Advances in Intelligent Systems and Computing

Volume 1326

The series "Advances in Intelligent Systems and Computing" contains publications on theory, applications, and design methods of Intelligent Systems and Intelligent Computing. Virtually all disciplines such as engineering, natural sciences, computer and information science, ICT, economics, business, e-commerce, environment, healthcare, life science are covered. The list of topics spans all the areas of modern intelligent systems and computing such as: computational intelligence, soft computing including neural networks, fuzzy systems, evolutionary computing and the fusion of these paradigms, social intelligence, ambient intelligence, computational neuroscience, artificial life, virtual worlds and society, cognitive science and systems, Perception and Vision, DNA and immune based systems, self-organizing and adaptive systems, e-Learning and teaching, human-centered and human-centric computing, recommender systems, intelligent control, robotics and mechatronics including human-machine teaming, knowledge-based paradigms, learning paradigms, machine ethics, intelligent data analysis, knowledge management, intelligent agents, intelligent decision making and support, intelligent network security, trust management, interactive entertainment, Web intelligence and multimedia.

The publications within "Advances in Intelligent Systems and Computing" are primarily proceedings of important conferences, symposia and congresses. They cover significant recent developments in the field, both of a foundational and applicable character. An important characteristic feature of the series is the short publication time and world-wide distribution. This permits a rapid and broad dissemination of research results.

Indexed by DBLP, EI Compendex, INSPEC, WTI Frankfurt eG, zbMATH, Japanese Science and Technology Agency (JST), SCImago.

All books published in the series are submitted for consideration in Web of Science.

More information about this series at http://www.springer.com/series/11156

Miguel Botto-Tobar · Henry Cruz ·
Angela Díaz Cadena
Editors

Artificial Intelligence, Computer and Software Engineering Advances

Proceedings of the CIT 2020 Volume 1

Springer

Editors
Miguel Botto-Tobar
Eindhoven University of Technology
Eindhoven, Noord-Brabant, The Netherlands

Henry Cruz
Universidad de las Fuerzas Armadas ESPE
Sangolquí, Ecuador

Angela Díaz Cadena
University of Valencia
Valencia, Valencia, Spain

ISSN 2194-5357 ISSN 2194-5365 (electronic)
Advances in Intelligent Systems and Computing
ISBN 978-3-030-68079-4 ISBN 978-3-030-68080-0 (eBook)
https://doi.org/10.1007/978-3-030-68080-0

This Springer imprint is published by the registered company Springer Nature Switzerland AG
The registered company address is: Gewerbestrasse 11, 6330 Cham, Switzerland

Preface

The XV Multidisciplinary International Congress on Science and Technology (CIT 2020) was held in Quito, Ecuador, on October 26–30, 2020, proudly organized by Universidad de las Fuerzas Armadas ESPE in collaboration with GDEON. CIT is an international event with a multidisciplinary approach that promotes the dissemination of advances in science and technology research through the presentation of keynote conferences. In CIT, theoretical, technical, or application works that are research products are presented to discuss and debate ideas, experiences, and challenges. Presenting high-quality, peer-reviewed papers, the book discusses the following topics:

- Artificial Intelligence
- Computational Modeling
- Data Communications
- Defense Engineering
- Innovation, Technology, and Society
- Managing Technology and Sustained Innovation and Business Development
- Modern Vehicle Technology
- Security and Cryptography
- Software Engineering

CIT 2020 received 234 submissions written in English by 660 authors coming from 22 different countries. All these papers were peer reviewed by the CIT 2020 Program Committee consisting of 236 high-quality researchers. To assure a high-quality and thoughtful review process, we assigned each paper at least to three reviewers. Based on the peer reviews, 70 full papers were accepted, resulting in an 30% acceptance rate, which was within our goal of less than 40%.

We would like to express our sincere gratitude to the invited speakers for their inspirational talks, to the authors for submitting their work to this conference, and the reviewers for sharing their experience during the selection process.

October 2020

Miguel Botto-Tobar
Henry Cruz Carrillo
Ángela Díaz Cadena

Organization

General Chairs

Miguel Botto-Tobar	Eindhoven University of Technology, The Netherlands
Henry Cruz Carrillo	Universidad de las Fuerzas Armadas (ESPE), Ecuador

Organizing Committee

Miguel Botto-Tobar	Eindhoven University of Technology, The Netherlands
Henry Cruz Carrillo	Universidad de las Fuerzas Armadas (ESPE), Ecuador
Ángela Díaz Cadena	Universitat de Valencia, Spain

Steering Committee

Miguel Botto-Tobar	Eindhoven University of Technology, The Netherlands
Ángela Díaz Cadena	Universitat de Valencia, Spain

Publication Chair

Miguel Botto-Tobar	Eindhoven University of Technology, The Netherlands

Program Chairs

Life Sciences

María Emilia Medina — Universidad de las Fuerzas Armadas (ESPE), Ecuador

Earth and Construction Sciences

Débora Simón-Baile — Universidad de las Fuerzas Armadas (ESPE), Ecuador

Energy and Mechanics

Vinicio Lara Padilla — Universidad de las Fuerzas Armadas (ESPE), Ecuador

Electrical and Electronic

Lenín Granda Gutiérrez — Universidad de las Fuerzas Armadas (ESPE), Ecuador

Computer Science

Rodrigo Fonseca Carrera — Universidad de las Fuerzas Armadas (ESPE), Ecuador

Human and Social Science

Edgardo Romero Frometa — Universidad de las Fuerzas Armadas (ESPE), Ecuador

Security and Defense

Antonio Bravo Calle — Universidad de las Fuerzas Armadas (ESPE), Ecuador

Economic and Administrative Sciences

Patricio Herrera Enríquez — Universidad de las Fuerzas Armadas (ESPE), Ecuador

Social

Ana Guamán — Universidad de las Fuerzas Armadas (ESPE), Ecuador

Local Committee

Research Department

Wendy Castillo — Universidad de las Fuerzas Armadas (ESPE), Ecuador

Sede Latacunga

José Bucheli Andrade — Universidad de las Fuerzas Armadas (ESPE), Ecuador

Program Committee

Adamid Beltrán	Centro Nacional de Tecnología Agropecuaria y Forestal "Enrique Álvarez Córdova" (CENTA), El Salvador
Agustín J. González V.	Universidad Técnica Federico Santa María, Chile
Alain Hernández Santoyo	Universidade Federal de Alfenas (UNIFAL), Brazil
Alba Arevalo Vergel	Universidad Francisco de Paula Santander, Colombia
Alex Cardoso Bastos	Universidade Federal do Espírito Santo, Brazil
Alex Jiménez Vélez	Universidad Politécnica de Catalunya, Universidad de las Fuerzas Armadas (ESPE), Ecuador
Alexei Pastolni	Southwest State University, Russia
Alexi Delgado	Universidad de Ciencias y Humanidades (UCH), Perú
Alicia Carabali	Escuela Politécnica Nacional, Ecuador
Álvaro Can Chulim	Universidad Autónoma de Nayarit, México
Álvaro Eulalio Villamizar	Universidad de Pamplona, Colombia
Ambrocio Teodoro Esteves Pairazamán	Universidad Cesar Vallejo, Perú
Ana Cecília Nunes	Pontifícia Universidade Católica do Rio Grande do Sul, Brazil
Anderson Steven Peña Sabogal	Universidad Nacional de Colombia
Andrea Sotomayor	Instituto Nacional de Investigaciones Agropecuarias INIAP, Ecuador
Andrés Erazo	Division of Biomedical Engineering, University of Saskatchewan, Canada
Andrés Felipe Besoain Pino	Universidad de Talca, Chile
Andrés Santiago Martínez León	Universidad Estatal del Suroeste de Russia, Russia

Andrés Tirado Espín	Universidad Nacional de Loja, Ecuador
Angela Jimenez	Universidade Federal do Paraná-UFPR, Brazil
Arturo Montenegro-Ramírez	Universidad Técnica de Ambato, Ecuador
Beatriz F. Giraldo Giraldo	Universitat Politècnica de Catalunya, Spain
Beltrán Castañon Cesar Armando	Pontificia Universidad Católica del Perú, Perú
C. Bachelier	Novatec Solar GmbH, Herrenstrasse, Germany
Caja Rivera Rocío Marilyn	University of South Florida, USA
Carles Riba	Universidad Politécnica de Catalunya, Spain
Carlos Almeida	Escuela Politécnica Nacional, Ecuador
Carlos Arturo Raymundo Ibañez	Universidad Peruana de Ciencias Aplicadas, Perú
Carlos Barrera	Instituto Tecnológico Metropolitano de Medellín, Colombia
Carlos de Castro Lozano	Universidad de Córdoba, Spain
Carlos Matovelle	Universidad Católica de Cuenca, Ecuador
Carlos Smith	University of South Florida, USA
César Jesús Pardo Calvache	Universidad del Cauca, Colombia
Cesar Puente Montejano	Facultad de Ingeniería, Universidad Autónoma de San Luis Potosí, México
Christian Byron Castro Miniguano	Universidad Técnica de Ambato, Ecuador
Cristina Pérez Rico	Escuela Politécnica Nacional, Ecuador
Daniel Garcés	Escuela Superior Politécnica del Litoral (ESPOL), Ecuador
Daniel Gorman	Universidad de São Paulo, Brasil
Darwin Mena	Universidad Santo Tomas, Colombia
Diana Endara	Escuela Politécnica Nacional, Ecuador
Diego Alejandro García Cárdenas	Universidad de Pamplona, Colombia
Diego Ávila	Escuela Superior Politécnica del Chimborazo (ESPOCH), Ecuador
Diego Barragán	Universidad Técnica Particular de Loja, Ecuador
Diego Cajas	Pontificia Universidad Católica del Ecuador, Ecuador
Diego Donoso Vargas	Universidad de las Américas, Ecuador
Dominik Schlipf	Universidad de Stuttgart, Germany
Dustin Tahisin Gómez Rodríguez	Universitaria Agustiniana, Colombia
Edinzo Iglesias	Universidad Técnica del Norte, Ecuador
Eduardo Pinos	Universidad Politécnica Salesiana, Ecuador
Edwin Paccha	Universidad de Chile, Chile
Efraín Freire	Instituto Nacional de Biodiversidad (INABIO), Ecuador

Efstathios Stefos	Pontificia Universidad Católica del Ecuador, Ecuador
Esteban Fuentes	Universidad Tecnológica Indoamérica, Ecuador
Fidel David Parra Balza	Universidad Israel, Ecuador
Francisco Ampuero	Universidad del Azuay, Ecuador
Francisco Edgar Castillo Barrera	Universidad Autónoma de San Luis Potosí, México
Francisco Flores	Universidad de las Fuerzas Armadas, Ecuador
Francisco Maciá-Pérez	Universidad de Alicante, Spain
Francisco Rossomando	Instituto de Automática-Universidad Nacional de San Juan-INAUT, Argentina
Franklin Cordova Buiza	Universidad Continental, Universidad Privada del Norte, Perú
Gabriel Enrique Pico Mera	Universidad Técnica de Manabí, Ecuador
Germán Harvey Alférez Salinas	Universidad de Montemorelos, México
Germania Vayas Ortega	Universidad Rey Juan Carlos, España
Ghassan Al-Kindi	Sohar University, Omán
Gilberto Reynoso-Meza	Pontifícia Universidade Católica do Paraná, Brasil
Gloria Díaz Londoño	Universidad Nacional de Colombia, Colombia
Gustavo Scaglia	Universidad Nacional de San Juan-INAUT, Argentina
Héctor Hugo Laverde Morales	Fundación Universitaria Agraria de Colombia, Colombia
Héctor Manuel Ortega Escobar	Colegio de Postgraduados, México
Henry Daniel Vera Ramírez	SENA-SENNOVA, Colombia
Henry Rivera Puga	Fuerza Aérea Ecuatoriana, Ecuador
Holger Benavides	Universidad Técnica Particular de Loja, Ecuador
Holger Geovanny Cevallos	Escuela Superior Politécnica del Litoral, Ecuador
Hugo Gamez Cuatzin	Centro de Ingeniería y Desarrollo Industrial–CIDESI, México
Hugo Leiva	Universidad Yachay Tech, Ecuador
Idalberto Macías Socarrás	Universidad Estatal Península de Santa Elena, Ecuador
Iglesias Navarro Iván	Universidad Técnica del Norte, Ecuador
Iker García Garizabal	Escuela Superior Politécnica del Litoral (ESPOL), Ecuador
Ilaria Bendato	University of Genoa, Italia
Ismael Minchala Ávila	Universidad Politécnica Salesiana, Ecuador
Ivana Maia	Instituto Federal Maranhão, Brasil
Jaime Alberto Buitrago	Universidad del Quindío, Colombia

Jaime Cadena	Escuela Politécnica Nacional, Ecuador
Jaime Jaramillo	Universidad Técnica Particular de Loja, Ecuador
Javier Augusto Vera Solano	Universidad de Antioquia, Colombia
Javier Castillo	Universidad Santiago de Cali, Colombia
Javier Sánchez Guerrero	Universidad Técnica de Ambato, Ecuador
Jesús Olivero-Verbel	Universidad de Cartagena, Colombia
Jesús Ramos Martín	Universidad Regional Amazónica (IKIAM), Ecuador
Jesús Rodríguez Manzano	Imperial College London, UK
Jianhui Wong	Southern Methodist University, USA
João Canavilhas	University of Beira Interior, Portugal
John Forero C.	Universidad Distrital Francisco José de Caldas, Colombia
John Ramiro Agudelo Santamaría	Universidad de Atioquía, Colombia
Jonathan Coronel León	Escuela Superior Politécnica del Litoral (ESPOL), Ecuador
Jordi Rafael Palacios González	Sergio Arboleda University, Colombia
Jorge Andrés Rosales Acosta	Escuela Politécnica Nacional, Ecuador
Jorge Herrera-Tapia	Universidad Laica Eloy Alfaro de Manabí, Ecuador
Jorge Luis Díaz Rivera	Universidad Central de Bayamón, Puerto Rico
Jorge Nogared Cardoso	Universidad del Sur de Santa Catarina (UNISUL), Brasil
José Alfredo Facio Umaña	Autonomous University of Coahuila, México
José Javier Capera Figueroa	Universidad Iberoamericana, México
José Julián Ñáñez Rodríguez	Universidad del Tolima, Colombia
José Manuel Menéndez	Universidad Politécnica de Madrid, Spain
José Raúl Castro	Universidad técnica Particular de Loja, Ecuador
Juan Carlos Guerri	Universidad Politécnica de Valencia, Spain
Juan Carlos Solano	Universidad Nacional de Loja (UNL), Ecuador
Juan Contreras	Escuela Naval Almirante Padilla ENAP, Colombia/University of Saskatchewan, Canada
Juan Leonardo Espinoza Abad	Universidad Católica de Cuenca, Ecuador
Juan Martínez	Universidad Rey Juan Carlos, Spain
Juan Romero Mediavilla	Comando de Operaciones Aéreas y Defensa, Departamento TIC-COAD, Ecuador
Juan Sanago	Universidad Católica de Cuenca, Ecuador
Juan Sanez	Federal University of Parana, Brazil

Julian Triana Dopico	Universidad de Leuven, Belgium
Julio Bastos Arrieta	Universitat Politècnica de Catalunya, Spain
Julio Nolberto Pérez Guerrero	Escuela Superior Politécnica de Chimborazo, Ecuador
Julio Ontiveros-Ravell	Universidad Politécnica de Madrid, Spain
Kheireddine Choutri	University Professional Blida 1, Algeria
Kléber Anibal Chiluisa Chiluisa	Universidad de las Fuerzas Armadas (ESPE), Ecuador
Kristina Elizabeth Bravo Huivin	Universidad Privada del Norte, Perú
Kruzkaya Ordóñez	Universidad técnica Particular de Loja, Ecuador
Leonardo A. Bermeo Varón	Universidad Santiago de Cali, Colombia
Leonardo Goyos	Universidad de las Fuerzas Armadas (ESPE), Ecuador
Leonardo Pachano	Instituto Francés del Petróleo (IFP Énergies nouvelles), France
Lilian Susana Molina Benavides	Universidad Estatal Península de Santa Elena, Ecuador
Luis Alberto Otero González	Universidad de Santiago de Compostela, Spain
Luis Amilcar Olvera Vera	Universidad de Guayaquil, Ecuador
Luis Chamba Eras	Universidad Nacional de Loja, Ecuador
Luis Manuel Vilchez Blázquez	Centro de Investigación en Computación del IPN, México
Luis Morales	Escuela Politécnica Nacional, Ecuador
Luis Santiago Choto	Escuela Superior Politécnica del Litoral (ESPOL), Ecuador
M. R. Ali	King Fahd University of Petroleum and Minerals (KFUPM), Arabia Saudita
Mahdi Shariati	Tôn Đức Thắng University, Vietnam.
Marcelo Zambrano	Universidad Politécnica de Valencia–España/Universidad Técnica del Norte, Ecuador
María Alejandra Zuluaga	EURECOM, Francia
María Dolores Sánchez Fernández	Investigador Universidad de A Coruña, Spain
María Gabriela Intriago Cedeño	Universidad Técnica de Manabí, Ecuador
María Hallo	Escuela Politécnica Nacional, Ecuador
María José Abasolo	Universidad Nacional de La Plata, Argentina
Mario César Maya Rodríguez	Instituto Politécnico Nacional (IPN), México
Marlena León-Mendoza	Universidad de las Américas, Ecuador
Mary Vergara	Universidad Politécnica Salesiana, Ecuador

Mauricio Sánchez	Universidad Nacional Mayor de San Marcos, Perú
Mercy J. Borbor Córdova	Escuela Superior Politécnica del Litoral (ESPOL), Ecuador
Meritxell Gisbert Traveria	Universidad de Barcelona, Spain
Michael Giraldo	Universidad EAFIT, Colombia
Miguel Antonio Morales Chan	Galileo Universidad, Guatemala
Miguel Araque	Universidad Politécnica Salesiana, Ecuador
Miguel Flores	Escuela Politécnica Nacional, Ecuador
Miguel Lapo Palacios	Universidad Técnica de Manabí, Ecuador
Miriam Viviana Ñañez Silva	Universidad Nacional de Cañete, Perú
Mónica Delgado	Escuela Politécnica Nacional, Ecuador
Naga Raju Maddela	Universidad Técnica de Manabí, Ecuador
Nayeth Solórzano Alcívar	Escuela Superior Politécnica del Litoral (ESPOL), Ecuador
Néstor Darío Duque Méndez	Universidad Nacional de Colombia, Colombia
Ney Raúl Balderramo Vélez	Universidad Técnica de Manabí, Ecuador
Nisia Rizzo	Universidade do Estado da Bahia, Brasil
Orlando Vicente Lizaldes Espinosa	Universidad Nacional de Loja, Ecuador
Oscar Campo	Universidad Autónoma de Occidente, Colombia
Oscar Gonzales	Escuela Superior Politécnica del Litoral (ESPOL), Ecuador
Oscar Raúl Mancilla Villa	Universidad e Guadalajara, México
Ovais Khan	Texas A&M University, USA
Pablo Carrera	Colegio de Administración y Economía-CADE, Ecuador
Pablo Londoño	Universidad Central del Ecuador, Ecuador
Paola Guzmán	Universidad Politécnica de Valencia-Instituto de Telecomunicaciones y Aplicaciones Multimedia (iTEAM), Spain
Paúl Palmay Paredes	Escuela Superior Politécnica del Chimborazo (ESPOCH), Ecuador
Paul Romero	Escuela Superior Politécnica del Chimborazo (ESPOCH), Ecuador
Pedro Ortiz Oblitas	Universidad Nacional de Cajamarca, Perú
Pere Ariza	Institute of Environmental Science and Technology, Universitat Autònoma de Barcelona (UAB), Barcelona, Spain
Priscila Cedillo	Universidad Católica de Cuenca, Ecuador
Rafael Vásquez	Universidad Pontificia Bolivariana, Colombia
Raúl Fernández Concepción	Universidad de Pinar del Río, Cuba

Raúl Manzanilla	Universidad Yachay Tech, Ecuador
Ricardo Alfredo Cajo Diaz	Escuela Superior Politécnica del Litoral (ESPOL), Ecuador/University of Genoa-DIME, Italy
Rodrigo P. Ramos	Universidade Federal do Vale do São Francisco, Brasil
Romel Ángel Erazo Bone	Universidad Estatal Península de Santa Elena, Ecuador
Ronnie Araneda Cabrera	Universidad de A Coruña, Spain
Rosa Elena Ordoñez Vivero	Universidad Laica Vicente Rocafuerte de Guayaquil, Ecuador
Rosario Violeta Grijalva Salazar	Universidad César Vallejo, Perú
Rosero Chandi Carlos Xavier	Universidad Politécnica de Catalunya, Spain
Rubén Darío Rojas Sulbarán	Universidad de Los Andes, Venezuela
Rubén Hidalgo	Escuela Superior Politécnica del Litoral (ESPOL), Ecuador
Ruth Zamora	Universidad Técnica de Ambato, Ecuador
Saba Infante	Universidad Yachay Tech, Ecuador
Salomon Quito	Universidad Central del Ecuador, Ecuador
Samiha Fadloun	Ecole Nationale Supérieure D'Informatique, Argelia
Sandra Cabrera	Universidad Católica de Cuenca, Ecuador
Santiago Cortes	Universidad Antonio Nariño, Colombia
Santiago Ferrandiz Bou	Universidad Politécnica de Valencia, Spain
Santiago Gómez	Western University of Ontario, Canada
Sara Fernández-López	Universidad de Santiago de Compostela, Spain
Sayedeh Parastoo Saeidi	UTE, Ecuador
Segundo Edwin Cieza Mostacero	Universidad Cesar Vallejo, Perú
Simon Klančnik	University of Maribor, Slovenia
Sipiran Mendoza Iván Anselmo	Pontificia Universidad Católica del Perú, Perú
Sonia Contreras Ortiz	Universidad Tecnológica de Bolívar, Colombia
Stephen Thomas	University of Greenwich, UK
Tania Aleyda Acosta Hurtado	Escuela Politécnica Nacional, Ecuador
Teobaldis Mercado Fernández	Universidad de Córdoba, Colombia
Tito Raúl Vargas	Universidad Santo Tomás, Colombia
Turgay Kıvak	Duzce University, Turkey
Vasilica-Maria Margalina	Universidad Técnica de Ambato, Ecuador
Venkata Subba Reddy Gangireddygari	University of South Africa, Sudáfrica
Vera Gelashvili	Rey Juan Carlos University, Spain

Organizing Institutions

ESPE
UNIVERSIDAD DE LAS FUERZAS ARMADAS
INNOVACIÓN PARA LA EXCELENCIA
1922
ECUADOR

Gdeon
Impulsando la investigación

Contents

Artificial intelligence, Communications, Security and Cryptography, and Software Engineering

Prediction of Parkinson's Disease Severity Based on Gait Signals Using a Neural Network and the Fast Fourier Transform

J. D. Suquilanda-Pesántez[1], M. C. Zambonino-Soria[1], D. E. López-Ramos[1], M. G. Pineda-Molina[1], N. S. Milán[1], M. C. Muñoz[1], G. F. Villalba-Meneses[1,2], and D. Almeida-Galárraga[1(✉)]

[1] School of Biological Sciences and Engineering, Yachay Tech University, Urcuquí 100119, Ecuador
{jefferson.suquilanda,marjorie.zambonino,diana.lopez, maria.pineda,nataly.milan,michelle.munoz,gvillalba, dalmeida}@yachaytech.edu.ec

[2] Department of Design and Manufacturing Engineering, University of Zaragoza, María de Luna 3, 50018 Zaragoza, Spain

Abstract. Parkinson's disease (PD) is a chronic and progressive movement disorder affecting the nervous system of patients. PD alters the capacity to sustain a steady rhythm in the gait leading to impaired motor skills. The diagnosis of PD is a complex task that encompasses motor and non-motor symptoms. More than 50% of patients suffer from freezing in their legs during walking, generally defined as Freezing of Gait (FOG). Hence, gait evaluation is essential for physicians to detect the disease and track its severity. In this context, intelligent learning algorithms have been recently used to facilitate the diagnostic process. However, an accurate diagnosis is still required. This paper presents an effective and novel approach to diagnose PD using gait analysis, wearable sensors, and deep learning techniques. To achieve this goal, an Artificial Neural Network multiclassifier system (ANN) has been developed. The proposed model predicts the severity of PD by detecting the stage of the disease. It uses a further analysis of movement alteration, supervised learning, and clinical assessment based on Hoehn & Yahr (H&Y) scores. The experimental results indicate a high efficiency of the proposed method in tracking PD severity. The method can make a distinction between idiopathic PD patients and healthy individuals, and classify them in the levels 0, 2, 2.5 and 3 according to the symptoms progress and level of disability. The model achieves an outstanding performance with an accuracy of 97%. Thus, this study emphasizes the use of a prediction learning algorithm that can be helpful and valuable for clinical examinations.

Keywords: Parkinson's disease · Fast Fourier Transform · Artificial Neural Network · Gait analysis · Multiclassifier system

M. Botto-Tobar et al. (Eds.): CIT 2020, AISC 1326, pp. 3–18, 2021.
https://doi.org/10.1007/978-3-030-68080-0_1

1 Introduction

Neurological disorders are currently the prominent cause of disability in the world. Parkinson's disease (PD) is ranked as the second most frequent neurodegenerative disorder after Alzheimer's. The Parkinson Disease Foundation [1] considers there are about 7 to 10 million cases worldwide, and the numbers continue to grow exponentially due to the direct relation with the aging of the population. This chronic and progressive disease is caused mainly by the deterioration of dopamine-producing nerve cells in the brain. Dopamine is an essential neurotransmitter that controls coordinated muscle function and movement coordination. The four main symptoms of PD are tremors, resistance to movement, postural instability, and decreased spontaneous and voluntary movement known as Bradykinesia [2]. Tremor is a common symptom exhibited by approximately 70% of PD patients at an Early Stage (ESPD), which leads to a comfortable and frequent fall [3]. Therefore, impaired gait and tremor occurrences are notable symptoms in different stages of the disease. These disturbances are classified as continuous and episodic. The first classification is due to reduced stride length, gait speed, increased double support time, stride variability, and left-right asymmetry. The second classification involves festination and freezing [4].

The Freezing of Gait (FOG) is one of the most abnormal and critical conditions that affect the autonomy and quality of life of PD patients. As the disease progresses, there is a greater deficit in the amplitude and control of the frequency of the steps. Other factors that may contribute to gait abnormalities include rhythmicity and skewness [5]. Thus, gait dynamics have been used in previous studies to distinguish patients with Parkinson's and normal subjects based on certain gait parameters such as kinetics, kinematics, and spatio-temporal, and systems such as video motion analysis system and Local Field Points (LFP) [6–12].

ANNs are computational structures inspired by how biological neural processing works. They are highly parallelized dynamic systems that work interactively to solve specific problems such as analyzing signals and pattern classification [13]. The topology of ANNs consists in that each input neuron receives a numerical input from the variables of the database. Then, they are multiplied by factors, known as connection weights. The product is summed up to become the net input and finally be processed in the inner layers of the network through an activation function to generate the output information [14]. The backpropagation algorithm is the method to compute the error gradient for a feed-forward network. It uses a supervised learning technique of a predefined set of pairs of inputs and outputs. The output is compared with the target output, and the error is computed for each exit neuron [15]. Thus, ANNs have been used in clinical settings to predict the effectiveness of instrumental evaluation and increase diagnostic sensitivity and specificity. They are also used in biochemical analysis, medical imaging, and drug development [14].

The application of ANNs in predicting biological outcomes has been increasing in the last decade to diagnose several disorders such as PD. Various researches have reported different analysis methods, including machine learning [16–18], deep learning [19, 20], and hybrid neural network [21]. Abdulhay et al. [2] applied machine learning techniques to diagnose PD by using peak detection and pulse duration analysis of various gait features. This method achieved an accuracy of 92.7%. Chang et al. [22] presented a successful Support Vector Machine (SVM) classifier to distinguish PD patients from control subjects requiring only kinematic gait data and center of pressure (CoP) that means the point of force application. This method suggested the relevance and efficiency of machine learning in increasing the accuracy of predicting PD severity. In [23] the gait analysis was divided into three lines: kinetics, kinematics and electromyography to evaluate the muscle activity during walking. Zeng et al. [24] presented a PD classification approach using gait data via deterministic learning. The approach revealed that the time series of Vertical Ground Reaction Force (VGRF) could be used to detect PD patients. Here, the sensor operation and gait signal approach used small, low-power, and low-cost wearable sensors [25].

Cole et al. [25] presented a dynamic algorithm in pattern classification based on dynamic neural networks (DNN), dynamic support vector machines (DSVMs) and Hidden Markov Model (HMM) frameworks. The dataset was gathered while eight ill patients and four healthy subjects performed spontaneous and unrestricted quotidian activities. The study showed significant experimental results in reducing error rates. Mannini et al. [26] proposed a similar algorithm to build an SVM classifier for post-stroke and Huntington's disease. In [27], the authors used a K-Nearest Neighbor (KNN) classifier to obtain probability density parameters as a feature for each class. After that, those probabilities with other statistics aspects including standard deviation (SD), root-mean-square-deviation (RMSD) and t-student test values were inputs to train an SVM.

Moreover, Hoang et al. [28] developed a gait rating model using 1D and 2D Convolutional Neural Network (CNN). The used VGRF data was preprocessed into images to optimally extract the characteristics of one-dimensional time series data optimally. Lee et al. [29] classified idiopathic PD subjects and healthy controls through an ANN with weighted fuzzy membership functions (NEWFM) by using gait characteristics and wavelet. Both studies [28, 29] demonstrates the significance of preprocessing and information extraction methods to obtain a good performance of the neural network.

However, gait study can become a difficult task since it is influenced by physiological conditions such as age, health, and weight [19, 30]. Therefore, PD has been systematically tested from a clinometric point of view by applying global assessments such as the Hoehn & Yahr (H&Y) categorical scale or the Unified Parkinson's Disease Rating Scale (UPDRS) [31]. For example, Jane et al. [20] built a Q-back propagated time delay neural network (Q-BTDNN) for a temporal classification system to predict the severity of gait disturbances in PD patients considering the H&Y scale. This approach performed an effective clinical decision-making system (CDMS) by analyzing the instabilities in the postures and walking patterns. The most recent study performed in [19] applied deep learning techniques to build an intelligent Parkinson identification model for analyzing gait data using a 1D Convolutional Neural Network (1D-CNN). The authors developed a Deep Neural Network (DNN) classifier according to the UPDRS, achieving an accuracy of 85.3% in Parkinson's severity prediction. The disadvantages of UPDRS is the time-consuming requirement to measure all the items and the shortcomings of the scale's usefulness in the ESPD [31].

In this context, this paper proposes an expert Artificial Neural Network (ANN) classifier that is trained using a Backpropagation (BP) learning technique that tracks the severity of Parkinson's. The trained model uses gait analysis and the H&Y scale to identify the stage of the disease and predict its progress. The main contribution of this work is the use of the Fast Fourier Transform (FFT) in the preprocessing of data, the segmentation of the gait signals, and the use of distinct extracted features as inputs for the feed-forward algorithm. The rest of this paper is organized as follows: Section 2 describes the material and methodology used to build the proposed algorithm. Section 3 presents the experiments and the results. Finally, Sect. 4 deals with the concluding remarks.

2 Materials and Methods

This work evaluates the use of an ANN classification model for VGRF data to correctly classify the severity of PD considering H&Y categorical scale. The methodology workflow is composed by four phases (see Fig. 1). Firstly, the data is collected. Secondly, it is normalized and preprocessed to diagnose PD. Thirdly, the most discriminative features in the frequency domain are extracted from Fast Fourier Transform. Finally, the proposed classification algorithm is constructed to predict the severity of the disease. Hence, this study presents an efficient ANN classification model trained and tested.

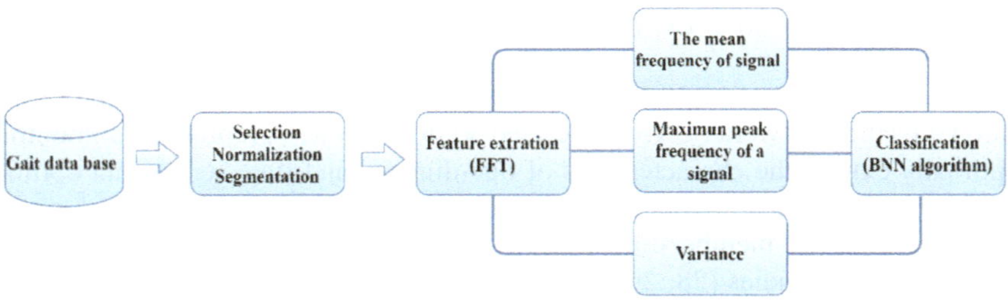

Fig. 1. Block diagram of the general proposed method.

2.1 Dataset Description

This work considers a Parkinson's Disease Database from Physionet [32], which contains data from 166 subjects divided between an idiopathic PD group (93 subjects) and a control group (73 subjects). The database is composed by three studies in which the patients executed different styles of walking. These include treadmill walking (Si Group), Rhythmic-auditory stimulation (RAS) walking (Ju Group), walking on a ground level using a walker, unassisted walking on a ground level, and dual tasking walking (Ga Group). This database stores VGRF measured in Newtons of the walking (gait) observed every 0.01 (s) for 2 (min) on level ground with 100 Hz as sampling rate (see Table 1). The data was obtained by computerized force-sensitive wearable sensors that measure the force walking and the variations of the stride-to-stride [25]. The records were measured by 16 sensors (8 for each shoe) integrated into the insole of two shoes (see Fig. 2).

Moreover, some demographic features were recorded such as gender, age and weight (Kg), and measures that indicates how Parkinson's disease symptoms progress such as

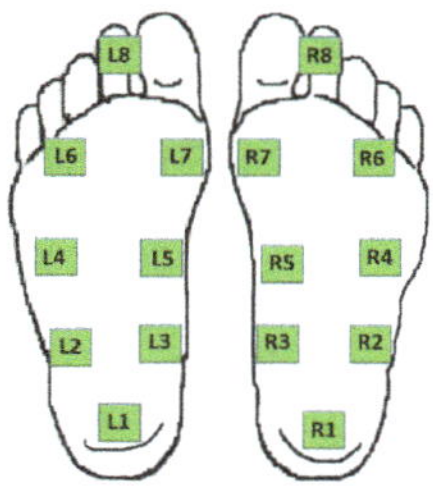

Fig. 2. Position of force sensors in left foot (L1–L8) and right foot (R1–R8).

H&Y scale. The stages are 1.0 (unilateral involvement), 1.5 (unilateral and axial involvement), 2.0 (bilateral involvement without impairment of balance), 2.5 (mild bilateral disease), 3.0 (moderate bilateral disease with postural instability), and the two most severe stages 4.0 and 5.0 that indicate a high walking difficulty or the patients' need for a wheelchair [33]. In the dataset, there were 10 patients with a *3* H&Y score; 55 patients with a 2 score; and 28 patients with 2.5 scores. This means that most of the subjects in the idiopathic PD group were classified at ESPD or as moderate disease severity. Thus, this data described a walking sequence of each subject during a period of time.

Table 1. Overall description of dataset.

Column	Description	Units
1	Time	Seconds (s)
2–9	VGRF measurements from sensors in left foot (L1–L8)	Newton (N)
10–17	VGRF measurements from sensors in right foot (R1–R8)	Newton (N)

2.2 Pre-processing of Data

The entire dataset was selected, ordered, and processed in MATLAB 2019b to calculate the most significant frequency parameters. These features characterize each patient with regard to their PD stage. The pre-processing of data includes two phases that are data selection and standardization.

Data Selection. Out of the total 2-min walk, the records of the first 30 s and the final 30 s were not taken into consideration to reduce the beginning and end-up effects of the walk [34]. Consequently, only 1 min of each patient was taken. The study group Ju was dismissed because it only contains 53-s records. The final dataset considered for this work was labeled by the disease severity stage of each patient to track the progress of PD. The experimental subjects were classified into four categories according to H & Y scores: 0 for control subjects, 2 for mild levels, 2.5 for moderate levels, and 3 for severe levels. Table 2 describes the number of subjects per category.

Table 2. Description of the dataset selected for this study.

Study	Subjects	Subtotal	Total
Ga	PD	27	45
	Control	18	
Si	PD	34	62
	Control	28	
Total			107

Normalization. Datasets were normalized to the body weight of the patients to avoid the influence of this parameter on VGRF measurements [19]. The normalization was performed following the below formula:

$$Normalized\ VGRF(\%) = \frac{VGRF(N)}{Body\ Weight(N)}(100\%) \tag{1}$$

Upon the normalization, the dataset was segmented to extract the proper frequency-domain features.

2.3 Information Extraction (Features)

For an appropriate windowing of each Sensor Signal (SS), a segmentation of 1.5 s (a total of 40-time segments) from the 1 min data was considered. The selection of features from the frequency domain provides representative information of SS. According to Sarbaz et al. [35], the differences between PD people and people in good health are relevant, mostly in relation to mean and variance values in gait patterns. In addition, it asserts that normal individuals show regular behaviors and PD patients irregular behaviors with irregularly distributed spectra in some regions. Then, considering these literature findings, the power spectra was used to obtain three significant features from each segment of SS: maximum amplitude, mean, and variance. The three parameters are concatenated to form an array (1×48). These concatenations compose the feature dataset in frequency that will be utilized to carry out training and evaluation of the performance of the presented ANN system (see Fig. 3).

2.4 Model Structure of the Proposed Artificial Neural Network

The Keras library and Colab notebook were used to construct the ANN. The implementation of the network was done in python language. The structure of the ANN is shown in Fig. 4. Where five layers are presented. The first one is the input layer with 48 data inputs corresponding to the 3 features extraction of each sensor, resulting in a 1×48 vector. There are three hidden layers: the initial one has 120 neurons, the next one has 60, and the third one has 20. All of them belong to the class "Dense" and have Relu activation function. They also use the BP technique and reach a minimum error of

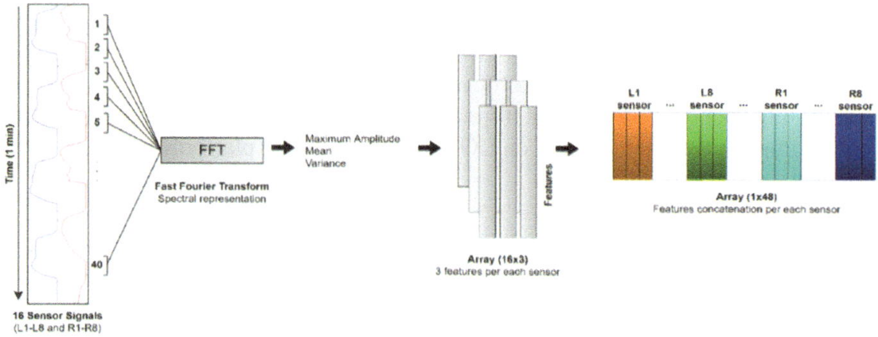

Fig. 3. Preprocessing of 16 SS. This process includes segmentation of 1.5 s (40-time segments), the frequency analysis by FFT, and the extraction of features.

classification by adjusting the network's weights. The last layer (output layer) has four possible outputs that correspond to the labels (stages) of the disease: 0, 2, 2.5, and 3. Then, the decision is performed by the Softmax function.

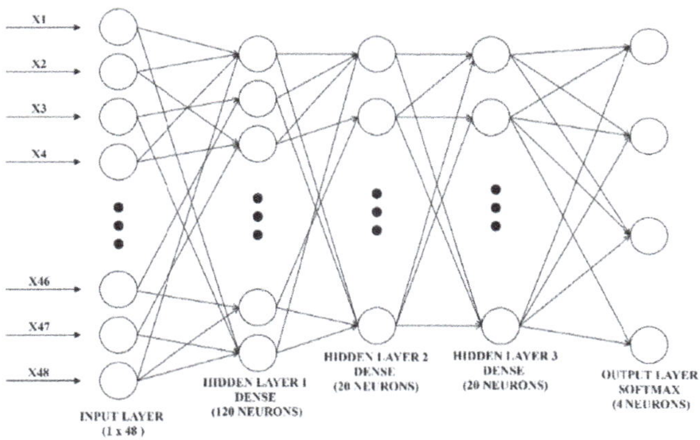

Fig. 4. Architecture of the ANN. The first layer is composed of 48 inputs; the three hidden layers have 120, 60 and 20 nodes; the output layer produces the classification process.

2.5 Performance of the Proposed Model

To obtain an adequate ANN capable of learning and making predictions, the model must fit into three stages named training, validation and testing. Then, the algorithm itself must be divided it into training (70%), validation (15%), and testing (15%) sub-datasets. The training stage is when the model learns the patterns from the gait signals; validation was used to minimize overfitting, and both the predictive power and the accuracy level of the network were corroborated by testing. The evaluation of the proposed model enabled to verify an adequate classification by analyzing some parameters such as confusion matrix,

F1 scores, selectivity, precision, and sensitivity. These metrics allowed determining the efficiency of the method to correctly classify the severity of PD into four stages (0, 2, 2.5 and 3). The experimental part was developed on a personal computing device which possesses Intel(R) Core(TM) i5-4210 processor, speed of 2.40 GHz and 8 GB RAM. The algorithms were implemented on COLAB, a computing platform based on Python language.

3 Results and Discussion

This section presents the experimental results and a discussion as follows: (i) the evaluation of the algorithm, (ii) the prediction of PD stages by the ANN system, (iii) a comparison with similar researches and (iv) the study limitations.

3.1 Evaluation of the Backpropagation Algorithm

Figure 5 shows the training and validation curves, which converge rapidly and grow with a similar trend, because they are directly proportional. This indicates that the algorithm learns significant characteristics without excessive adjustment. As the epochs begin to run, the precisions and losses are measured against the validation set. As demonstrated in Fig. 6. Demonstrates, the precision of the model increases from approximately 0.65 to around 1.00 throughout the 150 epochs. This accomplishes an accuracy of 0.9720, which equals to 97.20%. The accuracy is a relevant parameter corresponding to the proportion of the segments from the data entered (patient walking) in the training or validation set.

On the other hand, the cross-entropy loss function or cost function is a quantitative measure that allows evaluating the difference or deviation between the expected and real outputs. This measures the errors made by the network when predicting the stage at which a patient is categorized or, if by default, the patient belongs to the control group. Figure 6 shows that as the epochs or the number of times that the dataset passes through the network increases, an initial trend is detected at 5 and 25 epochs. Here, the loss reaches values of less than 0.3. From about 25 epochs on, a minimal trend is shown to increase loss as the number of epochs also increases. This is because the network begins to become overtrained and is no longer useful enough in predicting the stages of new patients.

3.2 Prediction of Parkinson's Stages

The ANN computational model presents a global understanding of a specific brain function in PD and normal subjects. To study gait behavior and its disorders it was necessary to take into account of the physiological findings to design a model that allows the understanding of disease to give a prediction about the earlier stages to diagnose and control PD. Table 3 shows a cross-validation summary of the predicted results based on precision, recall, and F1 for the total patients of each class, depending on the stage of PD. F1 (%) scores are relatively similar in all four classes, demonstrating that the back-propagated learning algorithm is very consistent and valid to predict the stages. Level 2 is the most successful level for the performance showing 99% accuracy. Level 0

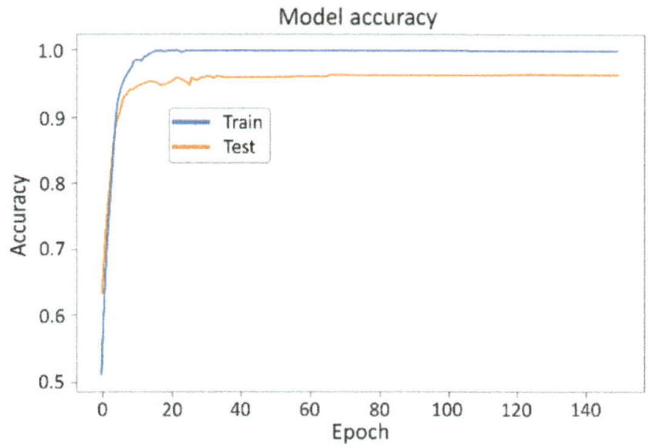

Fig. 5. Training and testing model accuracy curve represented by the blue line and orange line respectively for Parkinson's detection.

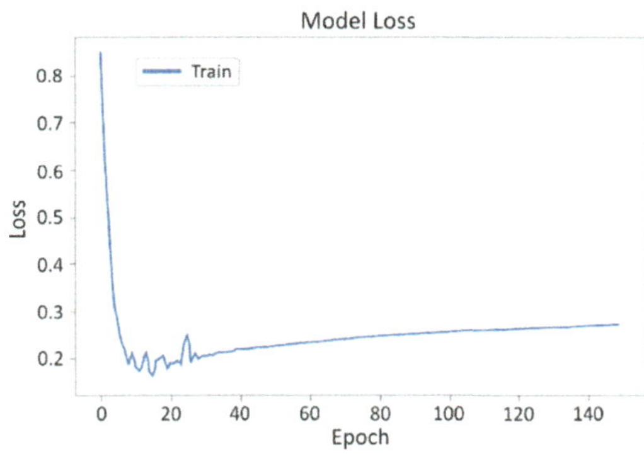

Fig. 6. Loss curve of the algorithm for Parkinson's detection.

is the most consistent due to it contains all the control subjects. One leading explanation for this trend may be the number of data (gait features) per level. Hence, considering the weighted average, the proposed ANN has an efficient performance, reaching an overall accuracy of 97% equivalent to the average recall.

Furthermore, the Confusion Matrix shown in the Fig. 7 presents the experimental results regarding the identification of the Parkinson's stages.

The Confusion Matrix demonstrates high efficiency and performance of the model by correctly classifying almost all subjects according to the severity level. Thus, it provides relevant information about the minimum error range in the classifier. In the present case, the model has four possibilities corresponding to PD stages. Level 0 is the control group; level 1 represents stage 2; level 2 satisfies stage 2.5, and level 3 corresponds to stage 3. Basically, the matrix configuration allows keeping a count of the true and false negative and the right and false-positive predictions of a classifier, as shown by the confusion matrix for PD prediction. The confusion matrix presents a total of 288 TP for the control group; 238 TP for level 2; 69 TP for level 2.5; and 69 TP for level 3. This confirms that

Table 3. Cross-validation summary for PD stages prediction on the ANN model.

Class	Precision (%)	Recall (%)	F1 (%)	n (subjects)
0	98	98	98	295
2	96	99	97	241
2.5	97	97	97	71
3	100	83	91	35
Accuracy	–	–	97	642
Macro average	98	94	96	642
Weighted average	97	97	97	642

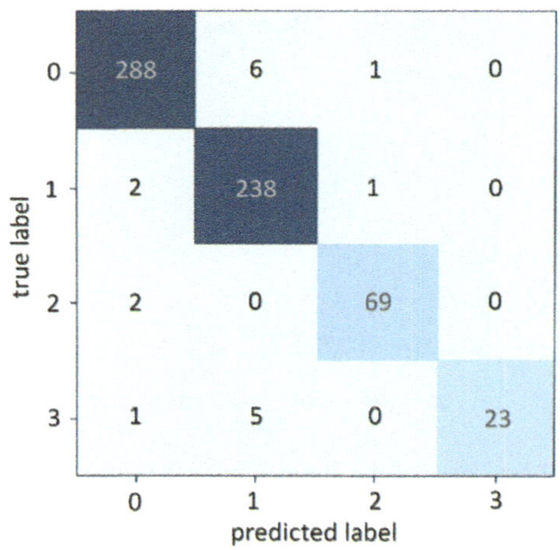

Fig. 7. Confusion matrix for stages of Parkinson's detection.

the BP algorithm is consistent with the system accuracy of the ANN reported on Table 3, showing a well-small rank of error. Therefore, due to the efficiency and performance of the classification model based on the PD stages, the system can be used in clinical decision supporting systems for the treatment of Parkinson's patients.

3.3 Comparison with Similar Studies

In the literature, there are some studies [4, 12, 16–21, 36, 37] that have used speech signals and gait analysis to identify Parkinson's Disease, but just a few are related to the prediction of the severity by using clinical assessments (see Table 4). Their accuracy varies from 74% to 96% and the two most used rating scales to analyze the PD progression are H&Y scale and UPDRS.

Table 4. Recent PD detection and diagnostics studies found in literature.

Method	Learning technique	Clinical scale	Accuracy (%)
Proposed model	Back-propagation learning algorithm	H & Y	97
Demand driven stimulation (DDM) by a Multi-Layer Perceptron [12]	Discrete wavelet transform and artificial neural network	N/a	83.2
Classification algorithms: NB, RF, DT, RT [16]	Automated machine learning classifiers	UPDRS	96.11
NB, RF, C4.5, MLP, AdaBoost with C4.5 and Bagging with C4.5 [17]	Supervised machine learning methods to build a FOG-detection chain	N/a	95
MLP, RF, NB, Bayes Network [18]	Machine learning	N/a	76.9 to 88.9
1D CNN (1D-Convnet) to build a DNN classifier [19]	Deep learning	UPDRS	87.3
Q-BTDNN [20]	Q-learning and Back-propagation (Q-BP)	H & Y	91.53
Hybrid Long Short-Term Memory (LSTM) and CNN [21]	Deep Learning that capture spatio-temporal characteristics of gait dynamics	H & Y	96.2
NEWFM [4]	NEWFM to classify coefficients of wavelet transform	N/a	74.32
Marquardt algorithm for small- and medium sized networks [36]	Lavenberg-Marquardt	N/a	83.29
SVM using n-fold cross validation [37]	Quadratic programming (QP) optimization	UPDRS	85 to 92

*NB: Naïve Bayes, RF: Random Forest, MLP: Multilayer Perceptron, DT: Decision Tree, RT: Random Tree.

To a proper comparison, the studies that use gait data from patients with different severity levels, and in which the experimental work was performed with an intelligent learning technique were chosen for this analysis. Moreover, since the model presented in this study is a multiclassification system, it will be compared with other multiclass models (see Table 5). Then, the studies by Zhao et al. [21] and El Maachi et al. [19] were chosen because they are the most recent studies that use deep learning algorithms and present high accuracies. The two selected studies, including the proposed model in this paper, performed a multi-category classification (see Table 5) and reported an investigation based on predicting the severity of PD from several VGRF measures recorded by the sensors.

Table 5. Comparison of Cross-validation results to detect Parkinson's disease.

Method	Learning	Scale	N° of Classes	Precision (%)	Recall (%)	F1 (%)
Proposed BNN model	DL	H&Y	4	97	97	97
Hybrid LSTM+CNN [21]	DL	H&Y	4	–	96.2	–
Deep 1D-Convnet [19]	DL	UPDRS	5	87.3	85.3	85.3

*The weighted average values of each study are considered. DL: Deep Learning.

Zhao et al. [21] employed a deep learning architecture conformed by a two-channel parallel network that includes a Hybrid Long Short-Term Memory (LSTM) and CNN. In this study, some spatio-temporal characteristics were analyzed over a while. The authors concluded that CNN+LSTM is better than only LSTM system at detecting PD, demonstrating that the hybrid model is more efficient. On the other hand, El Maachi et al. [19] built a DNN classifier using 1D convolutional neural network (1D-Convnet). The authors presented a model with an accuracy of 98.7% using UPDRS scale, proving the adequate training and efficiency to detect potential PD patients and predict the disease.

In contrast, with the two previous studies, the ANN classifier used in this study is not a hybrid model. Nevertheless, some frequency, statistical and demographic parameters were analyzed. This study's preprocessing and segmentation procedure can be adapted to more gait input data without difficulty due to the proposed flexible algorithm based on the multiclass supervised deep learning. It is important to note that all the three studies have reported neural networks that can learn intrinsic characteristics from gait signals to generalize to unobserved patients, a valuable parameter for clinical diagnosis.

In addition, Zhao et al. [21], El Maachi et al. [19] and the present method take the majority classification of all the segmented walk samples to detect the final stage classification of the patient. Thus, the main difference is the accuracy and efficiency between the models, where our ANN model also achieves a high value. The use of H&Y scale in this method favors the assessment of impairments in ESPD. One of the results of this study shows that each stage presents different gait patterns in patients, who can advance from one H&Y score to the next. Furthermore, since H&Y is a widely used clinical scale by professionals, it facilitates the implementation of the model in some current procedures for rehabilitation and PD treatments. Then, this work confirms the validity of the utilization of H&Y as a measure of PD progression analyzed in [38, 39].

In the present study, the specific spatial kernels enable the model to learn automatically and allow the best distinction of gait data between the Parkinson and control group. Compared with machine learning methods, DNNs are used to study complex signals through successive layers with non-linear activation because of their capability [19]. The major strength of applying the FFT in our ANN model is the efficiency of the preprocessing of data in classifying patients and standard cases. It must be noted that this study includes the following steps: a) a stride time interval relevant and specific to each subject's gait was chosen; b) the frequency-relevant features were extracted considering the main differences in gait behavior between healthy and PD cases; c) the weights were adjusted in a feed-forward algorithm to construct an appropriate ANN; d) H&Y scores were used to design an efficient classification of the PD levels. Thus, the proposed model can be presented as a standard solution for PD severity prediction. It can also be used as a medical tool for the treatment and prevention of gait impairments.

3.4 Study Limitation

Although the multiclassifier system provides significant accuracy and efficiency, few limitations are presented. A more in-depth biostatistical data analysis can help to extract more significant gait features based on biomechanical gait studies that are needed to understand the severity of FOG symptoms in Parkinson's patients. In that way, more significant characteristics of each stage of the disease will improve the training of ANN and generate better severity predictions in a patient with PD. However, considering the physiological plausibility of the model, it can be used as a baseline to automatically predicts the disease severity. This study even shows that once the proposed model is well-trained, the predicted severity is directly obtained from features collected by sensor data, which resolves any subjectivity concerns of the specialists. Therefore, this study provides further personalized treatment approach for each patient and simplifies the diagnosis procedure. The future work will consider a foot strike profile, lifting and landing parameters in walking, and the differences between stride and swing time variability.

4 Conclusions and Future Perspectives

This work proposes a model that predicts the severity of PD by monitoring and detecting the stages via gait analysis. An accurate ANN multiclass classification model based on a backpropagation algorithm was constructed. The proposed method is a preliminary study that classifies the patients following their symptom progression by associating significant gait features with H&Y scores. This multi-category method enabled the model to track the severity of the disease, differently from previous methods that only provide a binary detection. The model is trained by four levels of severity because obtaining VGRF measures from patients, who are located between the most critical stages, is quite difficult due to the fact that they can barely walk. Each PD patient should have an individualized evaluation and treatment plan to determine more effective strategies for medications, training, and follow-up examinations. The experimental findings reveal the high accuracy and efficiency of the present ANN model, standing as a relevant primary template in predicting the PD progression and providing improvements in the intelligent gait analysis approaches.

As future work, the ANN multiclassifier can achieve a higher prediction accuracy by implementing the following characteristics: (i) the extraction of more gait features by considering a complex domain line of work; (ii) the use of multiple data sources with patients in more levels of disease, and (iii) the use of fused features such as skeleton data or biochemical data. The results reported in this paper present a valuable model for the computer-assisted medical diagnosis that simplifies the procedure and provides a personalized treatment to help the patients in assessing and reducing FOG occurrences.

References

1. Parkinson's Disease Foundation (n.d.). https://www.pdf.org
2. Abdulhay, E., Arunkumar, N., Narasimhan, K., Vellaiappan, E., Venkatraman, V.: Gait and tremor investigation using machine learning techniques for the diagnosis of Parkinson disease. Futur. Gener. Comput. Syst. **83**, 366–373 (2018). https://doi.org/10.1016/j.future.2018.02.009
3. Perumal, S.V., Sankar, R.: Gait and tremor assessment for patients with Parkinson's disease using wearable sensors. ICT Express. **2**, 168–174 (2016). https://doi.org/10.1016/j.icte.2016.10.005
4. Frazzitta, G., Pezzoli, G., Bertotti, G., Maestri, R.: Asymmetry and freezing of gait in parkinsonian patients. J. Neurol. **260**, 71–76 (2013). https://doi.org/10.1007/s00415-012-6585-4
5. Lo, A.C., Chang, V.C., Gianfrancesco, M.A., Friedman, J.H., Patterson, T.S., Benedicto, D.F.: Reduction of freezing of gait in Parkinson's disease by repetitive robot-assisted treadmill training: a pilot study. J. Neuroeng. Rehabil. **7**, 1–8 (2010). https://doi.org/10.1186/1743-0003-7-51
6. Cho, C.W., Chao, W.H., Lin, S.H., Chen, Y.Y.: A vision-based analysis system for gait recognition in patients with Parkinson's disease. Exp. Syst. Appl. **36**, 7033–7039 (2009). https://doi.org/10.1016/j.eswa.2008.08.076
7. Daliri, M.R.: Automatic diagnosis of neuro-degenerative diseases using gait dynamics. Meas. J. Int. Meas. Confed. **45**, 1729–1734 (2012). https://doi.org/10.1016/j.measurement.2012.04.013
8. Khorasani, A., Daliri, M.R., Pooyan, M.: Recognition of amyotrophic lateral sclerosis disease using factorial hidden Markov model. Biomed. Tech. **61**, 119–126 (2016). https://doi.org/10.1515/bmt-2014-0089
9. Krystkowiak, P., Blatt, J.-L., Bourriez, J.-L., Duhamel, A., Perina, M., Blond, S., Guieu, J.-D., Destée, A., Defebvre, L.: Effects of subthalamic nucleus stimulation and levodopa treatment on gait abnormalities in Parkinson disease. Am. Med. Assoc. **60**, 80–84 (2015)
10. Rodger, M., Young, W., Craig, C.: Synthesis of walking sounds for alleviating gait disturbances in Parkinson's disease. IEEE Multimed. **22**, 1 (2015). https://doi.org/10.1109/mmul.2015.21
11. Su, B.L., Song, R., Guo, L.Y., Yen, C.W.: Characterizing gait asymmetry via frequency sub-band components of the ground reaction force. Biomed. Signal Process. Control. **18**, 56–60 (2015). https://doi.org/10.1016/j.bspc.2014.11.008
12. Camara, C., Isasi, P., Warwick, K., Ruiz, V., Aziz, T., Stein, J., Bakštein, E.: Resting tremor classification and detection in Parkinson's disease patients. Biomed. Sig. Process. Control. **16**, 88–97 (2015). https://doi.org/10.1016/j.bspc.2014.09.006
13. Soleimanian, F., Mohammadi, P.: A case study of Parkinson's disease diagnosis using artificial neural networks. Int. J. Comput. Appl. **73**, 1–6 (2013). https://doi.org/10.5120/12990-9206
14. Nguyen, T., Malley, R., Inkelis, S.H., Kuppermann, N.: Comparison of prediction models for adverse outcome in pediatric meningococcal disease using artificial neural network and logistic regression analyses. J. Clin. Epidemiol. **55**, 687–695 (2002). https://doi.org/10.1016/S0895-4356(02)00394-3
15. Betancourt, H.E., Armijos, D.A., Martinez, P.N., Ponce, A.E., Ortega-Zamorano, F.: Portable expert system to voice and speech recognition using an open source computer hardware. In: Proceedings of the 2018 2nd European Conference on Electrical Engineering and Computer Science, EECS 2018, pp. 564–568 (2018). https://doi.org/10.1109/EECS.2018.00110
16. Tripoliti, E.E., Tzallas, A.T., Tsipouras, M.G., Rigas, G., Bougia, P., Leontiou, M., Konitsiotis, S., Chondrogiorgi, M., Tsouli, S., Fotiadis, D.I.: Automatic detection of freezing of gait events in patients with Parkinson's disease. Comput. Meth. Programs Biomed. **110**, 12–26 (2013). https://doi.org/10.1016/j.cmpb.2012.10.016

17. Mazilu, S., Blanke, U., Calatroni, A., Gazit, E., Hausdorff, J.M., Tröster, G.: The role of wrist-mounted inertial sensors in detecting gait freeze episodes in Parkinson's disease. Pervasive Mob. Comput. **33**, 1–6 (2016). https://doi.org/10.1016/j.pmcj.2015.12.007
18. Ertuğrul, Ö.F., Kaya, Y., Tekin, R., Almali, M.N.: Detection of Parkinson's disease by shifted one dimensional local binary patterns from gait. Exp. Syst. Appl. **56**, 156–163 (2016). https://doi.org/10.1016/j.eswa.2016.03.018
19. El Maachi, I., Bilodeau, G.A., Bouachir, W.: Deep 1D-Convnet for accurate Parkinson disease detection and severity prediction from gait. Exp. Syst. Appl. **143**, 113075 (2019). https://doi.org/10.1016/j.eswa.2019.113075
20. Jane, N., Nehemiah, K., Arputharaj, K.: A Q-backpropagated time delay neural network for diagnosing severity of gait disturbances in Parkinson's disease. J. Biomed. Inform. **60**, 169–176 (2016). https://doi.org/10.1016/j.jbi.2016.01.014
21. Zhao, A., Qi, L., Li, J., Dong, J., Yu, H.: A hybrid spatio-temporal model for detection and severity rating of Parkinson's disease from gait data. Neurocomputing **315**, 1–8 (2018). https://doi.org/10.1016/j.neucom.2018.03.032
22. Chang, D., Alban-hidalgo, M., Hsu, K.: Diagnosing Parkinson' s Disease From Gait, pp. 1–4. Stanford (2015)
23. Chen, P.H., Wang, R.L., Liou, D.J., Shaw, J.S.: Gait disorders in Parkinson's disease: assessment and management. Int. J. Gerontol. **7**, 189–193 (2013). https://doi.org/10.1016/j.ijge.2013.03.005
24. Zeng, W., Liu, F., Wang, Q., Wang, Y., Ma, L., Zhang, Y.: Parkinson's disease classification using gait analysis via deterministic learning. Neurosci. Lett. **633**, 268–278 (2016). https://doi.org/10.1016/j.neulet.2016.09.043
25. Cole, B.T., Roy, S.H., De Luca, C.J., Nawab, S.H.: Dynamical learning and tracking of tremor and dyskinesia from wearable sensors. IEEE Trans. Neural Syst. Rehabil. Eng. **22**, 982–991 (2014). https://doi.org/10.1109/TNSRE.2014.2310904
26. Mannini, A., Trojaniello, D., Cereatti, A., Sabatini, A.M.: A machine learning framework for gait classification using inertial sensors: application to elderly, post-stroke and Huntington's disease patients. Sensors (Switzerland) **16**, 134 (2016)
27. Wang, X.: Machine Learning for Gait Classification (2017)
28. Hoang, N.S., Cai, Y., Lee, C.W., Yang, Y.O., Chui, C.K., Heng Chua, M.C.: Gait classification for Parkinson's disease using stacked 2D and 1D convolutional neural network. In: International Conference on Advanced Technologies for Communications, October 2019, pp. 44–49 (2019). https://doi.org/10.1109/ATC.2019.8924567
29. Lee, S.H., Lim, J.S.: Parkinson's disease classification using gait characteristics and wavelet-based feature extraction. Exp. Syst. Appl. **39**, 7338–7344 (2012). https://doi.org/10.1016/j.eswa.2012.01.084
30. Borrione, P., Tranchita, E., Pierpaolo, S., Parisi, A.: Effects of physical activity in Parkinson's disease: a new tool for rehabilitation. World J. Methodol. **4**, 133 (2014). https://doi.org/10.5662/wjm.v4.i3.133
31. Post, B., Merkus, M.P., de Bie, R.M.A., de Haan, R.J., Speelman, J.D.: Unified Parkinson's disease rating scale motor examination: are ratings of nurses, residents in neurology, and movement disorders specialists interchangeable? Mov. Disord. **20**, 1577–1584 (2005). https://doi.org/10.1002/mds.20640
32. Goldberger, A.L., Amaral, L.A., Glass, L., Hausdorff, J.M., Ivanov, P.C., Mark, R.G., Mietus, J.E., Moody, G.B., Peng, C.K., Stanley, H.E.: PhysioBank, PhysioToolkit, and PhysioNet: components of a new research resource for complex physiologic signals. Circulation **101**, E215-220 (2000). https://doi.org/10.1161/01.cir.101.23.e215
33. Linna, M., Taimela, E., Apajasalo, M., Reijo, M.: Probabilistic sensitivity analysis for evaluating cost-utility of entacapone for Parkinson's disease. Exp. Rev. Pharmacoecon. Outcomes Res. **2**(2), 91–97 (2002). https://doi.org/10.1586/14737167.2.2.91

34. Modestino, E.J., Reinhofer, A.M., Blum, K., Amenechi, C., O'Toole, P.: Hoehn and yahr staging of Parkinson's disease in relation to neuropsychological measures. Front. Biosci. Landmark. **23**, 1370–1379 (2018). https://doi.org/10.2741/4649
35. Sarbaz, Y., Gharibzadeh, S., Towhidkhah, F., Banaie, M., Jafari, A.: A gray-box neural network model of Parkinson's disease using gait signal. Basic Clin. Neurosci. **2**, 33–42 (2011)
36. Hagan, M.T., Menhaj, M.B.: Training feedforward networks with the Marquardt algorithm. IEEE Trans. Neural Netwo. **5**, 989–993 (1994). https://doi.org/10.1109/72.329697
37. Khan, T., Westin, J., Dougherty, M.: Classification of speech intelligibility in Parkinson's disease. Biocybern. Biomed. Eng. **34**, 35–45 (2014). https://doi.org/10.1016/j.bbe.2013.10.003
38. Zhao, Y.J., Wee, H.L., Chan, Y.H., Seah, S.H., Au, W.L., Lau, P.N., Pica, E.C., Li, S.C., Luo, N., Tan, L.C.S.: Progression of Parkinson's disease as evaluated by Hoehn and Yahr stage transition times. Mov. Disord. **25**, 710–716 (2010). https://doi.org/10.1002/mds.22875
39. Staffen, W., Mair, A., Unterrainer, J., Trinka, E., Ladurner, G.: Measuring the progression of idiopathic Parkinson's disease with [123I] β-CIT SPECT. J. Neural Transm. **107**, 543–552 (2000). https://doi.org/10.1007/s007020070077

An Analysis of Deep Learning Architectures for Cancer Diagnosis

Galo Enrique Valverde Landivar, Jorge André Bedor Caballero, Daniel Humberto Plua Moran, Miguel Angel Quiroz Martinez(✉), and Maikel Yelandi Leyva Vazquez

Department of Computer Science, Universidad Politécnica Salesiana Sede Guayaquil, Chamber 227 y 5 de junio, Guayaquil, Ecuador
{gvalverde,dplua,mquiroz}@ups.edu.ec, jbedor@est.ups.edu.ec, mleyvaz@gmail.com

Abstract. It was analyzed the reference information on Deep Learning applications in the areas of diagnosis and prediction of different types of cancer. The problem is to perform the analysis and obtain the criteria to select a Deep Learning architecture for cancer diagnosis. The objective is to carry out an analysis of Deep Learning architectures and select a model to apply training and tests that assist in the diagnosis of cancer. It was used as a method the exploratory research and deduction to analyze the reference information on Deep Learning theories and architectures applied in cancer diagnosis; it also describes the reasons for selecting a model, scope, proposal, configuration parameters and structure for a CNN network. It resulted in the Impact of Deep Learning in cancer diagnosis, Training the CNN network, and Testing the CNN network. It was concluded that the 9-layer CNN Simple model used for training and testing on a data set of 8801 breast cancer images, has good properties and generates quantitative results for image classification; in the adopted model, a precision rate was obtained on the data set that reached 85.67% in training and 85.87% in tests; the quality of the model in classification tasks is 86%; this indicates the good stability and efficiency of the model.

Keywords: Deep learning · Cancer diagnosis · Machine learning · Architectures

1 Introduction

Deep Learning (DL) is an extensive area within Machine Learning (ML), ML is an area within Artificial Intelligence (AI); DL is used in various environments of human life because its algorithms process nonlinear information with hierarchical architectures that use trend characteristics and classification; this learning is also called representation learning [1].

DL is used in areas such as scam detection in financial applications, financial data succession forecasting, predictive and diagnostic analysis, medical image processing, power system scanning, and rendering systems [2].

M. Botto-Tobar et al. (Eds.): CIT 2020, AISC 1326, pp. 19–33, 2021.
https://doi.org/10.1007/978-3-030-68080-0_2

Imaging or biopsy is used to detect cancer; in the biopsy, the cells are benign or malignant, the disadvantage is the irregular shape of both types of cells and manual analysis by the health professional; algorithms with precise results are used in the images to determine if the cancer is benign or malignant based on the characteristics [3].

Deep Learning is an ideal universal learning approach to solving problems in different areas [1]; it is a network framework used for schematic recognition, imaging, computer vision, and bioinformatics; the architecture uses hidden input and output layers that are connected through a weight matrix [4]; it consists of multilayer networks of artificial neurons that identify relevant characteristics of the input data to detect and classify them, large data volumes of data may be labeled or unlabeled [5].

According to [1, 6] there are 4 types of algorithms in DL (Fig. 1):

- Supervised: Deep Neural Networks (DNN), Convolutional Neural Networks (CNN), Recurrent Neural Networks (RNN), Long Short Term Memory (LSTM), Gated Recurrent Units (GRU)
- Semi-supervised: Deep Reinforcement Learning (DRL), Generative Adversarial Networks (GAN)
- Unsupervised: Auto-Encoders (AE), Restricted Boltzmann Machines (RBM)
- Reinforcement: Bayesian Deep Learning (BDL)

Some characteristics of DL [1] are:

- Universal learning vision because it applies to any domain
- Robust because the optimal characteristics are assimilated by the algorithms
- General, because the algorithm is used in different environments
- Scalable because it supports multiple layers and nodes

DL is used in organ imaging recognition and fractionation, lesion segmentation or detection, localization of organic structures, tissue analysis, image restoration, evaluation of radiotherapy treatment [5]. The types of cancer reviewed by the references are breast [3], prostate, brain tumor [7], oral cavity [8], bladder, lung, lymphoma [9], lung, skin,

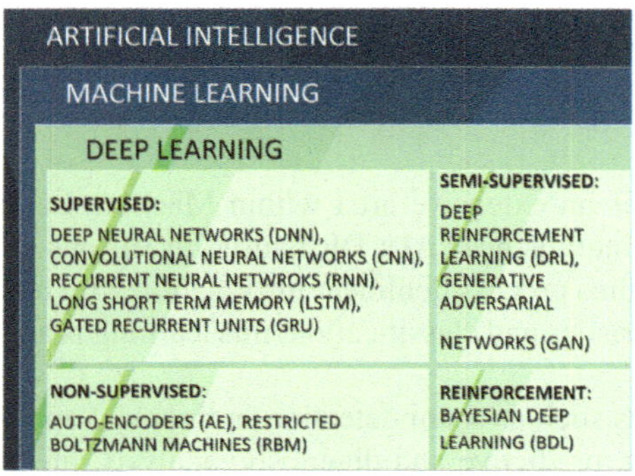

Fig. 1. Classification of Deep Learning algorithms. DL is an area that belongs to Machine Learning, and ML is within Artificial Intelligence. According to the references, there are four types of algorithms in DL [1].

brain, colorectal, neck, uterine liver [10]. Breast cancer is the most common worldwide with 2.4 million cases and 523,000 deaths annually [9].

The problem is to perform the analysis and obtain the criteria to select a Deep Learning architecture for cancer diagnosis.

Why is a deep learning architecture analysis necessary for cancer diagnosis?

To know about supervised architectures, models, or classification algorithms and select the most appropriate to adopt a diagnostic system; and determine an algorithm and architecture has the greatest potential to perform well in a variety of medical data sets for a type of cancer.

Other motivations are: different algorithms and DL architectures that can be applied to a set of medical data, execution of training and tests on robust algorithms, algorithms recognize patterns [2]; challenge in alternatives to obtain patterns inherent to the data set [11].

The objective is to carry out an analysis of Deep Learning architectures and select a model to apply training and tests that assist in the diagnosis of cancer.

As a method are applied the exploratory research and deduction to analyze the reference information on DL theories and architectures applied in cancer diagnosis.

2 Materials and Methods

In Materials the references on DL applications in the areas of cancer diagnosis and prediction are described; in Methods, it describes the scope, the proposal, the configuration, and structure parameters for a CNN network.

2.1 Materials

1) **Deep Neural Networks (DNN):** It consists of hidden multilayers with multiple hidden links [1]. DNN was applied in use for prediction of pathologies and took another network as input data [5]; it tested DNN with multiple image types with an average of 86.4% [6]; was used to classify breast cancer through a molecule classifier model [12].
2) **Convolutional Neural Networks (CNN):** Its architecture has three layers: convolution, maximum grouping, and classification, the first two are at the low and middle level of the network; here the output nodes of the convolution and maximum grouping layers extract the attributes of the input images through procedures on the input nodes [1]; this advanced neural networks have an advantage in the work of image classification according to experiences [7]; a signal passes through the network without loops or loops, it contains hidden layers with functions each [10]. CNN was applied in: Detection, segmentation, classification and reconstruction of images [2]; in a model for detection and classification of malignant cells in mammary images, the model obtained 97.67% of average accuracy [3]; in organ cancer detection, organ injury, patient response to therapies, image reconstruction, and image classification [5], different DL applications of cancer imaging were reviewed including segmentation of tumor lesions, cancer classification, and survival prognosis [6]; an image classifier was modeled to diagnose, the training precision rate

is 80.15% and the test precision rate is 78.15%, it was deduced that the model is continuously trained [7]; images of cancer in oral cavities were classified, here the filters are based on recognition of patterns of the visual cortex or texture, the results of the data set are precision of 88.3% [8]; the document has 63 studies of different cancers on CNN, this shows that it is widely used in cancer detection [10]; in tumor images, subclass classification and prediction [13], are performed, image classification, samples, lymph node detection, lung node detection [11]; language diagnosis from image classification [14]; and diagnosis of tumor in the brain through magnetic resonance imaging [15]; categorization in detection, classification, segmentation and prediction of survival, for the study of colon cancer [16]; macro classification, classification and detection of breast cancer images [17]; new model of artificial CNN for the diagnosis of gastric cancer [18]; model for survival prognosis in gastric cancer [19].

3) **ResNet:** It is a model that adds more non-linear capacity that going from connected to less connected models, it uses deep and superficial networks for error control in training [9]. ResNet was applied in: lung classification and segmentation of lesions in entire image with 89% of the tests [5]; Pre-trained and improved neural networks were applied for four types of cancer with an average precision of 99.8% [9].
4) **Recurrent Neural Networks (RNN):** It analyzes successive data such as text or voice because its internal memory state retains information on previous data locations [5]; this network allows the operation in a succession of vectors through time, in architecture, the inputs of the hidden layers are the output of hidden layers [1]. RNN was applied in: ventricular quantification with CNN and Bayesian tasks [5]; used in identification systems, series analysis, classifications, voice and audio processing, sound prediction, determining heart failure, data traffic analysis, and in driver behavior prediction [1].
5) **Long Short Term Memory (LSTM):** This architecture improves RNN memory retention and is better at various tasks such as captioning images, voice confirmation, and machine translation [5]; the information is updated in the state of the cell or gate and can be processed temporarily, there is a version with the gates closed [1]. LSTM was applied to obtain diagnostic images that may be 3D [5].
6) **Gated Recurrent Units (GRU):** It is an LSTM variation that works with recurring networks, has a low computation cost, is simpler and lighter; this technique harmonizes the input and output inputs towards an update gate in conjunction with the cell status. GRU was applied in combination with LSTM and audio tagging [1].
7) **Deep Reinforcement Learning (DRL):** It is an approach that experiments to act with ordinary sense from the unknown ecosystem, it is applied to computing, engineering and mathematics, robotics, control of power stations, neuroscience, economics, and decision making; this learning is a practice on what to do and how to correspond between a situation and an action. DRL was applied in medical care through devices, optimizing learning policies, observation, and action, energy management, improving facial images [1].
8) **Generative Adversarial Networks (GAN):** The two neural networks compete for a zero-sum game, here the competitors are the discriminator and the generator; In the generation of images, the generator starts with a Gaussian echo to conceive images and the discriminator determines the quality of the images generated [1]. It has two

networks in its architecture; the first network is a generator that delivers data instances of the previously trained data, the second network is discriminating that establishes the probability of the generated candidates corresponding training; These two networks are trained in conjunction with back expansion to obtain accurate samples [5]. GAN was applied in: image processing to improve resolution, segmentation, synthesizing, image blending, image manipulation in real-time audio processing; in medical images, it is used to eliminate noise, segment tumors, MRI analysis [1]; imaging, feature recognition, faulty data, three-dimensional image manipulation [2].

Table 1 presents as a summary the references with their revised architectures, the reference [4] is not in the table because it was only used to obtain the concept of DL; it contains the architectures studied by these references, with the type of cancer that they reviewed with their training data and quantitative results; Architectural analysis references do not specify results or cannot be averaged because the authors reviewed between different architectures.

Table 1. Summary of references.

Ref	Application	Architectures	Data	Results	Cancer
[1]	Architecture analysis	RNN, LSTM, GRU, DRL, GAN	Images	No specified	Named: Breast cancer
[2]	Architecture analysis	CNN, ResNet, LSTM, GAN	Test	No specified	Named: Skin cancer, brain tumor
[3]	Detect and classify breast cancer	CNN	Images	97.52% accuracy	Studied: Breast cancer
[5]	Architecture analysis	DNN, CNN, ResNet, RNN, LSTM, GAN	Images	No specified	Named: Lung nodule, tissues, interstitial, calcium score, ventricular, liver tumor, liver fibrosis, fatty liver disease, brain
[6]	Segmentation, classification, and prediction of images	DNN	Molecular Imaging	86.4% accuracy	Studied: Prostate cancer, brain tumor, lung tumor, breast cancer
[7]	Diagnosis and classification in prostate cancer	CNN	Magnetic resonance	78.15% accuracy	Studied: Prostate Cancer

(continued)

Table 1. (*continued*)

Ref	Application	Architectures	Data	Results	Cancer
[8]	Diagnosis and classification in oral cancer	CNN	Images	88.3%	Studied: Cancer in oral cavities
[9]	Classification of Breast Cancer	ResNet	Images	99.8%	Studied: Breast, Bladder, Lung Cancer and Lymphoma
[10]	Detection of different types of cancer	CNN	Biopsy images	No specified	Studied: Breast cancer, lung cancer, skin cancer, prostate cancer, brain cancer, colon cancer Named: cervix, bladder cancer, liver cancer
[11]	Architecture analysis	CNN	Test	No specified	Named: Breast Cancer and Brain Cancer
[12]	Classification of molecules	DNN	Images	96.3% accuracy	Study of breast cancer
[13]	Classify benign or malignant breast tumors	CNN	Histopathological images	95.4% accuracy	Studied: Breast cáncer
[14]	Disease progression in tongue	CNN	Tongue images	90.5% Accuracy	Studied: Tongue Cancer Named: Diabetes, breast cancer, gastritis
[15]	Classify images and diagnosis of brain tumor	CNN	Magnetic resonance	97.2% accuracy	Studied: Brain tumor
[16]	Architecture analysis	CNN, LSTM	Images	No specified	Study of colon cancer
[17]	Architecture analysis	CNN, RNN, LSTM, RestNet	Images	No specified	Study of breast cancer

(*continued*)

Table 1. *(continued)*

Ref	Application	Architectures	Data	Results	Cancer
[18]	Diagnosis of cancer	CNN artificial	Images	95% accuracy	Study of gastric cancer
[19]	Classification of cells	CNN	Images	95% accuracy	Study of gastric cancer
[20]	Architecture analysis	DNN, CNN, RNN, GAN	Test	No specified	Named: brain cancer, prostate cancer, skin cancer, rectal cancer, bladder cancer, liver tumor, cervical cancer, lung cancer, pediatric cancer, head, and neck cancer

2.2 Methods

1) **Reasons to adopt CNN:**

- A significant number of references cite the usefulness and advantages of CNN.
- CNN generates an output correlated with a quantitative medical measure [5].
- It has a variety of versions and applications in deep learning because these form multiple layers of stronger neurons, they are general in large data sets; architectural versions are the field of computer vision that analyze two-dimensional and more dimensional images on color images [13].

2) **Scope for the proposed CNN network**

- Adopt and apply a CNN network for cancer diagnosis
- The platform used is in Intel Core i3-6006U 2.00 GHz CPU, 8 GB of RAM, 64-bit operating system and 512 GB SSD
- Training and testing phases are carried out in 100 iterations
- The data set is 8801 breast cancer images, 80% (7040 images) is for training and 20% (1761 images) is for testing
- Python software and SciPy libraries are used that contain modules for optimization, linear algebra, integration, interpolation, special functions, image processing, and resolutions.
- 0 is for Negative result and 1 is for Positive result
- For the execution of the software, random samples are taken, not in sequence, it is mixed; Fig. 2 first image of the iterations
- The shape of the arrays is $21 \times 28 \times 3$; 3 is the number of colors each image has: RGB (red, green, blue) with values from 0 to 255

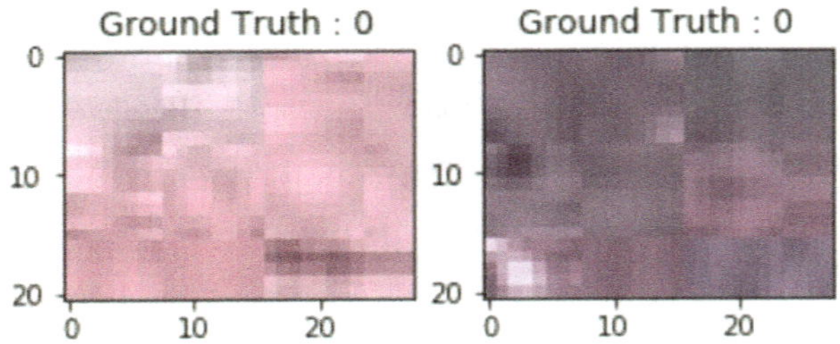

Fig. 2. Sample of the first image for 100 iterations (epoch).

- The shape data for the training of 7040 images are: (7040, 21, 28, 3)
- The form data for the 1761 image tests are: (1761, 21, 28, 3)
- Data is subdivided back into 80-20 for training and testing with a validation subset; results from 7040 images in 5632 images (80%) and 1408 images (20%)
- Training is now (5632, 21, 28, 3) and the validation subset is (1408, 21, 28, 3)

3) **Network configuration parameters**

- Learning rate value is 0.001 (INIT_LR = 1e−3)
- Number of iterations to the data set for the second group (epochs = 100)
- Number of images obtained from memory (batch_size = 64)

4) **Structure of the CNN system**

It is used a CNN trained model, this model has good results in biomedical, the network has 9 layers in total; these layers are Conv2D, LeakyReLU, MaxPooling2, Dropout, Flatten and Dense. The names, parameter numbers, and other characteristics of the layers are in Table 2.

Table 2. System layer model.

Layer	Type	Output	Parameters
conv2d_6	Conv2D	(None, 21, 28, 32)	896
Leaky_re_lu_11	LeakyReLU	(None, 21, 28, 32)	0
max_pooling2d_6	MaxPooling2	(None, 11, 14, 32)	0
dropout_11	Dropout	(None, 11, 14, 32)	0
flatten_6	Flatten	(None, 4928)	0
dense_11	Dense	(None, 32)	157728
leaky_re_lu_12	LeakyReLU	(None, 32)	0
dropout_12	Dropout	(None, 32)	0
dense_12	Dense	(None, 2)	66

The model layers are explained below:

- Conv2D: First layer of neurons (2-dimensional convolutional), here the 21 × 28 × 3 images enter; with 32 filters (kernel) of size 3 × 3 that detect certain characteristics of the image.
- LeakyReLU: This function is for the activation of neurons; It also transforms the read values, multiplies the negatives by a corrective coefficient and leaves the positive this enter.
- MaxPooling: (2 × 2) reduces the incoming image from 21 × 28 in half, (11 × 14) to maintain unique features detected.
- DropThis: this technique intervenes by removing network nodes to avoid overfitting.
- Flatten: for the 32 filters and creates a layer of 32 traditional neurons, that is, Dense.
- The output layer with 10 neurons with Softmax activation, for correspondence with the "hot encoding".

3 Results

3.1 Impact of Deep Learning in Cancer Diagnosis

Table 3. Profit of architectures by references.

Architectures	Ref	Training	Dataset
Deep Neural Networks (DNN)	[5]	1 reference	95 patients
	[6]	1 training	Not specified
	[12]	100 iterations	1975 patients
	[20]	8 references	Not specified
Convolutional Neural Networks (CNN)	[2]	17 references	No specified
	[3]	1000 iterations	8000 images
	[5]	23 references	No specified
	[6]	16 training	No specified
	[7]	6000 iterations	10056 images
	[8]	7 training	11000 images
	[10]	63 references	No specified
	[13]	1 training	7909 images
	[11]	19 references	No specified
	[14]	1 training	1548 images
	[15]	7 training	253 images
	[16]	127 references	No specified
	[17]	31 references	118968 images
	[18]	2 training	572 patients
	[19]	100 iterations	640 patients
	[20]	62 references	No specified

(continued)

Table 3. (*continued*)

Architectures	Ref	Training	Dataset
ResNet	[5]	665 training	166 nodules
	[5]	15 training	8 patients
	[9]	3000 iterations	7909 images
	[2]	3 references	No specified
	[17]	3 references	11675 images
Recurrent Neural Networks (RNN)	[1]	17 references	No specified
	[5]	1 reference	145 cases
	[17]	1 reference	736 images
	[20]	1 reference	No specified
Long Short Term Memory (LSTM)	[1]	4 references	No specified
	[5]	562 training	141 cases
	[2]	6 reference	6 proposals
	[16]	3 references	No specified
	[17]	2 references	830 images
Gated Recurrent Units (GRU)	[1]	3 references	No specified
Deep Reinforcement Learning (DRL)	[1]	11 references	No specified
Generative Adversarial Networks (GAN)	[1]	26 references	No specified
	[5]	1 references	182 cases
	[2]	6 references	No specified
	[17]	2 references	19157 images
	[20]	3 references	No specified

References were tabulated to obtain points of view or approaches of the algorithms, architectures, dataset, and types of cancer studied. Table 3 is grouped by architectures, to obtain the references, the number of training sessions, and the number of data sets that were processed by each article.

3.2 Training the CNN Network

A trained model gets better-extracted features and increases its efficiency. Loss values are the deviation between the training value and the actual value. The training and validation introduces the images, pixels, arrays, colors, filters and the network regulates itself (learns) with the best weights for the 158690 interconnections to distinguish the 2 diagnoses: positive or negative. The training dataset is 7040 images (80% of 8801 images) in 100 iterations, this validation subset is divided: for training with 5632 images (80% of 7040); for testing with 1408 images (20% of 7040).

The set of loss values for training has a downward trend, Fig. 3 presents the minimization of loss in the training process and has 0.3681 units of minimum loss in iteration number 97.

Figure 4 the training precision can be kept close to 0.85 after iteration number 45; the maximum precision is 0.8567 in iteration number 92; the model in this phase is sufficiently trained and stable.

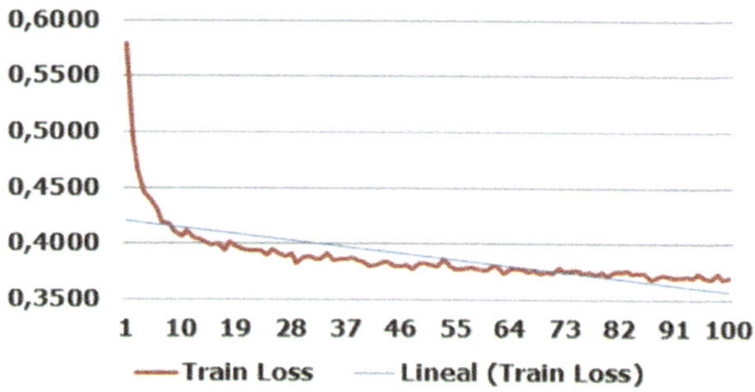

Fig. 3. Loss values in training versus iterations.

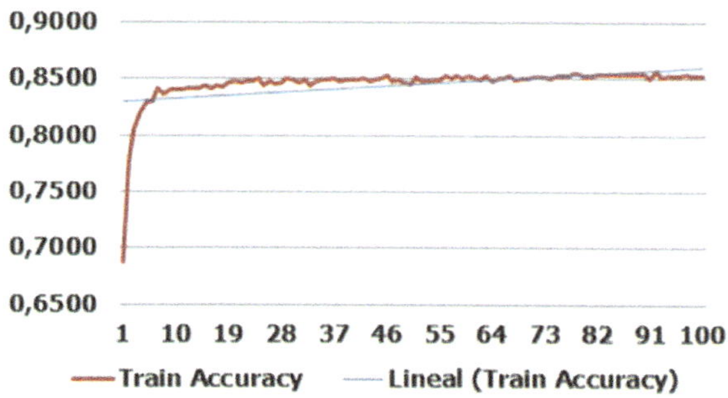

Fig. 4. Training accuracy versus iterations.

The set of loss values for testing has a downward trend, Fig. 5 presents loss minimization in the testing process and has 0.3612 units of minimum loss in iteration number 97.

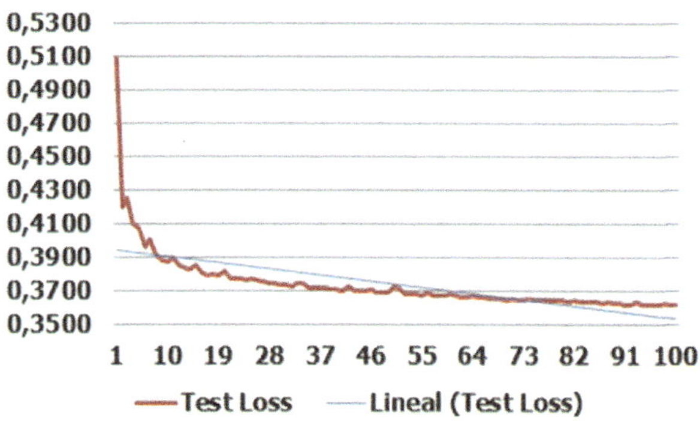

Fig. 5. Loss values in tests versus iterations.

The precision is the degree of closeness of the measurements of several images, here the precision curve is a measurement between the trained and the real value; Fig. 6 shows the precision curve, the horizontal axis is the iteration number and the vertical axis is the percentage of precision; the precision remains between 0.85 and 0.86 after iteration number 47; that is, the model is stable and has better efficient training.

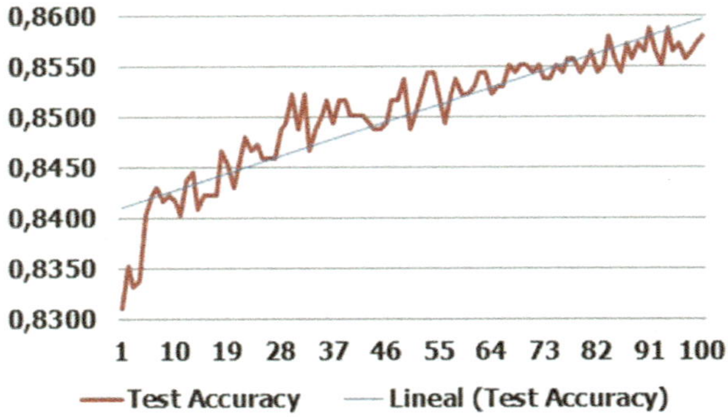

Fig. 6. Test accuracy versus Iterations.

3.3 Testing the CNN Network

The test is 1761 images (20% of 8801 images). Accuracy of 38.86% was achieved in 100 complete iterations of the training set; in the validation set, it reaches 84.78% accuracy. Accuracy measures the percentage of cases that the model has hit. The classification of new images can now be predicted; any image in the test set is used, the prediction gets the classes as an integer. The total number of classes obtained is 1761. The number of correct labels in 100 iterations is 1493 in this test.

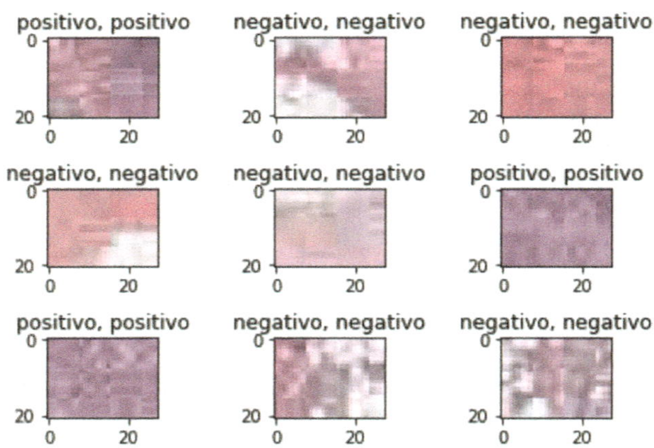

Fig. 7. Prediction of an image of the data set.

According to the training of the data set and after obtaining characteristics in the test phase, the diagnosis and prediction of the image (Fig. 7) is negative. For these tests in

100 iterations: 1147 images were found to be benign cancer and 614 images were found to be cancer, the accuracy of the model is 86% (Table 4).

Table 4. Prediction for 100 iterations.

Class type	Accuracy	Recall	F1-score	Images
Benign	0.85	0.96	0.90	1147
Malignant	0.89	0.68	0.77	614
Accuracy			0.86	1761
Macro avg	0.87	0.82	0.84	
Weighted avg	0.86	0.86	0.86	

The comparative analysis of the model used shows Class Type for the type of Benign or Malignant cancer and its respective precision, recovery, and F1 score; Precision metric measures the quality of the ML model in classification tasks; Recall metric informs us of the quantity that the ML model is capable of identifying; F1-Score value is used to combine the precision and recall measurements into a single value, making it easy to compare the combined performance of precision and completeness between various solutions.

4 Discussion

- It was proposed to use the CNN algorithm in breast cancer for image classification based on information from the references; 16 of the 20 references carry out trainings on CNN.
- The references and the results obtained in the applied model demonstrate the great value in the clinical area; This research is of reference value for other types of cancer.
- Only a simple 9-layer CNN network structure was used, the dataset is 8801 breast cancer images, the convolution nuclei are $21 \times 28 \times 3$ pixels, and the iterations parameter is 100.
- In the training phase of the CNN network, the loss values indicate the deviation in training is 0.3681 units and in tests it is 0.3612 units; the accuracy values for classification in training is 85.67% and in tests it is 85.87%; this indicates the good stability and efficiency of the model.
- Our research agrees with the other references in the following: Breast cancer data training and testing with references [3, 6, 9, 10, 13, 17]; in the training and testing phase, we obtained quantitative results to assess the adopted architecture, in the same way as [3, 7, 8, 13–15]; we applied CNN architecture to breast cancer as [3, 6, 9, 13, 17]; all the references of our research analyzed or revised the CNN architecture, except [1, 4, 9, 12].
- The results show that the adopted algorithm achieves an acceptable performance of detection and classification precision in breast cancer images

5 Future Work and Conclusions

As future work, a Deep Learning Analysis was proposed to diagnose cancer images with noise.

It was concluded that the 9-layer CNN Simple model used for training and testing on a data set of 8801 breast cancer images, has good properties and generates quantitative results for image classification; in the adopted model, a precision rate was obtained on the data set that reached 85.67% in training and 85.87% in tests; the quality of the model in classification tasks is 86%; this indicates the good stability and efficiency of the model.

Acknowledgment. This work has been supported by the GIIAR research group and the Salesian Polytechnic University of Guayaquil.

References

1. Alom, M.Z., Taha, T.M., Yakopcic, C., Westberg, S., Sidike, P., Nasrin, M.S., Hasan, M., Van Essen, B.C., Awwal, A.A.S., Asari, V.K.: A state-of-the-art survey on deep learning theory and architectures. Electron **8**, 292 (2019). https://doi.org/10.3390/electronics8030292
2. Sengupta, S., Basak, S., Saikia, P., Paul, S., Tsalavoutis, V., Atiah, F., Ravi, V., Peters, A.: A review of deep learning with special emphasis on architectures, applications and recent trends. Knowl. Based Syst. **194**, 105596 (2020). https://doi.org/10.1016/j.knosys.2020.105596
3. Khan, S.U., Islam, N., Jan, Z., Ud Din, I., Rodrigues, J.J.P.C.: A novel deep learning based framework for the detection and classification of breast cancer using transfer learning. Pattern Recognit. Lett. **125**, 1–6 (2019). https://doi.org/10.1016/j.patrec.2019.03.022
4. Alakwaa, F.M., Chaudhary, K., Garmire, L.X.: Deep learning accurately predicts estrogen receptor status in breast cancer metabolomics data. J. Proteome Res. **17**, 337–347 (2018). https://doi.org/10.1021/acs.jproteome.7b00595
5. Sahiner, B., Pezeshk, A., Hadjiiski, L.M., Wang, X., Drukker, K., Cha, K.H., Summers, R.M., Giger, M.L.: Deep learning in medical imaging and radiation therapy. Med. Phys. **46**, e1–e36 (2019). https://doi.org/10.1002/mp.13264
6. Xue, Y., Chen, S., Qin, J., Liu, Y., Huang, B., Chen, H.: Application of deep learning in automated analysis of molecular images in cancer: a survey. Contrast Media Mol. Imaging **2017**, 9512370 (2017). https://doi.org/10.1155/2017/9512370
7. Liu, Y., An, X.: A classification model for the prostate cancer based on deep learning. In: Proceedings of the 2017 10th International Congress on Image and Signal Processing, BioMedical Engineering and Informatics, CISP-BMEI 2017, 1–6 January 2018 (2018). https://doi.org/10.1109/CISP-BMEI.2017.8302240
8. Aubreville, M., Knipfer, C., Oetter, N., Jaremenko, C., Rodner, E., Denzler, J., Bohr, C., Neumann, H., Stelzle, F., Maier, A.: Automatic classification of cancerous tissue in laserendomicroscopy images of the oral cavity using deep learning. Sci. Rep. **7**, 1–10 (2017). https://doi.org/10.1038/s41598-017-12320-8
9. Jannesari, M., Habibzadeh, M., Aboulkheyr, H., Khosravi, P., Elemento, O., Totonchi, M., Hajirasouliha, I.: Breast cancer histopathological image classification: a deep learning approach. In: Proceedings of the 2018 IEEE International Conference on Bioinformatics and Biomedicine, BIBM 2018, pp. 2405–2412 (2019). https://doi.org/10.1109/BIBM.2018.8621307

10. Hu, Z., Tang, J., Wang, Z., Zhang, K., Zhang, L., Sun, Q.: Deep learning for image-based cancer detection and diagnosis – a survey. Pattern Recogn. **83**, 134–149 (2018). https://doi.org/10.1016/j.patcog.2018.05.014
11. Chan, H.P., Samala, R.K., Hadjiiski, L.M., Zhou, C.: Deep learning in medical image analysis. Adv. Exp. Med. Biol. **1213**, 3–21 (2020). https://doi.org/10.1007/978-3-030-33128-3_1
12. Mohaiminul Islam, M., Huang, S., Ajwad, R., Chi, C., Wang, Y., Hu, P.: An integrative deep learning framework for classifying molecular subtypes of breast cancer. Comput. Struct. Biotechnol. J. **18**, 2185–2199 (2020). https://doi.org/10.1016/j.csbj.2020.08.005
13. Nawaz, M., Sewissy, A.A., Soliman, T.H.A.: Multi-class breast cancer classification using deep learning convolutional neural network. Int. J. Adv. Comput. Sci. Appl. **9**, 316–322 (2018). https://doi.org/10.14569/IJACSA.2018.090645
14. Wang, X., Liu, J., Wu, C., Liu, J., Li, Q., Chen, Y., Wang, X., Chen, X., Pang, X., Chang, B., Lin, J., Zhao, S., Li, Z., Deng, Q., Lu, Y., Zhao, D., Chen, J.: Artificial intelligence in tongue diagnosis: using deep convolutional neural network for recognizing unhealthy tongue with tooth-mark. Comput. Struct. Biotechnol. J. (2020). https://doi.org/10.1016/j.csbj.2020.04.002
15. Çinar, A., Yildirim, M.: Detection of tumors on brain MRI images using the hybrid convolutional neural network architecture. Med. Hypotheses **139**, 109684 (2020). https://doi.org/10.1016/j.mehy.2020.109684
16. Pacal, I., Karaboga, D., Basturk, A., Akay, B., Nalbantoglu, U.: A comprehensive review of deep learning in colon cancer. Comput. Biol. Med. **126**, 104003 (2020). https://doi.org/10.1016/j.compbiomed.2020.104003
17. Pang, T., Wong, J.H.D., Ng, W.L., Chan, C.S.: Deep learning radiomics in breast cancer with different modalities: overview and future. Exp. Syst. Appl. **158**, 113501 (2020). https://doi.org/10.1016/j.eswa.2020.113501
18. Sun, R.J., Fang, M.J., Tang, L., Li, X.T., Lu, Q.Y., Dong, D., Tian, J., Sun, Y.S.: CT-based deep learning radiomics analysis for evaluation of serosa invasion in advanced gastric cancer. Eur. J. Radiol. **132**, 109277 (2020). https://doi.org/10.1016/j.ejrad.2020.109277
19. Zhang, L., Dong, D., Zhang, W., Hao, X., Fang, M., Wang, S., Li, W., Liu, Z., Wang, R., Zhou, J., Tian, J.: A deep learning risk prediction model for overall survival in patients with gastric cancer: a multicenter study. Radiother. Oncol. **150**, 73–80 (2020). https://doi.org/10.1016/j.radonc.2020.06.010
20. Meyer, P., Noblet, V., Mazzara, C., Lallement, A.: Survey on deep learning for radiotherapy. Comput. Biol. Med. **98**, 126–146 (2018). https://doi.org/10.1016/j.compbiomed.2018.05.018

Identifying Similar Groups of Countries According to the Impact of Corona Virus (COVID-19) by a Two-Layer Clustering Method

Juan Riofrío(✉), Carlos Muñoz-Moncayo, Isidro R. Amaro, and Israel Pineda

Yachay Tech University, Hda. San José S/n Y Proyecto Yachay,
Urcuquí 100119, Ecuador
{juan.riofrio,carlos.munoz,iamaro,ipineda}@yachaytech.edu.ec

Abstract. This paper presents a new clustering algorithm to identify groups of countries. First, a layer of several clustering methods is applied to the original dataset. Then, after performing dimensionality reduction techniques like t-SNE or SOM on the resulting data, a second clustering layer (K-Means) is applied to identify the final clusters. This method is applied to a dataset from 163 countries, considering the following variables population, area, Gross Domestic Product (GDP), Gross Domestic Product adjusted for Purchase Power Parity (GDP-PPP), and COVID-19 related data (Confirmed, Recovered, and Deaths). The implementation with SOM dimensionality reduction outperformed the one with t-SNE for the considered dataset. We expect that using this information, countries can have an insight on which measures against COVID-19 replicate or avoid, based on the results in countries from the same cluster.

Keywords: Clustering · COVID-19 · K-Means · Self-Organizing Map · Two-layer clustering

1 Introduction

Since an abnormal amount of transmittable pneumonia cases, similar to SARS-CoV-1, was reported in Wuhan on December 2019, authorities of China began fighting what seemed to be an epidemic outbreak. Indeed, this disease was identified as Novel Coronavirus (COVID-19), caused by the extremely contagious virus SARS-CoV-2. Some of its common symptoms are fever, dry cough, and dyspnoea; in severe cases it can even cause pneumonia, sepsis, and septic shock, see e.g., [4]. On March 11, 2020, the WHO characterized COVID-19 as a pandemic; since then, unwavering efforts from countries worldwide have been made to control the spreading and develop a vaccine to prevent it. Science-based decisions taken by politicians around the world is a decisive factor in what course the human race will take against such a threat against its well-being. Therefore, to analyze how the virus spreads and its global impact is essential.

M. Botto-Tobar et al. (Eds.): CIT 2020, AISC 1326, pp. 34–48, 2021.
https://doi.org/10.1007/978-3-030-68080-0_3

Data Science, with its wide and versatile fields of study such as Big Data, Data Mining, and Machine Learning, is a well-suited discipline to analyze large datasets as the provided by the COVID-19 tracking worldwide. Clustering approaches have been proposed and shown to be of great significance for decision takers, see [10]. In this work, we present a novel method, based in a two-layer clustering, applied to data concerning the global impact of COVID-19. In the first layer of clustering, we use several well-known methods, such as Affinity Propagation, Agglomerative Clustering, BIRCH, K-Means, and Mini Batch K-Means. In the second layer, we use a combination of t-SNE/SOM and K-Means, as illustrated in Fig. 1. The variables considered in this work are the population, area, the number of confirmed cases, recovered cases, and deaths related to COVID-19 in 163 countries.

The remaining of this paper is structured as follows: Sect. 2 describes the dataset used and overviews the clustering methods considered in this work as well as the proposed experimental methodology. Section 3 gathers the results and discussion. Finally, the concluding remarks are drawn in Sect. 4.

2 Materials and Methods

In this section, we give an insight into the different stages of the algorithm. The first stage is fetching and preprocessing of data regarding COVID-19. Then, we review some clustering, comparison, and dimensionality reduction methods. Figure 1 presents the methodology and all the steps executed during the research.

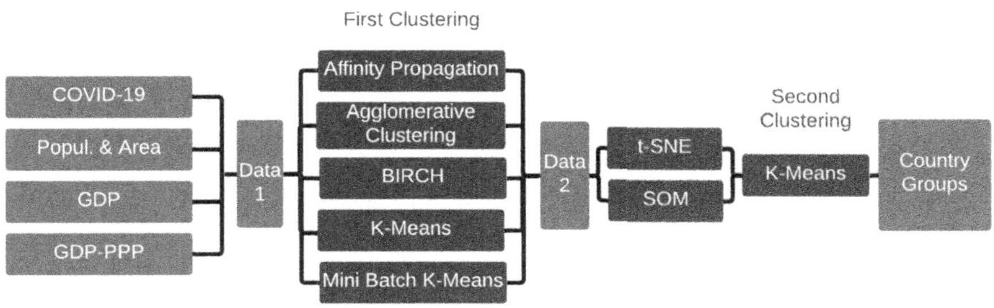

Fig. 1. Methodology followed during the research

2.1 Dataset

In the face of the Novel Corona Virus outbreak, the United Nations Office for the Coordination of Humanitarian Affairs (OCHA), in collaboration with the Johns Hopkins University Center for Systems Science and Engineering, gathered data from 266 countries related to the health crisis. We fetched this data from the Humanitarian Data Exchange, a platform managed by the OCHA [16]. For technical reasons, we reduced the number of observations to 163. The dataset contains the total number of *confirmed cases*, *deaths*, and *recoveries* of COVID-19 of each country. It has been updated each day from January 22 of 2020, to June 29 of 2020. In this study, we consider the data corresponding to *June 23, 2020*.

To develop a broader analysis, and since recent works, e.g., [3,6], suggest that a correlation between population density and impact of COVID-19 may exist, we considered the area and population of each country. This data was retrieved from [13], and its last update was done on April 18, 2020. Finally, as stated in the previous section, it can be expected that the economic status of a country influences the way COVID-19 affects it and vice versa. Therefore, we took into account the Gross Domestic Product (GDP) and the Gross Domestic Product adjusted for Purchase Power Parity (GDP-PPP). Data corresponding to GDP and GDP-PPP were fetched from the International Monetary Fund report [19] and [20], and last updated in April 2019 and April 2020, respectively.

2.2 Data Clustering

When searching for patterns within a voluminous data set, it is key to classify the observations in distinguishable groups. Data clustering is a widely used method to do so. Its main characteristic is that classification groups or *clusters* are not explicitly provided a priori. They depend on the data instead. The process of performing clustering depends on several factors, such as the number of clusters, the desired approach, and the nature of the data.

We want elements of a cluster to be similar and elements from different clusters to be dissimilar. A widely used approach is to choose some observations adequately, often referred to as *exemplar* of each cluster, and then assigning each observation to the cluster corresponding to the closest exemplar observation. To do this, it is required to quantify the distance between observations, we do so using a function denoted *similarity measure*.

Table 1. First rows of Data 1

Country	Population	Area (Km^2)	GDP (Billions)	GDP-PPP (Billions)	Confirmed	Recovered	Deaths
Afghanistan	38742911	652860	20.682	74.79	29481	9260	618
Albania	2878420	27400	17.21	38.08	2047	1195	45
Algeria	43685618	2381740	193.0	638.36	12076	8674	861
...	...	...	...	...	...	...	...

As shown in Table 1, after preprocessing, all of the variables considered in our first dataset, which we will refer as *Data 1*, are numerical. Therefore, standard similarity measures between entries such as Euclidean, Minkowski or Mahalanobis distance may be used [15]. With the help of the Machine Learning package for Python, Scikit-learn [11], we applied different clustering methods on Data 1. Let us briefly describe them, assuming we are dealing with n observations and

p variables, thus a data matrix $Y_{n\times p} = (y_{ij})$; we denote the similarity between observations y_i and y_j by $s_{ij} := s(y_i, y_j)$.

Affinity Propagation. In this method, the similarity is often computed by $s_{ij} = -\|y_i - y_j\|^2$, although it can be set according to different necessities. The main characteristic of Affinity Propagation is that all the observations are candidates to be exemplars from the beginning [5]. The process followed to define the exemplars. Therefore, the clusters rely on sending messages between observations, and the following iterative process does it:

1. For all i, j, the responsibility r_{ij}, is updated. Each r_{ij} is a real-valued indicator of how optimal is the observation x_j to be set as an exemplar for the observation x_i, compared to other candidate observations to be exemplars for x_i. It is computed by
$$r_{ij} = s_{ij} - \max_{j'\neq j}\{a_{ij'} + s_{ij'}\}, \tag{1}$$
where a_{ij} is the availability; it denotes how ideal it would be for x_i to choose x_j as its exemplar, considering how optimal is x_j to be the exemplar for other observations.
2. For all i, j the availability is updated by
$$a_{ij} = \min\left(0, r_{jj} + \sum_{i'\notin\{ij\}} \max(0, r_{i'j})\right), \quad i \neq j, \tag{2}$$
$$a_{jj} = \sum_{i'\neq j} \max(0, r_{i'j}). \tag{3}$$

This process is repeated until the members of each cluster remain fixed for several iterations or a limit number of iterations is exceeded. The final exemplars are those x_j that satisfy $r_{jj} + a_{jj} > 0$.

Agglomerative Clustering. This is a hierarchical method, in which to begin, every observation is set as a cluster and if not cut before, the whole dataset becomes a unique cluster at the end. Since the last is a trivial result, it is often required that the final number of clusters is given a priori. First, a distance, metric, or dissimilarity measure between observations needs to be defined. In our case, it is the Manhattan distance:

$$d_{ij} := \|x_i - x_j\|_1 = \sum_{k=1}^{p} |x_{ik} - x_{jk}|. \tag{4}$$

Then a linkage criteria or distance between clusters is defined, we use the complete linkage:

$$Lnk(A, B) := \max\{d_{ij} \,:\, x_i \in A,\, x_j \in B\}. \tag{5}$$

The agglomeration of clusters is done in an iterative process in which, at each step, two clusters are merged into one if the linkage between them is minimized. This is repeated until the desired number of clusters is reached. As can be expected, the complexity of this algorithm is considerable, $O(n^3)$ for time and $O(n^2)$ for memory.

BIRCH. This method works by building a Clustering Feature Tree (CFT) for the data, which is compacted into a set of Clustering Feature Nodes (CFN). The CFN has subclusters called Clustering Feature subclusters (CFS), and CFS can have CFN as children if they are not in terminal CFN (leaves). Each CFS carries the number of samples in a subcluster, the total sum of samples, the squared euclidean sum of samples, and the centroids. This is the information needed for clustering, and it is done to not handle the whole dataset, which is especially useful when dealing with voluminous datasets.

The algorithm provided by Scikit [11] requires two hyperparameters: A threshold to control the distance between a subcluster and a potential member observation, and a branching factor that limits the number of subclusters in a CFN. It works in the following way:

1. An observation is placed in the first CFN or root and then merged with one of its subclusters such that the radius is minimized and the threshold and branching constraints. If there are child nodes to this subcluster, this process is repeated until a leaf is reached; then, the properties of all crossed subclusters is updated.
2. If the radius of the subcluster exceeds the threshold and has a greater amount of subclusters than the branching factor, the observation is stored in a separate space. The two subclusters with greater distance between them are considered, and their subclusters are divided into two groups based on the distance between them.
3. If the separate node has a parent subcluster and the branching factor is exceeded, the parent is divided. If there is still no space, this process is repeated until the root is reached.

K-Means. This is the main method considered in this work; through it, we want to partition our dataset into k clusters, while minimizing the variance within each cluster or cluster's *inertia*, i.e., we want to minimize

$$\sum_{i=1}^{k} \sum_{x \in C_i} \|x - \mu_i\|^2, \tag{6}$$

Where each C_i is a cluster and μ_i its corresponding mean. The clustering is performed in the following way:

1. A centroid for each cluster is selected (k in total); in the simplest case these are observations.

2. Clusters are constructed by assigning each observation to its nearest centroid.
3. For each cluster C_i, the mean μ_i is computed and assigned as the new centroid.
4. For each cluster, the distance $d(\mu_i, \widetilde{\mu}_i)$ between the new centroid μ_i and the old centroid is computed $\widetilde{\mu}_i$.

Steps 2), 3) and 4) are repeated until

$$\sum_{i=1}^{k} d(\mu_i, \widetilde{\mu}_i) < \gamma, \tag{7}$$

For a threshold γ, fixed a priori. For congruence with 6, it is common to choose d as the squared euclidean distance. Even though the k-means algorithm has complexity $O(n^{pk+1})$ [8], some implementations aiming to maximize distance between initial centroids show a faster performance.

Mini Batch K-Means. This is a variation of the K-means algorithm aimed to make calculations in less time. It works in the following way:

1. k clusters and centroids are set as in k means.
2. A random subset of the data is taken, and each point is assigned to the nearest centroid.
3. For each point x from the mini-batch M, its assigned centroid c is updated using a convex combination of itself and the point:

$$c = (1 - \rho)c + \rho x. \tag{8}$$

4. Each observation is assigned to its nearest centroid.

Steps 2, 3, and 4 are repeated until the centroids do not have significant changes. This is done by letting ρ go to zero as the number of iterations increases. This method is faster than K-Means with a loss in the quality of the results.

Self-Organizing Map (SOM). A Self-Organizing Map (SOM) [9] is a type of artificial neural network that is trained through unsupervised learning and belongs to the category of competency learning networks. These types of artificial neural networks use a neighborhood function to preserve the topological properties of the input space. It is used as a grouping and visualization tool in exploratory data analysis. The primary goal of a SOM is to transform a high-dimensional input space into a smaller, discrete output space (dimensionality reduction) while preserving relationships in the data, but not actual distances [2]. SOM has three characteristic steps [7]:

1. Competition: For each sample, the neurons in the network compete with each other. The neuron that best represents the sample (most similar) based on a discriminant function is declared the winner.
2. Cooperation: The neighboring neurons of the winning neuron determine the neighborhood. These excited neurons learn (not equally) from the same entry.

3. Adaptation: The weight vectors of the winner neuron and its neighboring neurons are updated taking into account the learning rate and the distance to the winner neuron. Through this process neighborhood nodes become more similar to the entry sample.

T-Distributed Stochastic Neighbor Embedding (t-SNE). This is a non-linear dimensionality reduction method. It aims to represent h dimensional points x_i with l dimensional points y_i such that $l < h$ and the probability that y_i and y_j are close implies that x_i and x_j are close is maximized. To review the process to do so, let us assume that we have a set of n h-dimensional vectors $\{x_1, ..., x_n\}$. Then, for $i \neq j$, we compute

$$p_{i|j} = \frac{\exp(-\|x_i - x_j\|^2/2\sigma_i^2)}{\sum_{k\neq i} exp(-\|x_i - x_k\|^2/2\sigma_i^2)}, \tag{9}$$

and define $p_{i|i} = 0$. The similarity bewtween x_i and x_j will be given the condition probability $p_{i|j}$ that x_j is a neighbor of x_i, if neighborhood is proportional to the probability density under a Gaussian centered at x_i, see [18]. We also define

$$p_{ij} = \frac{p_{i|j} + p_{j|i}}{2n}. \tag{10}$$

The goal of t-SNE is to obtain a function $x_i \mapsto y_i \in \mathrm{R}^l$ preserving the similarities p_{ij}. For this purpose and in an analogous way, the similarity between y_i and y_j is computed:

$$q_{ij} = \frac{(1 + \|y_i - y_j\|^2)^{-1}}{\sum_{k\neq i}(1 + \|y_i - y_k\|^2)^{-1}}, \tag{11}$$

and $q_{ii} = 0$. The values of each y_i are obtained by minimizing the Kullback–Leibler divergence of the distribution P from the distribution Q, i.e.,

$$KL(P||Q) = \sum_{i\neq j} p_{ij} \log \frac{p_{ij}}{q_{ij}}. \tag{12}$$

This minimization process is done by gradient descent.

Rand Index. This is a method to measure the similarity between clustering methods, let us describe it in its most simple form. Let $X = \{x_1, ..., x_n\}$ and two partitions of X, $B = \{B_1, ..., B_r\}$, $C = \{C_1, ..., C_s\}$, we set:

- a, the number of pairs of elements from X that belong to B_i and C_j for some i, j.
- b, the number of pairs of elements from X that belong to different elements of B and C.
- c, the number of pairs of elements from X that belong to B_i for some i and belong to different elements of C.

- d, the number of pairs of elements from X that belong to C_i for some i and belong to different elements of B.

The Rand index is computed by

$$R = \frac{a + b}{a + b + c + d}. \tag{13}$$

Here, $a + b$ can be seen as the agreements between B and C and $c + d$ can be seen as the number of disagreements between B and C. R can take values between 0 and 1, with 0, meaning a complete dissimilarity between clusters and 1, indicating that the clusters are the same, see [14].

3 Results and Discussion

This section presents the main results derived from the research. Figure 1 shows the experimental approach consisting of two clustering steps applied to the data. The second clustering step is performed in the data obtained from the first clustering.

3.1 First Clustering

We performed this clustering step on the dataset obtained after joining and cleansing of the raw data. Five different clustering methods are applied to this dataset: affinity propagation, agglomerative, birch, K-means, and minibatch K-means. The number of desired clusters, in which the data will be clustered, is passed as an argument to all the methods except for affinity propagation, which identify the number of clusters on its own.

After applying all the methods, we obtained a new dataset that contains the countries and the group they belong to in each method. Table 3 presents the structure of the new dataset, and Table 2 shows the representative country for each cluster in all the clustering methods. Table 4 compares all the clustering methods using Adjusted Rand Index as a comparison measure.

3.2 Second Clustering

In this part of the research, we used two different methods, t-distributed Stochastic Neighbor Embedding (t-SNE) and Self-Organizing Map (SOM). This clustering step is applied to the data obtained from the previous step. We obtained a categorical dataset with five columns indicating the cluster to which each country belongs in every clustering method; before working with this dataset it is important to encode the categorical variables. For this purpose, we used one hot coding technique; in this technique, a single categorical column with d distinct values is transformed to d binary columns, each observation indicating if the sample belongs(1) or not(0) to a certain category, see e.g., [12]. The categorical

Table 2. Representative countries for each clustering method

Cluster	Affinity	Agglomerative	Birch	K-Means	MB K-means
0	Bangladesh	Philippines	Russia	Chile	Ethiopia
1	Brazil	Congo (Brazzaville)	Armenia	India	China
2	China	Brazil	Yemen	Brazil	Hungary
3	France	Burma	Philippines	Turkey	Bhutan
4	Honduras	Saudi Arabia	Turkey	US	Italy
5	India	Turkey	Iraq	Lithuania	Saudi Arabia
6	Indonesia	Sri Lanka	Zambia	Uganda	Brazil
7	Lithuania	Uganda	Brazil	Russia	Indonesia
8	Mali	Russia	Burma	China	India
9	Japan	US	China	Indonesia	Georgia
10	Peru	Japan	India	Philippines	Congo (Kinshasa)
11	Philippines	Bangladesh	Indonesia	Mozambique	Zambia
12	Russia	India	US	Italy	Bangladesh
13	Spain	Indonesia	Japan	Hungary	Spain
14	Turkey	United Kingdom	Portugal	Japan	Madagascar
15	US	China	United Kingdom	Bangladesh	Russia

Table 3. First rows of data 2

Country	Affinity	Agglomerative	Birch	K-Means	MB K-means
Afghanistan	10	4	5	6	5
Albania	7	1	1	5	9
Algeria	13	7	5	6	13
...	...	...	...	...	...

Table 4. Comparison of first clustering methods using Adjusted Rand Index

	Affinity	Agglomerative	BIRCH	K-means	MB K-means
Affinity	1.0	0.59	0.81	0.96	0.7
Agglomerative		1.0	0.61	0.57	0.39
BIRCH			1.0	0.84	0.59
K-means				1.0	0.72
MB K-means					1.0

dataset of shape 163×5 is transformed into a binary dataset of shape 163×80 as every column of the categorical dataset has 16 different groups.

To test the proposed technique, we also applied the second clustering to the original dataset used in the first clustering to see the differences in the obtained results.

T-SNE Results. First, we applied t-SNE to both datasets with a dimensionality reduction purpose. After obtaining the new two dimensions representing each dataset, a K-means clustering algorithm is applied to cluster the data in 10 different groups. Figure 2 presents the results obtained for the original dataset, while Fig. 3 presents the results obtained with the proposed technique of applying some clustering methods first and transforming the dataset (two-layer clustering).

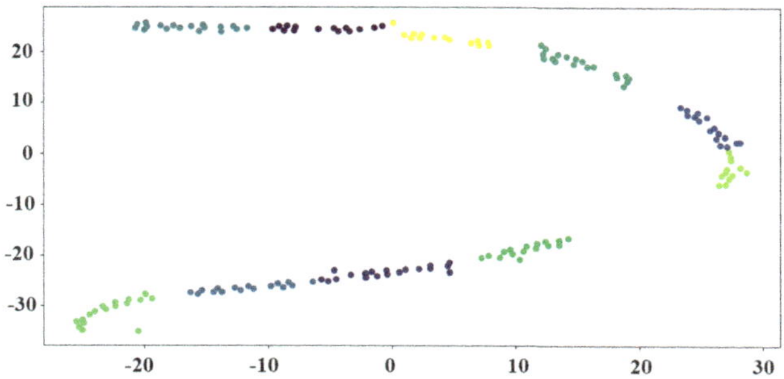

Fig. 2. Clusters after applying t-SNE method to Data 1 (1 layer t-SNE+ K-means)

It is easier to visually distinct some groups with the proposed method. Some samples are already grouped after applying t-SNE, making it easier for K-means to distinguish each cluster. The separation of the samples into groups is better, or at least the clusters are easier to identify with the proposed methodology, as shown in Fig. 3 in comparison with Fig. 2. To test this, we compute the mean variance of the distance from the centroid of the cluster to each sample within the cluster and the average distance among all the centroids in each approach. Table 5 presents the results of the comparison. A smaller variance is preferred, while a greater distance among centroids is desired.

When using t-SNE, the approach presented in this research has shown a good performance as it minimizes the variance of the distance from each sample to their corresponding centroid, yet the centroids are closer to each other in this approach. However, comparing one layer clustering t-SNE+K-means with two layers clustering t-SNE+K-means, we obtain an Adjusted Rand Index of 0.6154921500297419. This suggests that even though some observations are classified in similar clusters for both methods, there is still a great proportion of observations classified in a different way, which is congruent with the compactness of some clusters with our proposed method, as shown in Fig. 3.

SOM Results: We applied this method to both datasets, again with a dimensionality reduction purpose. After obtaining the new two dimensions that represent each dataset, we performed a K-means clustering algorithm to cluster the data in 10 different groups. Figure 4 presents the results obtained for the original dataset, while Fig. 5 presents the results obtained with the proposed technique of applying some clustering methods first and transforming the dataset.

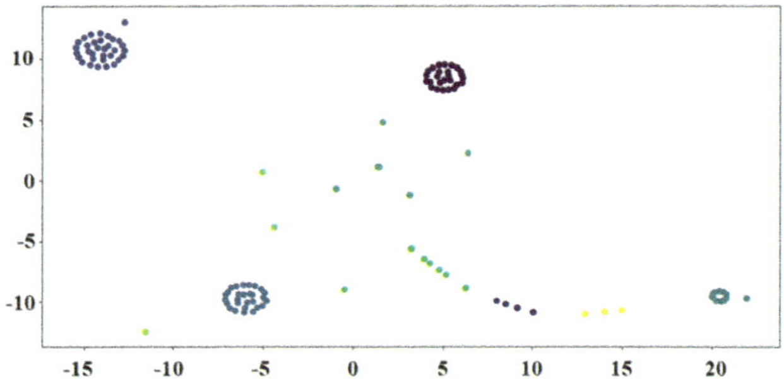

Fig. 3. Clusters after applying t-SNE method to Data 2 (2 layer t-SNE+ K-means)

Table 5. t-SNE approaches comparison

Metric	1 Layer	2 Layer
Mean variance of distance sample-centroid	1.908586	0.434635
Mean distance among centroids	332.712943	161.481747

This time the comparison of both approaches is easier to spot. When applied to the original dataset, SOM yields to a 2D map that does not have clusters as distinguishable as the one generated with the proposed technique. To test this, we compute the mean variance of the distance from the centroid of the cluster to each sample within the cluster, as well as the average distance among all the centroids in each approach. Table 6 presents the results of the comparison. It is important to notice that Fig. 5 has fewer plotted points. The reason for this is that many samples have the same coordinates. A single point in the 2D map may represent several samples, making it easier for K-means to cluster the data obtained with SOM as one point may cluster several countries with less disperse samples on the map. Fewer points are desirable as it shows the samples are already grouped by similarity.

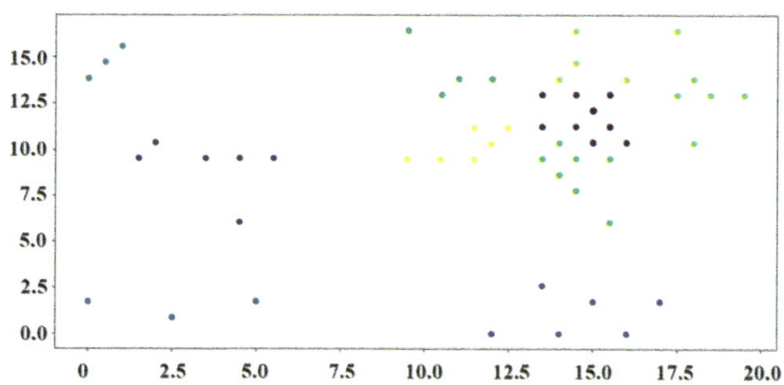

Fig. 4. Clusters after applying SOM method to Data 1 (1 layer SOM + K-means)

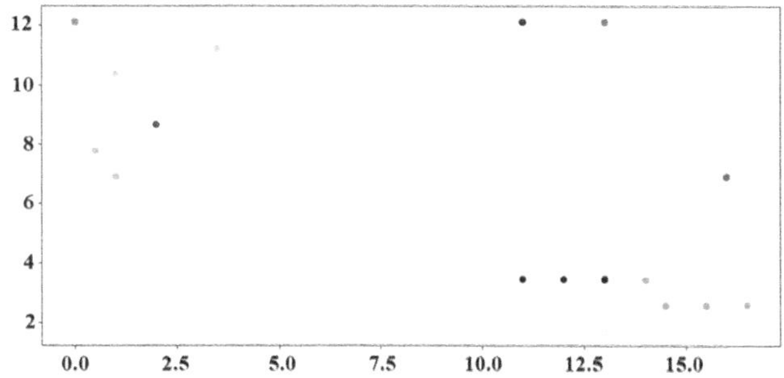

Fig. 5. Clusters after applying SOM method to Data 2 (2 layer SOM + K-means)

Table 6. SOM approaches comparison

Metric	1 Layer	2 Layer
Mean variance of distance sample-centroid	0.500746	0.069282
Mean distance among centroids	92.166437	87.062974

When using SOM, the approach presented in this research minimized the variance of the distance from each sample to their corresponding centroid, yet the centroids are slightly closer to each other in this approach. Comparing between clusters from 1 layer clustering SOM+K-means and 2 layer clustering SOM+ K-means, we get an Adjusted Rand Index of 0.2395; as can be seen in Figs. 4 and 5 this is an indicator that both methods perform clustering in a completely different way.

3.3 Final Results

We obtained the best results using the proposed method and SOM (Fig. 5), in the sense that samples were tightly grouped together, resulting in the smallest mean variance of distance sample-centroid of the four studied methods. Table 7 shows the clusters formed with this approach. The countries are expected to take actions similar to other countries within their groups to fight coronavirus infection and spreading.

Table 7. Groups formed with the two-layer SOM clustering method

Group	Countries
1	Algeria, Argentina, Congo (Kinshasa), Germany, India, Iran, Iraq, Japan, Mexico, Russia, Spain, Turkey, US, Uganda, Ukraine
2	Albania, Armenia, Bosnia and Herzegovina, Botswana, Central African Republic, Congo (Brazzaville), Costa Rica, Croatia, Denmark, Eritrea, Finland, Gabon, Georgia, Guinea-Bissau, Ireland, Jamaica, Kuwait, Lesotho, Liberia, Lithuania, Mauritania, Moldova, Mongolia, Namibia, New Zealand, Norway, Oman, Panama, Qatar, Singapore, Slovakia, Slovenia, Uruguay
3	Australia, Cameroon, Cote d'Ivoire, Ghana, Madagascar, Mozambique, Nepal, Yemen
4	Austria, Azerbaijan, Belarus, Belgium, Benin, Bolivia, Bulgaria, Burundi, Czechia, Dominican Republic, El Salvador, Greece, Guinea, Haiti, Honduras, Hungary, Israel, Jordan, Lebanon, Libya, Nicaragua, Papua New Guinea, Paraguay, Portugal, Rwanda, Serbia, Sierra Leone, South Sudan, Sweden, Switzerland, Tajikistan, Togo, Tunisia, United Arab Emirates
5	Antigua and Barbuda, Bahrain, Barbados, Bhutan, Comoros, Cyprus, Djibouti, Dominica, Equatorial Guinea, Estonia, Fiji, Grenada, Guyana, Iceland, Latvia, Luxembourg, Maldives, Malta, Mauritius, Montenegro, San Marino, Seychelles, Suriname, Timor-Leste, Trinidad and Tobago
6	Burkina Faso, Cambodia, Chad, Chile, Ecuador, Guatemala, Kazakhstan, Malawi, Mali, Netherlands, Romania, Senegal, Somalia, Sri Lanka, Zambia, Zimbabwe
7	Bangladesh, Colombia, Egypt, Ethiopia, France, Indonesia, Philippines, South Korea, Thailand, United Kingdom, Vietnam
8	Afghanistan, Angola, Brazil, Canada, Malaysia, Morocco, Nigeria, Pakistan, Peru, Poland, Saudi Arabia, Uzbekistan
9	China, Niger, Taiwan.
10	Burma, Italy, Kenya, South Africa, Tanzania

4 Conclusions and Recommendations

This work presented a novel clustering method. This methodology consists in transforming the data using several clustering methods followed by a new clustering with dimensionality reduction using t-SNE and SOM over the new data to group the data and later apply K-Means to identify the clusters. According to the minimum mean variance of distance sample-centroid, the best results were found using SOM with the proposed method of two-layer clustering. It also showed good performance when considering the mean distance among centroids, even though it was not the best one in this category. This method showed a reduced amount of plotted points in the graph, as it groups the countries based on the first clustering similarities. After applying the dimensionality reduction, several countries have the same x and y coordinates, making it easier for K-Means algorithm to make groups. This technique is applied to COVID-19 data collected from 163 countries to group countries into clusters according to their similarities. With this information, countries can apply similar effective measures that worked in a particular country in the same cluster or, on the contrary, avoid applying measures that have shown poor performance in a given country within the same cluster.

The proposed method can be modified for future work to add different clustering methods in the first or second clustering layers. It is important to consider methods that work well with the selected dataset. To further study the influence of the variables in the clusters, and the contrast between the latter, a MANOVA Biplot could be performed [1]. The dataset can be expanded to consider more variables that may have significantly influenced the impact of COVID-19. For example, considering the public policies applied in the countries to stop the advance of the disease, as this information is directly related to the spreading of COVID-19. Another work that can be performed in the future is to compare the efficiency of the proposed method with other clustering algorithms.

References

1. Amaro, I.R., Vicente-Villardón, J.L., Galindo-Villardón, M.P.: Manova biplot for two way multivariate general linear models. Interciencia **29**(1), 26 (2004)
2. Asan, U., Ercan, S.: An introduction to self-organizing maps. Computational Intelligence Systems in Industrial Engineering. pp. 295–315. Springer (2012)
3. Carozzi, F., Provenzano, S., Roth, S.: Urban Density and COVID-19. Discussion Paper Series, IZA Institute of Labor Economics. http://ftp.iza.org/dp13440.pdf
4. European Centre for Disease Prevention and Control: Q & A on COVID-19. European Union. (2020). https://www.ecdc.europa.eu/en/covid-19
5. Frey, B.J., Delbert, D.: Clustering by passing messages between data points. Science. Am. Assoc. Adv. Sci. **315**(5814), 972–976 (2007)
6. Hamidi, S., Sabouri, S., Ewing, R.: Does density aggravate the COVID-19 pandemic? J. Am. Plann. Assoc. **86**(4), 495–509 (2020). https://doi.org/10.1080/01944363.2020.1777891
7. Haykin, S.: Neural Networks: a Comprehensive Foundation. Prentice-Hall, Inc., Upper Saddle River (2007)
8. Inaba, M., Katoh, N., Imai, H.: Applications of weighted Voronoi diagrams and randomization to variance-based k-clustering. In: Proceedings of 10th ACM Symposium on Computational Geometry, pp. 332–339 (1994) https://doi.org/10.1145/177424.178042
9. Kohonen, T.: The self-organizing map. In: Proceedings of the IEEE, vol. 78, no. 9, pp. 1464–1480. IEEE (1990)
10. Sanjay, K.: Monitoring Novel Corona Virus (COVID–19) Infections in India by Cluster Analysis. Annals Data Sci. 7(3), 417–442 (2020)
11. Pedregosa, F., et al.: Scikit-learn: machine learning in Python. JMLR **12**, 2825–2830 (2011)

12. Potdar, K., Pardawala, T.S., Pai, C.D.: A comparative study of categorical variable encoding techniques for neural network classifiers. Int. J. Comput. Appl. **175**(4), 7–9 (2017)
13. Prabhu, T.N.: Population by Country - 2020. Dataset. https://www.kaggle.com/tanuprabhu/population-by-country-2020/metadata
14. Rand, W.M.: Objective criteria for the evaluation of clustering methods.Journal of the American Statistical Association. American Statistical Association **66**(336), 846–850 (1971) arXiv:1704.01036. https://doi.org/10.2307/2284239 JSTOR 2284239
15. Rencher, A.C.: Methods of Multivariate Analysis. vol. 492, Wiley, New York (2003)
16. The Humanitarian Data Exchange, service provided by the OCHA. https://data.humdata.org/dataset/novel-coronavirus-2019-ncov-cases
17. The pandas development team: pandas-dev/pandas: Pandas. Zenodo. Version 1.0.5. (2020). https://doi.org/10.5281/zenodo.3509134
18. van der Maaten, L.J.P., Hinton, G.E.: Visualizing data using t-SNE. J. Mach. Learn. Res. **9**, 2579–2605 (2008)
19. World Population Review: GDP Ranked by Country 2020. https://worldpopulationreview.com/countries/countries-by-gdp
20. World Economic Outlook Database. International Monetary Fund. April, 2020. https://www.imf.org/external/pubs/ft/weo/2020/01/weodata/index.aspx

ToraxIA: Virtual Assistant for Radiologists Based on Deep Learning from Chest X-Ray

Martin Carnier[1], Raul Albertti[1], Liliana Gavidia[2], Erika Severeyn[3], and Alexandra La Cruz[4(✉)]

[1] Alumbra.ai™, Santiago, Chile
{martincarnier,raulalbertti}@alumbra.ai
[2] GBBA, Grupo de Bioingeniería y Biofísica Aplicada, Caracas, Venezuela
lgavidia@usb.ve
[3] Department of Thermodynamics and Transfer Phenomena, Universidad Simón Bolívar, Caracas, Venezuela
severeynerika@usb.ve
[4] Faculty of Engineering, Universidad de Ibagué, Ibagué, Tolima, Colombia
alexandra.lacruz@unibague.edu.co
https://www.alumbra.ai

Abstract. Misdiagnosis of pulmonary pathology may have several causes. Some of them are related to the unavailability of radiology specialists, the increasingly overwhelming number of Chest X-Ray images generated daily, or the respiratory disease by itself. Consequently, it might lead to radiologist burnout. Furthermore, an unusual location of pathological lung signs may lead to a misdiagnosis of severe pulmonary disease. Several pathological signs are observed from a Chest X-Ray and on different tissues: Heart and Pericardium, Hili, Mediastinum, Lungs, Pleura, and Chest wall. In the present study, a Deep Learning model is presented, which identifies several tissues and abnormal signs that may require a deeper evaluation. The model was trained using a total of 240,000 Chest X-Ray images, which can automatically identify several structures observed on a Chest X-Ray and some pathological signs of respiratory disease. The model reached 97% of accuracy. These results suggest that our model can be used as an automatic identifier of tissues and pathological signs from Chest X-Ray images. In consequence, it constitutes a meaningful and essential tool for preliminary studies of the severe respiratory disease since it can potentially decrease the number of false positives in the most common radiological examination ordered for patient evaluation. It also could be used for screening purposes to detect findings that may be useful in any pathology assessment, and in the decision-making process of conducting further image modality evaluations, such as CT.

Keywords: Severe respiratory disease · Artificial intelligence · Medical validation · Convolutional neural networks · Deep learning

M. Botto-Tobar et al. (Eds.): CIT 2020, AISC 1326, pp. 49–63, 2021.
https://doi.org/10.1007/978-3-030-68080-0_4

1 Introduction

Chest X-ray evaluation remains the primary tool by excellence for the screening of respiratory disease [4,23]. Therefore, it should be readily available to all patients with respiratory symptoms. The diagnosis of respiratory diseases requires specialized training for proper evaluation. Currently, the number of specialists evaluating respiratory disease is insufficient, and the few available are generally suffering fatigue due to excessive work. Besides, there is a highly increasing incidence of lung disease due to different unfavorable environmental factors and particularly under the COVID-19 Coronavirus pandemic. On the other hand, the overwhelming amount of clinical data and radiological images causes medical decision-making in uncertain conditions, which in turn leads to inaccuracy in medicine due to burnout in radiologists [6]. This seriously compromises patient safety and might shorten life expectancy, as well as increase health costs. Besides, it produces fatigue of the radiologist [6,17], and delays in report generation to proper diagnosis and treatment. In addition, the increasing demand for radiological services has been worsened with the COVID-19 pandemic.

Convolutional neural networks (CNN) have become an increasingly popular and effective method for analyzing diagnostic images [22]. In general, machine learning techniques have been used in the diagnosis of lung diseases. Particularly, the convolutional neural networks have been used in the diagnosis of tuberculosis, obtaining satisfactory results in the classification of subjects with tuberculosis with a sensitivity of 99% and specificity of 100% [12]. CNN has also been used in the classification of subjects with pneumonia [20] and with viral pneumonia due to COVID-19 [3]. CNN generally takes a set of training images, learns meaningful features from the data, and optimizes a set of parameters such as weights and bias to be able to distinguish and classify the data by their type [19]. Diagnostic image evaluation of respiratory disease is commonly carried out directly by an expert who observes if there are irregular patterns that do not exist in healthy lungs. This process consists of a visual recognition, based on the experiences of the specialist, the presence of pathological patterns suggestive of lung disease, and the comparison with the form, proportion, location, and definition of the visible organs considered as healthy. Since the information is unstructured, it is not possible to perform its analysis using traditional statistical models. As it is well known, CNN has become actively used for solving this kind of issue efficiently. In [21], several CNN were implemented and compared for differentiation between normal and abnormal frontal chest radiographs, including normal versus lung opacity classification, with best-reported values of AUC of 0.9804 ± 0.0032 (with an accuracy of $94.71 \pm 0.32\%$, a sensitivity of $92.20 \pm 0.34\%$ and a specificity of $96.34 \pm 0.31\%$).

This study proposes an application using a convolutional neural network model to operate as a predictive model for diagnosis support as a virtual assistant. This model allows for the detection of pathological signs suggestive of respiratory diseases from Chest X-rays. Some of the pathological signs detected by the application are pleural effusion, interstitial pattern, consolidation, atelectasis, pneumothorax, cardiomegaly, and nodule-mass. This model of artificial intelligence (AI) provides seamless integration with the pre-existing workflow without disrupting the regular procedure of the physicians, enhancing the specialized care that sometimes cannot always become easily accessible due to the limited availability of radiologists. The highly audited algorithm presented will help demystify the misleading or erroneous information physicians sometimes have about AI as a black-box, or that it constitutes a menace to their job. It is important to emphasize that AI will not replace radiologists. Radiologists who use AI will replace radiologists who do not. It is the opposite. Toraxia is a tool designed to be used as a virtual assistant that provides augmented radiology, making medical expertise transcend to function as an extension of themselves.

2 Methodology

All datasets used are available in the following public databases: PadChest [5], Chexpert [9], and SIIM-ACR [14]. The model was trained in two phases. The first training phase used 200,000 Chest X-Ray images from the open databases mentioned before. The second training phase considered COVID-19 cases, then 40,000 images were added, from the database MIMIC [10] and some others were from several health institutions from Latin America. Some cases where patients with COVID-19. In the first training phase, the result was not enough for localizing the area of the image where the pathological sign needed to be identified. We realized there were some mismatches in the annotations. Consequently, several radiologists and medical specialists were invited to participate in a manual annotation process, to audit the data that would be used in the second phase of the training. The dataset annotation process is explained in Sect. 2.2, while the model is described in Sect. 2.3.

2.1 Data Set

At an early stage, full datasets from PadChest, SIIM-ACR, and Chexpert, were used to train the model with 200,000 images. In the second stage, the database MIMIC was added to the training data set. In addition, other Chest x-ray images from local healthcare institutions in Chile, Venezuela, Colombia, Nicaragua, and Ecuador were added (see Table 1). A total of 240,000 images were used for training, validation, and testing: 70–73% of them were used for training, 20% for model validation, and 7–10% for testing (see Table 2). All these data were previously annotated by the expertise of a group of medical specialists and radiologists, in addition to the available annotations of the open databases already mentioned. The reports of the radiologists and the clinical history of the patients were used as well. The dataset was classified and annotated by three experts into the following classes: Interstitial pattern, Consolidation, Atelectasis, Pneumothorax, Pleural Effusion, Cardiomegaly, and Nodule/mass (see Table 3).

Table 1. Dataset source and annotation

Source of data	Image quantity	Annotation
Health centers from Chile, Venezuela, Colombia, Nicaragua, and Ecuador	10,000	By experts
PadChest	120,000	Already annotated
Chexpert	35,000	Already annotated
MIMIC	30,000	Already annotated
SIIM-ACR https://www.kaggle.com/c/siim-acr-pneumothorax-segmentation	45,000	Already annotated

Table 2. Dataset distribution per pathological sign for training, validation, and testing, used in the first and second training phases

Pathological signs	First training phase			
	Total used	Training	Validation	Testing
Interstitial lung pattern	43,000	30,100	8,600	4,300
Consolidation pattern	34,000	23,800	6,800	3,400
Atelectasis	15,000	10,500	3,000	1,500
Pneumothorax	45,000	31,500	9,000	4,500
Effusion	15,000	10,500	3,000	1,500
Cardiomegaly	30,000	21,000	6,000	3,000
Nodulo/mass	18,000	12,600	3,600	1,800
Total	**200,000**	**140,000**	**40,000**	**20,000**
		70%	**20%**	**10%**
	Second training phase			
Pathological signs	Dataset used	Training	Validation	Testing
Interstitial lung pattern	8,000	5,840	1,600	564
Consolidation pattern	8,000	5,840	1,600	564
Atelectasis	4,000	2,920	800	282
Pneumothorax	4,000	2,920	800	282
Effusion	6,000	4,380	1,200	423
Cardiomegaly	6,000	4,380	1,200	423
Nodulo/mass	4,000	2,920	800	282
Total	**40,000**	**29,200**	**8,000**	**2,818**
		73%	**20%**	**7%**

2.2 Annotation Process

Datasets from PadChest, Chexpert, MIMIC, and SIIM-ACR are already annotated. These annotations are regarded as our first expert. Several institutions and collaborators have provided collections of images with the respective radiology reports (see Acknowledgement section). This big dataset not annotated was then analyzed in a blinded study by several radiology experts. A group of two radiology experts was formed and each of them visualized a set of images without prior knowledge of the radiology report nor the classification. A set of images was then classified and annotated by at least three experts: two groups of specialists and the open annotated databases (PadChest, MIMIC, Chexpert, and SIIM-ACR) considered as the third expert. A total of twenty (20) collaborators between radiologists and specialists in the area of diagnosis of respiratory diseases participated in the annotation process. All of them between five (5), ten (10), and (15) years of experience analyzing Chest X-Ray and identifying pathological signs, and from different countries in Latin America.

The segmentation-based training process requires a mask annotation in addition to the previously available classification. This is why the annotation consisted of creating a mask that covered the pathological sign identified in the image. In addition, due to the existence of high variance resulting from these indirect classification methods, it was necessary to perform an audit process in order to select the gold standard dataset for testing. This process was carried out by physicians and radiologists. The image database, and its previous classification, were made available by a web site from Toraxia, so that the radiologist collaborators could access them for the auditing process.

Each image in the database was analyzed by two physicians. For them, the findings identified by at least two of the classifying elements (pre-sorting, physician 1, physician 2) have been determined to be definitive. In cases of positive signs, the final mask for each image and sign has been determined as the intersection of the masks generated by the individuals participating in the annotation process, resulting in a number of masks per pathological sign as shown in Table 3.

Table 3. Segmented and annotated images with pathological signs, selected as gold standard for testing the model

Pathological sign	Number of segmented and annotated images	Number of images used
Interstitial lung pattern	4,800	4,300
Consolidation pattern	3,750	3,400
Atelectasis	1,600	1,500
Pneumothorax	4,800	4,500
Effusion	1,600	1,500
Cardiomegaly	3,700	3,000
Nodule/mass	1,800	1,800

2.3 Deep Learning Model

U-Net [18] is a convolutional neural network used for biomedical image segmentation, that has been successfully used for semantic segmentation, and shown to exhibit great performance on medical images. It comprises encoding and a decoding stage. The encoder consists of a set of blocks that perform a max-pooling operation, in which the resolution of the image is halved, and the number of channels is doubled. The decoder encompasses transposed convolution blocks that are now used to reverse the process. A distinctive feature is that each reduction stage is concatenated with its equivalent in maximization, managing to maintain the spatial information that would otherwise be lost.

Res-Net [8] is a neural network architecture inspired by the pyramidal cells of the cerebral cortex that make use of shortcuts so that the layers can jump over others, as a solution to the problem of gradient vanishing. In this model, the binary cross-entropy function, a loss function specialized for binary classification problems, is often used and shown to be very effective for balanced classes. However, in our case, the level of balancing is not available for the location of pathological signs, so the use of this function was not encouraged. Therefore, the Dice loss function [26] was used instead, since it maximizes the values for the least present classes and faces the problem of imbalanced datasets.

The U-Net architecture allows us to integrate a sub-network into the encoder/decoder process [25]. In order to assist in the training process, several architectures can be implemented in specific segments of the model. For this use case, Res-Net18, Res-Net50, and VGG11 (AlbuNet) were considered for integration to U-Net. The following results were obtained for the detection of signs of interstitial and consolidation patterns during the training, validation, and evaluation process, with the following accuracy for each architecture (see Table 4). These results lead to consider Res-Net50 as the best coder option in this study.

Table 4. Results of testing different network architectures integrated to U-Net

Model	Parameters	Validation process (Accuracy)	Evaluation process (Accuracy)
Res-Net18	15M	0.83	0.85
Albu-Net	34M	0.78	0.79
Res-Net50	34M	0.88	0.93

The information system training and development process was carried out with all these implemented components, as detailed in the following process. To prepare the image training set, images are first selected by class, then a manual segmentation is carried out and a mask is generated. This mask is used to guide the semantic segmentation in our learning model. On the other hand, the training process is performed in three steps: the first one involves the first training with a learning rate (Lr) equal to 0.0001; then a second training with a smaller Lr is done. Finally, a validation process is carried out to estimate the model precision. Results from our model U-Net using Res-Net50 as a coder is shown in Table 5. The accuracy values obtained for segmentation per pathological sign are shown. These values are obtained from the Dice Loss Function [26] used for the coder in the validation process of the model. The results suggest that currently, our model is still not a very good identifier of Nodule/mass (0.69 of accuracy), as can be observed in Table 5.

Table 5. Precision values of the segmentation by pathological sign

Pathological sign	Dice loss function
Interstitial lung pattern	0.87
Consolidation	0.81
Atelectasis	0.72
Pneumothorax	0.84
Pleural effusion	0.84
Cardiomegaly	0.94
Nodule/mass	0.69

3 Experimental Results

3.1 Library Performance

Throughout the implementation phase of the proposed model, several combinations of platforms and hardware were evaluated for optimal training and testing process performance. Initially, the model was developed using Tensorflow [1] with Keras [7], a high-level library, on a Python Anaconda environment. The model was also developed using Pytorch library [16], and performance compared with each other. Table 6 presents the accuracy results obtained from each ResBlock in the validation and the evaluation (testing process). Results in Table 6 showed that Pytorch constituted the best selection to carry out the deep learning tasks related to the project and deployment to production. In the beginning, it was planned to use Ubuntu as the operating system to carry out the development and deployment of the project, but after a documentary review of the support offered by Nvidia for both operating systems and the hardware capabilities in terms of the driver options for each platform, it was concluded that Windows Server 2016 in an IBM cloud computing service was the most stable platform for the installation of all the required dependencies necessary for the deployment of the comprehensive system.

Table 6. Comparison of Keras/Tensorflow vs Pytorch processing

Process	CPU (AMD Ryzen 2700X)	GPU (Nvidia Tesla K80)	CPU/GPU Maximum consumption
Training with Keras/Tensorflow	86 h	4 h	100%/70%
Testing Keras/Tensorflow	45 s	8 s	100%/70%
Training with Pytorch	170 h	2,5 h	100%/100%
Testing Pytorch	190 s	2,5 s	100%/100%

3.2 Deep Learning Model Results

Table 7 shows the confusion matrix obtained from the first and the second training phase. The first training phase used 200,000 images for training, where 10% (20,000) were used for testing the model. The second training phase used 40,000 images for training, where 7% (20,000) were used for testing the model. Several statistical metrics were calculated from the confusion matrix for the first and the second training phase, respectively (see Table 8).

Table 7. The confusion matrix obtained in the first training phase with 20,000 images for testing. The second Confusion matrix is from the second training phase where there are two (2) classes: Positive, for pathological sign detection suggestive of COVID-19, Negative for no COVID-19 signs detected

First training phase

Predicted	True	
	Positive	Negative
Positive	8,788	1,606
Negative	1,512	8,094

Second training phase

Predicted	True	
	Positive	Negative
Positive	266	43
Negative	30	2,479

Table 8. Statistical metrics obtained form the first and the second training phases

Statistical metric	First training phase	Second training phase
Accuracy	0.8441	0.9741
Precision	0.8455	0.8608
Sensitivity (Recall)	0.8986	0.8986
Specificity	0.8384	0.9830
Negative predicted value	0.8426	0.9880
F1	0.8493	0.8793

As we can observe in Table 8, all values, including Accuracy, Precision, Recall, Negative Predicted Value, and F1, in both phases of training reached an excellent value over 0.8, and the Accuracy reached a value over 0.9 in the second training phase, which are good indicators of an excellent CNN model. Besides, every pathological sign identified can be visible and available to the specialists for deeper analysis.

Figure 1 and 2 show the results obtained from four (4) patients evaluated using Toraxia^{TM}. Figure 1 shows the patients previously and randomly selected. Each image is passed through our model, and consequently, the model generates a color map image for each pathological sign detected. In Fig. 2, every row corresponds to a pathological sign identified for each patient. The signs found are listed in the following order, from row one to row six: effusion, atelectasis, consolidation, cardiomegaly, pneumothorax, and interstitial pattern, and reported for each patient, which correspond to the pathological signs better classified by our model. An example of a report generated by Toraxia^{TM} is shown in Fig. 3. Colors are also a codification of the severity of the signs, and this represents a piece of important visual information for the radiologist. However, this is still under study. Currently, there are no official reports or in-depth studies about it.

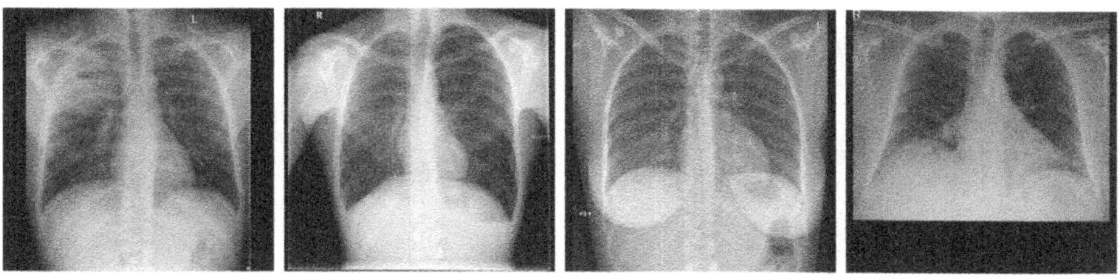

Fig. 1. Chest X-Ray of 4 patients with different pathological signs, which are shown in Fig. 2 as a result from Toraxia^{TM}

3.3 COVID-19 Dataset

Experimentally, our model was evaluated for COVID-19 disease, then, a dataset of patients with COVID-19 was introduced to Toraxia^{TM}. A patient with COVID-19 typically presents an interstitial pattern or consolidation sign [24]. Therefore, we trained our model in order to detect both signs. This training was performed in the second training phase (see Table 2). For this, the data were separated into two groups, one with COVID-19 images and the other with non-COVID-19 images. For our model, a suggested positive patient presents an interstitial pattern or consolidation sign and a negative otherwise. The results of this binary classification are shown in Table 8 and the corresponding confusion matrix in Table 7. The area under the ROC curve gave a value of $ROC_{AUC} = 0.99$. In this case in particular, the binary cross-entropy loss function was used for the validation process, since the classification made was binary.

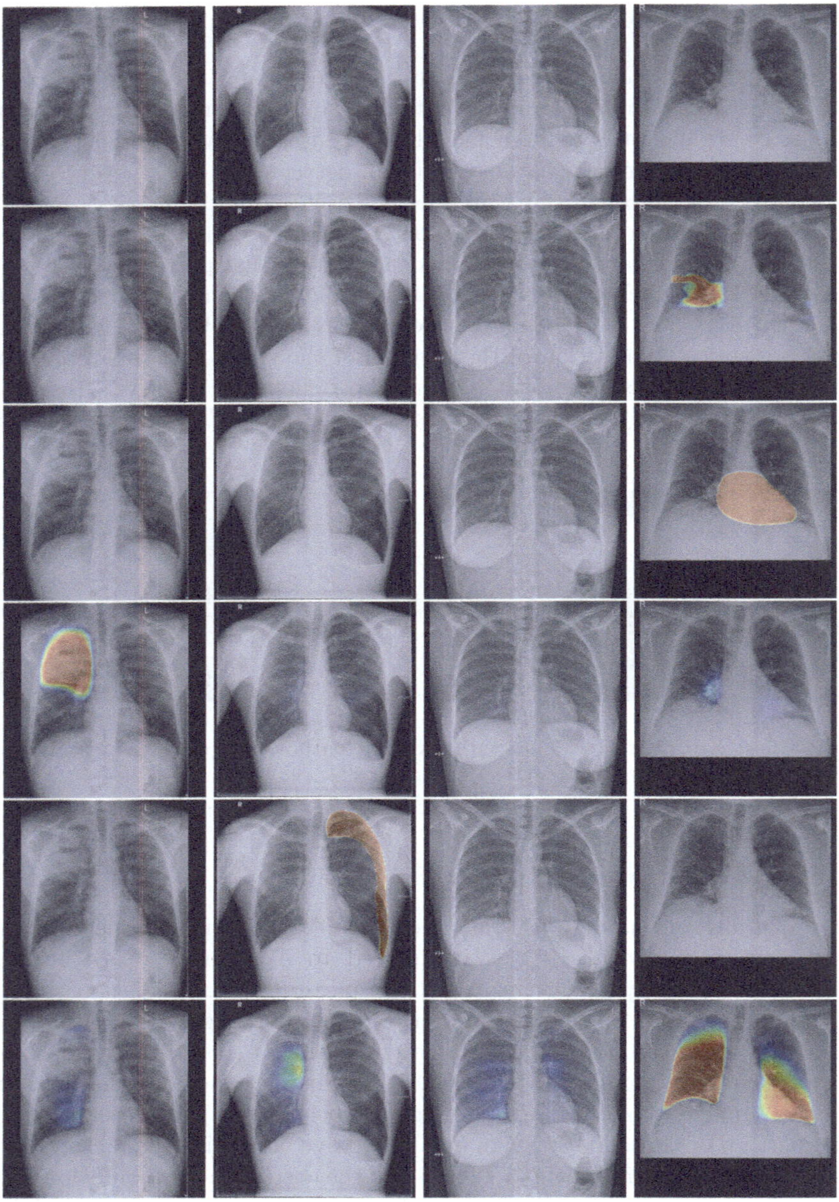

Fig. 2. Each column corresponds to a pathological sign identified as a result of applying Toraxia™ over each patient from Fig. 1

3.4 Toraxia™Output

Toraxia™automatically generates a report to radiologists and physician specialists (see Fig. 3). The report is divided into 4 sections (each section is indicated into a circle in Fig. 3): (1) the demographic data from the patient; (2) a detailed % of the area covered by the pathological sign in the left and the right lung; (3) an estimation of the detection accuracy, and finally in section (4), the color map of the signs identified in the image.

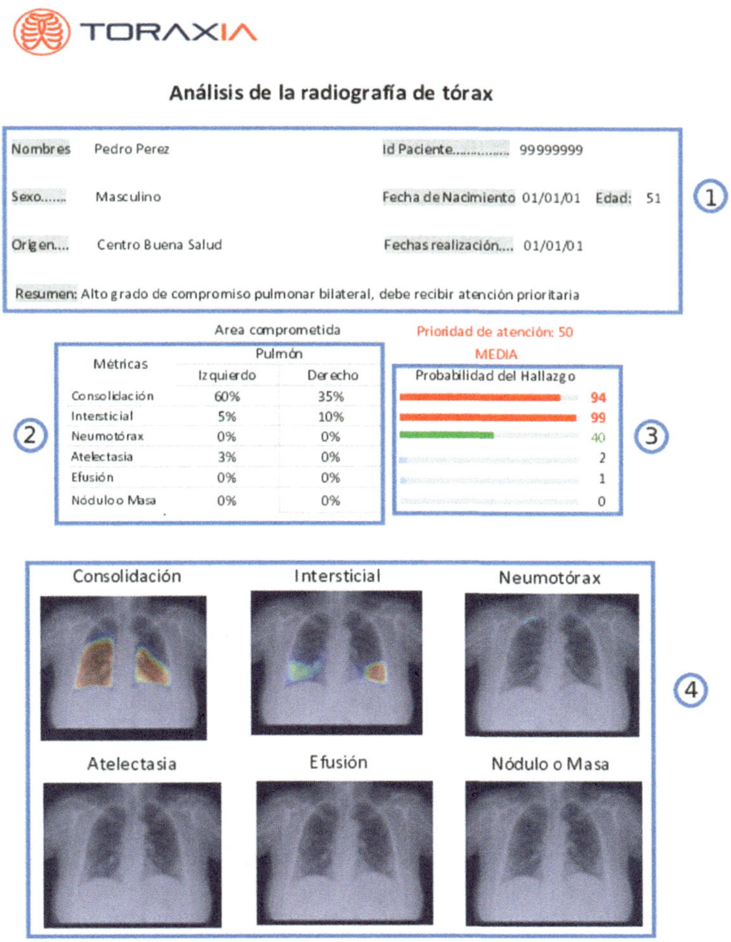

Fig. 3. Toraxia™report generated for one patient. The study result is obtained in 3 s.

4 Discussion

The method applied in this work exhibit excellent results for the detection of pathological signs from Chest X-rays, with a high probability of detecting subjects with and without the disease, as can be seen in the high values of positive predictive value and negative predictive value. The method yields a few false positives and false negatives. These results corroborate the validity of using convolutional neural networks for the diagnoses of pulmonary disease, as suggested by other studies that reference the pertinent use of CNN for the detection of pathologies such as Tuberculosis and pneumonia [2,13,15]. A comparative evaluation, validation, and training based on different cost functions has resulted in Dice loss function to be the optimal cost function for detection of pathological signs in anteroposterior chest radiographs, and it has proven to be suitable for optimization of the model with unbalanced classes typically representing pathological signs that, when present, occupy a specific area of the image.

The highest precision indexes during the training and evaluation process were achieved with the Res-Net50 codifier, thereby continuing its use on the present operation of the system. This type of architecture has been used in the training of neural networks for the diagnosis of cancer and it has turned out to be the most suitable for these cases, which corroborates what was found in this research [11]. The use of segmented images for the training of the model resulted in a better orientation for model classification and localization of pathological signs, enabling optimal performance of the algorithm while requiring a reduced number of images. This facilitates the manual evaluation and audit process, encouraging the participation of healthcare professionals in the calibration of the system.

The Pytorch library resulted in the best for this implementation, regarding response time and system resource optimization on the hardware level. This enables expansion of the model to include additional functions without significantly affecting its performance on assistance in image analysis with an added value proposition, thus maintaining timely execution parallel to the workflow execution in today's healthcare environment.

5 Conclusion and Future Work

Toraxia potentially constitutes an excellent virtual radiology assistant for automatic detection of pathological signs suggestive of respiratory disease, reducing the diagnostic error, increasing efficiency, and quality of care. It also reduces the burnout radiologists sometimes experience, when immersed in environments with limited availability of radiologists or highly increasing demand of radiology reporting, since it can provide precise quantitative information that the radiologist cannot usually calculate himself due to time constraints and the need for urgency evaluation. This information is derived from the automatic delineation of affected areas. In other words, it can reduce the time spent in identification, and characterization, which gains the radiologist valuable time for a more accurate interpretation and diagnosis. This represents a contribution to the degree of ground-glass opacity progression of COVID-19 disease. Toraxia can assist the radiologist to expand coverage of practice without losing diagnostic quality, and ultimately assist in the expansion of radiology services. Using U-Net with ResNet50 offered the optimal combination for the detection of pathological signs through semantic segmentation. This implementation has demonstrated that it is possible to continuously include medical criteria in the calibration process of the system. Based on this, a more in-depth evaluation is currently undergoing targeting calibration and localization of the algorithm, considering phenotypes and particular characteristics of populations, local medical criteria, and local radiology services in Latin American countries.

Currently, Toraxia is being tested by several health institutions in Latin America. Furthermore, we are working on adding semantic information to our dataset, to create a knowledge graph and on top of it, apply a Machine learning process for better support and results.

Acknowledgments. Full acknowledgement is given to Alumbra.ai^{TM}, CUI.DATE, Allm.net, IBM, IBM Cloud, InterSystemsTM, CORFO Openbauchef. Universidad de Chile, Centro Nacional en Sistemas de Información en Salud Chile, Inacap, Smart Salud, AGFATM. Special thanks to the following collaborators and their institutions, who either provided images to the training phase of the algorithm, time and effort segmenting and reviewing images used or are committed to the multinational study for calibration and localization of the algorithm considering phenotypes, local medical criteria and radiology services in Latin America. In Venezuela: Dr. Ivonne Rivas Gavidia. Radiologist. Instituto Clínico La Florida. President of Sociedad Venezolana de Radiología 2018–2020 (SOVERADI). Dr. Sergio Tovar. Radiologist. Vice President of the Board of Directors. Hospital de Clínicas Caracas. Dr. Oswaldo Ramos Núñez. Radiologist. Diagnostic Imaging Director. Centro Clínico La Sagrada Familia. President of Asociación Venezolana de Informática en Salud. Dr. Bernardo Lander. Radiologist. Chief of Radiology Service. Centro Médico de Caracas. Dr. Miguel Rocha and Dr María Elena Savino. Radiologists. Instituto Médico La Floresta, Dr. Anamel García. Radiologist. Instituto Clínico Unare. Dr. Eleonora Garcia Hernandez and Dr. Lisseth Diaz Liendo. Radiologists. Centro Médico de Caracas. Dr. Angel Riera García. Radiologist. Policlínica Metropolitana. Dr. Marialejandra Mendoza. Pulmonologist. Hospital Universitario de Los Andes. Researcher from the Department of Medicine. Universidad de Los Andes. Dr. Daniel Foldats. Internal Medicine. Inpatient Care Coordinator. Centro Médico Docente La Trinidad. In Colombia: Radiologists in Fundación Santa Fe de Bogotá: Dr. Javier Romero. President; Dr. María José Veloza and Dr. María Pinzón. In Hospital Universitario San Ignacio de Bogotá: Dr. Julio César Castellanos Ramírez. Specialist in Public Health and Healthcare Administration. Director. Dr. Margarita Manrique Andrade. Research Director. Dr. Carolina Gómez Rodríguez. Director of the Radiology Department. Dr. Ariel Emilio Cortés Martínez Researcher from Pontificia Javeriana de Bogotá. Dr. Nicolás Pérez Almanza. Public Health Specialist. Novatec Colombia. In Peru: Dr. Gustavo Adolfo Rondón Mendoza. Radiologist. Lima. in Chile: Benjamin Pizarro Galleguillos. PhD Candidate in Science and Radiology Specialty. Universidad de Chile.

References

1. Abadi, M., Agarwal, A., Barham, P., Brevdo, E., Chen, Z., Citro, C., Corrado, G.S., Davis, A., Dean, J., Devin, M., Ghemawat, S., Goodfellow, I., Harp, A., Irving, G., Isard, M., Jia, Y., Jozefowicz, R., Kaiser, L., Kudlur, M., Levenberg, J., Mané, D., Monga, R., Moore, S., Murray, D., Olah, C., Schuster, M., Shlens, J., Steiner, B., Sutskever, I., Talwar, K., Tucker, P., Vanhoucke, V., Vasudevan, V., Viégas, F., Vinyals, O., Warden, P., Wattenberg, M., Wicke, M., Yu, Y., Zheng, X.: TensorFlow: Large-scale machine learning on heterogeneous systems (2015), http://tensorflow.org/, software available from tensorflow.org
2. Anthimopoulos, M., Christodoulidis, S., Ebner, L., Christe, A., Mougiakakou, S.: Lung pattern classification for interstitial lung diseases using a deep convolutional neural network. IEEE Trans. Med. Imaging **35**(5), 1207–1216 (2016)
3. Apostolopoulos, I.D., Mpesiana, T.A.: Covid-19: automatic detection from x-ray images utilizing transfer learning with convolutional neural networks. Phys. Eng. Sci. Med. **43**, 635–640 (2020)
4. Bordley, W.C., Viswanathan, M., King, V.J., Sutton, S.F., Jackman, A.M., Sterling, L., Lohr, K.N.: Diagnosis and testing in bronchiolitis: a systematic review. Arch. Pediatr. Adolesc. Med. **158**(2), 119–126 (2004)
5. Bustos, A., Pertusa, A., Salinas, J.M., de la Iglesia-Vayá, M.: Padchest: a large chest x-ray image dataset with multi-label annotated reports. Med. Image Anal. **66**, 101797 (2020). https://doi.org/10.1016/j.media.2020.101797, http://www.sciencedirect.com/science/article/pii/S1361841520301614
6. Chetlen, A.L., Chan, T.L., Ballard, D.H., Frigini, L.A., Hildebrand, A., Kim, S., Brian, J.M., Krupinski, E.A., Ganeshan, D.: Addressing burnout in radiologists. Acad. Radiol. **26**(4), 526–533 (2019). https://doi.org/10.1016/j.acra.2018.07.001, http://www.sciencedirect.com/science/article/pii/S1076633218303416
7. Gulli, A., Pal, S.: Deep Learning with Keras. Packt Publishing Ltd., Birmingham (2017)
8. He, K., Zhang, X., Ren, S., Sun, J.: Deep residual learning for image recognition. In: 2016 IEEE Conference on Computer Vision and Pattern Recognition (CVPR), pp. 770–778 (2016)
9. Irvin, J., Rajpurkar, P., Ko, M., Yu, Y., Ciurea-Ilcus, S., Chute, C., Marklund, H., Haghgoo, B., Ball, R., Shpanskaya, K., Seekins, J., Mong, D., Halabi, S., Sandberg, J., Jones, R., Larson, D., Langlotz, C., Patel, B., Lungren, M., Ng, A.: Chexpert: a large chest radiograph dataset with uncertainty labels and expert comparison (2019)
10. Johnson, A.E., Pollard, T.J., Shen, L., Li-WeI, L.H., Feng, M., Ghassemi, M., Moody, B., Szolovits, P., Celi, L.A., Mark, R.G.: Mimic-iii, a freely accessible critical care database. Sci. Data **3**(160035) (2016). https://doi.org/10.1038/sdata.2016.35.
11. Kocher, M., Ruge, M.I., Galldiks, N., Lohmann, P.: Applications of radiomics and machine learning for radiotherapy of malignant brain tumors. Strahlentherapie und Onkologie: Organ der Deutschen Rontgengesellschaft (2020)
12. Lakhani, P., Sundaram, B.: Deep learning at chest radiography: automated classification of pulmonary tuberculosis by using convolutional neural networks. Radiology **284**(2), 574–582 (2017)
13. Liu, C., Cao, Y., Alcantara, M., Liu, B., Brunette, M., Peinado, J., Curioso, W.: Tx-cnn: detecting tuberculosis in chest x-ray images using convolutional neural network. In: 2017 IEEE International Conference on Image Processing (ICIP), pp. 2314–2318. IEEE (2017)

14. Norweck, J.T., Seibert, J.A., Andriole, K.P., Clunie, D.A., Curran, B.H., Flynn, M.J., Krupinski, E., Lieto, R.P., Peck, D.J., Mian, T.A., Wyatt, M.: Acr-aapm-siim technical standard for electronic practice of medical imaging. J Dig. Imaging **26**, 38–52 (2013). https://doi.org/10.1007/s10278-012-9522-2
15. Pasa, F., Golkov, V., Pfeiffer, F., Cremers, D., Pfeiffer, D.: Efficient deep network architectures for fast chest x-ray tuberculosis screening and visualization. Sci. Rep. **9**(1), 1–9 (2019)
16. Paszke, A., Gross, S., Massa, F., Lerer, A., Bradbury, J., Chanan, G., Killeen, T., Lin, Z., Gimelshein, N., Antiga, L., Desmaison, A., Kopf, A., Yang, E., DeVito, Z., Raison, M., Tejani, A., Chilamkurthy, S., Steiner, B., Fang, L., Bai, J., Chintala, S.: Pytorch: an imperative style, high-performance deep learning library. In: Wallach, H., Larochelle, H., Beygelzimer, A., Alché-Buc, F., Fox, E., Garnett, R. (eds.) Advances in Neural Information Processing Systems, vol. 32, pp. 8024–8035. Curran Associates, Inc. (2019). http://papers.neurips.cc/paper/9015-pytorch-an-imperative-style-high-performance-deep-learning-library.pdf
17. Reiner, B.I., Krupinski, E.: The insidious problem of fatigue in medical imaging practice. J. Dig. Imaging **25**(1), 3–6 (2012)
18. Ronneberger, O., Fischer, P., Brox, T.: U-net: Convolutional networks for biomedical image segmentation. In: Medical Image Computing and Computer-Assisted Intervention (MICCAI). LNCS, vol. 9351, pp. 234–241. Springer, Heidelberg (2015). http://lmb.informatik.uni-freiburg.de/Publications/2015/RFB15a, (available on arXiv:1505.04597 [cs.CV])
19. Simonyan, K., Zisserman, A.: Very deep convolutional networks for large-scale image recognition. CoRR **abs/1409.1556** (2015)
20. Sirazitdinov, I., Kholiavchenko, M., Mustafaev, T., Yixuan, Y., Kuleev, R., Ibragimov, B.: Deep neural network ensemble for pneumonia localization from a large-scale chest x-ray database. Comput. Electric. Eng. **78**, 388–399 (2019)
21. Tang, Y., Peng, Y., Yan, k., Bagheri, M., Redd, B., Brandon, C.J., Lu, Z., Han, M., Xiao, J., Summers, R.M.: Automated abnormality classification of chest radiographs using deep convolutional neural networks. npj Digit. Med. **70**(3) (2020). https://doi.org/10.1038/s41746-020-0273-z
22. Taylor, E., Haven, K., Reed, P., Bissielo, A., Harvey, D., McArthur, C., Bringans, C., Freundlich, S., H., I.R.J., Perry, D., F., W., Milne, D., Modahl, L., Huang, Q.S., Gross, D., Widdowson, M.A., Grant, C.: A chest radiograph scoring system in patients with severe acute respiratory infection: a validation study. BMC Med. Imaging **15**(61) (2015). https://doi.org/10.1186/s12880-015-0103-y
23. Van Cleeff, M., Kivihya-Ndugga, L., Meme, H., Odhiambo, J., Klatser, P.: The role and performance of chest x-ray for the diagnosis of tuberculosis: a cost-effectiveness analysis in nairobi, kenya. BMC Infect. Dis. **5**(1), 1–9 (2005)
24. Vancheri, S.G., Savietto, G., Ballati, F., Maggi, A., Canino, C., Bortolotto, C., Valentini, A., Dore, R., Stella, G.M., Corsico, A.G., Iotti, G.A., Mojoli, F., Perlini, S., Bruno, R., Preda, L.: Radiographic findings in 240 patients with covid-19 pneumonia: time-dependence after the onset of symptoms. Eur. Radiol. **30**, 1–9 (2020). https://doi.org/10.1007/s00330-020-06967-7
25. Zhou, Z., Siddiquee, M.M.R., Tajbakhsh, N., Liang, J.: Unet++: redesigning skip connections to exploit multiscale features in image segmentation. IEEE Trans. Med. Imaging **39**(6), 1856–1867 (2020)
26. Zou, K.H., Warfield, S.K., Bharatha, A., Tempany, C.M.C., Kaus, M.R., Haker, S.J., William, M.W., Jolesz, F.A., Kikinis, R.: Statistical validation of image segmentation quality based on a spatial overlap index. Acad. Radiol. **11**(2), 178–189 (2004). https://doi.org/10.1016/s1076-6332(03)00671-8

Classification of Andean Chocho (Lupinus Mutabilis Sweet) by Shape and Color Using Artificial Vision

Paul Freire Diaz(✉), Ximena López-Mendoza, Byron Casignia, Andrés Santiago Cisneros Barahona, and María Isabel Uvidia Fassler

Universidad Nacional de Chimborazo, Riobamba, Ecuador
jpfreire@unach.edu.ec

Abstract. In Ecuador, the classification of the cooked grain of Lupinus Mutabilis Sweet, Andean lupine, according to its color and shape is done manually. This work shows an alternative classification using artificial vision. The system is developed on Matlab's Simulink visual programming platform, running on a Raspberry Pi B + board, incorporated into a test bench that maintains constant lighting during image acquisition. The image processing is carried out under the HSV color model, which allows to intuitively choose the defined ranges of its three components of hue, saturation and value to discriminate areas in the image based on their color. To determine the acceptance by color of the grain, the average value of the components is compared with thresholds that are calculated experimentally. The grains are evaluated by shape based on their eccentricity. The classification system indicates suitable grains according to their shape and color with an accuracy of 81.83% compared to those selected by human operators.

Keywords: Andean lupine · HSV color model · Raspberry Pi · Simulink · Artificial vision

1 Introduction

Lupinus Mutabilis Sweet is a native legume of the Andean countries that is cultivated from Ecuador to Bolivia [1]. In Ecuador, known as "chocho", it is a widely consumed food; with a protein percentage higher than that of soy, of more than 50%, it has fewer carbohydrates than other legumes [2, 3]. Cooked without the shell it is used in combination with products with low protein content [4].

In agricultural products, color is a parameter that allows evaluating their suitability. Thus, the consumer will accept or reject the product according to what they observe [5, 6]. In lupine suitable for human consumption, color defects are identifiable only in cooked grain. The presence of spots or irregularities has traditionally been evaluated through visual inspection by qualified personnel [7, 8] to eliminate beans with green, black cotyledons, or with brown or black spots on the cover [1]. Where these defects are

M. Botto-Tobar et al. (Eds.): CIT 2020, AISC 1326, pp. 64–78, 2021.
https://doi.org/10.1007/978-3-030-68080-0_5

visualized, regardless of its size, is known as an "imperfect grain" [2]. The manual classification process requires a great deal of time and is subject to a subjective appreciation of characteristics by the operator, based on empirical experience [8].

Artificial Vision emulates the way human vision works, relying on algorithms such as neural networks, fuzzy logic, statistical classifiers, among others [9], to extract specific attributes, such as color and shape, among others, using computer systems [10]. It is intended to automate repetitive inspection tasks, product quality controls, object inspection without physical contact at high speed [9].

The RGB model represents a color image combining red (R), green (G) and blue (B) in each pixel, assigning each a value in a range from 0 to 255. This model implies the superposition of the components to form a single image [11]. The main drawback of using this model in artificial vision lies in the difficulty of finding the ranges of the three RGB components that form a specific tone since they are very similar to each other [12].

The HSV model represents the image through 3 layers or channels, "Hue" (H), "Saturation" (S) and "Value" (V). Hue defines a pure color in a range, measured in sexagesimal degrees, from 0 to 360° [13]. Saturation is determined as the white color contained in a color in a range from 0% to 100%, which can also be expressed normalized from 0 to 1; and the Value channel indicates how light or dark the color also becomes in a range from 0% to 100% or from 0 to 1, normalized [14].

Artificial vision techniques have been successfully tested with difficult-to-define and relative visual parameters, such as the shape of cells or fruit, regardless of the position or size of the object being inspected. In Constante et al. [15], images of strawberries are acquired through a webcam or a database stored in a computer to be analyzed first using a process algorithm and then training a neural network.

Machine vision is used to classify elements of more regular shapes such as the Tommy type handle [16]. Size and color are the selection parameters chosen of the mango, to infer how ripe it is.

Recognition processes require devices capable of processing the information from a digital camera in real-time. Arduino or Raspberry Pi are the processes that can be done by a computer or by hardware platforms. As shown in Othman et al. [17] in the systems area they used artificial vision for the design of an embedded system for detection and measurement of objects in real-time with Raspberry PI, which presents a great performance at low cost.

2 Materials and Methods

This section refers to the stages of project development. The first section describes the Artificial Vision System (AVS), its software and hardware components, and the test bench for image capture. The following describes the process known as image segmentation and finally the extraction of characteristics for classification.

For the system tests, lupine grains of different sizes and shapes are used, separating in the first instance the grains categorized as imperfect from those suitable for marketing, according to the criteria of a person dedicated to the classification and sale of this product. From the group of suitable grains, the characteristics of shape and color are extracted. In Fig. 1 the algorithm followed for the classification is represented.

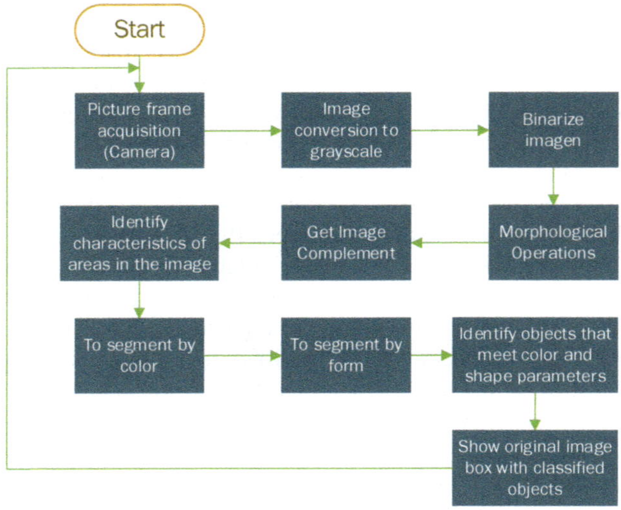

Fig. 1. Stages followed in the lupine classification process by artificial vision.

2.1 Artificial Vision System

The system hardware consists of a Raspberry Pi B + board, with a compatible camera board v1.3 attached (Table 1), configured to receive the video with a size of 320 × 240 pixels, in RGB format and a sampling time of 1/10 s.

Table 1. Characteristics of Raspberry Pi Camera Board v1.3

Compatible with both the Model A and Model B Raspberry Pi	5 MP omnivision 5647 camera module
Still Picture Resolution: 2592 × 1944	Video: Supports 1080 p @ 30 fps, 720 p @ 60 fps and 640 × 480 p 60/90 Recording
15-pin MIPI Camera Serial Interface	Size: 20 × 25 × 9 mm

The images resulting from the process are displayed on a 17-inch LCD monitor connected to the board's HDMI port (Fig. 2 a).

To guarantee optimal results in the acquisition of the analyzed images, through space free of variation in the intensity of illumination and reflections, since variations in illumination affect the interpretation of colors by the camera [16, 18]. In Fig. 2 b, the diagram of the test bench of the lupine classification system is shown, formed by a wooden box with a 210 × 300 mm base and a height of 200 mm. It has a front cover that allows the grains to be analyzed and a perforation in the central part of the upper cover. On this cover is located the Raspberry Pi card with the camera aligned to the hole. Inside there are two strips of led luminaires, powered by a 12-V DC voltage source.

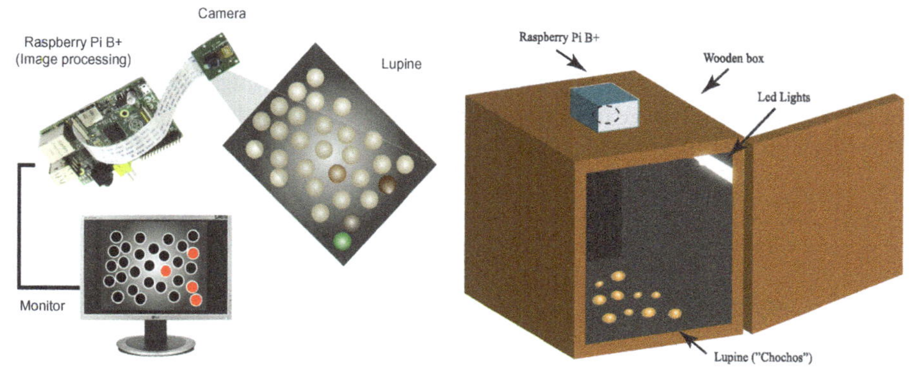

Fig. 2. a) Outline of the lupine grading system test bench b) Description of the proposed classification system for lupine grains

For the development of the system, the Simulink graphical programming environment was used, part of the Matlab software, which allows modeling and simulating dynamic systems using block diagrams. From this diagram, Simulink compiles, generates, and embeds the system code on the Raspberry board.

2.2 Image Segmentation

The process known as segmentation consisted of getting different objects or areas of an image to have values that allow them to be discriminated according to their characteristics. There are several general steps for the analysis of an image, among which are the binarization of the image and the morphological operations [16].

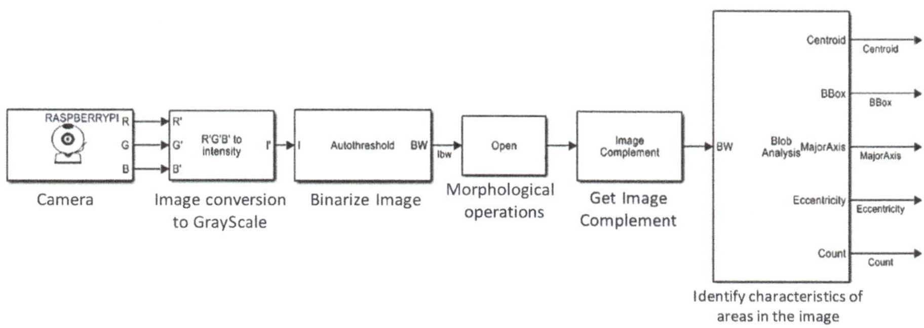

Fig. 3. Section of the Simulink Block diagram that performs the segmentation of an image obtained from the camera

The sequence of Simulink blocks that the system uses for this effect is shown in Fig. 3, at the end of which there will be matrices with the coordinates of the centroid, the bounding box, the length of the greatest axis, and the eccentricity of each the detected regions.

Grayscale Conversion

In the GrayScale or Intensity space, the color of each point is stored with a value between 0 and 255, which indicates its "degree of darkness". The following equation defines the conversion of an RGB image to Grayscale, where R′, G′, B′ are the components of each of the pixels in the image [19].

$$\text{Intensity} = \begin{bmatrix} 0.299 & 0.587 & 0.114 \end{bmatrix} \begin{bmatrix} R' \\ G' \\ B' \end{bmatrix} \tag{1}$$

Image Binarization

The goal of this step is to obtain identifiable regions for analysis. When transforming the grayscale image to a binary image, it will contain values of 1 and 0 in the matrix that describes it [20, 21]. The AutoThreshold block applies the Otsu method [22]. For the binarization process of the grayscale image, in the first tests with the Threshold scaling factor below 1, the system recognizes the neighboring objects as one. It was determined through several experiments that the Threshold scaling factor = 1.27 achieved the best results in this case.

Morphological Operations

In the proposed algorithm, the morphological operation "open" is used, which executes an "erode" operation and then a "dilate" operation [23], using the same structuring element (strel), which is a binary element valued by the neighborhood in both two dimensions and multidimensional, in which pixels equal to 1 are included in the operation and pixels with a value of 0 are not. The central pixel of this element is called the origin and identifies the pixel in the image subjected to this process [24].

Image Complement

By the parameters chosen for the binarization process, an image is obtained with the objects found in black on a white background. In the following stages of the system, you must have an image in which the areas to be searched are shown in white, so the inversion of the image is necessary. The process of obtaining the complement of a binary image consists of replacing the value of 1 in the pixels by 0, and vice versa. In Fig. 4 the results of the processes described are shown.

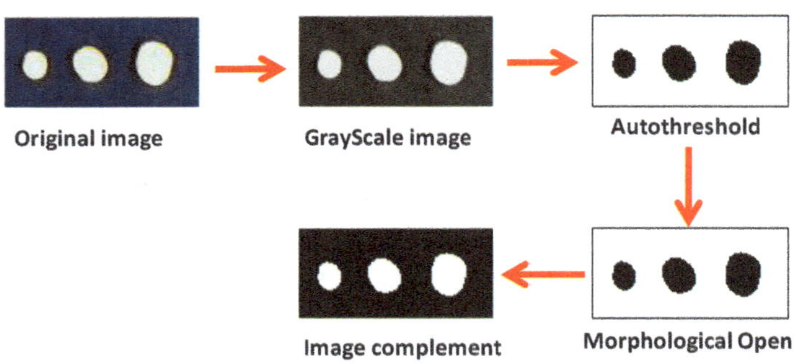

Fig. 4. Image segmentation process

The regions or areas with contiguous pixels are called connected components, objects, or "Blobs" [25] and for their analysis, the Blob Analysis block of the Simulink Computer Vision System Toolbox library is used, which calculates statistics of the regions obtained after image binarization [26].

2.3 Feature Extraction

A sample of lupine grains chosen at random from a portion of cooked grains, without prior categorization, is separated, according to the criteria of a person dedicated to the classification and sale of this product, into two groups: grains classified as imperfect and those considered suitable for marketing. The characteristics of color and shape are extracted from suitable grains (Fig. 5) to determine the thresholds of these variables for automatic classification.

The value of the color of each grain is taken as the average of the value of the H, S and V components of the pixels that represent it within the image. In this way, if there are spots on the grain, the average value is affected, falling outside the established thresholds for its acceptance.

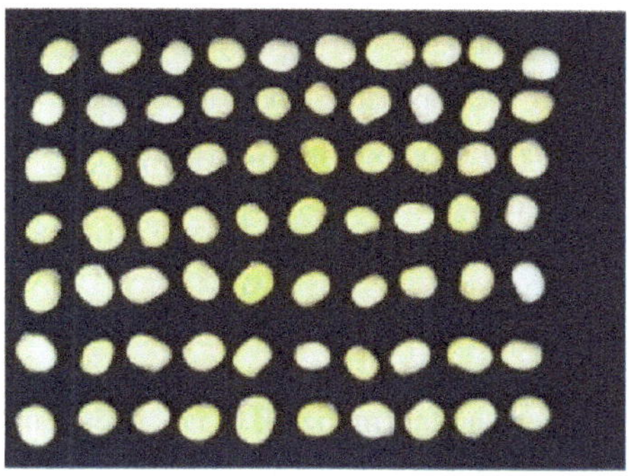

Fig. 5. Sample of lupine grains classified as suitable.

Extraction of the Characteristics of the Color Variable

The originally captured image is in RGB space, but it is transferred to HSV space, since it is easier to separate the objects by their colorimetry, defining ranges of interest based on a single layer of data compared to the RGB model, where they have three [13]. The conversion is carried out internally by the Color Space Conversion block through the equations:

$$H = \begin{cases} \left(\frac{G'-B'}{MAX-MIN}\right)/6, & si\ R' = MAX \\ \left(2 + \frac{B'-R'}{MAX-MIN}\right)/6, & si\ G' = MAX \\ \left(4 + \frac{R'-G'}{MAX-MIN}\right)/6, & si\ B' = MAX \end{cases} \tag{2}$$

$$S = \frac{MAX - MIN}{MAX} \tag{3}$$

$$V = MAX \tag{4}$$

Where MAX and MIN represent the maximum and minimum values of the RGB components and H, S and V will vary from 0 to 1, where 1 represents the maximum saturation value [27].

According to the Ecuadorian standard, the lupine suitable for consumption must be white to cream in color [28], so the values of tint or "pure" color will be approximately in the range of 60° to 120° of the H component, that is to say, between yellow and light green.

Each color component is processed separately. The "Average Area H", "Average Area S" and "Average Area V" user function blocks require the inputs described in Table 2:

Table 2. Description of the inputs of the functions "Average Area H", "Average Area S" and "Average Area V"

Input variable	Type	Description
Component	Matrix	H, S or V component of the original image
Nblobs	Entire	Number of regions or blobs detected
Blob	Matrix	Matrix of dimension M × 4 with the information of the smallest rectangle that contains the detected region. M is the number of regions detected

The outputs of the functions are normalized, so it is necessary to apply gains to them to obtain the real values of the components (Fig. 6).

With the inputs described, for each component the function in Fig. 7 executes the following algorithm:

1. Using the coordinates of each of the regions detected in the binarized image, a submatrix is delimited in the "component" matrix
2. For each submatrix, the average of the values greater than 0 contained is calculated, that is, those that contain color information.
3. The averages found are delivered as output in a vector called "prom_area" of (t × 1) elements, where t represents the number of "blob" to which the information belongs.

The result of the algorithm is described in Fig. 7.

To determine the thresholds between which the values of the components must fluctuate, only suitable grains are placed on the test bench. The maximum and minimum values of each vector "prom_area" are determined, which will be the threshold values. The values of the color components obtained by applying this process are shown in Table 3.

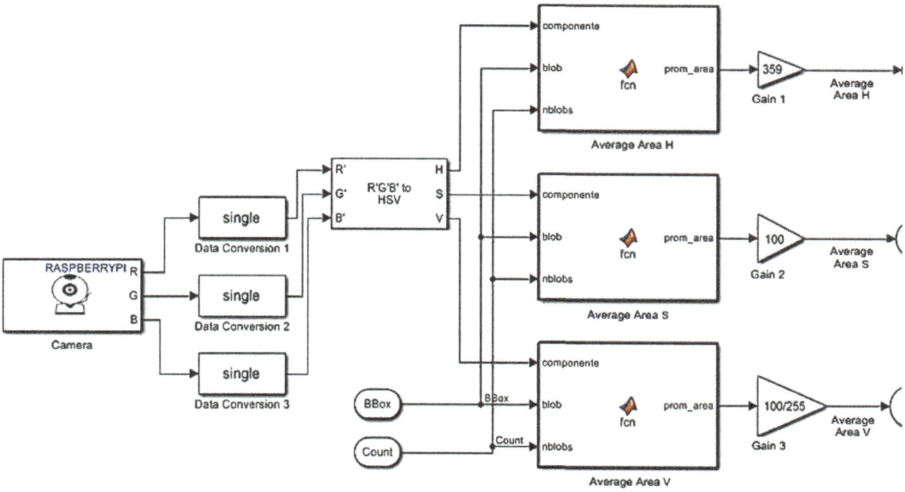

Fig. 6. Simulink block structure that averages the value of the H, S and V components of the objects detected in the image.

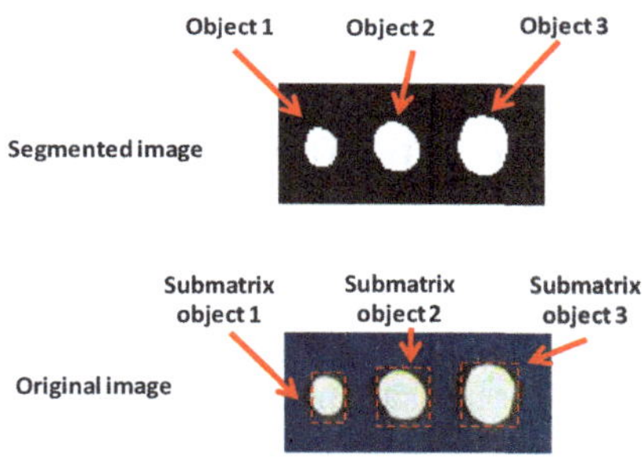

Fig. 7. The segmented image where the objects are identified in the order, with their coordinates, to delimit submatrices in the original image

Table 3. Average of the maximum and minimum values of the HSV components found in images of suitable grains

Thresholds	H (Hue)	S (Saturation)	V (Value)
Minimum	60	17	70
Maximum	124,25	34	90

The maximum and minimum values found for the components H, S and V are entered into the system in the form of constant blocks. The blocks that fulfill this function are shown in Fig. 8.

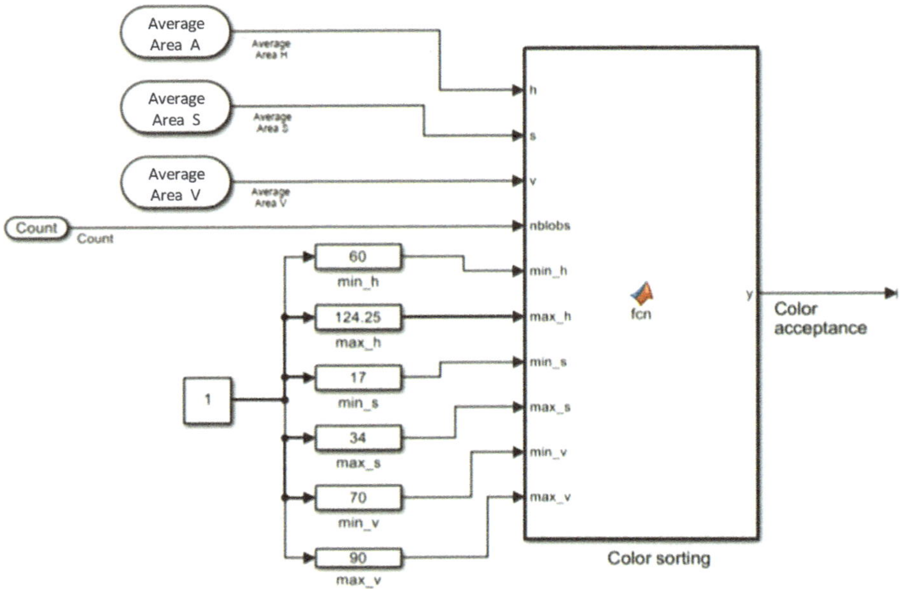

Fig. 8. System section that determines the acceptance of a detected element based on its color

Extraction of the Characteristics of the Variable Shape

At this stage, the shape of the object is determined through the value of its eccentricity, delivered by the Blob Analysis block (Fig. 9). Eccentricity relates the distance between the foci of the ellipse and the length of its principal axis. To obtain the thresholds of the shape variable, the images obtained from corn kernels with grains manually classified as suitable were used which, after measurement, were shown to have eccentricity values less than or equal to 0.7.

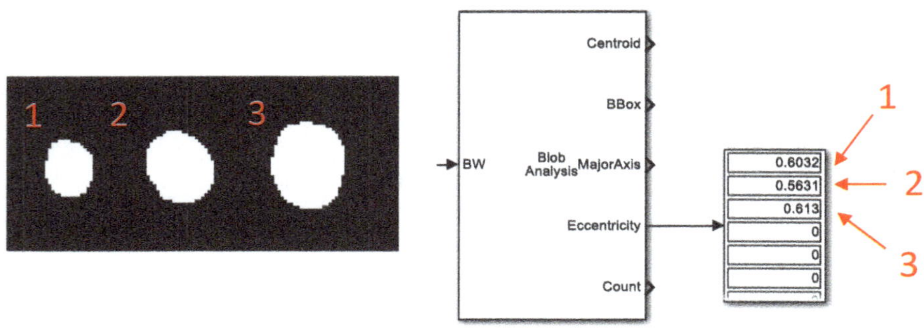

Fig. 9. The eccentricity of three objects identified by the Blob Analysis Block

2.4 System Tests

If the detected object meets the defined eccentricity and color parameters, the system draws a blue circle on the image of the object, otherwise, a red circle is displayed on the original image. To test the algorithm, images are processed with grains considered unsuitable by the human operator located within the test bench (Fig. 10).

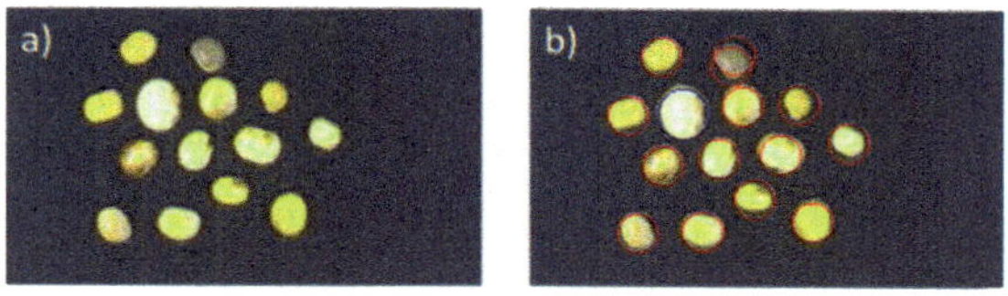

Fig. 10. Imaging tests containing only grains considered defective. a) Original image b) Processed image

With a sample containing both fit and imperfect grains, the system returns the image shown in Fig. 11.

Fig. 11. Output of the lupine grains sorting system

3 Results and Discussion

To evaluate the results of the system, a sample of 1066 lupine beans randomly chosen from a batch of unbroken, unsorted beans was used. The sample had independent manual classifications made by two people (operators) who work in this activity. In each case, the optimal grains were separated from the defective ones.

The classified grains were fed into the machine vision system in batches, as the area captured by the camera can contain a maximum of 80 grains. Batches containing only grains classified as defective by the operator or batches with grains considered optimal were placed in the system. In this way, the results of the manual can be compared with those of the proposed system. The results are shown in Tables 4, 5, 6 and 7. A compilation of the data is shown in Table 8.

Table 4. Results of the artificial vision system for the grains classified as defective by the operator 1.

Batch	Grains with defects detected by operator 1	Artificial vision system		Coincidence operator/artificial vision
		Optimal grains	Grains with defects	
1	70	18	52	52
2	66	10	56	56
Total	136	28	108	108

Table 5. Results of the artificial vision system for the grains classified as defective by the operator 2.

Batch	Grains with defects detected by operator 2	Artificial vision system		Coincidence operator/artificial vision
		Optimal grains	Grains with defects	
1	70	19	51	51
2	74	13	61	61
Total	144	32	112	112

Table 6. Results of the artificial vision system for the optimal grains by the operator 1.

Batch	Optimal grains detected by operator 1	Artificial vision system		Coincidence operator/artificial vision
		Optimal grains	Grains with defects	
1	70	59	11	59
2	79	64	15	64
3	70	62	8	62
4	70	63	7	63
5	80	66	14	66
6	80	65	15	65
7	79	67	12	67
8	80	70	10	70
9	80	64	16	64
10	70	58	12	58
11	70	62	8	62
12	70	61	9	61
13	32	26	6	26
Total	930	787	143	787

Table 7. Results of the artificial vision system for the optimal grains by the operator 2

Batch	Optimal grains detected by operator 2	Artificial vision system		Coincidence operator/artificial vision
		Optimal grains	Grains with defects	
1	79	66	13	66
2	70	65	5	65
3	70	57	13	57
4	70	58	12	58
5	80	73	7	73
6	80	75	5	75
7	70	54	16	54
8	80	71	9	71
9	80	74	6	74
10	70	56	14	56
11	70	55	15	55
12	70	65	5	65
13	33	20	13	20
Total	922	789	133	789

Table 8. Comparison of the results of the manual classification of a sample of 1066 grains of lupine with that carried out by the proposed system.

Type of grain detected	Manual detection		Average detection by operators	Number of right answers		Average of right answers
	Operator 1	Operator 2		Artificial vision system vs operator 1	Artificial vision system vs operator 2	
Optimal grains	930	922	926	787	789	788
Grains with defects	136	144	140	108	112	110
Total Sample	1066	1066				

Sorting performance is measured using the Eq. 5:

$$Accuraccy = \left(\frac{AGO}{2GOVA} + \frac{AGD}{2GDVA} \right) \times 100 \tag{5}$$

Where:
AGO = Average of right answers optimal grains.
AGD = Average of right answers grains with defects.
GOVA = Average of optimal grains detected by the human operators.
GDVA = Average of defective grains detected by the human operators.
Evaluating Eq. (5) with the results obtained is obtained for the system:

$$Accuraccy = 81{,}83\%$$

4 Conclusions

An artificial vision classification system was developed for lupins, which indicates the grains suitable for human consumption with an accuracy of 81.83% concerning those selected by a human operator. Having fixed values as thresholds for the maximum and minimum values of the color components caused a grain with a value slightly outside the ranges to be discarded, as well as the defective grains to be considered valid. Another factor that influenced the result is that grain defects may appear on the opposite side to that captured by the camera, causing a false positive to occur. There is a difference between the number of grains considered optimal by operator 1 and operator 2 because each has a slightly different tolerance criterion for the lupine to be considered suitable.

Using a suitable combination of binarization threshold and structuring element values, it was possible to obtain a binarized image with well-defined areas. Because valid lupine grains are practically elliptical, eccentricity allows classifying a grain by its shape, accepting those that have a value less than or equal to 0.7, which was determined experimentally. The algorithm used required few computational resources, so the result runs optimally on a low-cost device like the Raspberry Pi. However, the results obtained could be improved using other techniques such as fuzzy logic or neural networks.

References

1. García Zambrano, J.L.: Gestión de la producción asociativa del chocho (Lupinus Mutabilis Sweet) y su incidencia en el nivel de ingresos de los habitantes productores de la comunidad Sarachupa (2014). https://repositorio.uta.edu.ec/handle/123456789/8567
2. Baldeón Salgado, P.E.: Procesamiento del Chocho (Lupinus Mutabilis Sweet) para la obtención de leche y yogurt como alimentos alternativos de consumo humano (2012). https://repositorio.ug.edu.ec/bitstream/redug/1928/1/TESIS_CHOCHO_PDF.pdf
3. Suca, G.R.A., Suca, C.A.A.: Potencial del tarwi (Lupinus mutabilis Sweet) como futura fuente proteínica y avances de su desarrollo agroindustrial. Rev. Per. Quím. Ing. Quím. **18**, 55–71 (2015)
4. Caicedo, V.C., Peralta, I.E., Villacrés, P.E., Rivera, M.M.: Poscosecha y mercado de chocho (Lupinus Mutabilis Sweet) en Ecuador, Quito (2001)
5. Bhargava, A., Bansal, A.: Fruits and vegetables quality evaluation using computer vision: a review. J. King Saud Univ. Comput. Inf. Sci. (2018). https://doi.org/10.1016/j.jksuci.2018.06.002
6. Fischer, G., Martínez, O.: Calidad y madurez de la uchuva (Physalis peruviana l.) en relación con la coloración del fruto. Agron. Colomb. **16**, 35–39 (1999)

7. Blasco, J., Cubero, S., Gómez-Sanchís, J., Moltó, E.: Avances en visión artificial automática de productos hortofrutícolas. Poscosecha **288**, 48–51 (2010)
8. Bonilla Gonzalez, J.P., Prieto Ortiz, F.A., Perez Rodriguez, C.P.: Mass and volume estimation of passion fruit using digital images. IEEE Lat. Am. Trans. **15**, 275–282 (2017). https://doi.org/10.1109/tla.2017.7854623
9. Vasco Orozco, A., Tamayo Ocampo, A.M.: Visión artificial, una alternativa a la sensórica convencional en la implementación de los sistemas de control. IngEam **3**, 1–8 (2016)
10. Parra, P., Negrete, T., Llaguno, J., Vega, N.: Determination of the degree of fermentation of cocoa through different techniques of artificial vision. In: 16th LACCEI International Multi-Conference for Engineering, Education, and Technology "Innovation in Education and Inclusion", pp. 19–21 (2018). https://doi.org/10.18687/LACCEI2018.1.1.163
11. López, I.E., López, G.A., Argueta, R.A., Galvez, J.A., Cardona, M.N.: Desarrollo de software para extracción de propiedades de cristales de azúcar utilizando visión artificial – SugarVision. In: CHILECON 2015 - 2015 IEEE Chilean Conference on Electrical, Electronics Engineering, Information and Communication Technologies, pp. 835–840 (2016). https://doi.org/10.1109/Chilecon.2015.7404669
12. Russ, J.C.: The Image Processing Handbook (Google eBook) (2007). https://doi.org/10.1201/9780203881095. 817 p
13. Méndez, H., Jalomo, J., Ordóñez, F., García, J.: Procesamiento de Imagen para Seguimiento de Objetos por Control PI de un Robot Manipulador Didáctico. Memorias del Congr. Nac. Control Automático, pp. 43–48 (2012)
14. Barba-Guamán, L., Calderon-Cordova, C., Quezada-Sarmiento, P.A.: Detección de objetos en movimiento a través de la umbralización del color (2017)
15. Constante, P., Chang, O., Pruna, E., Escobar, I.: Artificial vision techniques for Strawberry's industrial classification. IEEE Lat. Am. Trans. **14**, 2576–2581 (2016). https://doi.org/10.1109/TLA.2016.7555221
16. Romero Acero, Á., Marín Cano, A., Jiménez Builes, J.A.: Sistema de clasificación por visión artificial de mangos tipo Tommy. Rev. UIS Ing. **14**, 21–31 (2015)
17. Othman, N.A., Salur, M.U., Karakose, M., Aydin, I.: An embedded real-time object detection and measurement of its size. In: 2018 International Conference on Artificial Intelligence and Data Processing, IDAP 2018, pp. 1–4 (2019). https://doi.org/10.1109/IDAP.2018.8620812
18. Omid, M., Soltani, M., Dehrouyeh, M.H., Mohtasebi, S.S., Ahmadi, H.: An expert egg grading system based on machine vision and artificial intelligence techniques. J. Food Eng. **118**, 70–77 (2013). https://doi.org/10.1016/j.jfoodeng.2013.03.019
19. García, P.P.: Reconocimiento de imágenes utilizando redes neuronales artificiales (2013). https://eprints.ucm.es/23444/1/ProyectoFinMasterPedroPablo.pdf
20. Hernandez-Juarez, S., Mejia-Rodriguez, A.R., Arce-Santana, E.R., Scalco, E., Rizzo, G.: Evaluación Cuantitativa del Desempeño de Métodos de Segmentación Aplicados a Imágenes Médicas para el Análisis de Estructuras Anatómicas de Interés. Memorias del Congr. Nac. Ing. Biomédica, pp. 374–377 (2015). https://doi.org/10.24254/CNIB.15.66
21. Mendoza, L.E., Meza, E.F., Gualdron, O.E.: Segmentación y Parametrización Automática de Imágenes Iridológicas. Rev. Ing. Biomédica **10**, 13–21 (2016)
22. Otsu, N.: A Threshold Selection Method from Gray-Level Histograms. IEEE Trans. Syst. Man. Cybern. **9**, 62–66 (1979). https://doi.org/10.1109/TSMC.1979.4310076
23. Mathworks: Morphological erosion of binary pixel data - Simulink - MathWorks América Latina. https://la.mathworks.com/help/visionhdl/ref/erosion.html?s_tid=doc_ta
24. Vélez Serrano, J., Moreno Díaz, A., Sánchez Calle, Á., Sánchez-Marín, J.: Visión por computador (2019)
25. Mathworks: Medir las propiedades de las regiones de imagen - MATLAB regionprops - MathWorks América Latina. https://la.mathworks.com/help/images/ref/regionprops.html

26. Mathworks: Compute statistics for labeled regions - Simulink - MathWorks América Latina. https://la.mathworks.com/help/vision/ref/blobanalysis.html
27. Mathworks: Convert color information between color spaces - Simulink - MathWorks América Latina. https://la.mathworks.com/help/vision/ref/colorspaceconversion.html?searchHighlight=colorspaceconversionblock&s_tid=doc_srchtitle
28. Instituto Ecuatoriano de Normalización INEN: Norma Técnica Ecuatoriana Voluntaria INEN 2 390:2004 Leguminosas. Grano desamargado de chocho. Requisitos (1989). https://www.normalizacion.gob.ec/buzon/normas/2390.pdf

Deep Learning for Edge Computing: A Survey

Joffre Luis Leon Veas, Luis Bryan Cordero Solis, Galo Enrique Valverde Landivar, and Miguel Angel Quiroz Martinez(✉)

Department of Computer Science, Universidad Politécnica Salesiana Sede Guayaquil, Chamber 227 y 5 de junio, Guayaquil, Ecuador
{jleon,gvalverde,mquiroz}@ups.edu.ec, lcordero@est.ups.edu.ec

Abstract. A compendium of Deep Learning algorithms that are applied in Edge Computing through IoT devices that generate a large amount of data and representative image types was analyzed. The problem is the lack of proposals for the analysis and recommendation of Deep Learning algorithms that use data from devices in an Edge Computing. The objective is to carry out a survey as a research technique for the collection of data on Deep Learning algorithms for Edge Computing in different application areas, and determine which algorithm is most used and efficient. The applied methodology is exploratory research for the study of problems associated with the treatment and processing of data in different areas, gradual approach and deduction based on information from selected references on Deep Learning and Edge Computing. This research resulted in an Impact of Deep Learning on Edge Computing, a Deep Learning Algorithm Summary and an Edge Computing Summary resulted. It was concluded that according to the analysis carried out, the best algorithm is DNN due to its high degree of precision 98% in different areas; DNN stands out for the complexity of its structure that uses multiple layers (input, hidden and output) for the processing and training of the neurons that make it up.

Keywords: Deep Learning · Edge Computing · Machine learning · Data analysis · Survey

1 Introduction

The Internet of Things (IoT) today is one of the most widely used technologies in the personal, home and business fields; many of the devices are connected to the network and these generate thousands of amounts of data for communication, concurrent access and storage; since this information is transmitted to the cloud for processing.

However, these requests by simultaneous clients and workgroups to a centralized server generate very long response times and high energy consumption; in the case of medical systems, automated systems, among others, require efficiency and higher performance.

Deep Learning (DL) is considered a set of essential algorithms in the field of Artificial Intelligence; because its algorithms for neural network training help in achieving high performance and precision (Fig. 1); therefore, said algorithms are applied to assist

M. Botto-Tobar et al. (Eds.): CIT 2020, AISC 1326, pp. 79–93, 2021.
https://doi.org/10.1007/978-3-030-68080-0_6

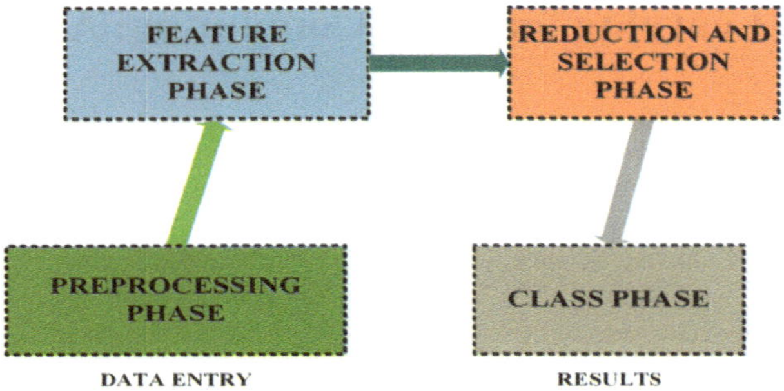

Fig. 1. Phases of operation of Deep Learning models.

and optimize tasks of recognition of human activities, health sections, optimization of advanced autonomous systems [1].

Edge Computing (EC) is a term related to the IoT, here a large number of devices connected wirelessly to the network generate millions of data that will be processed in a decentralized and distributed way (Fig. 2); it is used to avoid high energy consumption, latency and high costs; this mesh network of micro servers helps to achieve very high transmission speeds and, above all, to strengthen information security [1].

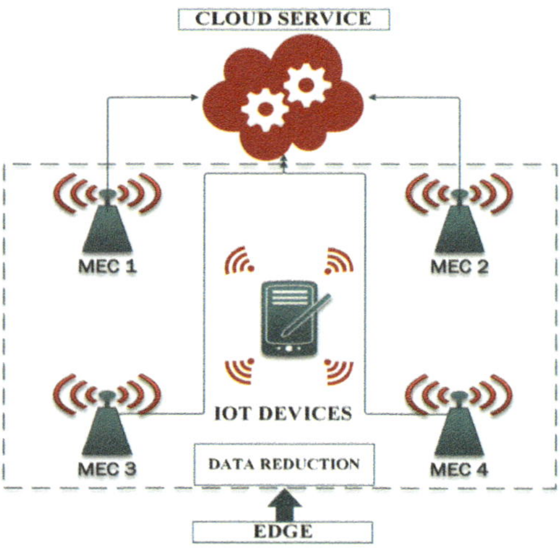

Fig. 2. Edge-Cloud Computing architecture

Generally, a state-of-the-art computing structure such as EC does not have edge servers capable of handling a high rate of requests by DL algorithms used to extract features from large data clusters at the input layer [2]. It should be noted that edge devices have limited storage resources due to the size of convolutional neural networks; therefore, there are studies that propose modifying the prediction layer of these networks so that the devices on the edge process the data obtained in the input layer [3]. Edge Computing and Image Recognition through Deep Learning converge for more accurate results and processing of that data in reduced times [2, 4, 5] (Fig. 3).

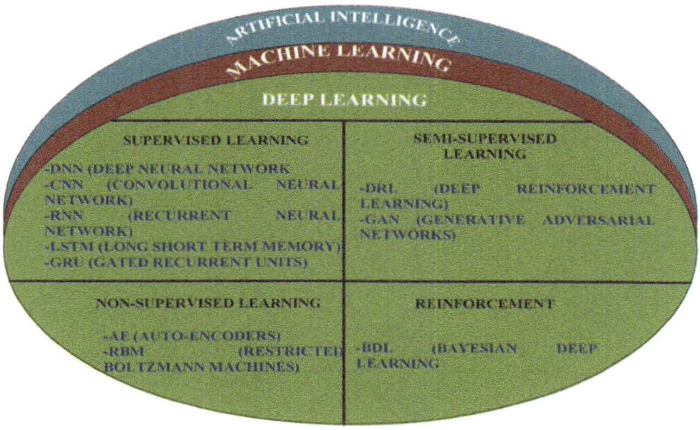

Fig. 3. Classification of Deep Learning algorithms [6].

The problem is the lack of proposals for the analysis and recommendation of Deep Learning algorithms that use data from devices in an Edge Computing.

Why analyze Deep Learning research information for Edge Computing?

To understand information recognition and classification algorithms that are directed or developed in connected device environments.

Our motivation is to achieve future work that allows the use of Deep Learning algorithms with improvements in their level of sensitivity and accuracy of results in the face of new technologies.

The objective is to carry out a survey as a research technique for the collection of data on Deep Learning algorithms for Edge Computing in different application areas, and determine which algorithm is most used and efficient.

The applied methodology is exploratory research for the study of problems associated with the treatment and processing of data in different areas, gradual approach and deduction based on information from selected references on Deep Learning and Edge Computing from Scopus and IEEE databases.

2 Materials and Methods

The proposals made by the authors of the references with topics related to Deep Learning and Edge Computing are described below.

2.1 Materials

The authors of [1] discussed reducing data transmission to the cloud to reduce latency and consumption on IoT devices; they explained that DL analyzes data with multilayer algorithms that recognize the structure of objects and their attributes, due to the reduction of linear dimensionality for data sent to the cloud; they also highlighted that the Edge-Cloud architecture manages a set of encoders that help the analysis and efficient saving of data obtained by IoT devices against human activity.

The authors of [2] proposed a DL model for IoT devices and data in EC to collect and send data efficiently; they emphasized that DL handles multiple layers here features will be extracted from large amounts of data in lower layers, some of this data will be processed in intermediate layers, and the rest in upper layers.

The authors of [3] proposed a DL model that analyzes a person's face and recognizes their age and gender through EC devices; they discussed new possible methods that help reduce the size of neural networks and increase their performance and precision so that they can be processed and distributed by edge devices.

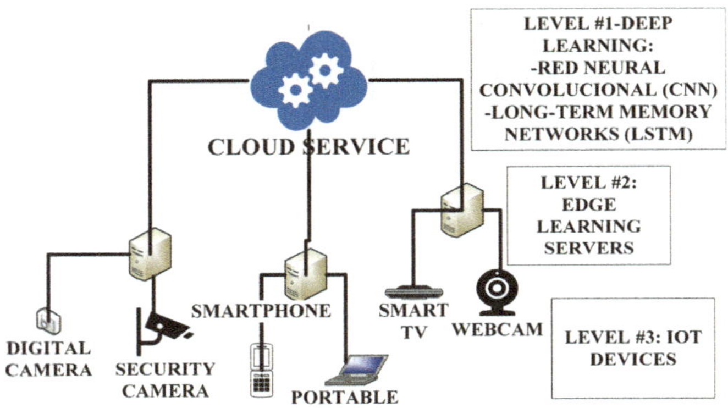

Fig. 4. 3-layer Edge Learning model.

Figure 4 is a representation of the model called Edge Learning made up of 3 levels supported with CNN and LSTM algorithms for data classifiers; this model improves performance and reduces network traffic by 80% and runtime by 69% to run DL training tasks on Edge servers [4].

The authors of [5] proposed an intelligent surveillance architecture, that optimizes face detection and improves recognition in a local processing system and cloud backup; they also highlighted the implementation of 4 renowned layers for the operation of the algorithm, with the ability to differentiate face patterns recognized by a group of devices applied to the IoT; they developed image manipulation in Python and this information is exported in BD MySQL to get accurate recognition.

The authors of [7] implemented an IoT device for the transmission of video images, with periodic programming to guarantee energy savings with captures of interest; they also used image recognition through deep neural networks to reduce the content capture period and establish an algorithm that filters transmission to the cloud; they applied a set of equations that demonstrate low power consumption by ruling out images that are outside the range of interest and increase the battery life time.

The authors of [8] proposed to highlight the main challenges and limitations reflected by DL models in devices that act at the edge of the network; they also proposed parameters and techniques for reducing layers and scales that would help improve the performance of algorithms to save memory, storage and refine results; they used different gradients and parallelisms to achieve much more precise approximations compared to operations with dynamic characteristics.

The authors of [9] referenced a set of embedded vehicle assistance systems supported by the accuracy of object recognition provided by DL and the efficient power of perimeter processing; they demonstrated how the learning and edge devices that are built into these vehicles help optimize communication and performance; they further stated that edge devices can perform real-time video processing to free up memory, CPU cores and highlight more advanced autonomous driving systems.

The authors of [10] proposed a model that shows that large volumes of data generated in IoT can be processed by perimeter servers to guarantee quality of service with reduced response times; they explained how a DL algorithm can allocate resources and dimension the traffic of requests that are made to the perimeter servers by multiple devices in an optimal way.

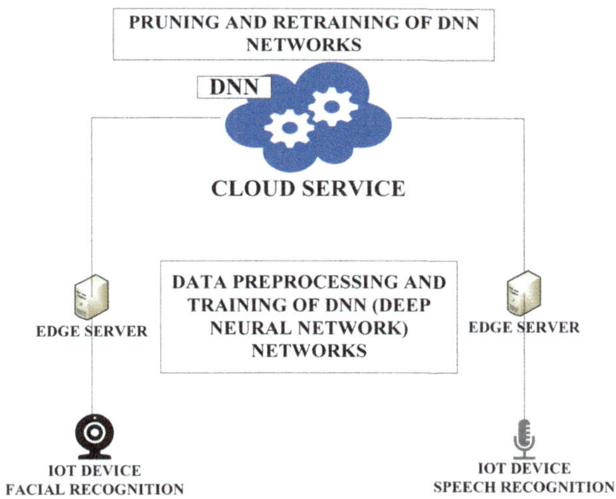

Fig. 5. Deep Neural Network model on EC devices.

Figure 5 shows the data preprocessing phase on Edge servers obtained by IoT recognition devices; the authors of [11] indicated that the model shows how noise could be suppressed in new data during retraining of neural networks after being pruned; it should be noted that this process would optimize the size of the network to improve performance and data processing on edge devices.

The authors of [12] discussed the advantages of Mobile Edge Computing against the high demand for mobile applications and technologies that need to transmit data for processing on distributed micro servers; they explained that the DL model contains two components that help detect an application's threat to network security; they concluded that the different deep learning models have mobile application detection and preprocessing algorithms that can be malicious.

The authors of [13] designed a DL model that identifies and classifies mosquitoes in rural spaces by employing a trap device that captures video frames and processes them in EC; they also highlighted the efficient work of reducing the size of the transmission before sending the data to the central cloud server, the images obtained are processed and the classification begins.

This article [14] detected anomalies and security problems that can arise during the use of IoT applications and devices through different DL models; they were referenced to health monitoring systems here algorithms allow detecting anomalies presented by a patient in real time using IoT devices; in addition, they cited anomaly detection models in different fields of application; each adopts high precision, and thanks to EC helps detect anomalies in real time.

The authors of [15] proposed an architecture to accelerate image recognition through DL networks and processing in the EC field for better performance; they also demonstrated through a block diagram the memory units, convolution processing and functions that make up the DV700 hardware architecture.

The authors of [16] proposed a linear regression model to predict the processing time for tasks generated by applications on mobile devices and divided into subtasks; they also designed central cloud monitoring software to analyze resource utilization, emphasizing that downloading of mobile applications on various MEC servers is more efficient.

The authors of [17] analyzed a process that classifies images assisted by recognition algorithms, this gets accurate results after analyzing thermographic data and rendering textures using mathematical processes; they emphasized that a classification model used for detection requires large amounts of data to guarantee its sensitivity, maximum performance and precision of results; they proposed that the algorithms applied to the analysis of minimal data from large image clusters obtained by thermography help in the prevention of advanced breast cancer.

2.2 Methods

Table 1 shows a summary of the algorithms that make up DL, its focus on each of the areas and fields of application, in addition to the precision results obtained in tests performed by the references themselves.

Table 1. Reference summary.

Ref.	Area	Proposed method	Edge Computing	Accuracy %
[1]	Analysis of data (Auto-Encoders)	Use of Edge-Cloud Computing technologies to analyze large volumes of data	Data traffic reduction at the edge	95,45%
[2]	Management and Productivity (CNN)	CNN multilayer models to reduce data size in intermediate layers	Run CNN algorithms on EDGE servers to improve efficiency	–

(continued)

Table 1. (*continued*)

Ref.	Area	Proposed method	Edge Computing	Accuracy %
[3]	Facial recognition (CNN)	Algorithms for classifying and predicting the age and gender of a person using data captured by IoT devices	Adapt age and gender recognition and estimation models to edge teams	88,4% 89,5%
[4]	Data traffic analysis (LSTM)	Deep learning tasks with high-resource teams	Reduce data traffic with Edge Learning	90%
[5]	Facial recognition (DNN)	Face recognition algorithm connecting Edge-Cloud technology	Facial recognition model through indoor cameras	98,93%
[7]	Video surveillance (DNN)	Transmission of images of interest after a classification process with DL algorithms	IoT camera with periodic on and off programming for low energy consumption	–
[8]	Segmentation of medical images (GAN)	Comparison between generative and discriminating gradients	Support in image preprocessing	97%
[9]	Real-time data analysis (CNN)	Algorithm that analyzes images, videos in real time	High power moving edge device	–
[10]	Management and Productivity (Auto-Encoders)	Distributed deep learning model	Multiple EDGE servers process tasks from one IoT device	93%
[11]	Face and voice recognition (DNN)	Pruning and retraining of neural networks to obtain improvements in recognition	Process DNN algorithms that take up less memory and storage space on a restructured network	93,76%
[12]	Security (RBM)	Model that captures non-linear relationships between the characteristics of mobile applications	Mobile computing environment to detect threats and guarantee security with DL algorithms	–
[13]	Classification and recognition (CNN)	It uses a precision function and the region of interest factor for the classification process of two data sets	IoT device to trap mosquitoes and capture video images for pre-processing and feature extraction	98%

(*continued*)

Table 1. (*continued*)

Ref.	Area	Proposed method	Edge Computing	Accuracy %
[14]	Health, Monitoring systems (DNN)	Proposed method to IDS for IoT networks	Distributed model to detect threats and anomalies at the edge	98,75%
[15]	Image recognition (CNN)	Floating point average precision model for recognition applications	IoT device to speed up image recognition	–
[16]	Management and Optimization (LSTM)	Model to manage the distribution of subtasks to MEC nodes according to their capacity	Parallel processing of tasks on multiple neighboring Edge mobile nodes	–
[17]	Diagnosis of breast cancer (DNN)	Classification of thermographic images	Computer assisted diagnosis	95,45%

Types of algorithms in DL [6] below:

- Supervised: Deep Neural Networks (DNN), Convolutional Neural Networks (CNN), Recurrent Neural Networks (RNN), Long Short Term Memory (LSTM), Gated Recurrent Units (GRU)
- Semi-supervised: Deep Reinforcement Learning (DRL), Generative Adversarial Networks (GAN)
- Unsupervised: Auto-Encoders (AE), Restricted Boltzmann Machines (RBM)
- Reinforcement: Bayesian Deep Learning (BDL)

3 Results

In this phase the following results were obtained:

- Impact of Deep Learning on Edge Computing.
- Summary of Deep Learning algorithms.

3.1 Impact of Deep Learning on Edge Computing

Table 2 shows the precision results obtained from the different DL algorithms; after applying video and image recognition methods through pre-processing, extraction and selection of characteristics, and classification of results; in this context, the DNN and CNN algorithms achieve precision results greater than 98% in recognition tasks. In the case of DNN algorithms, a facial recognition model is used through indoor camera captures supported with Raspberry Pi3 perimeter servers; for a much more efficient detection and energy saving and together with the processing layer, obtain results of 98.93% in less time. For CNN algorithms, its effectiveness is applied to machine vision tasks with 98% precision results compared to DNN algorithms; through CNN the learning and training of the neural network is developed after classifying and segmenting the images of a data set. Amidst these powerful algorithms, Auto-Encoder algorithms stand out with 93% accuracy results. Compared to DNN, Auto-Encoders are self-supervised learning algorithms that have different inputs to encode data and after being processed in the hidden layer, comparative linear outputs are obtained.

Table 2. Results of precision of DL algorithms.

Ref.	Model	Data set	Type of data	Accuracy %
[1]	AE	Collection of data generated by accelerometers, gyroscopes and proximity sensors through smart devices	The final data set resulted in 561 characteristics describing human activities	95,45%
[3]	CNN	Set of facial images with age and gender labels where a total of 62329 records are considered	Restructuring of image size and dimensioning of facial images at 96 × 96 pixels	88,4%
[4]	LSTM	Data generated by multiple IoT devices, where a total of 60,000 records are considered	The final dimension of the images is equivalent to 28 × 28 pixels	90%
[5]	DNN	Set of images for facial recognition through indoor IoT cameras. A total of 4932 images were used for detection and recognition	Capture of video surveillance images	98,93%
[8]	DNN	Recognition of human characteristics and activities, image classification	Different applications are used that can classify images of 155 frames per second (fps)	97%
[10]	AE	Dataset and service requests generated by multiple IoT devices	Download of joint works and allocation of resources at the edge of the network	93%
[11]	DNN	Dataset generated by multiple IoT devices	Variation in pruning methods from original data for retraining of neural networks	93,76%
[13]	CNN	Mixed and non-mixed data set by images captured on video for segmentation, extraction and selection of characteristics	Analysis of white pixels in images of the object captured by 350 × 350 pixel smart devices	98%

(continued)

Table 2. (*continued*)

Ref.	Model	Data set	Type of data	Accuracy %
[14]	DNN	Physical activity monitoring dataset	Images processed in the Python programming language	98,75%
[17]	DNN	Classification of thermographic images for diagnosis of breast cancer	Image size restructuring and grayscale RGB color conversion. The final dimension of the images is equivalent to 320 × 240 pixels	95,45%

The references presented in Table 2 showed the quantitative efficiency based on the model or architecture, in DL the models train and test with images; therefore, the references are comparable.

3.2 Deep Learning Algorithm Summary

Table 3 shows a summary of each of the algorithms that make up DL; supervised, semi-supervised, unsupervised, and reinforced learning algorithms are denoted; among the most important are the DNN and CNN for using high precision and efficient training methods in the different layers that make it up; both algorithms apply the linear regression method and the classification phase to obtain the desired results as a result of input data pairs; these algorithms are applied to IoT devices that generate data after recognition of human activities, to be then preprocessed on EDGE servers [1].

Table 3. Summary of algorithms in Deep Learning.

Models	Ref.	Edge Computing distributed connected devices
Deep Neural Networks (DNN)	[5, 7, 8, 11, 14, 17]	Support in image preprocessing to eliminate noise and train the weight vector of neurons. Allocation of resources and increase in the optimal performance of the architecture of a distributed system
Convolutional Neural Networks (CNN)	[1–4, 8, 9, 11, 13, 15]	
Long Short Term Memory (LSTM)	[4, 8, 16]	
Gated Recurrent Units (GRU)	[6]	
Generative Adversarial Networks (GAN)	[8]	
Auto-Encoders (AE)	[1, 10]	
Restricted Boltzmann Machines (RBM)	[12]	

Table 4 shows a summary of the set and type of data used to train neural networks according to the DL algorithm used in that field; it should be noted that the information from different databases was acquired to carry out extraction, processing and segmentation stages of characteristics; for the training of neural networks, thousands of records were used to categorize the false positives and negatives of the results obtained in laboratory or open field tests; many of the application areas use methods of reducing the images captured by IoT devices and proceed to the preprocessing phase on Edge Computing servers.

Table 4. Dataset summary.

Ref.	Image type	Use	Records	Access
[3]	Facial images	Age and gender estimation using facial recognition	62329	IMDB-WIKI
[5]	Facial images	Precision for facial recognition uses a preprocessing model that reduces power consumption and increases performance	4932	DynamoDB (MySQL)
[13]	Video images	Extraction and Classification of wild mosquitoes according to their characteristics after recognition	Mixed set – 100 Aedes mosquitoes (29 wild, 71 random lab) – 100 Culex mosquitoes (47 wild, 53 random lab)	Laboratory images and images captured by video with a recognition device
[4]	28 × 28 pixel elements and images	Neural network training to reduce traffic and increase processing efficiency	60000	MNIST

Table 5 shows the average of the precision results of the DL algorithms versus tests carried out with a previously established data set for each case; DNN algorithm stands out with an average precision of 96.72% in different areas of application and task execution; in addition to achieving a sensitivity of 98% and a specificity in the results of 93%, making it the highest performing supervised learning algorithm; these algorithms are related in their neural network and layer division, so the results in the output layer are compared linearly with the data in the input layer after being processed in the hidden layer.

Table 5. Average of precision results obtained in study tests according to the DL algorithm.

	DNN	CNN	LSTM	Auto-Encoders
Precision	96,72%	93,75%	90%	93%
References	[5, 11, 14, 17]	[3, 13]	[4]	[10]

3.3 Edge Computing Summary

Table 6 shows a comparison of study sources with qualitative and quantitative results between different DL algorithms and their application on EDGE servers through data generated by IoT devices; these results are distributed according to the area of application of Edge Computing technology:

- Processing devices for task distribution, data analysis and optimization of energy consumption in high-performance hardware.
- Performance improvements for resource allocation and optimization.
- Recognition to capture facial images or elements for extraction and selection of characteristics.

Table 6. Edge Computing summary

Edge Computing	Results	Algorithm	Ref.
Processing devices	Qualitative result	(LSTM)	[16]
		CNN	[15]
		DNN	[7]
			[14]
	Quantitative result	Auto-Encoders	[1]
		CNN	[13]
		DNN	[17]
			[8]
Improves performance	Qualitative result	CNN	[2]
		(LSTM)	[4]
	Quantitative result	Auto-Encoders	[10]
		CNN y LSTM	[4]
		DNN	[11]
Recognition	Qualitative result	CNN	[9]
	Quantitative result	CNN	[3]
		DNN	[5]

In the summary, the DNN and CNN algorithms for the precise and efficient development of these processes on EDGE servers can be highlighted.

4 Discussion

Based on the summaries made in the results phase, we establish that DL algorithms executed on decentralized micro servers are necessary resources in the field of productivity of organizational tasks, data security, data mining analysis, recognition of facial features and computer-assisted medical diagnostics.

These results are related to that the references maintain; in [1] and [10] here they highlighted the distribution of resources and data analysis on IoT devices; in [14] they established techniques for intrusion and anomaly detection systems in different areas and fields of study; in [3] and [5] architectures were established for easy recognition through IoT cameras and it uses DL algorithms as classifiers by characteristic features.

These authors express that the data set used for recognition tests must be very broad; in the layer that performs data processing near the source, it is necessary to extract a significant group of characteristics to increase the level of precision of the results; architectures only consider image or video data in frames per second to perform these tests with devices with high processing resources.

A theoretical consequence of this research is to establish data reduction parameters in the processing layer through EC servers that uses DL algorithms that generally need to sensitize large data in these devices; there are proposed strategies for retraining neural networks that could generate smaller and more efficient architectures from new data.

The results of this research agree with: [11] for the retraining of the neural network; [16] sets division of subtasks into different Edge nodes; [2] defines the learning of data generated by IoT devices in Edge Computing.

5 Future Work and Conclusion

As future work we proposed to design a training model for neurons to use the minimum amount of resources from Edge servers.

It was concluded that according to the analysis carried out, the best algorithm is DNN due to its high degree of precision 98% in different areas; DNN stands out for the complexity of its structure that uses multiple layers (input, hidden and output) for the processing and training of the neurons that make it up.

Acknowledgment. This work has been supported by the GIIAR research group and the Salesian Polytechnic University of Guayaquil.

References

1. Ghosh, A.M., Grolinger, K.: Deep learning: edge-cloud data analytics for IoT. In: 2019 IEEE Canadian Conference of Electrical and Computer Engineering, CCECE 2019, pp. 1–7 (2019). https://doi.org/10.1109/CCECE.2019.8861806

2. Li, H., Ota, K., Dong, M.: Learning IoT in edge: deep learning for the Internet of Things with edge computing. IEEE Netw. **32**, 96–101 (2018). https://doi.org/10.1109/MNET.2018.1700202
3. Giammatteo, P., Fiordigigli, F.V., Pomante, L., Di Mascio, T., Caruso, F.: Age gender classifier for edge computing. In: Proceedings of 2019 8th Mediterranean Conference on Embedded Computing, MECO 2019, pp. 8–11 (2019). https://doi.org/10.1109/MECO.2019.8760160
4. Huang, Y., Ma, X., Fan, X., Liu, J., Gong, W.: When deep learning meets edge computing. In: Proceedings of International Conference on Network Protocols, ICNP 2017, October 2017. https://doi.org/10.1109/ICNP.2017.8117585
5. Kaşkavalci, H.C., Gören, S.: A Deep learning based distributed smart surveillance architecture using edge and cloud computing. In: Proceedings of 2019 International Conference on Deep Learning and Machine Learning in Emerging Applications, Deep-ML (2019). https://doi.org/10.1109/Deep-ML.2019.00009
6. Alom, M.Z., Taha, T.M., Yakopcic, C., Westberg, S., Sidike, P., Nasrin, M.S., Hasan, M., Van Essen, B.C., Awwal, A.A.S., Asari, V.K.: A state-of-the-art survey on deep learning theory and architectures. Electronics **8** (2019). https://doi.org/10.3390/electronics8030292
7. Lim, J., Seo, J., Baek, Y.: CamThings: IoT camera with energy-efficient communication by edge computing based on deep learning. In: 2018 28th International Telecommunication Networks and Application Conference, ITNAC 2018, pp. 1–6 (2019). https://doi.org/10.1109/ATNAC.2018.8615368
8. Voghoei, S., Hashemi Tonekaboni, N., Wallace, J.G., Arabnia, H.R.: Deep learning at the edge. In: Proceedings of 2018 International Conference on Computational Science and Computational Intelligence, CSCI 2018, pp. 895–901 (2018). https://doi.org/10.1109/CSCI46756.2018.00177
9. Hochstetler, J., Padidela, R., Chen, Q., Yang, Q., Fu, S.: Embedded deep learning for vehicular edge computing. In: Proceedings of 2018 3rd ACM/IEEE Symposium on Edge Computing, SEC 2018, pp. 341–343 (2018). https://doi.org/10.1109/SEC.2018.00038
10. Wang, H., Chen, X., Xu, H., Liu, J., Huang, L.: Joint Job offloading and resource allocation for distributed deep learning in edge computing. In: Proceedings of 21st IEEE International Conference on High Performance Computing and Communications; 17th IEEE International Conference on Smart City; 5th IEEE International Conference on Data Science and System, HPCC/SmartCity/DSS 2019, pp. 734–741 (2019). https://doi.org/10.1109/HPCC/SmartCity/DSS.2019.00109
11. Chandakkar, P.S., Li, Y., Ding, P.L.K., Li, B.: Strategies for re-training a pruned neural network in an edge computing paradigm. In: Proceedings of 2017 IEEE International Conference on Edge Computing, EDGE 2017, pp. 244–247 (2017). https://doi.org/10.1109/IEEE.EDGE.2017.45
12. Subramaniam, P., Kaur, M.J.: Review of security in mobile edge computing with deep learning. In: 2019 Advances in Science and Engineering Technology International Conferences, ASET 2019, pp. 1–5 (2019). https://doi.org/10.1109/ICASET.2019.8714349
13. Huang, L.P., Hong, M.H., Luo, C.H., Mahajan, S., Chen, L.J.: A vector mosquitoes classification system based on edge computing and deep learning. In: Proceedings of 2018 Conference on Technologies and Applications of Artificial Intelligence, TAAI 2018, pp. 24–27 (2018). https://doi.org/10.1109/TAAI.2018.00015
14. Sharma, B., Sharma, L., Lal, C.: Anomaly detection techniques using deep learning in IoT: a survey. In: Proceedings of 2019 International Conference on Computational Intelligence and Knowledge Economy, ICCIKE 2019, pp. 146–149 (2019). https://doi.org/10.1109/ICCIKE47802.2019.9004362
15. Ohbuchi, E.: Low power AI hardware platform for deep learning in edge computing. In: 2018 IEEE CPMT Symposium Japan, ICSJ 2018, pp. 89–90 (2019). https://doi.org/10.1109/ICSJ.2018.8602619

16. Kim, K., Lynskey, J., Kang, S., Hong, C.S.: Prediction based sub-task offloading in mobile edge computing. In: International Conference on Information Networking, pp. 448–452, January 2019. https://doi.org/10.1109/ICOIN.2019.8718183
17. Raghavendra, U., Gudigar, A., Rao, T.N., Ciaccio, E.J., Ng, E.Y.K., Rajendra Acharya, U.: Computer-aided diagnosis for the identification of breast cancer using thermogram images: a comprehensive review. Infrared Phys. Technol. **102** (2019). https://doi.org/10.1016/j.infrared.2019.103041

SIR Model Parameter Fitting of SARS-CoV-2 Basic Reproduction Number in Venezuela and Ecuador Epidemic

Alexandra La Cruz[1(✉)], Erika Severeyn[2], Jesús Velásquez[2], Héctor Herrera[3], and Sara Wong[4]

[1] Faculty of Engineering, Universidad de Ibagué, Ibagué, Tolima, Colombia
alexandra.lacruz@unibague.edu.co

[2] Department of Thermodynamics and Transfer Phenomena, Universidad Simón Bolívar, Caracas, Venezuela
{severeynerika,jmvelasquezf}@usb.ve

[3] Department of Biological and Biochemical Process Technology, Universidad Simón Bolívar, Caracas, Venezuela
haherrera@usb.ve

[4] Department of Electronics and Circuits, Universidad Simón Bolívar, Caracas, Venezuela
swong@usb.ve

Abstract. A novel coronavirus, formerly called SARS-CoV-2 by the International Committee on Taxonomy of Viruses, caused an outbreak of atypical pneumonia in Wuhan, Hubei province by December of 2019 and then rapidly spread out in the whole China. In March 2020, the SARS-CoV-2 spread to Venezuela and Ecuador. The basic reproduction number (R_0) is a value that allows measuring the transmission rate of a virus. The SIR model is the simplest compartmental epidemiological model (Susceptible, Infectious, and Recovered). The SIR model can be used to estimate R_0 by fitting the curve of the infected compartment to the experimental curve of infected subjects per day. This work aims to study the projection of the R_0 of SARS-CoV-2 in Venezuela and Ecuador. For this purpose, five experiments took place by adjusting the SIR model curve of the infected compartment to experimental data at five-time intervals (the first 14, 28, 42, 56, and 178 days for Ecuador data and 165 days for Venezuela data). The R_0 fell from 6.923 to 1.003 (between experiments 2 and 3) for Venezuelan data, and from 3.622 to 1.022 (between experiments 1 and 2) for Ecuadorian data. These differences imply that R_0 depends on the preventive measures implemented to face the pandemic. On the other hand, the number of infections in Ecuador is higher compared to the infection numbers in Venezuela in each experiment. Venezuela is experiencing a complex economic crisis that limits the ability of its citizens to mobilize due to the lack of gasoline, monetary liquidity, and failures in basics services, suggesting that the economic activity of the countries will also influence the number of infections.

M. Botto-Tobar et al. (Eds.): CIT 2020, AISC 1326, pp. 94–108, 2021.
https://doi.org/10.1007/978-3-030-68080-0_7

Keywords: SARS-CoV-2 · Parametric fitting · Mathematical epidemiological model

1 Introduction

A novel coronavirus caused an outbreak of atypical pneumonia in Wuhan, Hubei province in December of 2019, and then rapidly spread out in the whole of China [13,39]. The virus formerly called SARS-CoV-2 (severe acute respiratory syndrome coronavirus 2) by the International Committee on Taxonomy of Viruses. In February 13, 2020, there were over 60,000 reported cases (including more than 1,000 death report) in China [19]. Since January 2020, the central government of China implemented preventive measures to avert the spreading of SARS-CoV-2 [9,37]. Many cities in Hubei province applied a lookdown strategy, and many measures were implemented, such as tracing close contacts, quarantining infected cases, promoting social consensus on self-protection (wearing a face mask in a public area), and social distance [4,22].

Despite all the measures implemented by the Chinese government to contain the spread of SARS-CoV-2, by March 2020 the virus had spread to North America, Europe, and South America [8,24]. By the time of writing (October 9, 2020), the United States of America (with 7,863,045 of cases), India (with 6,964,074 of cases), and Brazil (with 5,035,744 of cases) are the countries with the highest number of infections in the world [36]. Besides, the latest findings have shown that SARS-CoV-2 is an inflammatory infection in which mortality may increase in certain types of populations, such as the elderly [18], patients with diabetes or associated metabolic diseases [7,28,31], patients with cancer [16], among others.

On March 13, 2020, Venezuela confirmed the first case of SARS-CoV-2, and by March 16, 2020, the Venezuelan government adopted measures of social distancing and preventive quarantine [20]. In Ecuador, the first registered case of SARS-CoV-2 was on February 29, 2020. As a consequence, on March 1, 2020, the Ecuadorian government suspended the celebration of massive events in the cities of Babahoyo and Guayaquil, since the first case of coronavirus confirmed in this region [25]. Both located in the southwest of the country. On March 12, 2020, the government of Ecuador decrees the suspension of classes, suspension of massive events, and the implementation of quarantine of international passengers to prevent the spread of SARS-CoV-2 [30].

Several epidemiological models can describe and predict the spread of disease. Among of these models is the SIR model [3,34]. The SIR model is a standard epidemiological compartmental model that assumes that individuals are in one of three possible compartments: Susceptible individuals (S), infected (I), or recovered (R) [23,35]. The basic reproduction number (R_0) is a measure that allows studying the propagation capacity of any contagious disease [10]. From the SIR model, the R_0 can be calculated from the parametric adjustment of the experimental curves of contagion per day and the curve of infected individuals given by the SIR model [12].

This work aims to study the different projections of the R_0 value of the SARS-CoV-2 in Venezuela and Ecuador. For this purpose, the SIR model has been fitted to different periods of the existing data on infected cases. In the following section, the methodology developed in this research is explained. Results and discussion are presented in Sects. 3 and 4. Finally, conclusions and proposals for future work are presented in Sect. 5.

2 Methodology

2.1 Database

The database was collected from [36], the data used corresponds to the number of cases of SARS-CoV-2 infection per day. This database started reporting active cases since February 29 for Ecuador and March 13 for Venezuela. In both cases (Venezuela and Ecuador), the database cover from the day when the first case was reported [36].

2.2 SIR Model

The SIR (Susceptible-Infected-Recovered) model, developed by Ronald Ross, William Hamer, and others in the early 20th century, consists of a system of three coupled nonlinear ordinary differential equations, which does not possess an explicit formula solution [2]. The SIR model corresponds to the following three equations:

$$\frac{dS}{dt} = -\frac{\beta SI}{N} \tag{1}$$

$$\frac{dI}{dt} = \frac{\beta SI}{N} - \gamma I \tag{2}$$

$$\frac{dR}{dt} = \gamma I \tag{3}$$

where:

- β is the infection rate (the rate of the population susceptible become infected).
- γ is the removal rate (the rate of infected population recovering or dying), with the inverse of γ being the recovery period of a susceptible subject since the moment infection begins until the negative diagnostic test result.
- S is the susceptible population per day.
- I is the infected population per day.
- R is the recovered population per day.

The basic reproduction number is a theoretical parameter that provides some information about the speed with which a disease can spread in a given population. The R_0 can be calculated from the relation between β and γ from SIR model [2]:

$$R_0 = \frac{\beta}{\gamma} \tag{4}$$

2.3 Parameter Fitting

The fitting of γ and β parameters of the SIR model to the experimental data was performed following the procedure shown in Fig. 1 [26]. The vectors of γ and β were the inputs to the algorithm. In principle, these vectors were limited, taking as data the R_0 between 2 and 3.1 and the recovery time between 8 and 21 days, both reported in the literature [17,32]. However, these limitations prevented to achieve the minimization of the objective function, therefore very high and very low values of γ and β were established as limits, and their values were plotted as a function of the sum of absolute error to verify the minimum absolute error. Vectors were constructed with an equidistance of 10^{-4}.

Combinations were made with each of the values of the vectors γ and β. With each combination, the curve of infected subjects was found by solving the differential equation system of the SIR model using a six-stage of the fifth-order Runge-Kutta method. This curve was compared with the experimental curve of infected subjects found in [36] using the sum of absolute error as an objective function. Finally, the combination of γ and β that has the lowest sum of absolute error is the one chosen to find R_0.

In this study, five experiments were performed to look for the projection of the susceptible, infected, and recovered curves of the five-time intervals:

- Experiment 1: Active cases per day from the first day (February 29 for Ecuador and March 13 for Venezuela) to the fourteenth day were used for the projection.
- Experiment 2: Active cases per day from the first day (February 29 for Ecuador and March 13 for Venezuela) to the twenty-eighth day were used for the projection.
- Experiment 3: Active cases per day from the first day (February 29 for Ecuador and March 13 for Venezuela) to forty-second day were used for the projection.
- Experiment 4: Active cases per day from the first day (February 29 for Ecuador and March 13 for Venezuela) to fifty-sixth day were used for the projection.
- Experiment 5: Active cases per day from the first day (February 29 for Ecuador and March 13 for Venezuela) to two hundred and twenty-third day for Ecuador and two hundred and tenth day for Venezuela were used for the projection.

In each experiment, the procedure explained in Fig. 1 was performed for the curve of infected cases in Venezuela and Ecuador.

3 Results

Table 1 shows the projections for R_0, β, γ, and recovery period of the five experiments for Ecuador and Venezuela. In each case, the infected curve was fitted using the number of days of the experimental curve, for example, for the first

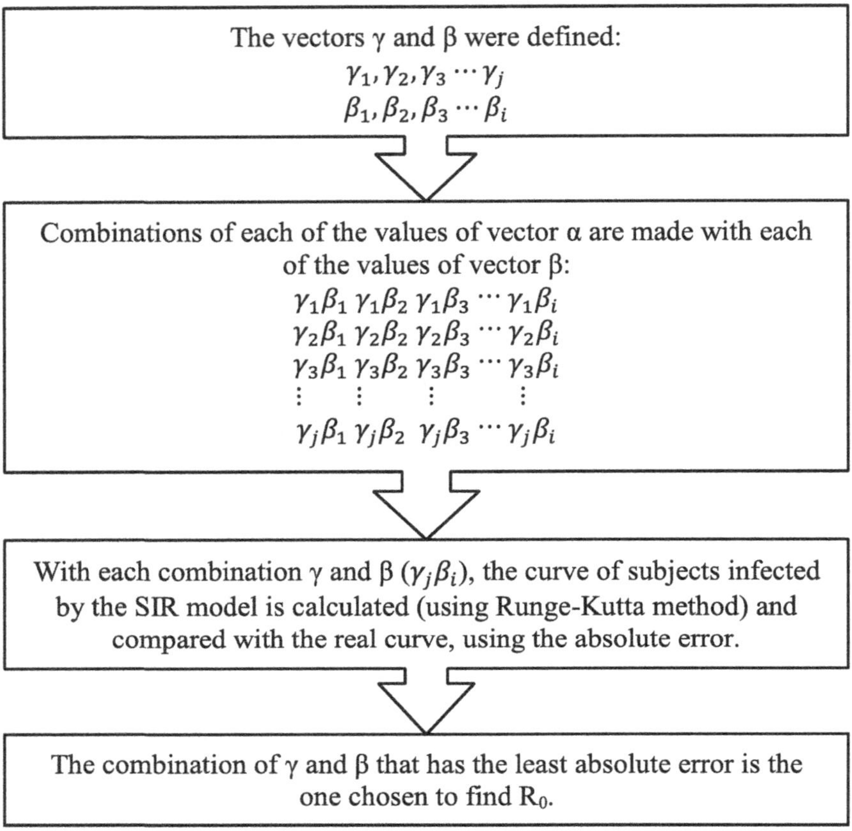

Fig. 1. Fitting of γ and β parameters of the SIR model to infected experimental curve.

experiment, the experimental curve was used from day one to fourteen. This was done in this way to observe how R_0 projection changes at different time points. Figure 2, 3, 4, 5, and 6 show the graphs of the fit of the SIR model curves for the Venezuelan data. Figure 7, 8, 9, 10, and 11 the SIR model curves for the Ecuadorian data. In the graphs (a) the SIR model curve fit graphs, the susceptible, infected, and recovered curves are shown on the left, and the infected curve is zoomed in on the right for more detail. On the other hand, in the graphs (b) the sum of the absolute error for the values of γ and β is showed.

4 Discussion

In the Table 1 the R_0 of experiments 1 and 2 of Venezuela data are reported between 5.833 and 6.923, this R_0 tendency changed in experiments 3, 4, and 5 where the R_0 fell to 1.003. Otherwise, the R_0 of experiment 1 for Ecuador data is 3.622 and since the experiment 2 to 5 this tendency changes and the R_0 fell

Table 1. Parameters projection to each experiment.

Country	Parameters	Exp.1	Exp.2	Exp.3	Exp.4	Exp.5
Ecuador	R_0	3.622	1.036	1.022	1.033	1.069
	γ	0.090	7.560	11.140	6.340	1.310
	β	0.326	7.830	11.390	6.550	1.400
	Recovery period [days]	11.111	0.132	0.090	0.158	0.763
	Sum of absolute error	4.803	59.857	265.286	876.179	8458.153
Venezuela	R_0	5.833	6.923	1.003	1.004	1.031
	γ	0.060	0.026	74.600	23.900	1.920
	β	0.350	0.180	74.800	24.000	1.980
	Recovery period [days]	16.667	38.462	0.013	0.042	0.521
	Sum of absolute error	17.481	40.153	38.769	45.734	1495.643

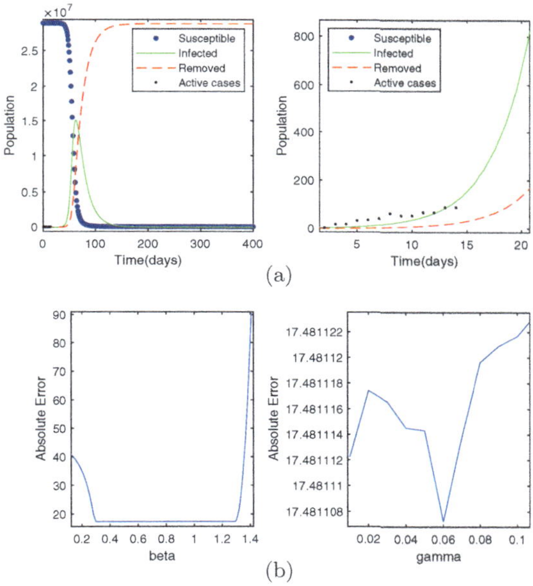

Fig. 2. Venezuela experiment 1: (a) SIR model simulation and zoom in on the experimental data interval. (b) β and γ vs. the sum of absolute error.

to 1.022. These findings suggest that social distancing, stipulated on March 12 for the Ecuador government and March 16 for the Venezuela government, had a significant impact on the decrease of R_0 [5].

It is important to note that SARS-CoV-2 has had a higher number of infections in Ecuador than in Venezuela, despite that the R_0 found in Ecuador data is lower. This could be due to the Venezuelan government implemented radical quarantine measures three days after the first case appeared, besides, the Venezuelan economic movement is very slowed down due to the lack of gasoline and the failure of basic services, all those conditions limit the movement of the

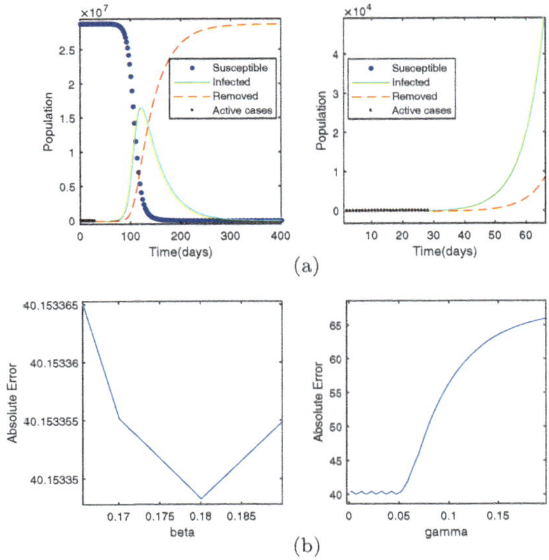

Fig. 3. Venezuela experiment 2: (a) SIR model simulation and zoom in on the experimental data interval. (b) β and γ vs. the sum of absolute error.

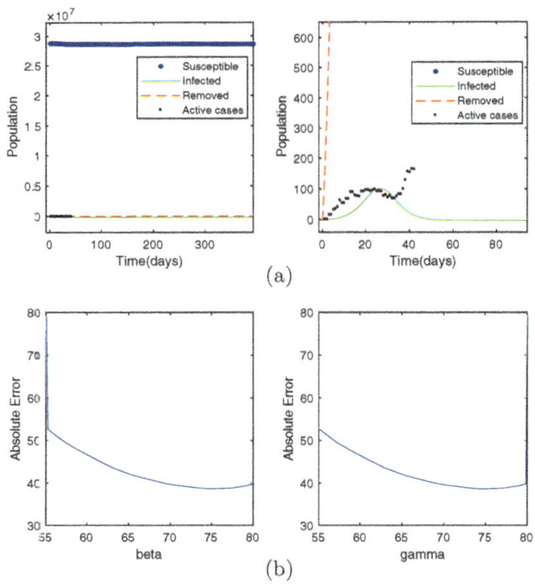

Fig. 4. Venezuela experiment 3: (a) SIR model simulation and zoom in on the experimental data interval. (b) β and γ vs. the sum of absolute error.

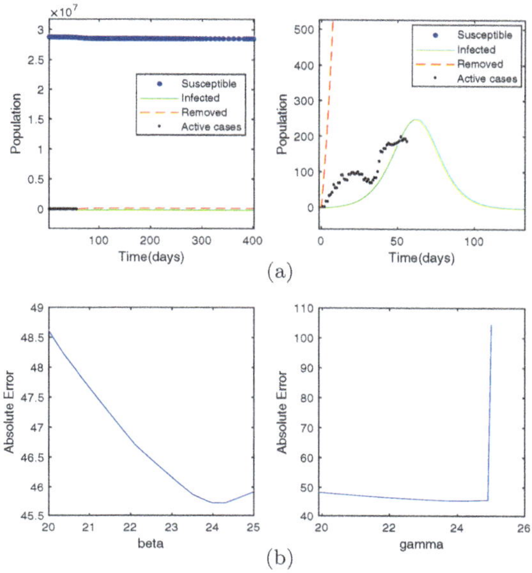

Fig. 5. Venezuela experiment 4: (a) SIR model simulation and zoom in on the experimental data interval. (b) β and γ vs. the sum of absolute error.

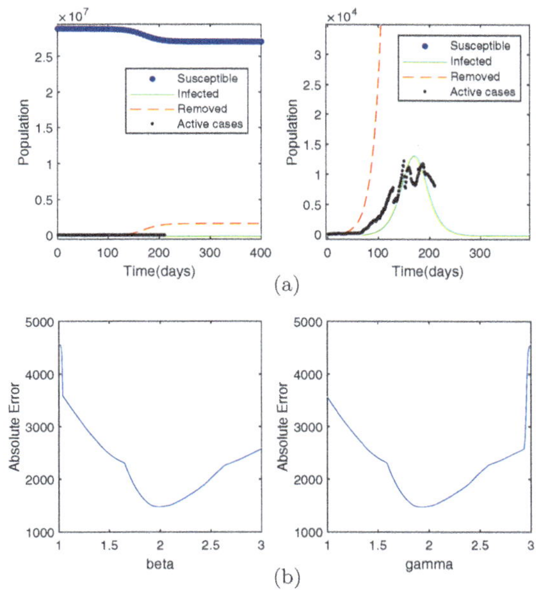

Fig. 6. Venezuela experiment 5: (a) SIR model simulation and zoom in on the experimental data interval. (b) β and γ vs. the sum of absolute error.

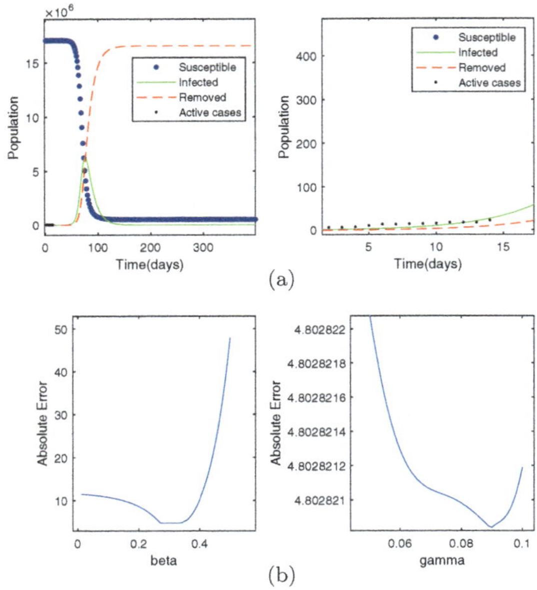

Fig. 7. Ecuador experiment 1: (a) SIR model simulation and zoom in on the experimental data interval. (b) β and γ vs. the sum of absolute error.

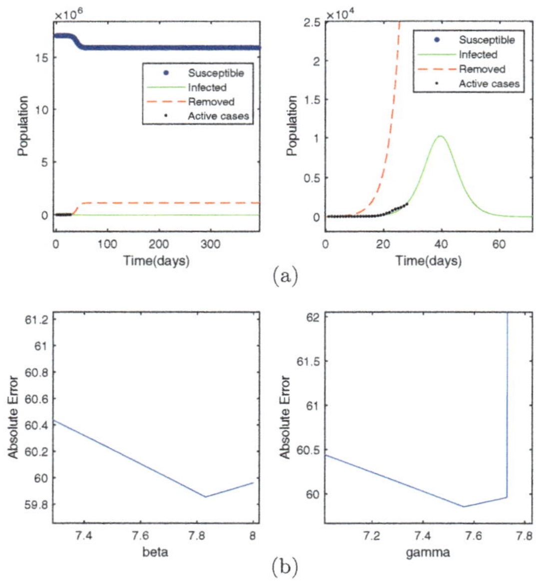

Fig. 8. Ecuador experiment 2: (a) SIR model simulation and zoom in on the experimental data interval. (b) β and γ vs. the sum of absolute error.

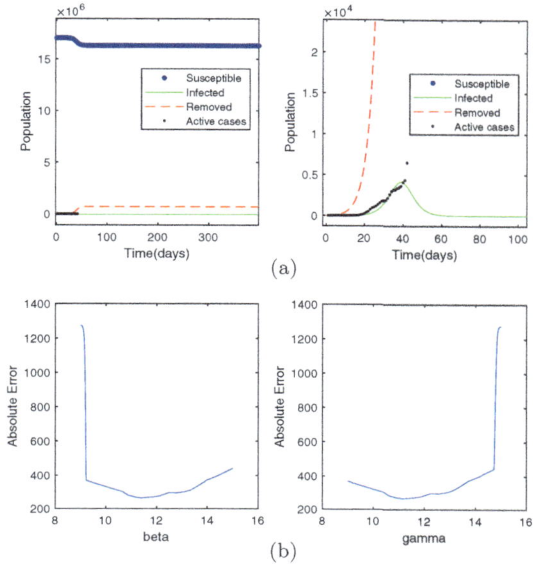

Fig. 9. Ecuador experiment 3: (a) SIR model simulation and zoom-in on the experimental data interval. (b) β and γ vs. the sum of absolute error.

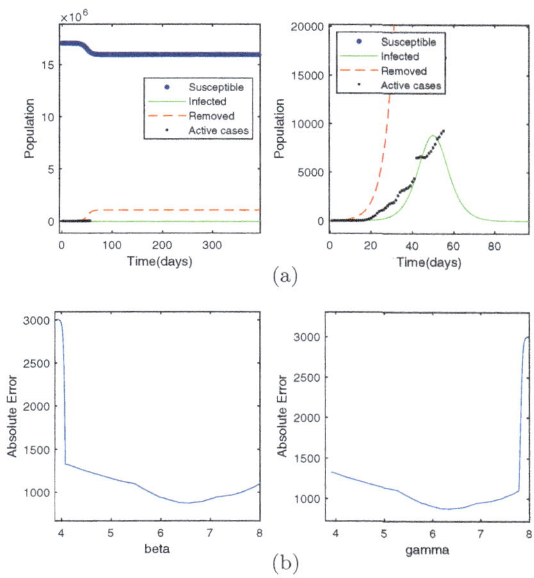

Fig. 10. Ecuador experiment 4: (a) SIR model simulation and zoom in on the experimental data interval. (b) β and γ vs. the sum of absolute error.

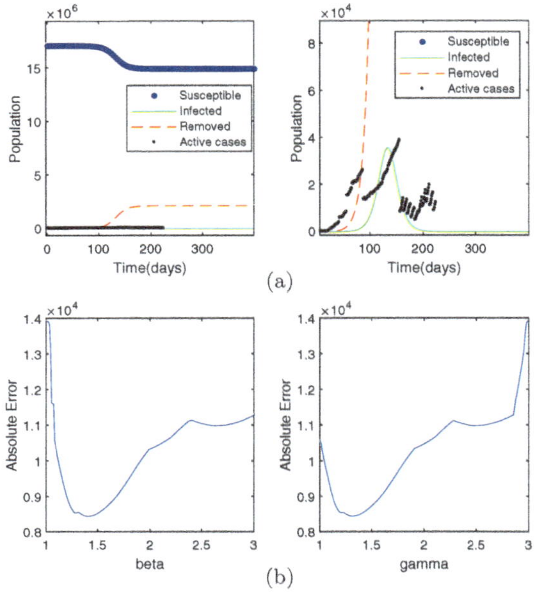

Fig. 11. Ecuador experiment 5: (a) SIR model simulation and zoom in on the experimental data interval. (b) β and γ vs. the sum of absolute error.

population. Instead, the Ecuadorian government implemented social distance measures for twelve days before the quarantine, during those days the virus was able to spread.

On the other hand, the R_0 found with the data from Venezuela (experiments 1 and 2) doubles the R_0 found in Ecuador (experiment 1), besides, the R_0 found in Venezuela took 25 days to drop after the establishment of quarantine, while in the case of Ecuador the R_0 decreased after 13 days of the establishment of social distancing and two days of the establishment of quarantine. This could suggest they are two different strains of the virus [21, 29], the strain that spread in Venezuela has similarities to the strain that spread in Spain, France, and Italy (more contagious and with a longer recovery time) [1]; and the strain that spread in Ecuador resembles the one that spread in China at the beginning of the pandemic that presented a R_0 between 2 and 3.1 [33, 38]. In any case, it should be taken as a starting point that SARS-CoV-2 has an R_0 between 2 and 7 for future projections and preventive health measures.

In Venezuela and Ecuador data a directly proportional relationship was observed between R_0 and the projection of the maximum number of infected subjects and between R_0 and the duration of the epidemic [6]. In Fig. 2 and 3 the maximum number of active infected cases in Venezuela is 1.5×10^7 subjects approximately with a duration that ranged between 200 and 350 days; in Fig. 4, 5, and 6 the R_0 fell and the maximum number of active infected decreased between 100 and 1.5×10^4 subjects approximately with a duration between 60 and 250 days. Likewise, the maximum number of active cases of infection in Ecuador (Fig. 7) is 6×10^6 subjects with a duration of 200 days approximately, after the R_0 decreased (Fig. 8, 9, 10, and 11) the maximum number ranged from 5×10^3 to 4×10^4 subjects and a duration between 60 and 200 days.

The sum of absolute error increases as the number of days studied increases in both data sets (Venezuela and Ecuador), this may be due to two reasons, the first reason is that when the experimental data has more days the error of the additional day is added to the sum of the absolute error values; and the second is that the model being used does not include certain aspects such as the amount of population varies every day by the number of people who naturally (not connected to SARS-CoV-2) died and were born, there is also a percentage of those recovered (14%) who could be re-infected by the disease [15], this type of population should be returned to the group of susceptible, and the disease has a latency period where the person has the virus without presenting symptoms this is not contemplated by the SIR model used in this research [11,15].

From the model, the average number of days a patient recovers can be calculated, as explained in the methodology section [14,27]. Table 1 shows that the recovery period ranged from 16.667 to 38.462 days in experiments 1 and 2 for the Venezuela data set, and 11.111 for Ecuador experiment 1. On the other hand, in experiments 3, 4, and 5 the recovery period ranged from 0.013 and 0.521 days for the Venezuela data set, and 0.090 and 0.763 days for the Ecuador data set. This could be due to that the probability of infection decreased as a consequence of establishing social distancing, and the dynamics of natural SARS-CoV-2 transmission are modified by the new conditions. It can also be verified that the recovery period found in the Venezuela dataset ($\approx$38 days) is higher than that found in Ecuador and China (between 8 and 21 days) [17] this could suggest that the strain that spread in Ecuador is different of the one spread in Venezuela.

5 Conclusions

In this study the variations in R_0 were assessed, using five-time intervals as experimental values for the fitting of the infected curve. The decrease in R_0 was observed after the establishment of social distancing, suggesting that the distancing measures implemented in Venezuela and Ecuador were able to decrease the ability of the infection to spread. This makes distance measurements very important in flattening the infected curve and preventing the collapse of the health care system of the countries around the world.

The differences in R_0 values for the five experiments suggest that R_0 depends not only on the nature of the infection but also on the economic activity of the country and the preventive measures taken against the pandemic. Therefore, the R_0 could be different in each country. For future works, R_0 values will be calculated for other countries with different economic developments and with different cultures.

The SIR model used in this work did not contemplate some known characteristics of SARS-CoV-2, for example, there have been cases of reinfection, therefore there is a percentage of recovered patients who return to the susceptible group, also the disease has a latency period where the person has the virus but does not suffer from it, however, they can transmit it, as is the case of the child population that can carry the virus without presenting symptoms. As future work, we will propose the inclusion of all these characteristics in the model to achieve a better fit with the experimental data in each country.

Acknowledgment. This work was funded by the Research and Development Deanery of the Simón Bolívar University (DID) and the Research Direction of the Ibagué University. Full acknowledgment is given to Tiberiu Harko, author of "Exact analytical solutions of the Susceptible-Infected-Recovered (SIR) epidemic model and of the SIR model with equal death and birth rates" (BioInfo Publications^{TM}).

References

1. Andersen, K.G., Rambaut, A., Lipkin, W.I., Holmes, E.C., Garry, R.F.: The proximal origin of sars-cov-2. Nat. Med. **26**(4), 450–452 (2020)
2. Anderson, R.M.: Discussion: the kermack-mckendrick epidemic threshold theorem. Bull. Math. Biol. **53**(1), 1–2 (1991)
3. Bjørnstad, O.N., Finkenstädt, B.F., Grenfell, B.T.: Dynamics of measles epidemics: estimating scaling of transmission rates using a time series SIR model. Ecol. Monogr **72**(2), 169–184 (2002)
4. Chen, W., Wang, Q., Li, Y., Yu, H., Xia, Y., Zhang, M., Qin, Y., Zhang, T., Peng, Z., Zhang, R., et al.: Early containment strategies and core measures for prevention and control of novel coronavirus pneumonia in China. Zhonghua yu fang yi xue za zhi [Chin. J. Prevent. Med.] **54**(3), 1–6 (2020)
5. Dalton, C.B., Corbett, S.J., Katelaris, A.L.: Pre-emptive low cost social distancing and enhanced hygiene implemented before local covid-19 transmission could decrease the number and severity of cases. Med. J. Aust. **212**(10), 1 (2020)
6. Delamater, P.L., Street, E.J., Leslie, T.F., Yang, Y.T., Jacobsen, K.H.: Complexity of the basic reproduction number (r0). Emerg. Infect. Dis. **25**(1), 1 (2019)
7. Fadini, G., Morieri, M., Longato, E., Avogaro, A.: Prevalence and impact of diabetes among people infected with sars-cov-2. J. Endocrinol. Invest. **43**, 867–869 (2020)
8. Fauver, J.R., Petrone, M.E., Hodcroft, E.B., Shioda, K., Ehrlich, H.Y., Watts, A.G., Vogels, C.B., Brito, A.F., Alpert, T., Muyombwe, A., et al.: Coast-to-coast spread of sars-cov-2 during the early epidemic in the United States. Cell **181**, 990–996 (2020)
9. Gabutti, G., d'Anchera, E., Sandri, F., Savio, M., Stefanati, A.: Coronavirus: update related to the current outbreak of covid-19. Infect. Dis. Therapy **9**, 1–13 (2020)
10. Harko, T., Lobo, F.S., Mak, M.: Exact analytical solutions of the susceptible-infected-recovered (SIR) epidemic model and of the sir model with equal death and birth rates. Appl. Math. Comput. **236**, 184–194 (2014)
11. Hethcote, H.W.: Three basic epidemiological models. In: Applied Mathematical Ecology, pp. 119–144. Springer, Heidelberg (1989)
12. Hethcote, H.W.: The basic epidemiology models: models, expressions for r0, parameter estimation, and applications. In: Mathematical Understanding of Infectious Disease Dynamics, pp. 1–61. World Scientific (2009)

13. Jin, Y., Yang, H., Ji, W., Wu, W., Chen, S., Zhang, W., Duan, G.: Virology, epidemiology, pathogenesis, and control of covid-19. Viruses **12**(4), 372 (2020)
14. Korobeinikov, A.: Global properties of SIR and SEIR epidemic models with multiple parallel infectious stages. Bull. Math. Biol. **71**(1), 75–83 (2009)
15. Lan, L., Xu, D., Ye, G., Xia, C., Wang, S., Li, Y., Xu, H.: Positive RT-PCR test results in patients recovered from covid-19. Jama **323**(15), 1502–1503 (2020)
16. Liang, W., Guan, W., Chen, R., Wang, W., Li, J., Xu, K., Li, C., Ai, Q., Lu, W., Liang, H., et al.: Cancer patients in sars-cov-2 infection: a nationwide analysis in China. The Lancet Oncology **21**(3), 335–337 (2020)
17. Liu, Y., Gayle, A.A., Wilder-Smith, A., Rocklöv, J.: The reproductive number of covid-19 is higher compared to SARS coronavirus. J. Travel Med. **27**(2), 1–4 (2020)
18. Nikolich-Zugich, J., Knox, K.S., Rios, C.T., Natt, B., Bhattacharya, D., Fain, M.J.: Sars-cov-2 and covid-19 in older adults: what we may expect regarding pathogenesis, immune responses, and outcomes. Geroscience **42**, 1–10 (2020)
19. Novel, C.P.E.R.E., et al.: The epidemiological characteristics of an outbreak of 2019 novel coronavirus diseases (covid-19) in China. Zhonghua liu xing bing xue za zhi= Zhonghua liuxingbingxue zazhi **41**(2), 145 (2020)
20. Organization, W.H., et al.: Coronavirus disease 2019 (covid-19): situation report, 72 (2020). Accessed 20 June 2020
21. Ou, X., Liu, Y., Lei, X., Li, P., Mi, D., Ren, L., Guo, L., Guo, R., Chen, T., Hu, J., et al.: Characterization of spike glycoprotein of sars-cov-2 on virus entry and its immune cross-reactivity with sars-cov. Nat. Commun. **11**(1), 1–12 (2020)
22. Pan, A., Liu, L., Wang, C., Guo, H., Hao, X., Wang, Q., Huang, J., He, N., Yu, H., Lin, X., et al.: Association of public health interventions with the epidemiology of the covid-19 outbreak in Wuhan, China. Jama **323**(19), 1915–1923 (2020)
23. Pathak, S., Maiti, A., Samanta, G.: Rich dynamics of an sir epidemic model. Nonlinear Anal. Model. Control **15**(1), 71–81 (2010)
24. Rodriguez-Morales, A.J., Gallego, V., Escalera-Antezana, J.P., Méndez, C.A., Zambrano, L.I., Franco-Paredes, C., Suárez, J.A., Rodriguez-Enciso, H.D., Balbin-Ramon, G.J., Savio-Larriera, E., et al.: Covid-19 in Latin America: the implications of the first confirmed case in Brazil. Travel Med. Infect. Dis. **35**, 101613 (2020)
25. Ruiz, G.P.I., León, A.C.I.: Evolución de la enfermedad por coronavirus (covid-19) en ecuador. La Ciencia al Servicio de la Salud **11**(1), 5–15 (2020)
26. Schneckenreither, G., Popper, N., Zauner, G., Breitenecker, F.: Modelling sir-type epidemics by odes, pdes, difference equations and cellular automata-a comparative study. Simul. Model. Pract. Theory **16**(8), 1014–1023 (2008)
27. Side, S., Noorani, M.: A sir model for spread of dengue fever disease (simulation for South Sulawesi, Indonesia and Selangor, Malaysia). World J. Model. Simul. **9**(2), 96–105 (2013)
28. Simonnet, A., Chetboun, M., Poissy, J., Raverdy, V., Noulette, J., Duhamel, A., Labreuche, J., Mathieu, D., Pattou, F., Jourdain, M., et al.: High prevalence of obesity in severe acute respiratory syndrome coronavirus-2 (sars-cov-2) requiring invasive mechanical ventilation. Obesity **28**, 1195–1199 (2020)
29. Tang, X., Wu, C., Li, X., Song, Y., Yao, X., Wu, X., Duan, Y., Zhang, H., Wang, Y., Qian, Z., et al.: On the origin and continuing evolution of sars-cov-2. Natl. Sci. Rev. **7**, 1012–1023 (2020)
30. Vallejo, G.: Control de vectores de arbovirus entiempos de covid-19 (sars-cov2) enel ecuador. INSPILIP **4**(2), 1–10 (2020)
31. Velásquez, J., Herrera, H., Encalada, L., Wong, S., Severeyn, E.: Análisis dimensional de variables antropométricas y bioquímicas para diagnosticar el síndrome metabólico. Maskana **8**, 57–67 (2017)

32. Verity, R., Okell, L.C., Dorigatti, I., Winskill, P., Whittaker, C., Imai, N., Cuomo-Dannenburg, G., Thompson, H., Walker, P.G., Fu, H., et al.: Estimates of the severity of coronavirus disease 2019: a model-based analysis. Lancet Infect. Dis. **20**(6), 669–677 (2020)
33. Viceconte, G., Petrosillo, N.: Covid-19 r0: magic number or conundrum? Infect. Dis. Rep. **12**(1), 1–2 (2020)
34. Wang, H., Wang, X.S.: Traveling wave phenomena in a kermack-mckendrick sir model. J. Dyn. Differ. Eq. **28**(1), 143–166 (2016)
35. Wilkinson, R.R., Ball, F.G., Sharkey, K.J.: The relationships between message passing, pairwise, kermack-mckendrick and stochastic sir epidemic models. J. Math. Biol. **75**(6–7), 1563–1590 (2017)
36. Worldometers.info: Reported cases and deaths by country, territory, or conveyance. https://www.worldometers.info/coronavirus/, Accessed 25 Aug 2020
37. Wu, Z., McGoogan, J.M.: Characteristics of and important lessons from the coronavirus disease 2019 (covid-19) outbreak in china: summary of a report of 72 314 cases from the chinese center for disease control and prevention. Jama **323**(13), 1239–1242 (2020)
38. Zhao, S., Lin, Q., Ran, J., Musa, S.S., Yang, G., Wang, W., Lou, Y., Gao, D., Yang, L., He, D., et al.: Preliminary estimation of the basic reproduction number of novel coronavirus (2019-ncov) in China, from 2019 to 2020: A data-driven analysis in the early phase of the outbreak. Int. J. Infect. Dis. **92**, 214–217 (2020)
39. Zhou, P., Yang, X.L., Wang, X.G., Hu, B., Zhang, L., Zhang, W., Si, H.R., Zhu, Y., Li, B., Huang, C.L., et al.: A pneumonia outbreak associated with a new coronavirus of probable bat origin. Nature **579**(7798), 270–273 (2020)

Machine Learning and Color Treatment for the Forest Fire and Smoke Detection Systems and Algorithms, a Recent Literature Review

Henry Cruz[1(✉)], Tatiana Gualotuña[1], María Pinillos[2], Diego Marcillo[1], Santiago Jácome[1], and Efraín R. Fonseca C.[1]

[1] Universidad de Las Fuerzas Armadas - ESPE, Sangolquí, Ecuador
hocruz@espe.edu.ec
[2] Escuela Politécnica de Chimborazo, Riobamba, Ecuador

Abstract. Currently, the forest fires are worldwide problem which requires a complete solution. The consequences of it are related, not only, to the environmental and biodiversity destruction but also to human and economic losses. In this article, forest fire and smoke detection systems, methods and technics literature review are carried out. To accomplish this aim, a generic mobile and non-mobile monitoring and surveillance systems applied in this field analysis took place. From the review carried out, it has been found that for fire detection two types of methods have been used, one based on machine learning and another through color features extraction of images or video frames, on the other hand, the combination of both methods is used for the detection of smoke and fire or only smoke, this guarantees accurate rates of more than 90%. Finally, a detailed analysis of the use of these fire and smoke detection algorithms in the UAS/UAV's systems was carried out paying special attention on them due to the flexibility, versatility, and maneuverability abilities among others displayed by these systems that help prevent and mitigate forest fires.

Keywords: Fire forest · Smoke forest · Machine learning · Color features extraction

1 Introduction

In the last years, the large-scale forest fires have become in a difficult and defying task of prevention and control as they have caused economic, human losses, a significant reduction in biodiversity and an increase in the global warming. The forest fires problem is one to be solved by a global community to face the increasing wildfire forest tendency around the world. According to the U.S National Interagency File Center during the 2018–2019 period, around 5 '463.256 Ha were burnt causing direct and indirect economic losses of around 72.000 million US dollars [1]. On the other hand, the European Forest Fire Information System (EFFIS), shows that in 31 countries of the Oriental Europe and North Africa around 165.308 Ha have been burnt [2].

M. Botto-Tobar et al. (Eds.): CIT 2020, AISC 1326, pp. 109–120, 2021.
https://doi.org/10.1007/978-3-030-68080-0_8

It can be said that the forest fires are take place more often and are devastating. Along 2019 and part of 2020, wildfires forest has been reported in the four most populated continents of the planet including the Artic [3]. The most destructive ones have been in Australia, the amazon, Siberia, Africa from Sudan to Occidental Africa, California, Alberta among others, causing the burned of millions of hectares [4].

The environmental consequences stemmed from the forest fires increase are related to the millions of carbon monoxide (CO) released which lead to the increase in the environmental temperature, the endemic and non-endemic fauna and flora species destruction, the water sources reduction, the fertile ground impoverishment, the forest biomass reduction, etc., and represent a serious threat for national security [5–7]. The forest fires are causes by the nature and by the human being, these ones represented the 96% of the forest fires [3]. Among the natural causes of the forest fires are the high atmospheric pressure, lave, high environmental temperatures, warm seas, lightning, among others. The human action causes include unattended fires, cigarettes, electric nets, debris burning, and agriculture, among others.

Several organizational methods have been proposed and different technologies which look for predicting, preventing and face these destructive events have been applied to tackle the causes and reduce the forest fires. Currently, this kind of technologies have been developed for their application in detection, monitoring and surveillance mobile and non-mobile systems. Specially, the use of optical sensors have allowed the development of surveillance algorithms to be applied in surveillance towers, Unmanned Aircraft System (UAS) known as drones, Unmanned Aerial Vehicles (UAV's), remote sensing, sensors networks and others [8, 9].

The signs acquired by the optical sensors are processed by computer vision, machine vision and learning machine algorithms though supervised, non-supervised and reinforcement training. The algorithms process result, mainly, is the detection and location of the forest fire and the emission of warnings and information about it. However, the accuracy would depend on the interaction efficiency and effectivity between the software and hardware and the accomplishment of the minimum requirements for the information communication such as high definition cameras, communication links between the monitoring and surveillance centers, access to high speed data nets, etc.

In this work, a revision of the main and recent methods and techniques for the detection of forest fire and smoke is carried out considering the following aspects:

- Fire and forest smoke detection systems through optical sensors and the current trend of use will be defined.
- The main methods of automatic detection of fire and smoke in forests through image and video signals training and processing.
- The ability of algorithms to perform real-time detection and the accuracy achieved.
- The detection capacities through UAS/UAV's.

The interest of the paper presented is to know the latest studies on methods of detection of forest fire, where images and video frames are used as inputs for detection and classification of fire from the rest of the environment. The present manuscript is organized as follows. In Sect. 2 the systems which mainly apply optical sensors for the forest fire detection are revised. In Sect. 3, the forest fire detection though automatic learning algorithms and digital images through the color methods are revised. In Sect. 4 a revision about the forest smoke detection methods and technics are revised and finally, in Sect. 5 the main work conclusions and future work are displayed.

2 The Forest Fire Detection Systems Applied and Acquisition Information Methodology

2.1 The Main Systems Used to Detect Fire and Smoke in the Forest

Several systems have been developed for forest fire detection, however, in all of them the use of sensors for surveillance and monitoring is a common requirement. Sensors are used in fixed watchtowers and networks to measure the amount of carbon, atmospheric pressure, humidity, wind intensity, amounts of atmospheric gases, luminance intensity and ambient temperature, etc. Optical sensors are also used in watchtowers and thus allow forest monitoring and fire alerts are issued by human operators or automatic systems located in control centers. The main problems that come up with this kind of monitoring are related to the forest limited coverage, telecommunication nets high dependency for the information transmission, high costs linked to the increase of monitoring areas, energetic autonomy limited, high rate of false positives in the warning emission, non-stop maintenance, the temperature sensors require physical presence when confirming a breakdown and they are not helpful to find out the fire spread or its intensity into a particular area [8–10].

Remote sensing is another way to carry out the forest fires monitoring and surveillance, generically, satellite systems equipped with high performance sensors are used for these purposes. The satellite remote sensing has some advantages such as wide surveillance coverage, transmission in real time, fire location. Its disadvantages are the delay in the reception of the information, the high costs involved, they can show low levels of efficiency and practicality, GPS location systems show accuracy issues in relation to blind points on the terrestrial surface.

The forests surveillance and the forest fire detection through UAS/UVA‘s comes up as a suitable option to monitor, prevent and generate warnings in possible events of fire in the forests because drones provide flexibility, versatility, maneuverability, coverage capability, real time transmission, low delay, low costs so that it makes possible to provide solutions of prevention and prediction and immediate responses and several types of sensors can be installed in the drones systems [11–13].

2.2 Procedure for Acquiring and Classifying Information

The preliminary analysis carried out in item 2.1, allows to establish that the algorithms for automatic detection of fire and forest smoke can be applied to any of the three systems studied (fixed towers, satellite remote sensing, UAV’s/UAS). However, the analysis also accounts for the use of UAV’s/UAS can provide results of greater effectiveness and efficiency.

On the other hand, the interest is also to know the recent and current trend of development, use and combination of methods, techniques and algorithms that provide the best results of pressure in the detection thus how they have employment skills in real time. In this sense and under the considerations presented, the search focuses on the following aspects:

- Establish two search fields one based on smoke detection to prevent the increase of forest fire and another on detecting fire to mitigate the effects of fire.
- Classify the methods presented for fire and smoke detection, consider high performance and real-time detection.
- Determine the most effective fire and smoke detection methods for use in UAS/UAV's.

3 Forest Fire Detection Methods Through Machine Learning Algorithms and Color Features Extraction

The use of optical sensors is a requirements for any of the surveillance methods detailed to guarantee the efficiency and the automatization of the detection and the emission of warnings due to the fact that the obtained information can be processed through different algorithms applicable to several fire detection methodologies, which can be grouped in two: detection through machine learning algorithms and detection through the digital processing of images through the color treatment [10, 14]. The following factors are critical to locate the fire outlines or fronts: a) the places setting where the intensity of the fire is higher, b) the direction of the spreading fire, c) the versatile nature of the spreading fire, etc.

Technical literature shows that the fire detection works developed have several approaches and some of them depend on the information acquisition in the initiation process of the forest fire which consists of the environmental, heat, fire, radiation or smoke phases in which the optical sensors are passive elements of information acquisition [15]. The optical sensors do not add additional value to the information, this happens when the visual signal is processed by the machine learning and digital processing of images algorithms. In this way, it is possible to establish the fire size, obtain a fire segmentation from the rest of the forest area, the burning area, location, visual warnings. This processed information provides advantages in the prevention, treatment and mitigation of the forest fires [16]. The forest fire detection through optical sensors lets an appropriate action to face a fire event and avoid human losses happened from non-controlled situations [17].

3.1 Fire Forest Detection Using Machine Learning Methods

Actually, the gathering of information through optical sensors represents an important option in the automatic detection of forest fire, however, machine learning, digital image processing and the color treatment algorithms face common problems such as the generation of false positives given by the illumination variation, the low signals transmission rate, low quality resolutions and variability in the contrast which affects the video frames and images, which are basic elements for the corresponding process.

The convolutional neural network (CNN) is one of the most used machine learning methods in the detection and automatic classification of the forest fire. The existing works in the forest fire detection show, overall, changes in the architecture and training model, in example, Khan Muhammad et al. [18] reached a CNN computer-based efficient architecture and is based on the [19] SqueezeNet architecture. This architecture shows interesting performance results in relation to the images classification and detection tasks. The use of the CNN in the fire detection can significantly improve the detection accuracy what will help to reduce the damage caused in the forest ignition process.

Another analysis points out that the methods which apply CNN for the automatic detection of forest show two approaches of development; one focused on the binary learning which provides the classification through gathering of fire samples of the images and another focused on another learning and classification methods based on the complete fire shot and the whole image outline [14]. Recent works show improvements in the forest fire detection process combined with the classification process through the CNN, taking into account color characteristics and intensity and using different sensors signals providing Internet of Things (IoT), the accuracy reached is higher than 90% [20]. In the same way in [21], an improvement in the flame detection is shown using faster R-CNN (Regions with Convolutional Neural Network features) and including a color guide module, which classifies the flame when the value of blue is higher inside the RGB space. The false positives reduction in this method is reached taking into consideration the images training in the whole surveillance environment, the accuracy percentages reached up to 96,36%. In the work presented by Ayula S.B. et al., the fuzzy entropy optimized thresholding is applied combined with CNN what provides a high accuracy percentage because of the algorithm capacity of reducing the gaussian noise and blurriness removal. In this stage of the processing, the classification is multi-scale allowing the fire and smoke detection [22].

To use the algorithms based in CNN in the forest fire detection in real time, there are some problems to be solved. These inconveniences are related to the reduction in in the high computational charge, the limited capacity to predict forest fire events, the long delay times, the high quality of the images or video frames demand to reach accuracy [19, 23].

3.2 Fire Forest Detection Through Color Treatment

The use of algorithms based on the digital image processing through the color treatment is another option for the detection of the forest fire, different works establish relations among the different components in the RGB, HSV, YUV, YCbCr, CIELab spaces [24, 25].

In [26] the Fire Forest Detection Index (FFDI) is proposed, this index is the result of the relation and ponderation of the different color components in the RGB space and it is shown as 5r-4g-b. It is obtained as a result the fire contrast and its component of smoke from the rest of the forest setting, finally, the Otsu threshold is used to convert the image or frame into the binary system, this method shows an accuracy of up to 97,67% and a very low execution time in a way that can be applied in the detection in real time.

In [27] a comparative of the fire detection and smoke using different color spaces, specifically among HSV, RGB, YCbCr, LUV is shown. The conclusions are that the color spaces show better performances according to the environment, area to be covered and the detection system to be used, because of that a combination of the methods is proposed to reach better results. Other works show a combination of the forest fire detection with the analysis of textures, in fact in [28] the color model YCbCr is applied to extract the regions of interest (ROI), a posteriori the structures in 2D and in 3D in statistics and dynamics forms are evaluated, this allows the obtaining of characteristics such as shape, flickering area and movement and finally, a classifier based on Extreme Learning Machine (ELM) to separate the flame from the rest of the scene is used.

The automatic learning algorithms based on the digital processing of images through color is a common option to be applied. Actually, in [29] CNN and the local patch NN classifier are applied, both classifiers share the use of neuronal network to carry out the training of a images dataset that allows the forest fire detection, the accuracy rate is up to 93.1%, these results are better than the support vector machine (SVM) classificatory, this kind of detection can be used in drones because of the multi-scale application. In a similar way in [30] the combined use of the local binary pattern (LBP) for the extraction of characteristics from the image, the use of the support vector machine (SVM) to do the classification and a first separation of the forest fire from the rest of the scene and, finally, CNN are applied to identify forest fire early detection characteristics, the results show detections in real time for the use in UAVs.

3.3 Fire Forest Detection to UAV's/UAS

Precisely, the combination of the extraction of characteristics, classifiers and training methods provide a higher accuracy in the forest fire detection, however, one of the main inconveniences shown by these algorithms is the high computational load and the difficulty of implementations for surveillance in real time. In this way, an interesting detection with reduction in the computational time proposal is shown in [31], where the training through the YOLOv3 is applied, this neuronal network reaches a high performance with a 91% accuracy and processes the video frames with a rate of 30 frames per second (fps), what allows the surveillance in real time.

On the other hand, the digital processing of images based on the color have been used in the forest fire detection systems in UAS/UAV's because, among other reasons, of the high accuracy reached and, overall, the effectivity of use due to the process simplicity and the high load of work in real time. Actually, [26] shows that the FFDI application provides a 0.0447(s) time for frames with 960×540 pp size and the accuracy rate is of 96,82%. In the same way in [32] a forest fire detection system for UAVs is applied which combines the highlighting of color process, stream optics and remote sensing techniques. Also, in [11] the color processing through the de Normalized Difference Vegetation Index (NDVI) in the Lab and RGB space is applied to recognize the fire, the smoke and the movement change of the flame and the smoke. In a similar way, [33] shows a multicolor detection method based on the extraction of the fire and smoke saturation through the LBP in the color YCbCr and HSV spaces, giving weighted values to the different ROIs, a posteriori the extracted characteristics are trained and classified through the Artificial Neural Network (ANN), the final result is the recognition of the fire and the smoke with high rates of accuracy.

4 Forest Smoke Detection Methods and Techniques

The smoke is the product of the combination of gasses, solid and liquid particles and has static and dynamic characteristics. The former is related to the color, the texture and the irregularity and the latter to the movement, change of direction and speed. To obtain hints about the type and localization of the fire in the forest setting it is necessary to know the smoke location, analyze its color, density, high, volume, direction, etc. In fact, the forest fires are detected by analyzing the smoke static and dynamic characteristics. The smoke detection leads to solve some challenges related to the color, shadow, movement, and density variations.

4.1 Combined Methods to Smoke Forest Detection

In the initial stage, the forest fire before the flame, smoke is produced, because of that, the identification of the smoke can be an early warning indicator [34]. This variable is analyzed generally by two types of algorithms, those based on the color treatment and those based on the automatic learning. A third group is based on a combination of the two types of algorithms. In this way, the literature shows some methods for the smoke detection. One of the first works shows the smoke segmentation applying the contrast peaks in the spaces of color HSI, RGB, YCbCr, CIELab and the smoke separation from the rest of the scene is complimented by the Bayes classificatory [35]. In this same line, in [36] a combination of the extraction of the smoke characteristics in the image taking into consideration the HIS space considering uniquely the intensity value in 1/3 is carried out. A posteriori, the stream optics value is introduced so as to distinguish the smoke fluctuation movement, these two samples are supplies for the training through the Back-Propagation Neural Network to obtain the smoke detection from the rest of the scene.

The work presented by [37] the smoke detection is done taking into consideration the components contrast in the RGB space, where the Green value is higher than the previous ones, in this way the smoke is separated from the rest. On the other hand, a training through the Faster R-CNN for which three types of images are applied: 1) smoke and forest setting synthetic images separated, 2) real smoke and forest setting separated and 3) smoke and forest setting without being processed is carried out, the detection rate is higher than 93%. The Fig. 2 shows the input of three types of images for the training and the result about the smoke detection. In [38] the smoke is detected combining static characteristics of color, the texture and shape with the Robust AdaBoost classification method. In [39], the temporal signal in a CNN of two sources of information was used. This opened a new way for the hybrid techniques. The learning transference requires a data group smaller than for the training that can avoid the over adjustment as the example depicted in [40], where a CNN based on AlexNet is applied.

In [41] the smoke detection and location problem are faced through the development of a neuronal convolutional network of the RCNN y 3D CNN type. El RCNN uses the static spatial information to locate the smoke and the CNN 3D is used to detect the smoke with spatial and temporary characteristics. In this work RCNN and 3D CNN are used to develop a framework for the recognition of images. The use of these RCNN and 3D CNN for the image implies complexity in the execution time. On the other hand, in

[42] the smoke is detected trough saliency in the forms considering some aspects such as pixels or regions of smoke, from it, saliency maps are done and posteriori they are classified for the corresponding detection.

In [43] a training of different types of images of forest outlines with fire, smoke, smoke, and fire, without smoke and fire is applied, to do it that the deep learning convolutional neural network (DCNN) is applied. The training is quite complete with around 100 epochs, what reaches an accuracy of up to 92% in the fire and the smoke classification and detection. Another detection of smoke approach is based on the length of the movement in the video frames analysis, precisely in [44] a wavelet energy slope fitting is proposed to discriminate the continuous movement and similitude in the energy of the smoke concentration from the rest of the objects in the scene such as vehicles, people, clouds, frog, etc., taking into account a window of 40 frames processed in a continuously, the results curve shows an accuracy improvement in relation to other methods.

In [45], the movement is detected through the visual background extractor (ViBe). For that, smoke is extracted from the rest of the scene through the contrast caused by the smoke taking into consideration the components of the color in the HIS model, a posteriori the smoke texture characteristics are extracted through the adaptative scale local binary pattern (AS-LBP). This information goes through a process of normalization considering the use of a Gaussian and principal component analysis (PCA) filter. Finally, the analysis and classification are carried out through the hidden Markov model (HMM), the obtained results show that the reduction of the false positives and stability in the detection. In [46] a combination of methods is done for the smoke detection. It starts with the extraction from the foreground through the Gaussian Mixture Model which considers the movement of the smoke, in the next step, a segmentation of the smoke through the extraction from the areas with higher color contrast based on the HSV model takes place, for that a particular threshold is established, the segmentation of the objects in movement and the smoke specifically can be appreciated. Later on the segmented images are stacked in blocks, the characteristics are extracted and, finally, the SVM trains and classifies in this way, the smoke is identified from the video frames with the establishment of an average accuracy of 97.34% in the classification. Effective methods used to detect smoke through UAV's/UAS.

4.2 Smoke Detection Using Combined Methods Applied to UAS/UAV's

On the other hand, in [32] a smoke and fire detection method through the UAV's method is proposed. For that, the la luminance and chrominance in the Lab space color, the changes, speed, and the pixels movement are analyzed and evaluated through the two types of stream optics algorithms what allows the ROIs extraction. In [47], in the same way, the smoke color characteristics are extracted for the RGB model. The R and G values are weighed as higher in relation to the B one and the intensity levels are analyzed in the HIS model to establish the areas with higher light and gray. These supplies are used in the established threshold. This design is optimized through the Extended Kalman Filter (EKF) and the result is the smoke segmentation which compared with the Otsu threshold provides better results.

The detection of the smoke and the fire through the combination of methods and processes is proposed in [48]. It starts with video frames digital processing through the fire and the smoke segmentation considering the color concentration in the RGB space and the intensity in HIS. On the other hand, information of the heat concentration through infrared sensors is obtained, the fusion of the signals takes place to establish a map of detection where the intensities of the signals are established and with that the flame, the smoke and non-flame non-smoke are classified. This process is used to generate a feedback loop Fuzzy Neuron Network (FNN) and provides a higher accuracy in the classification process. In the works shown in [11, 12, 31, 33, 43] similar procedure are executed where the combination of the extraction of characteristics are based in the color treatment and, a posteriori, training and classification methods are applied to provide an accurate detection of smoke and fire. A summary of the most recent detection methods which combine methods and techniques in their application are displayed in Table 1.

Table 1. Summary of the most recent combined methods and systems applied for fire and smoke detection.

Method	Document	Fixed system	Designed to UAS/UAV	Use in real time	% Accuracy
FFD-Color features extraction-Optical flow	Sudhakar, S., et al. (2020)		X	No	----------
Color features extraction-NVI-Deep CNN	Kinaneva, D., et al. (2019)		X	Yes	-----------
FFDI Otsu Th	Cruz, H. et al. (2016)	X	X	Yes	96.82%
YOLOv3	Jiao, Z. et al. (2019)		X	Yes	91%
ANN SVM	Hossain, F. et al (2020)		X	No	84% flame 90% smoke
DCNN	Ashiquzzaman, A. et al. (2020)	X	X	Yes	92%
Color features extraction FNN	Sherstjuk, V. et al. (2020)	X		No	99%

5 Conclusions and Future Work

Forest fires around the world are taking place more frequently, which has negative consequences for the environment and for life in all its ways causing losses of all kinds. To face this situation, in the last years it is evident the increase in the study of solutions to prevent and mitigate these forest fires.

The literature review on the forest fire and smoke detection systems reflects an evolution from non-mobile systems to mobile ones, especially, the capacities shown by the UAV/UAS have allowed the surveillance and the monitoring to be more effective.

On the other hand, the methods and techniques, mainly, have focused on the proposal of integral solutions which lead to the interaction among them and include the fire and smoke detection, can be applied in real time and provide high accuracy levels.

It has been verified that there are combinations of methods which reach an accurate detection more than 90% in all cases. The digital image processing of the fire and the smoke through color features extraction is applied, these signals are used to the training process. Based on the study carried out, it can be said that in the detection and classification process, the application of CNN in its different versions leads to excellent results.

The future work has relation with use some variables related to signals acquisition in real time and combination environment characteristics to predict fire forest.

Acknowledgements. This work was partially supported by the PIS-04 "Detección temprana de incendios forestales basado en aprendizaje automático y en algoritmos de clasificación aplicados al espectro visible" research project of the Universidad de las Fuerzas Armadas ESPE, Ecuador and partially supported by the Work Group "Internet de las Cosas y Ciudades Inteligentes" of the Corporación Ecuatoriana para el Desarrollo de la Investigación y la Academia (CEDIA), Ecuador.

References

1. U.S Geological Survey. https://www.usgs.gov/news/wildfire-support-438-miles-above. Accessed 2 Aug 2020
2. The European Forest Fire Information System (EFFIS). https://effis.jrc.ec.europa.eu/reports-and-publications/annual-fire-reports/. Accessed 2 Aug 2020
3. Deutsche Welle Press. https://www.dw.com/es/incendios-forestales-el-cambio-clim%C3%A1tico-y-la-deforestaci%C3%B3n-aumentan-el-riesgo-global/a-51946276. Accessed 3 Jul 2020
4. Global Forest Fire Watch. https://www.globalforestwatch.org/. Accessed 07 Jul 2020
5. Bergeron, Y., Gauthier, S., Kafka, V., Lefort, P., Lesieur, D.: Natural fire frequency for the eastern Canadian boreal forest: consequences for sustainable forestry. Can. J. For. Res. **31**(3), 384–391 (2001)
6. Nolan, R.H., et al.: Causes and consequences of eastern Australia's 2019–20 season of mega-fires. Global Change Biol. **26**(3), 1039–1041 (2020)
7. Cruz, H.: La securitización ¿percepción o realidad tangible?. Un análisis multidimensional. Revista De La Academia Del Guerra Del Ejército Ecuatoriano **13**(1), 11 (2020)
8. Yuan, C., Zhang, Y., Liu, Z.: A survey on technologies for automatic forest fire monitoring, detection, and fighting using unmanned aerial vehicles and remote sensing techniques. Can. J. For. Res. **45**(7), 783–792 (2015)
9. Alkhatib, A.A.: A review on forest fire detection techniques. Int. J. Distrib. Sens. Netw. **10**(3), 597368 (2014)
10. Chowdary, V., Gupta, M.K.: Automatic forest fire detection and monitoring techniques: a survey. In: Singh, R., Choudhury, S., Gehlot, A. (eds.) Intelligent Communication, Control and Devices, pp. 1111–1117. Springer, Singapore (2018)
11. Sudhakar, S., Vijayakumar, V., Kumar, C.S., Priya, V., Ravi, L., Subramaniyaswamy, V.: Unmanned aerial vehicle (UAV) based forest fire detection and monitoring for reducing false alarms in forest-fires. Comput. Commun. **149**, 1–6 (2020)

12. Kinaneva, D., Hristov, G., Raychev, J., Zahariev, P.: Early forest fire detection using drones and artificial intelligence. In: 2019 42nd International Convention on Information and Communication Technology, Electronics and Microelectronics (mipro), pp. 1060–1065. IEEE (2019)
13. Moulianitis, V.C., Thanellas, G., Xanthopoulos, N., Aspragathos, N.A.: Evaluation of UAV based schemes for forest fire monitoring. In: Aspragathos, N., Koustoumpardis, P., Moulianitis, V. (eds.) International Conference on Robotics in Alpe-Adria Danube Region, pp. 143–150. Springer, Cham (2018)
14. Bu, F., Gharajeh, M.S.: Intelligent and vision-based fire detection systems: a survey. Image Vis. Comput. **91**, 103803 (2019)
15. Grishin, A.M., Perminov, V.A.: Mathematical modeling of the ignition of tree crowns. Combust. Explosion Shock Waves **34**(4), 378–386 (1998)
16. Ko, B.C., Ham, S.J., Nam, J.Y.: Modeling and formalization of fuzzy finite automata for detection of irregular fire flames. IEEE Trans. Circuits Syst. Video Technol. **21**(12), 1903–1912 (2011)
17. Dimitropoulos, K., Barmpoutis, P., Grammalidis, N.: Spatio-temporal flame modeling and dynamic texture analysis for automatic video-based fire detection. IEEE Trans. Circuits Syst. Video Technol. **25**(2), 339–351 (2014)
18. Muhammad, K., Ahmad, J., Lv, Z., Bellavista, P., Yang, P., Baik, S.W.: Efficient deep CNN-based fire detection and localization in video surveillance applications. IEEE Trans. Syst. Man Cybern. Syst. **49**(7), 1419–1434 (2018)
19. Iandola, F.N., Han, S., Moskewicz, M.W., Ashraf, K., Dally, W.J., Keutzer, K.: SqueezeNet: alexNet-level accuracy with 50x fewer parameters and$<$ 0.5 MB model size. arXiv preprint arXiv:1602.07360. (2016)
20. Cui, F.: Deployment and integration of smart sensors with IoT devices detecting fire disasters in huge forest environment. Comput. Commun. **150**, 818–827 (2020)
21. Chaoxia, C., Shang, W., Zhang, F.: Information-guided flame detection based on faster R-CNN. IEEE Access **8**, 58923–58932 (2020)
22. Avula, S.B., Badri, S.J., Reddy, G.: A Novel forest fire detection system using fuzzy entropy optimized thresholding and STN-based CNN. In: 2020 International Conference on COMmunication Systems & NETworkS (COMSNETS), pp. 750–755. IEEE (2020)
23. Mohammed, Z., Hanae, C., Larbi, S.: Comparative study on machine learning algorithms for early fire forest detection system using geodata. Int. J. Elect. Comput. Eng. **10**(5), 5507 (2020)
24. Vipin, V.: Image processing based forest fire detection. Int. J. Emerg. Technol. Adv. Eng. **2**(2), 87–95 (2012)
25. Patel, P., Tiwari, S.: Flame detection using image processing techniques. Int. J. Comput. Appl. **58**(18), 13–16 (2012)
26. Cruz, H., Eckert, M., Meneses, J., Martínez, J.F.: Efficient forest fire detection index for application in unmanned aerial systems (UASs). Sensors **16**(6), 893 (2016)
27. Pritam, D., Dewan, J.H.: Detection of fire using image processing techniques with LUV color space. In: 2017 2nd International Conference for Convergence in Technology (I2CT), pp. 1158–1162. IEEE (2017)
28. Prema, C.E., Vinsley, S.S., Suresh, S.: Efficient flame detection based on static and dynamic texture analysis in forest fire detection. Fire Technol. **54**(1), 255–288 (2018)
29. Zhang, Q., Xu, J., Xu, L., Guo, H.: Deep convolutional neural networks for forest fire detection. In: 2016 International Forum on Management, Education and Information Technology Application, Atlantis Press (January 2016)
30. Chen, Y., et al.: UAV Image-based forest fire detection approach using convolutional neural network. In: 2019 14th IEEE Conference on Industrial Electronics and Applications (ICIEA), pp. 2118–2123. IEEE (June 2019)

31. Jiao, Z., et al.: A deep learning based forest fire detection approach using UAV and YOLOv3. In: 2019 1st International Conference on Industrial Artificial Intelligence (IAI), pp. 1–5. IEEE (2019)
32. Yuan, C., Liu, Z., Zhang, Y.: Aerial images-based forest fire detection for firefighting using optical remote sensing techniques and unmanned aerial vehicles. J. Intell. Rob. Syst. **88**(2–4), 635–654 (2017)
33. Anim Hossain, F.M., Zhang, Y.M., Tonima, M.A.: Forest fire flame and smoke detection from UAV-captured images using fire-specific color features and multi-color space local binary pattern. J. Unmanned Veh. Syst. **8**(4), 285–309 (2020)
34. Perminov, V.: Numerical solution of Reynolds equations for forest fire spread. In: Sloot P.M.A., Hoekstra A.G., Tan C.J.K., Dongarra J.J. (eds) International Conference on Computational Science, pp. 823–832. Springer, Berlin (2002)
35. Krstinić, D., Stipaničev, D., Jakovčević, T.: Histogram-based smoke segmentation in forest fire detection system. Inf. Technol. Control **38**(3), 237–244 (2009)
36. Chunyu, Y., Jun, F., Jinjun, W., Yongming, Z.: Video fire smoke detection using motion and color features. Fire Technol. **46**(3), 651–663 (2010)
37. Zhang, Q.X., Lin, G.H., Zhang, Y.M., Xu, G., Wang, J.J.: Wildland forest fire smoke detection based on faster R-CNN using synthetic smoke images. Procedia Eng. **211**, 441–446 (2018)
38. Wu, X., Lu, X., Leung, H.: A video based fire smoke detection using robust AdaBoost. Sensors **18**(11), 3780 (2018)
39. Hu, Y., Lu, X.: Real-time video fire smoke detection by utilizing spatial-temporal ConvNet features. Multimedia Tools Appl. **77**(22), 29283–29301 (2018)
40. Yuan, F., Zhang, L., Xia, X., Wan, B., Huang, Q., Li, X.: Deep smoke segmentation. Neurocomputing **357**, 248–260 (2019)
41. Lin, G., Zhang, Y., Xu, G., Zhang, Q.: Smoke detection on video sequences using 3D convolutional neural networks. Fire Technol. **55**(5), 1827–1847 (2019)
42. Xu, G., Zhang, Y., Zhang, Q., Lin, G., Wang, Z., Jia, Y., Wang, J.: Video smoke detection based on deep saliency network. Fire Saf. J. **105**, 277–285 (2019)
43. Ashiquzzaman, A., Oh, S.M., Lee, D., Lee, J., Kim, J.: Context-aware deep convolutional neural network application for fire and smoke detection in virtual environment for surveillance video analysis. In: Zhang, Y.D., Senjyu, T., SO–IN, C., Joshi, A. (eds.) Smart Trends in Computing and Communications: Proceedings of SmartCom 2020, pp. 459–467. Springer, Singapore (2020)
44. Zhang, Y., Wang, H., Fan, X.: Algorithm for detection of fire smoke in a video based on wavelet energy slope fitting. J. Inf. Process. Syst. **16**(3), 557–571 (2020)
45. Tao, H., Lu, X.: Smoke vehicle detection based on multi-feature fusion and hidden Markov model. J. Real-Time Image Proc. **17**(3), 745–758 (2020)
46. Islam, M.R., Amiruzzaman, M., Nasim, S., Shin, J.: Smoke object segmentation and the dynamic growth feature model for video-based smoke detection systems. Symmetry **12**(7), 1075 (2020)
47. Yuan, C., Liu, Z., Zhang, Y.: Learning-based smoke detection for unmanned aerial vehicles applied to forest fire surveillance. J. Intell. Rob. Syst. **93**(1–2), 337–349 (2019)
48. Sherstjuk, V., Zharikova, M., Dorovskaja, I., Sheketa, V.: Assessing forest fire Dynamicsin UAV-based tactical monitoring system. In: Babichev, S., Lytvynenko, V., Wójcik, W., Vyshemyrskaya, S. (eds.) International Scientific Conference "Intellectual Systems of Decision Making and Problem of Computational Intelligence", pp. 285–301. Springer, Cham (2020)

Dimensionality Reduction Using PCA and CUR Algorithm for Data on COVID-19 Tests

Marco Enríquez, Samantha Naranjo(✉), Isidro Amaro, and Franklin Camacho

School of Mathematical and Computational Sciences, Yachay Tech University, Urcuquí, Ecuador
naranjosamantha1j@gmail.com

Abstract. In this paper we present the results of two well known analyses, Principal Component Analysis and CUR algorithm, conducted on data related to tests of coronavirus, which were performed from May 17th to June 26th, 2020 in Ibarra, Ecuador. We analyzed the effectiveness of CUR over PCA and found out that, for our data matrix, CUR is more effective than PCA whenever the control parameters of the CUR algorithm c and k are equal. Furthermore, the results of CUR algorithm suggest that the laboratory tests D-dimer, ferritin and PCR are the most important variables.

Keywords: CUR algorithm · Principal Components Analysis · COVID-19

1 Introduction

The World Health Organization declared COVID-19 outbreak as a pandemic on 11 March 2020 and people's lives have changed dramatically ever since. In Ecuador the number of cases is still increasing and, according to Ministerio de Salud Pública Ecuador, 182.523 tests have been performed until July 12th, 2020. There are two kinds of tests that can show us if a person is infected or not: reverse transcription polymerase chain reaction (RT-PCR) and serology tests (also know as rapid tests). The former is a diagnostic test that, from a respiratory sample of a person suspected of being infected, detects the presence of SARS-CoV-2 virus' RNA, and the latter, through a blood sample, is able to detect antibodies generated by the immune system.

Generally, people who want to know if they have COVID-19 do not ask clinical laboratories to do the RT-PCR test; they prefer to use the serology tests instead because those are faster and affordable. However, relying solely on these rapid tests can be dangerous. According to Ministerio de Salud Pública Ecuador [1], serology tests may yield a type II error, which means that it is possible that the laboratory concludes some patients are not infected when they are. In some

M. Botto-Tobar et al. (Eds.): CIT 2020, AISC 1326, pp. 121–134, 2021.
https://doi.org/10.1007/978-3-030-68080-0_9

cases, people ask laboratories for other tests, which will be named and described in Sect. 2, besides the serology ones. When serology tests are not very reliable or accurate, could the other tests help us detect or at least give more information about the presence of the virus? This question motivates our research.

The purpose of this research is to compare the results given by the application of two multivariate statistical techniques: Principal Components Analysis (PCA) and CUR algorithm (also known as CUR matrix decomposition), to a database related to COVID-19 in Imbabura, Ecuador. This database consists of some laboratory tests. Currently, only two of them are being used to determine if a person is infected with SARS-CoV-2. From this point these laboratory tests will be referred to as variables.

Then, our main objective is to use multivariate statistical techniques to reduce the dimension of the problem to determine which of those variables are more significant. In order to achieve this, we have to analyze our data set using PCA and CUR algorithm, and then interpret the results obtained by both techniques. Moreover, we will determine which technique is more effective to reduce the dimension for our data set. Our hypothesis is that CUR algorithm is more effective than PCA. The effectiveness of these techniques will be measured with the Frobenius Norm.

If we successfully fulfill our main objective, then we will be able to determine which variables play a role in the detection of the SARS-CoV-2, especially when serology tests are not reliable. This research is important because it provides lab technicians with valuable information about the variables that can be used to diagnose this disease.

We took for granted our interaction with other species and as a result the new coronavirus has affected our lifestyle dramatically. For this reason, research on this topic has increased a lot recently. In particular, the multivariate statistical analysis techniques, such as regression tree, cluster analysis or PCA, have been used for the study of COVID-19 disease. For example, Rahaman Khan and Hossain, in [2], used these techniques for a COVID-19 data from 133 countries. For each country they considered the following variables: total confirmed cases, new confirmed cases, total deaths, total recovered patients, total active cases, total seriously critical patients, infection rate per million, death rate per million, total tests conducted, and test rate per million. Another example is the research of Kumar et al. [3]. They use correlation and PCA to study the relationship of coronavirus spread with socio-economic factors of different states of India using the Rstudio software[1]. Nonetheless, we were not able to find research on the use of CUR algorithm in some aspect of the COVID-19 disease. Hence, our research takes a different approach because our COVID-19 data is local from Ibarra, Ecuador, and the use of multivariate statistical analysis techniques, such as CUR algorithm is focused in clinical laboratory tests.

In this paper, in Sect. 2, we first describe our data matrix and its variables, then we give a brief description of PCA and CUR algorithm and how to determine which of these techniques is more effective. Second, in Sect. 3, we present

[1] https://rstudio.com/.

the main results of our work, which was done using R language[2]. Finally, in Sect. 4, we summarize the article.

2 Materials and Methods

In this section, first we describe the data set used, including information about the number of individuals and the type of variables. Then we explain the two techniques used for the analysis and the effectiveness of one against another.

2.1 Description of Data

The data set used in this research[3] was obtained from a clinical laboratory in Ibarra, Ecuador, called Laboratorio Clínico - Clínica Ibarra. The tests of 256 individuals were performed from May 17th, 2020 to June 26th, 2020. The variables in this data are: *Edad, IGG, IGM, Dimero D, Ferritina, PCT* and *PCR*, but only two of them are specialized for COVID-19. Since this clinical laboratory is private, we asked its owner permission to use this data. We decided not to include the ID of the individuals because of privacy matters.

First, let's explain the variables of our data set that are used to confirm or rule out the new strain of coronavirus: *IGG* and *IGM*. According to Portillo et al. [4], "rapid tests would be a solution that would speed up the identification and isolation of infected individuals, estimating more accurately the number of infections". For that reason, in this study we consider the variables *IGG* and *IGM*, which are the serology tests. They both detect the immune response of the individual and thus the COVID-19 IgG and COVID-19 IgM antibodies. The Laboratorio Clínico - Clínica Ibarra applies a technique known as immunofluorescence for these tests.

The difference between *IGG* and *IGM* is that *IGM* reveals the immune response against active infection, while *IGG* indicates the presence of non-current infections; i.e., those that have already developed a secondary immune response. The time it takes to get the results, from when the patient's blood is drawn, is the same for *IGG* and *IGM*, 1 h. That is why they are called rapid tests.

The results of these tests are qualitative, but they can be expressed in numerical form against a cut off index (COI). The range of *IGG* and *IGM* is greater than or equal to 0. According to Laboratorio Clínico - Clínica Ibarra, both test are negative, which means that the patient does not have COVID-19 when their values are less than or equal to 0.9. If the values of *IGG* and *IGM* are between 0.9 and 1.1, the result of the tests are indeterminate; i.e., it is not possible to conclude whether or not the patient has COVID-19. If the values of *IGG* and *IGM* are greater than or equal to 1.1, then the result of the test is positive and the patient has COVID-19.

[2] http://www.r-project.org/.
[3] If you need the data matrix, you can email the corresponding author.

Now, let's briefly explain the other five variables. The variable *Edad* is the age of the individual and we considered it because in medical research, it has been shown to be an important variable, particularly in coronavirus studies. The variable *Dimero D* is a test that looks for the D-dimer in the blood, a protein fragment that is produced when a blood clot dissolves in the body. The method used in this test is heterogeneous immunoassay and its reference value range is [0, 500] ng/mL. The variable *Ferritina* is a test that measures the level of ferritin in the blood. Ferritin is a protein that stores iron in cells. The immunofluorescence method was used to identified this protein and its reference value range is [30, 350] ng/mL. The variable *PCT* is the procalcitonin test that measures the level of procalcitonin in the blood. A high level could be a sign of a serious bacterial infection, such as sepsis. As for this method, immunofluorescence was used. *PCT* is negative when is less than 0.5 ng/mL; the risk of severe sepsis and/or septic shock is low if *PCT* is between from 0.5 to 2 ng/mL, moderate if *PCT* is between from 2 to 10 ng/mL and high if *PCT* is greater or equal than 10 ng/mL. Finally, the variable *PCR*, also known as Ultrasensitive CRP (that must not be mistaken for RT-PCR), tell us if C-reactive protein is secreted by a current infection or inflamation. The method used for this test is the automated photometry. Ultrasensitive CRP is measured in mg/L. The reference values for this test is [0, 5] mg/L.

2.2 Techniques for the Analysis

In many disciplines and data analysis applications, the use of data matrices with a huge number of columns or rows is common. But sometimes, the studies with a large dimension of the data matrix has dissimilar or contradictory results. In such cases, it is necessary to employ some dimensionality reduction method in order to avoid wrong conclusions. In this research, we use only two of these techniques: PCA and CUR algorithm. So, before to review them, let's first briefly recall two main definitions.

For a matrix $M \in \mathbb{R}^{n \times p}$ the Frobenius norm is defined as

$$\|M\|_F = \sqrt{tr(MM^t)}, \tag{1}$$

where $tr(A)$ is the trace of a square matrix $A \in \mathbb{R}^{p \times p}$ defined by $tr(A) = \sum_{i=1}^{p} A_{i,i}$, the sum of its diagonal elements.

Also, let's recall Singular Value Decomposition (SVD). Given $X \in \mathbb{R}^{n \times p}$, there exists orthogonal matrices $U = [u^1 u^2 \cdots u^n] \in \mathbb{R}^{n \times n}$ and $V = [v^1 v^2 \cdots v^p] \in \mathbb{R}^{p \times p}$, where $\{u^i\}_{i=1}^{n} \subseteq \mathbb{R}^n$ and $\{v^i\}_{i=1}^{p} \subseteq \mathbb{R}^p$ are such that

$$U^T X V = \Sigma = diag(\sigma_1, \cdots, \sigma_\rho), \tag{2}$$

where $\Sigma \in \mathbb{R}^{n \times p}$, $\rho = min\{n, p\}$, $\sigma_1 \geq \sigma_2 \geq \cdots \geq \sigma_\rho \geq 0$, and $diag(\cdot)$ represents a diagonal matrix with the specified elements on the diagonal.

Principal Component Analysis. This is one of the most popular multivariate statistics technique to reduce the data dimension and boils down to the SVD.

We use PCA in our data set because it not only reduces the data dimension but also give us information about the behavior of the variables. According to Rencher and Christensen [5], the idea of this method is to create a new kind of variables called principal components (PC). The principal components associated with the largest singular values have strong optimality properties, and they can often be quite useful as a tool to summarize and identify major patterns of the data. The ith, $i = 1, \ldots, p$, principal component is defined by

$$PC_i = X\boldsymbol{v_i} = v_{1,i}\text{col}(1, X) + v_{2,i}\text{col}(2, X) + \cdots + v_{p,i}\text{col}(p, X), \tag{3}$$

where $v_{j,i}$ is the jth component of the ith right singular vector the data matrix $X \in \mathbb{R}^{n\times p}$, which contains n observations of a random vector $\boldsymbol{\mathcal{X}} = (\mathcal{X}_1, \cdots, \mathcal{X}_p)^T$.

Note that $\mathcal{X}_i$ is the random variable whereas $\text{col}(i, X)$ is a collection of n observations of this random variable. Thus the columns of our new data matrix D with lower dimension are the observations of the random variables

$$\mathcal{PC}_i = v_{1,i}\mathcal{X}_1 + v_{2,i}\mathcal{X}_2 + \cdots + v_{p,i}\mathcal{X}_p. \tag{4}$$

However, the ability to interpret the random variables $\mathcal{X}_i$, or equivalently the columns of X, in no way guarantees an ability to interpret linear combinations of them.

CUR Algorithm. The CUR matrix decomposition is also a dimensionality reduction method that was developed by Mahoney and Drineas [6]. We use this technique in our data set because it can solve the disadvantage of PCA with the interpretation. The idea behind CUR is to decompose the data matrix X into the product of three matrices C, U and R such that

$$X \approx CUR, \tag{5}$$

where the C matrix contains a small number of columns from X, the R matrix contains a small number of rows of X and the matrix U is the product of their Moore-Penrose pseudo inverses with X; that is $U = C^+XR^+$. The role of U is to guarantee that the product CUR is close to X. Since matrix C is built from the original variables of X, CUR matrix decomposition can be easily interpreted by practitioners of the field from which the data is drawn.

Although CUR matrix decomposition and CUR algorithm refer to the same, in this paper we establish a difference between them. Here, the R code implementation of the CUR matrix decomposition is referred to as CUR algorithm. This implementation is in the rCUR package, which is available for free in an article of Bodor et al. [7][4]. In the rCUR package, a primary function that makes the decomposition called 'CUR' is implemented. The inputs of this function are the data matrix X, the column parameter c, the row parameter r and the rank parameter k. The function 'CUR' returns a CURobj-class object.

[4] We are using the version 1.3 of rCUR that was updated on 2012-07-02.

Recall that C matrix contains a small number of columns from X. To select the columns of X to include in C (and similarly for the row selection of R), we have to compute an importance score for each columns of X. Then, by using the rank parameter k and the data matrix X, the CUR algorithm randomly samples a small number of columns from X using that score as an importance sampling probability distribution. That is, for the ith column X, 'CUR' associates a number l_i such that the collection $\{l_i\}_{i=1}^{p}$ forms a probability distribution. These l_i are called the *normalized statistical leverage scores*. If $V_{i,j}$ is the ith element of the jth right singular vector X, then the *leverage scores* equal

$$l_i = \frac{1}{k}\sum_{j=1}^{k} V_{i,j}^2, \tag{6}$$

for all $i = 1, \ldots, p$. The function 'leverage' retrieve column leverage scores from CURobj-class.

Using the leverage scores the algorithm makes a pass over the columns and selects the ith columns with a probability $\min\{cl_i, 1\}$. So the algorithm takes the leverage scores and scales them by the column parameter. If $cl_i < 1$ then the algorithm chooses that column as important with probability cl_i. If $cl_i > 1$ then the algorithm always chooses that column as important, i.e., with probability 1. Once the algorithm has chosen a set of columns it creates a matrix C whose columns are those selected columns. The function 'getC' retrieve matrix C from CURobj-class.

According to Drineas, Mahoney and Muthukrishnan [8], with probability at least 99%, this choice of columns satisfies the following

$$\|X - PcX\|_F \leq (1 + \epsilon/2)\,\|X - X_k\|_F\,, \tag{7}$$

where Pc denotes a projection matrix onto the column space of C and ϵ is an error parameter. The proof of (7) (see [8]) depends crucially on the use of (6). Thus, by using (7), Mahoney and Drineas [6] prove with probability at least 98% the following

$$\|X - CUR\|_F \leq (2 + \epsilon)\,\|X - X_k\|_F\,, \tag{8}$$

where $X_k = \sum_{i=1}^{k} \sigma_i u^i v^{iT}$.

In [7], Bodor et al. make a suggestion to improve efficiency by switching off the computation of components that are not used. So we make that the row parameter r be the set to almost the actual numbers of rows of X. The function 'getR' retrieve matrix R from CURobj-class.

Hence, based on Hunt [9], the steps of the CUR algorithm can be written as follows:

1. Get data matrix X and the two parameters c and k.
2. Calculate the leverage scores $\{l_i\}_{i=1}^{p}$ using X and k.
3. Make a pass over the columns of the matrix and keep the ith columns with a probability $\min\{cl_i, 1\}$.

4. If we have chosen at least one column, return a matrix C consisting of those chosen columns. If we have not chosen any column, then go back to step 3.

Similarly to PCA, the way in which one decides how to choose dimensionality we should reduce the data is not a question CUR algorithm can answer. However, one of the conclusions of Hunt [9] is that, if we want to use CUR algorithm in place of retaining m principal components, we should run the function CUR with $c = m$ and $k = m$.

2.3 Effectiveness

The performance of PCA and CUR algorithm depends on the data matrix X and on the control parameters c and k of the CUR algorithm. So, we have to know which of both techniques is better for our data set. To reduce the dimensionality of the data matrix implies to throw away part of the data. Hence, to measure this loss, Hunt [9] propose a statistic to measure the performance of CUR algorithm relative to PCA by "computing how much variance we lose, or equivalently, how much variance we retain" for each method.

Recall that PCA re-expresses the data in terms of a new data matrix D. By using (4), we construct D, the comprised matrix of k principal components of X. Thus, the effectiveness of CUR algorithm relative to PCA, by using the Frobenius norm, is

$$e = \frac{\|C\|_F^2 - \|D\|_F^2}{\|X\|_F^2}. \tag{9}$$

Since "neither CUR algorithm nor PCA can do better than retaining 100% more variance than the other", then $-1 < e < 1$. Therefore, if $e < 0$ then $\|C\|_F^2 < \|D\|_F^2$ which means that CUR algorithm retains less of the variance than PCA. Similarly, if $e > 0$ then $\|C\|_F^2 > \|D\|_F^2$ which means that CUR algorithm retains more of the variance than PCA.

3 Results and Discussions

In this section we summarize the results that were found after running PCA and CUR algorithm on our data matrix in R language[5].

3.1 Principal Component Analysis on Data Matrix

Implementation of PCA in R is straightforward. We have applied PCA on the original data matrix X, which has 7 columns and 256 rows. In order to run PCA, R program uses the correlation matrix. In our case, the determinant of the correlation matrix is 0.02606074 which is relatively close to 0. This means that the variables of our data are correlated. We can see that the cumulative proportion of the third principal component amounts to 0.8023 (see Table 1) of

[5] If you want to replicate the results, you can email the corresponding author.

the total variance. Since the criteria we are using to retain components is to account 80% of the variance, we take PC1, PC2, PC3. We get the same result if instead of using the previous criteria, we take principal components such that their eigenvalues are greater than or equal to 1 (see Fig. 1). In any case, we retain three principal components.

Table 1. Summary of PCA on the original data matrix

	PC1	PC2	PC3	PC4	PC5	PC6	PC7
Standard deviation	1.8355	1.1247	0.9908	0.8408	0.59853	0.4443	0.34868
Proportion of variance	0.4813	0.1807	0.1403	0.1010	0.05118	0.0282	0.01737
Cumulative proportion	0.4813	0.6620	0.8023	0.9032	0.95443	0.9826	1.00000

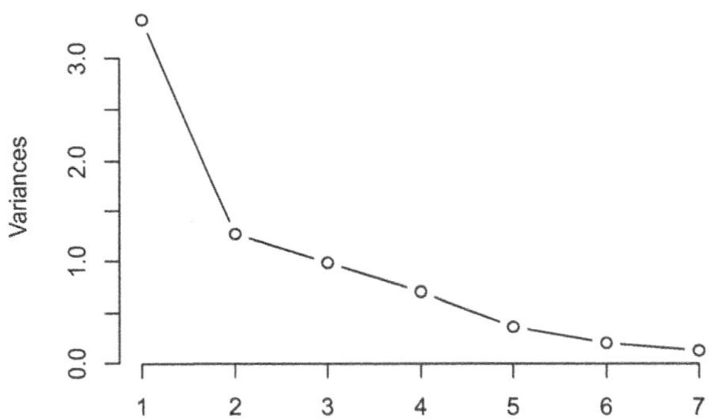

Fig. 1. Scree plot: the numbers in the abscissa represent the PC

PCA also gives us the coordinates of each of the seven variables using the principal components, which are given in Table 2. These coordinates, along with the coordinates of each individual and the retained PC (see Table 3), are used to get the plot shown in Fig. 2.

In Fig. 2 we can see that people on the left hand side of the plot are similar to each other and that they are people who tested negative for COVID-19. We can also notice that all of the seven variables are positively correlated; i.e. if one of the variables is high and far from its normal value, then it is more likely that the other variables have high values as well. On one hand, there is a high positive correlation among *Ferritina*, *Dimero D* and *IGG*; similarly among *PCT*, *PCR* and *IGM*. On the other hand, *Edad* has a high correlation among *Ferritina*, *Dimero D* and *IGG*, but it has a relatively low correlation among *PCT* and *PCR*. Moreover, the correlation between *Edad* and *IGM* is significantly low; that is, it value is rather close to zero. This shows that the higher the age the higher the values of these variables. We can further notice that people on the

Table 2. Coordinates of each variable in terms of PC

	PC1	PC2	PC3	PC4	PC5	PC6	PC7
Edad	0.0516	0.2320	0.9687	−0.0206	0.0584	−0.0086	0.0344
IGM	0.3161	−0.6276	0.1367	0.3134	−0.1580	−0.5612	0.2218
IGG	0.4502	0.2878	−0.1151	−0.4238	−0.1915	−0.0039	0.6962
Dimero D	0.4588	0.3574	−0.0779	−0.1447	−0.2774	−0.4098	−0.6242
Ferritina	0.2807	0.4338	−0.1113	0.8193	0.0977	0.1342	0.1471
PCT	0.4516	−0.1460	−0.0403	−0.1705	0.8554	0.0296	0.1066
PCR	0.4486	−0.3633	0.0976	0.0139	−0.3415	0.7057	−0.2052

Table 3. Matrix D obtained from PCA: we only show the first three and last three rows of the matrix

	PC1	PC2	PC3
1	−0.4112	2.5987×10^{-1}	−0.5766
2	−0.7519	-1.3912×10^{-1}	0.2170
3	−1.2401	-4.3013×10^{-1}	−0.4711
		⋮	
254	−1.0526	−0.6707	−0.3653
255	−0.4939	−0.0055	−0.0225
256	6.6718	2.1108	−0.6041

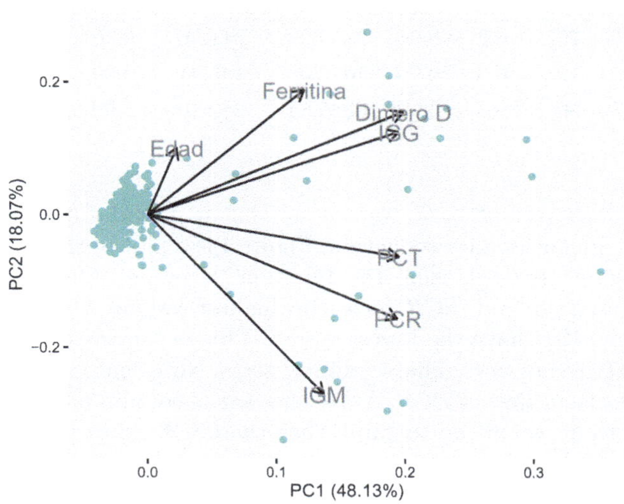

Fig. 2. PCA on original matrix

upper right of the graph have high values of *Ferritina*, *Dimero D* and *IGG* while people on the bottom right have high values of *PCT*, *PCR* and *IGM*.

Up till now, we have retained three principal components. Nonetheless, we still do not know which variables we should get rid off, which is precisely why we need to implement CUR algorithm.

3.2 CUR Algorithm on Data Matrix

Unlike PCA, we must provide CUR algorithm the values of c and k. To improve the efficiency of CUR algorithm with respect to PCA, we ran the function 'CUR' with input parameters $c = k = 3$, following the suggestion of Hunt [9] explained before. Recall that the data matrix is decomposed into three matrices C, U, and R. We can see that the function 'getR' returned a matrix R of dimension 134×7 (see Table 4); that is, 119 individuals of the original matrix were omitted because they do not provide as much information (low row leverage scores) as the other entries. In fact, this can be seen in Fig. 3, which is the plot of the row leverage scores.

Table 4. Matrix R for our data matrix: we only show the first three and last three rows of the matrix

	Edad	IGM	IGG	Dimero D	Ferritina	PCT	PCR
4	52	1.6	0.00	864.20	246.50	0.50	50.12
5	33	1.2	21.60	2456.50	368.45	1.50	145.20
8	34	0.1	0.00	321.45	302.20	0.20	0.22
				⋮			
250	70	0.2	0.00	385.62	318.56	0.20	1.50
251	12	0.3	0.00	107.20	145.41	0.09	2.60
256	50	0.5	36.20	6285.84	380.10	2.00	35.10

When it comes to matrix C, we can see that it provides us with more information about dimensionality reduction. Figure 4 is a plot of the column leverage scores that were retrieved using the function 'leverage'. On one hand, *Edad*, *Dimero D*, *Ferritina* and *PCR* have the highest values. On the other hand, *PCT*, *IGM* and *IGG* have the lowest values. This fact proves that the variables *IGM* and *IGG* are not very reliable clinical tests. Note that the leverage score of *Edad* is greater than that of *PCR*. As we choose the column parameter $c = 3$, the most natural thing would be to think that the CUR algorithm will choose the variables with highest leverage score: *Edad*, *Dimero D* and *Ferritina*. However, the function 'getC' returns a 256×3 matrix, whose columns are *Dimero D*, *Ferritina* and *PCR* (see Table 5). The variable *Edad* was omitted by the algorithm. This happens because when PCA was performed *Ferritina*, *Dimero D*, and *PCR*

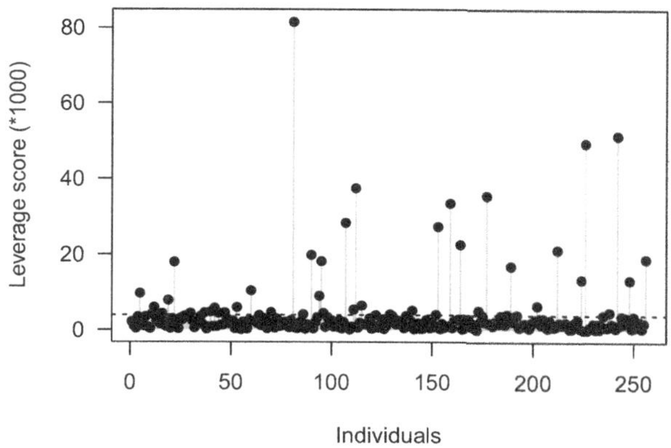

Fig. 3. Plot of row leverage scores

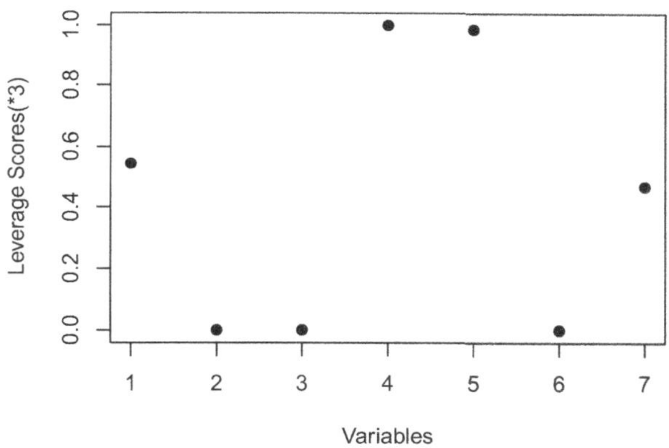

Fig. 4. Plot of column leverage scores. Variable *Edad* is represented by 1, *IGM* by 2, *IGG* by 3, *Dimero D* by 4, *Ferritina* by 5, *PCT* by 6 and *PCR* by 7

turned out to be more important for the first PC while *Edad* is more important for the second PC.

In order to show that the choice of the three variables was correct, we run a PCA on the matrix C and the matrix given by the four variables (see Fig. 5 and Fig. 6), and computed the cophenetic correlation coefficients. With the former matrix we got a coefficient of 0.995, but with the later the coefficient is 0.955. This implies that *Dimero D*, *Ferritina* and *PCR* are the most important variables. Furthermore, on Fig. 5 we see that *Ferritina*, *Dimero D* and *PCR* are positively correlated and that the proportion of variance is high. Unlike Fig. 5, Fig. 6 shows that when we add *Edad* the variance is lower and that this variable is barely related with the other three variables.

Table 5. Matrix C for our data matrix: we only show the first three and last three rows of the matrix

	Dimero D	Ferritina	PCR
1	265.85	246.15	4.57
2	145.65	102.23	4.63
3	41.90	45.00	0.00
		⋮	
254	165.20	45.00	0.90
255	87.10	216.80	3.68
256	6285.84	380.10	35.10

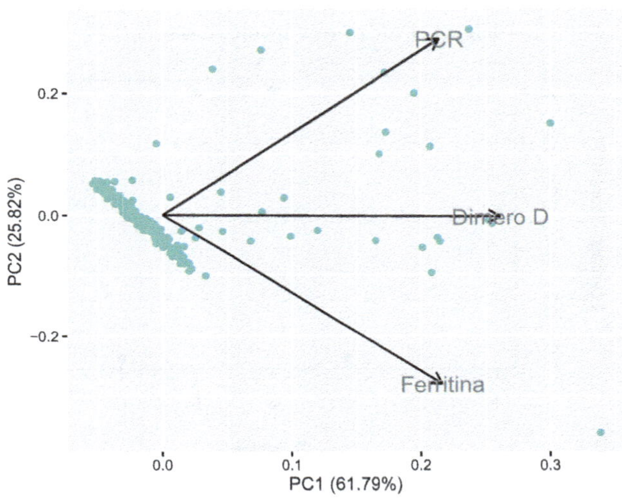

Fig. 5. Plot of PCA on matrix C

3.3 Effectiveness

We have already run PCA and CUR algorithm on the data matrix. It remains to see which method was more effective. Recall that here matrix X is the data matrix and D is the matrix obtained by performing PCA. Using (9) we get that $e = 0.9992723$, which is rather close to 1. This implies that, for our data matrix, CUR algorithm excels in dimensionality reduction, i.e., it is better than PCA.

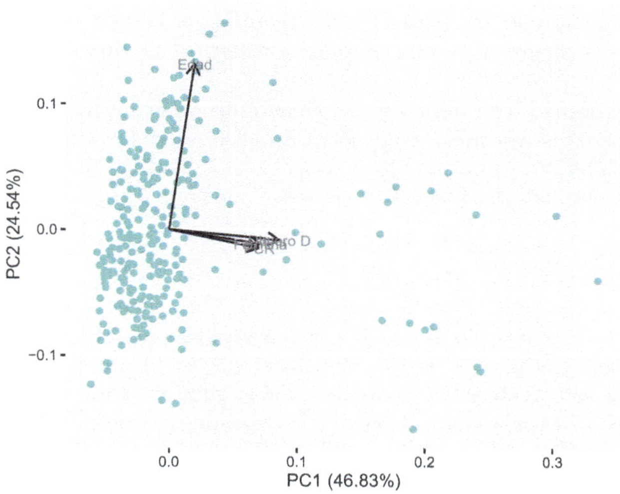

Fig. 6. Plot of PCA on the matrix with four variables: *Edad*, *Dimero D*, *Ferritina* and *PCR*

4 Conclusions

Throughout this paper, we focused on dimensionality reduction techniques and how we can use them to contribute more information about the detection of the virus SARS-CoV-2, which yields in COVID-19 disease. We considered two multivariate statistical techniques, PCA and CUR algorithm. In order reduce the dimension, PCA creates new variables called principal components while CUR algorithm removes the less important variables based on their leverage scores. Unlike PCA and whenever it is possible, CUR algorithm allows us to have a better interpretation of the results because matrix C is built from the original variables.

After applying PCA to our data matrix, we found out that it is possible to retain three principal components, but they do not give us much information about the variables. Then we ran CUR algorithm with input parameters $c = k = 3$. Three variables were chosen by the CUR algorithm. This variables are *Dimero D*, *Ferritina* and *PCR*. Furthermore, for our data matrix, CUR retains more variance than PCA. This means that CUR algorithm was more effective than PCA.

This exploratory data analysis led us to the conclusion that the laboratory tests D-dimer, ferritin and CRP are the most significant in our data set and they come in handy in the detection of SARS-CoV-2. For that reason and taking into account that COVID-19 is a new disease, we suggest for clinical research related to the detection of SARS-CoV-2 to consider them. We did not appraise other values of c and k other than the number of retained principal components.

Analyzing how the effectiveness of both techniques behaves when those values vary and using a bigger data matrix may constitute the object of future studies.

Acknowledgements. The data set was thankfully provided by Laboratorio Clínico - Clínica Ibarra. This research would not have been possible without their help. The authors also thank Vicecacillería de Invesigación of Yachay Tech University for its partial support through the Project Decisiones Cualitivas.

References

1. Avilés, E., et al.: Protocolo de uso de pruebas rápidas para detección de anticuerpos contra SARS-COV-2/COVID-19. Ministerio de Salud Pública Ecuador (2020). https://www.salud.gob.ec/wp-content/uploads/2020/04/Protocolo-de-uso-de-pruebas-r%C3%A1pidas-para-detecci%C3%B3n-de-anticuerpos-contra-Sars-Cov-2Covid-19_v2_20_04_2020.pdf. Accessed 21 Aug 2020
2. Rahaman Khan, H., Hossain, A.: Countries are clustered but number of tests is not vital to predict global COVID-19 confirmed cases: A machine learning approach. medRxiv (2020). https://doi.org/10.1101/2020.04.24.20078238. https://www.medrxiv.org/content/early/2020/04/29/2020.04.24.20078238
3. Kumar, A., Rani, P., Kumar, R., Sharma, V., Purohit, S.R.: Data-driven modelling and prediction of COVID-19 infection in India and correlation analysis of the virus transmission with socio-economic factors. Diabetes Metab. Synd. Clin. Res. Rev. **14**(5), 1231–1240 (2020). https://doi.org/10.1016/j.dsx.2020.07.008. http://www.sciencedirect.com/science/article/pii/S187140212030254X
4. Portillo, R.V., et al.: Pruebas rápidas para COVID-19, la mejor alternativa para Ecuador. Bionatura (2020). https://doi.org/10.21931/RB/2020.05.03.21
5. Rencher, A., Christensen, W.: Methods of Multivariate Analysis, 3rd edn. John Wiley & Sons, Inc., New York (2012). https://doi.org/10.1002/9781118391686
6. Mahoney, M.W., Drineas, P.: Cur matrix decompositions for improved data analysis. Proc. Nat. Acad. Sci. **106**(3), 697–702 (2009). https://doi.org/10.1073/pnas.0803205106. https://www.pnas.org/content/106/3/697
7. Bodor, A., Csabai, I., Mahoney, M.W., Solymosi, N.: rCUR: an R package for CUR matrix decomposition. BMC Bioinf. **13**(1), 1–6 (2012)
8. Drineas, P., Mahoney, M.W., Muthukrishnan, S.: Relative-error cur matrix decompositions. SIAM J. Matrix Anal. Appl. **30**(2), 844–881 (2008). https://doi.org/10.1137/07070471X
9. Hunt, G.J.: CUR: An interpretable alternative to principal components analysis. Ph.D. thesis, Drew University (2013)

Ensuring Traceability and Orchestration in the Food Supply Chain

Mauricio Espinoza-Mejía[1(✉)], Víctor Saquicela[1], and Victoria Abril-Ulloa[2]

[1] Computer Science Department, University of Cuenca, Cuenca, Ecuador
{mauricio.espinoza,victor.saquicela}@ucuenca.edu.ec
[2] Faculty of Medical Sciences, University of Cuenca, Cuenca, Ecuador
victoria.abril@ucuenca.edu.ec

Abstract. Ensure traceability and orchestration of the participants in the food supply chain can help improve the food production and reduce the distribution of unsafe or low-quality products. This article provides some insights about the design of a high-level architecture to support a semantic blockchain platform that ensures traceability and orchestration of the food systems. The design involves two layers that cover: (a) the decentralized orchestration of the participants, (b) the semantic modeling of the data and processes involved, and (c) the storage and integrity of the data. To deal with the platform design, it is analyzed the operation and attributes of the food supply chain management and it is discussed how the combination of Semantic and Blockchain technologies can address the platform features.

Keywords: Linked data · Blockchain · Food supply chain · Smart contracts

1 Introduction

According to Food and Agriculture Organization of the United Nations (FAO), the current food supply chain will be subject to radical changes in the next 10 to 15 years, driven by changing consumer demand, technological advances, trade dynamics, and other factors [1]. In this perspective, it is essential to examine the opportunities offered by technology, in order to develop low-cost solutions that can be implemented in countries with emerging economies, where the food supply chain is dominated by farmers and food companies small or medium.

The contribution of technology will undoubtedly improve the reality of these countries, where: (i) the compliance and security processes of the food supply chain continues being registered on paper or in private databases, making it difficult to identify a product at any stage of the supply chain [2]; and (ii) the industries of these countries are in the initial stages with respect to the use of applications with disruptive technologies, which causes a limited awareness of the relevance, development, implementation and benefits associated with new trends [3].

M. Botto-Tobar et al. (Eds.): CIT 2020, AISC 1326, pp. 135–149, 2021.
https://doi.org/10.1007/978-3-030-68080-0_10

Several experts agree that, in the case of developing countries, ensuring the traceability and orchestration of the food supply chain can provide a significant opportunity for the consumers, producers, and operators of these systems [4,5]. In fact, the lack of technological support in these value drivers can cause that the access to data to be costly, complex, subject to fraud, corruption, or error, causing threats of foodborne illness and/or financial loss [6].

Under this premise, the integration of technologies such as Semantic Web and Blockchain -the first with a high degree of maturity and the second still with the need for further research- can help mitigate these problems. This article explores the current situation and the capabilities of both technologies, as a means to design a platform that guarantees the traceability and improves the orchestration of the activities performed by the participants involved in the food supply chain. These aspects are vital to generating consumer confidence in the content and quality of the end-to-end products in the food supply chain.

The rest of the document is organized as follows: Sect. 2 discusses the operation and attributes of the food supply chain management, Sect. 3 defines the elements that underpin Semantic and Blockchain technologies. Section 4 identifies different technological settings to support food supply chain management. Section 5 proposes the design of a platform based on the integration of both technologies and this is followed by a conclusion in Sect. 6.

2 Characteristics of the Food Supply Chain

Food supply chains are complex networks that base their operation in the participation of different actors such as farmers, distributors, wholesalers and retailers, grocery stores, and consumers.

In general, the phases that characterize a food supply chain include: production, processing, distribution, wholesale and retail, and consumption [7]. These phases can be combined to support two main types of food supply chains: agricultural products or processed food products. In the first case, all stages of the supply chain leave intact the intrinsic characteristics of the cultivated or produced product in the field; whereas, in the second case, raw materials are used to generate consumer products with higher added value, making use of conservation and conditioning processes [8].

Information Technology has brought significant improvements to the food supply chain management in terms of automation of some specific tasks of the food production line such as: cleaning, packaging, and cold storage [9]. However, there are still some challenges that have not been resolved; by example, how to improve the transparency and integrity of the data at each stage of the process, which can generate consumer confidence.

Inspired by several studies (see, for example, [9–11]), in the following are described some functional attributes of the food supply chain management that could be addressed with the use of technology:

- *Data and Process Governance*. The term governance in the supply chain refers to the management of the participants and their actions. It is common for some participants (especially individual producers) to be skeptical about the data sharing processes and the actions they can take within the supply chain. The usual questions are: who will own the data, who will have access to it, and how it will be used. To avoid any inconvenience, it is necessary to implement standards that guarantee protection over private data and to perform mechanisms of supervision and control that establish the level of leadership over the participants in the supply chain.

- *Inter-organizational Relationship.* In a supply chain, the relationships between organizations are generally reflected through partnerships or buyer-seller relationships. In the first, the relationship is established between a manufacturing company and its suppliers; while, in the second, strategic relationships are established between independent companies. Encouraging inter-organizational information exchange is a key element to achieve a competitive advantage in the supply chain. This can be achieved by establishing agreements where it is defined: what can be done, how it can be done, and what happens if something is not done according to what has been agreed.
- *Interoperability.* Effective interoperability between the participants of the supply chain depends mainly on the degree of standardization of the production process. In the case of new production processes involving separate and independent companies, it is necessary to define the mechanism used to control the identification and monitoring of a product throughout the supply chain. Business standards can be used to expand visibility and enable the exchange of information between partners in the supply chain.
- *Decentralization.* Implementing decentralization processes in a food supply chain helps to improve the efficiency of these systems, as each participant makes decisions based on local information, which encourages working cooperatively. The decentralization also prevents that local decision-makers acting opportunistically, which can affect to other participants in the chain. The challenge is to move the current food supply chains towards a decentralized model that takes full advantage of each participant's available skills and assets but as a whole, the integrity of shared information to be preserved at all times.
- *Integration Between Stakeholders.* Different data collection systems can be found among participants in a supply chain: file-based systems, relational databases, or devices such as sensors, smartphones, and global positioning systems. The challenge in the current supply chains is to be able to generate a unified vision of the data, independent of the model, infrastructure, or system used by each participant. There is a need of integrating to supply chains, data management models that facilitate the retrieval and integration of information, which can encourage furthermore consumer confidence and decision-making.
- *Predictive Analysis.* The ability to be proactive and anticipate the future is a very important factor in supply chains. Improving the quality of food products, optimizing issues related to logistics, or establishing which food products are most in demand, are some of the cases that need to incorporate predictive analysis in the food supply chain.

3 Technology Review

In this section the generic elements that underpin the Semantic Web and Blockchain technologies are analyzed. These elements can be integrated to improve the traceability and orchestration of the food supply chains.

- *Semantic Web Elements.* The Semantic Web is a collection of technologies and standards that allow the creation, publication, and consumption of machine-interpretable data, the creation of vocabularies and the writing of rules to handle data according to the decentralized model of the Web[1]. The main elements that characterize this technology are:
  - *Ontologies* is the fundamental pillar of the Semantic Web and they formally describe shared conceptualizations of a domain, using representation languages such as RDF Schema[2] and OWL[3].
  - *Linked Data* allows the publication of data without any centralization and most importantly, once published, the data can be used by anyone who knows its location. In addition, the use of URIs[4] to identify resources allows the union of separate data sources, making them interoperable. To achieve these goals, the usual data representation standard is RDF[5] and to query the data SPARQL[6].
  - *Social Linked Data (Solid)* is a set of rules and tools for developing decentralized social applications. This platform combines some Web standards such as: WebId[7], Linked Data Platform[8], RDF and SPARQL to offer an approach where it is possible to discover and exchange information in a decentralized way [12]. An important feature on the Solid platform is that users have the freedom to choose where their data resides, including control of who is allowed to access it.
- *Blockchain Elements.* A blockchain is a type of distributed ledger, which consists of an ordered list of records whose complete content is shared by all nodes across the network. Two features define distributed ledgers, they are: *immutable* that is, once a record has been added, it cannot be deleted or edited; and *decentralized*, which indicates that there is no central authority to act as verifier [13]. Of course, a blockchain ensures all of these properties through blocks of transactions.
  From a technical point of view, a blockchain is a data structure that consists of a linked list of blocks, where each block refers to the previous one and contains information about transactions and data. The first block in the chain is called the genesis block, so it does not have a parent block. The core elements of this technology are briefly described in the following paragraphs:

[1] https://www.w3.org/standards/semanticweb/.
[2] https://www.w3.org/TR/rdf-schema/.
[3] https://www.w3.org/OWL/.
[4] https://www.w3.org/TR/uri-clarification/.
[5] https://www.w3.org/RDF/.
[6] https://www.w3.org/TR/sparql11-overview/.
[7] https://www.w3.org/wiki/WebID.
[8] https://www.w3.org/TR/ldp/.

- *Block* consisting of a header and a body. The header of a block provides the technical notions to form a blockchain and includes information such as: block version number, hash of the previous block, data hash tree, timestamp, or nBits. The body of the block moreover stores the transactions.
- *Transaction* is the fundamental unit of a blockchain and allows sharing and exchanging information between nodes as equals. This exchange of information from one node to another is performed by means of files generated by a source node and transmitted to the entire network for validation.
- *Node* element in the network (user or computer) that can perform various functions depending on the role it assumes. The responsibility and authority of each node depend on the governance used in the blockchain. Some nodes may have the function of creating new blocks (e.g., miners), reading/writing data, or participating in the consensus mechanism used to add new transactions in a block and incorporate a block to chain.
- *Consensus Protocol* is the backbone of a blockchain, as it allows all parts of the network to reach a common agreement. The choice of the consensus algorithm depends on the type of blockchain used; that is, not all consensus mechanisms are suitable for all types of blockchain. There are different ways to implement the consensus mechanism, the most common being: Proof-of-Work (PoW) and Proof-of-Stake (PoS). A further explanation of how works each of these consensus algorithms, is beyond the scope of this article, however, interested readers may refer to [14,15] for more information.
- *Smart Contract* is a way of establishing an agreement between two parties without the need for a physical intermediary. From a technical point of view, it is a program that is capable of executing and enforcing itself, autonomously and automatically. Its feature is not available on all blockchain platforms, but it is now becoming a highly desirable feature due to the flexibility and power it provides to applications.

4 Technological Proposal

This section offers different technological alternatives to support the food supply chain management. For each of the characteristics described in Sect. 2 are defined options that involve the combination of technical elements described in Sect. 3:

- *Data and Process Governance.* The governance in the food supply chain is critical to facilitate decision making between the participants. The way in which governance is established depends on the type of blockchain used:
  - *Public Blockchain:* The governance style in this case is permissionless, this means that anyone can participate in the consensus process, so, there is no need for a reliable intermediary to validate the transactions. Due to a large number of participants, it is almost impossible to alter transactions,

but performance is usually limited and latency low. The security of a public blockchain is determined through the consensus protocol which helps to confirm that a transaction is true. The most representative general purpose platform of this type is Ethereum[9].

- *Private Blockchain:* Only specific organizations can participate in the consensus process and a permission is required to access blockchain. This type of governance is known as permissioned, as it is controlled by a single organization. Access to transactions can be public or private, but the organizations serving as type of administrator are who decide on the inclusion of future participants or access to simple observers. Due to the smaller number of participants, transactions may be manipulated, but in return good levels of efficiency can be achieved. The security depends on the honesty of the entities validating transactions. Hyperledger Fabric[10] or Hyperledger Sawtooth[11] are examples of these blockchain types.
- *Federated Blockchain:* Another option for the food supply chain management is to use a federated or consortium blockchain, which applies the best of private and public blockchains. Governance is built by a small number of organizations, which can participate in decisions. The rest of the features such as access permissions, the immutability of transactions and the level of efficiency are similar to the private blockchain. An example of a consortium created by IBM from the Hyperledger Fabric platform is FoodTrust[12], which is used for the management of agricultural products.

- *Inter-organizational Relationship.* Although smart contracts are not an essential requirement of blockchain technology, they can be integrated as substitutes of the conventional legal contracts to encourage inter-organizational information exchange. The following options are possible:
  - *Only Smart Contracts:* Despite the relative immaturity of smart contracts, there are some applications in the field of the food supply chain where these elements are used to facilitate monetary and information exchange [16,17]. Table 1, compares the features of smart contracts by platform type, considering: the execution environment, the language used to implement business logic, and the model used to store data on the blockchain.
  - *Smart Contracts + Ontologies:* To reduce the conceptual ambiguity and improve the inference and automation of the operations executed by smart contracts, ontologies can be used [18].
    In the work proposed in [19], an ontology is used to change the logic of a smart contract, adapting it to unexpected situations. Although smart contracts are immutable, this proposal implements the logic of the contract in a source of information external to the blockchain. In this way, the rules defined in a domain ontology allow that the contract logic to be changed in unexpected situations.

[9] https://ethereum.org/.
[10] https://www.hyperledger.org/projects/fabric.
[11] https://sawtooth.hyperledger.org/.
[12] https://www.ibm.com/products/foodtrust.

Table 1. Features of smart contracts by blockchain platform type.

	Ethereum	Hyperledger fabric hyperledger sawtooth
Execution environment	Virtual machine ethereum	Docker container
Language	Solidity, serpent, Lisp like language	Java, Golang
Data model	Based on accounts	Key-value

- *Interoperability.* Accepted multi-sectoral global standards for identifying, capturing, and sharing product, service, and location information allow organizations to ensure interoperability. The options available for the food supply chain are:
  - *Blockchain + Standards:* The Electronic Product Code Information Service (EPCIS[13]) of GS1[14] is currently one of the main standards proposed for data exchange between organizations for tracking and tracing purposes in the supply chain.
    The proposal of combining GS1 standards like EPCIS and CBV[15] with Blockchain guarantees the visibility of information about a product, which in turn improves the transparency of the entire system. This quality is described in [20] as an essential factor in the implementation of a food product traceability system.
    It is necessary to comment that the EPCIS standard define only the structure of the data to be recorded. The semantics of data are informally defined; therefore, the output of the application generates silos of EPCIS event data for each participant in the supply chain. This feature of the EPCIS standard is not a problem if there is no need to perform discovery and validation services built from each participant's event repository.
  - *Blockchain + Standards + Ontologies:* As stated in [21] standards like EPCIS and CBV are not designed to share data between a multiplicity of supply chain participants. In order to link the events data of each participant, a semantic model called EPCIS Event Model (EEM) is defined in [22], it allows representing the life cycle of a product as it progresses through the entire supply chain.
    Another contribution is the OntoPedigree ontology proposed in [23]. This model defines conceptual entities to formalize the notion of linked pedigrees, which are real-time EPCIS events in the supply chain and encapsulated as linked data to enhance data visibility. Also, to facilitate the creation of linked pedigrees, the authors propose the LinkedEPCIS library[16] to capture EPCIS events as linked data.

[13] https://www.gs1.org/standards/epcis.
[14] https://www.gs1.org/.
[15] https://www.gs1.org/standards/epcis/epcis-cbv/1-0.
[16] https://github.com/nimonika/LinkedEPCIS.

These semantic models can be specialized and extended to define food domain-specific traceability chains.

- *Decentralization.* The decentralization process in the food supply chain must allow that each participant can decide where their data are stored and who can access it. Only those alternatives that ensure data integrity are proposed below:
  - *Only Blockchain:* The data stored on a blockchain are by design immutable and decentralized. Before opting for this decentralization alternative, it is necessary to consider that i) the storage of large amounts of data within blockchain may be inefficient because the information has to be replicated in all the nodes of the network; and ii) the distribution and replication structure of a blockchain can cause inconveniences regarding data sovereignty and privacy.
    One of the most efficient options for performing decentralized storage in blockchain, is the use of smart contracts [24], as this model provides a firm guarantee of data integrity. However, this option has a disadvantage the transaction cost, which may vary depending on the size of the contract.
    Regarding data access and privacy, this depends exclusively on the type of governance selected. At the beginning of this section, the access restrictions for each type of blockchain were described.
  - *Blockchain + Distributed Storage:* Another option is to store the data outside the blockchain, only registering indexes that point to the external content to maintain the origin of the data. This would cause data privacy and sovereignty to be maintained locally at each node, and blockchain be kept smaller. Furthermore, the authenticity of data outside the network can be guaranteed with the use of cryptographic techniques and digital summaries (hash functions) that can be stored within the blockchain.
    The proposal described in [25] states that one option to reduce costs is to incorporate a distributed storage platform into the blockchain. This alternative avoids the need that each participant to keep the hash of the registered content. In this model, the data are stored on the distributed storage platform and a content hash and a time stamp are registered on blockchain to ensure the integrity of the data. Some distributed storage alternatives are currently available such as: Interplanetary File System (IPFS)[17], Swarm [26], and FileCoin[18].
    In order to control the access to sensitive data of the participants, in [27] the authors propose the use of smart contracts to manage an access control list. In this way, participants can register the data and grant or revoke access to it. In addition, the IPFS platform provides access to the contract, while enforcing the permissions stored in the access control list.
  - *Blockchain + Solid:* This combination ensures the data integrity by maintaining a hash stored in the blockchain, which can be used to confront the externally stored data in Solid [28]. With this option, each participant in

[17] https://ipfs.io/.
[18] https://filecoin.io/.

the supply chain can store different data types within the Solid space such as: personal, financial, or logistical data, which can be managed through a personalized privacy control.

According to Solid architecture, the mechanism used to control the data access is through a control list and the use of a WebId public key. A WebID allows identifying a person, company, organization, or other agent using a URI.

- *Integration Between Stakeholders.* The integration process among the participants in the chain supply varies depending on the data collection mechanism used in each node of the network. The following options are valid in order to use the data in the production of knowledge for decision-making:
  - *Only Blockchain:* This option could be used in a food supply chain that does not require integrating external data sources to those data already stored in the blockchain itself. Although this alternative is viable and functional, it is necessary to take into account that the retrieval and integration of data on the blockchain can vary in complexity depending on the structure used to store data and transactions. By example, the Ethereum blockchain system provides a query language named Ethereum Query Language (EQL) [29], which allows users to write SQL-like queries to retrieve data from this platform.
  - *Linked Data:* The previous option solves several types of applications, however, in some scenarios there is the need of generating a shared knowledge repository from the data stored externally. In fact, it is highly unlikely that the local data sources (with domain specific data) have the information with the level of granularity required by users of any application.

    To generate a structured and interconnected knowledge repository between the participants of the supply chain, it is required that the data stored outside blockchain use the Linked Data standards. Thus, the data collection methods implemented at each node should result in an RDF graph.

    In [30] the authors extend the data persistence component of blockchain systems by adding a data-graph service-layer. In the proposal, the data are collected as objects commonly used in object-oriented programming and transformed into RDF statements. To perform the transformation process the following rules are used: i) an instance of a class is mapped as an RDF-Subject, ii) the properties of that class are mapped as RDF-Predicates, and iii) the values of those properties are mapped as RDF-Objects. This linked data generation alternative offers great flexibility especially when the scenario requires adding new data types and relationships between network participants. To integrate the data sources and to create a shared knowledge repository SPARQL federated queries can be used.
  - *Linked Data + Ontologies:* An additional alternative to the previous option is to use an ontology to describe a chained graph, similar to the structure used by a blockchain. In this way, it is possible to describe all

the participating entities in a blockchain system, which makes it easier the integration of data, but it is necessary to create and maintain a domain ontology for each domain of application. As described in [31], at the intersection between semantic technology and blockchain, different ontologies have been proposed to describe the structure of a chain. GraphChain [32], BLONDiE[19] and EthOn[20] are some examples of these semantic models. From these ontologies, a custom model can be designed to semantically represent a linked data structure to support the integration of data from the participants in the food chain supply.

- *Linked Data + Ontologies + Ontology Based Data Access Techniques:* This alternative is useful when participants locally already manage multiple data collection systems and non-synchronized terminologies. Without the application of a set of good practices for data curation and linking through ontologies, the extraction and integration of information could require the use of scripts, text files and other manual processes, which involve loss of time and risk of inconsistent data [33].
  In addition to the use of ontologies to provide a global vision of the multiple data sources registered at each node of the supply chain, it would be necessary to use techniques that describe the relationships between the global schemes used as integration mechanisms and the local data sources. In [34] the authors provide some techniques, algorithms, and systems that can be used to address the challenges of data integration using ontologies. Once the data sources have been homogenized by means of one or more ontologies, it is possible to use a mechanism similar to those discussed in the previous options.

– *Predictive Analysis.* At the time of writing this article, none of the most prominent blockchain platforms such as: Ethereum, Hyperledger Fabric, or Hyperledger Sawtooth offers the ability to manage Big Data and support predictive analysis, which is a facet important in today's world of supply chains. As mentioned in [35], the discussions to incorporate this capability into blockchain try: i) use new consensus protocols to improve the scalability of blockchains platforms, and ii) merge blockchains with other infrastructures that support analysis of large volumes of data such as Hadoop[21].

5 Platform Architecture

This section proposes the design of a generic platform that makes use of semantic and blockchain technologies to support the food supply chain management. The emphasis of the platform is to ensure the traceability and orchestration of the supply chain participants. The design covers two layers as can be seen in Fig. 1.

[19] https://github.com/hedugaro/Blondie.
[20] http://ethon.consensys.net/.
[21] https://hadoop.apache.org/.

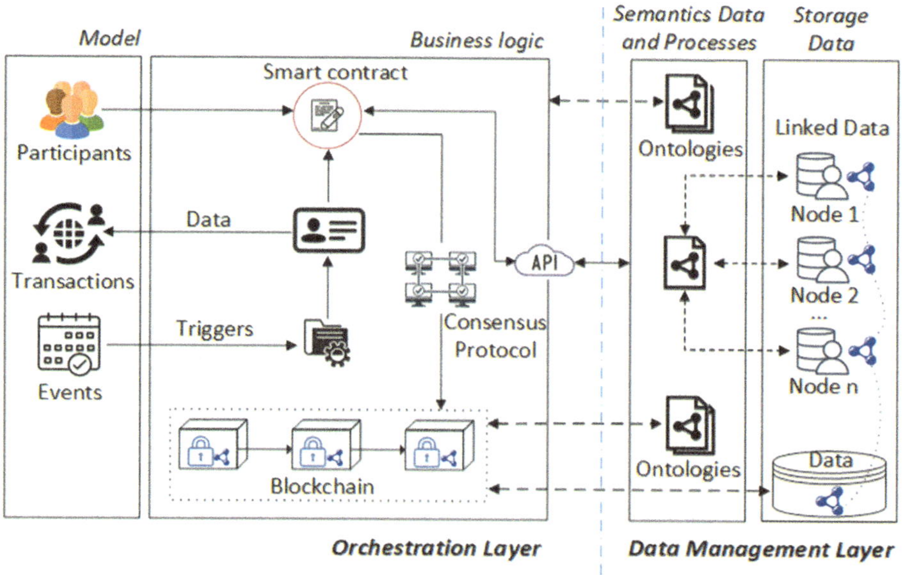

Fig. 1. Generic architecture to support the food supply chain management.

- *Orchestration Layer:* It is in charge of managing the decentralized orchestration of all the participants in the supply chain and validating the transactions executed according to the proposed business model.
- *Data Management Layer:* This layer is in charge of the data and process semantic modeling to ensure a common language among the participants. In addition, this layer has as aim the governance and integrity data and the semantic integration of diverse data sources.

It is important to show to the reader that the platform design proposed in this work does not address the use of new technologies that are being incorporated into the Industry 4.0 scope, which includes a variety of technological proposals for the development of a digital and automated manufacturing environment. However, issues related to the incorporation of technologies for tracking, monitoring, and locating of a product through the use of sensors can be incorporated into the platform as a data source belonging to any node in the network.

The platform design also does not address the incorporation of predictive analytic, which is a desirable feature in supply chains (see Sect. 2). In fact, issues related to the management, analysis, and visualization of large volumes of data to support predictive analytic in supply chains are currently relevant research topics in the scientific community.

An Example: Processed Food Management

Imagine a production process that involves multiple organizations that are part of a private consortium. This consortium seeks as part of a marketing strategy to increase the sales of its products. For this, a mobile application has been generated, which offers consumers information on the products developed by the consortium. For simplicity, some tasks and participants in the supply chain are omitted from the example, such as grocery stores, and logistics and transportation.

First, the *data and process governance* in the chain could be implemented with the use of a *private blockchain* since in this way the administration and control of the chain would be in charge of the business organization that leads to food production. This would also restrict access to other companies that are not part of the consortium. To fulfill this purpose, the *Hyperledger Fabric* platform can be used.

The identification and sharing of information on products in the processed food supply chain (*interoperability*) and the *integration between stakeholders*, can be supported by the combination of different ontologies (e.g., Blondie, EEM and OntoPedigree) as illustrated in Fig. 2. The core element in ontology is the concept of `Pedigree`, which captures certified information for a stakeholder in the supply chain. The property `hasTransactionInfo` defines the relationship between the pedigree and the conceptual entity capturing transaction information related to the sale, movement, return or transfer of goods. In order not to lose the link between the data provided by each participant, the entity `ChainInfo` stores the information of each node and the property `hasProductInfo` links these data to each pedigree to ensure consistency.

To support the *inter-organizational relationship* and promote the secure exchange of information between the participants, it can use the *smart contracts*. In this case, the use of ontologies is not necessary since it is not expected to alter the established business logic. The logic of the *smart contract* should implement at least two functions: the storing of a hash with RDF content and pedigree information on the blockchain and the retrieving of a hash, from a URI for performing verification processes.

As it is a private chain, it is understood that there would be no data sharing problems or sensitive information between the partners. In this way, *decentralization* could be implemented in the *data management layer*, using a distributed file system such as *IPFS*, thus avoiding overloading the blockchain.

Once the technological configuration to support the processed food supply chain management has been described, the following explains how the exchange of information between the participants would take place:

1. A farmer through a mobile application store the type of crop or the procedure used for planting; with this information, the system generates an RDF graph and publishes the graph via HTTP. The initial part of the Linked Pedigree is formed and their status is set as `Initial` on the ontology entity `PedigreeStatus`. The information of the event and the data is sent to the

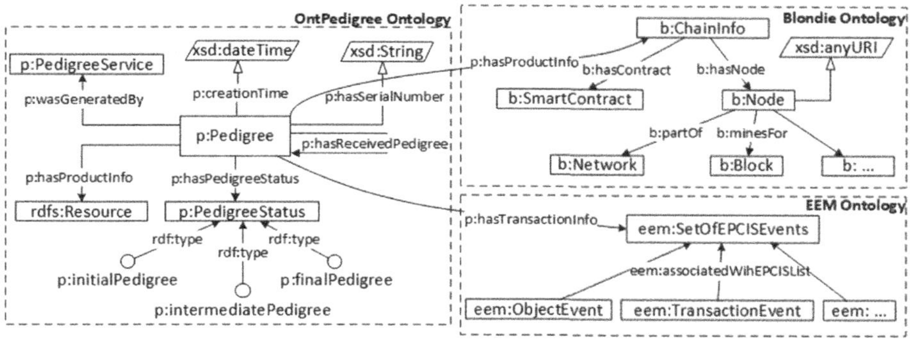

Fig. 2. Interlinked Ontologies to support data and process semantic modeling.

IPFS storage network, where a hash is generated. The hash is associated with the core pedigree information and it is stored on blockchain using a storing request to the smart contract.

2. Food processing companies and wholesalers, bid for crops or for processed products, respectively, following the steps of storing and publishing of data previously described. The property `hasReceivedPedigree` specifies the respective previous linked pedigree part by its URI.
3. Consumers track product information by scanning the QR code printed on the packaging, which contains the URI of a linked pedigree part. This link contains a reference to the previous parts, which consumers can deference to obtain the product information. In any moment the customer can to request to the smart contract information that verify the integrity of the information provided.

6 Conclusions

In this document, the operation of food supply chains was briefly described and some functional characteristics of the management of these systems were analyzed. In addition, they were analyzed the elements that characterize semantic and blockchain technologies with the aim of offering technological alternatives to support the attributes of the food supply chain management. With this information, it was established the design of a generic platform that covers a wide range of applications.

In order to support the platform operation, the design proposes two base layers; these layers support the decentralized orchestration of the participants, the semantic modeling of the data and processes involved, and the storage and integrity of the data. These functionalities seek to ensure the traceability and orchestration of the participants in the food supply chain.

The relationship between the attributes that characterize the food supply chain management and the layers of the platform can be used as a framework to help developers in selecting the most appropriate technologies in each case.

As future work, it is intended to end with the implementation and testing of the proposed scenario, as it is part of an ongoing research project.

References

1. FAO: Food systems for healthy diets (2018). http://www.fao.org/3/CA2797EN/ca2797en.pdf
2. Trienekens, J., Wognum, H., Beulens, P.M., van der Vorst, J.G.A.J.: Transparency in complex dynamic food supply chains. Adv. Eng. Inf. **26**(1), 55–65 (2012)
3. Queiroz, M.M., Telles, R., Bonilla, S.H.: Blockchain and supply chain management integration: a systematic review of the literature. Supply Chain Manage. Int. J. **25**(2), 241–254 (2019)
4. Lee, H.L., Mendelson, H., Rammohan, S., Srivastava, A.: Technology in Agribusiness: Opportunities to drive value. Technical report, Stanford Business (2017)
5. World Economic Forum In collaboration with McKinsey & Company: Innovation with a Purpose: Improving Traceability in Food Value Chains through Technology Innovations (2019)
6. Aung, M.M., Chang, Y.S.: Traceability in a food supply chain: safety and quality perspectives. Food Control **39**(4), 172–184 (2014)
7. Kamilaris, A., Fonts, A., Prenafeta-Boldú, F.X.: The rise of blockchain technology in agriculture and food supply chains. Trends Food Sci. Technol. **91**, 640–652 (2019)
8. van der Vorst, J., da Silva, C.A., Trienekens, J.H.: Agro-industrial Supply Chain Management: Concepts and Applications (2007)
9. Zhong, R., Xu, X., Wang, L.: Food supply chain management: systems, implementations, and future research. Ind. Manage. Data Syst. **117**, 2085–2114 (2017)
10. Bourlakis, M., Matopoulos, A.: 24 - Trends in food supply chain management. In: Delivering Performance in Food Supply Chains. Woodhead Publishing (2010)
11. Sufiyan, M., Haleem, A., Khan, S., Khan, M.I.: Analysing attributes of food supply chain management: a comparative study. In: Advances in Industrial and Production Engineering. Springer, Singapore (2019)
12. Sambra, A.V., Mansour, E., Hawke, S., Zereba, M., Greco, N., Ghanem, A., Zagidulin, D., Aboulnaga, A., Berners-Lee, T.: Solid : A Platform for Decentralized Social Applications Based on Linked Data. Technical report, MIT (2016)
13. Rauchs, M., Glidden, A., Gordon, B., Pieters, G., Recanatini, M., Rostand, F., Vagneur, K., Zhang, B.: Distributed Ledger Technology Systems. A Conceptual Framework. Cambridge Judge Business School, University of Cambridge, Cambridge Centre for Alternative Finance (2018)
14. Zheng, Z., Xie, S., Dai, H., Chen, X., Wang, H.: An overview of blockchain technology: architecture, consensus, and future trends. In: IEEE International Congress on Big Data (BigData Congress) (2017)
15. Zhang, R., Xue, R., Liu, L.: Security and privacy on blockchain (2019). https://arxiv.org/pdf/1903.07602.pdf
16. Kumar, M.V., Iyengar, N.C.S.N.: A framework for blockchain technology in rice supply chain management. Adv. Sci. Technol. Lett. **146**, 125–130 (2017)
17. Provenance: From Shore to Plate: Tracking Tuna on the Blockchain. Technical report, Provenance.org (2018). https://www.provenance.org/tracking-tunaon-the-blockchain#overview

18. de Kruijff, J., Weigand, H.: Ontologies for commitment-based smart contracts. In: On The Move to Meaningful Information Systems, vol. 10573–10574. Springer Verlag, Germany (2017)
19. Mohsin, F., Zhao, X., Hong, Z., de Mel, G., Xia, L., Seneviratne, O.: Ontology aided smart contract execution for unexpected situations (2019)
20. Lin, Q., Wang, H., Pei, X., Wang, J.: Food safety traceability system based on blockchain and EPCIS. IEEE Access **7**, 20698–20707 (2019)
21. Solanki, M., Brewster, C.: Enhancing visibility in EPCIS governing agri-food supply chains via linked pedigrees. Int. J. Semant. Web Inf. Syst. **10**(3), 45–73 (2014)
22. Solanki, M., Brewster, C.: Representing supply chain events on the web of data. In: DeRiVE@ISWC (2014)
23. Solanki, M., Brewster, C.: OntoPedigree: modelling pedigrees for traceability in supply chains. Semant. Web **7**(5), 483–491 (2016)
24. Chowdhury, N., Ramachandran, M., Third, A., Mikroyannidis, A., Bachler, M., Domingue, J.: Towards a blockchain-based decentralised educational landscape. In: The Twelfth International Conference on Mobile, Hybrid, and On-line Learning (eLmL 2020) (2020)
25. Third, A., Domingue, J.: Linked data indexing of distributed ledgers. In: 26th International Conference on World Wide Web Companion. WWW2017 (2017)
26. Hartman, J.H., Murdock, I., Spalink, T.: The Swarm scalable storage system. In: 19th IEEE International Conference on Distributed Computing Systems (1999)
27. Steichen, M., Fiz, B., Norvill, R., Shbair, W., State, R.: Blockchain-based, decentralized access control for IPFS. In: 2018 IEEE International Conference on Internet of Things and IEEE Green Computing and Communications and IEEE Cyber, Physical and Social Computing and IEEE Smart Data (2018)
28. Third, A., Domingue, J.: LinkChains: trusted personal linked data. In: Blockchain-Enabled Semantic Web (2019)
29. Bragagnolo, S., Rocha, H., Denker, M., Ducasse, S.: Ethereum query language. In: 1st International Workshop on Emerging Trends in Software Engineering for Blockchain. ACM, New York (2018)
30. Naim, B.A., Klas, W.: Knowledge graph-enhanced blockchains by integrating a graph-data service-layer. In: Sixth International Conference on Internet of Things: Systems, Management and Security (IOTSMS) (2019)
31. Braun, C., Käfer, T.: Verifying the integrity of hyperlinked information using linked data and smart contracts. In: Semantic Systems. The Power of AI and Knowledge Graphs. Springer, Cham (2019)
32. Sopek, M., Gradzki, P., Kosowski, W., Kuziski, D., Trójczak, R., Trypuz, R.: GraphChain: A distributed database with explicit semantics and chained RDF graphs. In: The Web Conference 2018 (2018)
33. Solórzano, J., Saquicela, V.: Integración de fuentes de datos heterogéneas: aplicando Tecnologías Semánticas y un Bus de Servicio Empresarial. Master's thesis, Universidad de Cuenca, Ecuador (2017)
34. Corcho, Ó., Priyatna, F., Chaves-Fraga, D.: Towards a new generation of ontology based data access. Semanti. Web **11**(1), 153–160 (2020)
35. Sahoo, M.S., Baruah, P.K.: HBasechainDB – a scalable blockchain framework on hadoop ecosystem. In: Supercomputing Frontiers. Springer, Cham (2018)

Network Neutrality: The Case of Five South American Countries

Roberto Triviño[1](✉), Antonio Franco-Crespo[1,2], and Rafael-Leonardo Ochoa-Urrego[3]

[1] Universidad de las Fuerzas Armadas, Sangolquí 17110, Ecuador
rdtrivino@espe.edu.ec
[2] Escuela Politécnica Nacional, Quito, Ecuador
[3] Universidad Central, Bogotá, Colombia

Abstract. There is an ongoing debate on the principles of Network Neutrality in many countries where have been implemented and refers to the concerns that Internet service providers could discriminate (block or prioritizes) against other providers (content, services, or applications), and how to design the regulation to avoid it. In the last years, policymakers in South America have grappled with the design process to implement laws that could protect the Internet as an open environment, ensure access and quality for users. The objective of this work is to present an overview of the Network Neutrality policies progress in five South American countries using a qualitative empirical approach and highlight the incorporated issues that make them ambiguous for effective enforcement.

Keywords: Internet · Net neutrality · South America · Content · Applications

1 Introduction

Over the last 25 years, the Internet has become an indispensable tool for users around the world, as well as being a fundamental facilitator of innovation, economic growth, [1] and social benefits. The Internet has become a determining technology of the information age in the same way that the electric motor was the vector of technological transformation during the industrial era [2]. Its structure has boosted the development of various Internet applications and services, also has spurred the proliferation of the Internet to become a global communications infrastructure that reaches 59% of the world population (4.54 billion people) [3]. The fundamental driver of this unique phenomenon has been the open ideological nature of the Internet, an environment where all data flows and applications are treated in an isonomic way, without discrimination, conform to the Network Neutrality - NN principle. The term was introduced by Wu [4] and refers to the principle that all Internet data should be accessed and handled equally without limitation or discrimination, regardless of the source, destination, content, service, application, type, or even device.

In the last years, many bandwidth-demanding applications and services have been developed in concert with the Internet. Those are more sensitive to latency than earlier

M. Botto-Tobar et al. (Eds.): CIT 2020, AISC 1326, pp. 150–161, 2021.
https://doi.org/10.1007/978-3-030-68080-0_11

ones, with different types of content (e.g., voice over Internet Protocol, Internet Protocol Television, audio, and video streaming). Therefore, there is a need for network management techniques. Based on those developments, the relationship between actors in the Internet ecosystem has changed. Many National Regulation Agencies - NRA believed that a new regulative framework would be required to guard user's rights and the Internet as a public, open platform to access information, content, and innovation [5].

Nowadays, Internet Service Providers - ISPs have become gateways for content and applications, as they control final access to consumers. The role played by those who manage the access infrastructure generates a great debate on the application of the principles of openness and non-discrimination at the transport level, and if it should be regulated [6, 7] to safeguard users and providers (content and application – CAPs).

Around the world, government authorities have to analyze the conditions offered by ISPs, the responsibilities they have with their customers, free competition, and innovation. Thus, NN has raised from frustrated regulatory proposals to regulatory policy in many advanced and developing nations [8]. However, unlike developed countries where there are studies about the subject, little has been written about the NN policies implemented in South America. Although there are a couple of works about NN implementation for South American countries [9, 10], they are neither comprehensive nor up-to-date.

Therefore, a study about NN implementation processes in South America is relevant to understand the interpretation of NN and its relationship with providers. Also, to describe aspects such as exceptions to the NN principle, ambiguity in the legal language, and traffic shaping that can affect competence and innovation. And since most of the studies come from developed countries, there is a need to explore implementing rules from other regions like South America, where expanding Internet access is a priority. Thus, as the States must ensure the efficacy of this policy through suitable legislation, we compared the differences between the regulations/laws in each of the researched countries. The rest of the paper is structured as follows. Section 2 presents the methodology. Section 3 shows a general view of the NN in the region. Section 4 discusses the NN policies in South America. Section 5 presents a policy comparison results and a brief discussion. Finally, Sect. 6 concludes the paper.

2 Methodology

The above is a general background of the NN debate about policies. According to Hooton [11], the basis of the critical arguments in the studies has been and continues to be theoretical hypotheses and speculations. And while the theory is relevant to contextualize, empirical research is required to determine its value in an applied policy context. He also mentions that the time is right to ground the debate and studies on NN in empirical aspects.

The objective of this work is to discuss the progress on NN policy implementations and its development process in five South American countries to see where they stand on the NN debate. The research focus is on the real-world context, so the study relies on an empirical approach. The chosen methodology is a multiple case study with a single-unit of analysis, which is appropriate for this study because of the complexity of NN, where the context affects the case in a real situation with many uncontrollable and observable events throughout the time. According to [12], multiple case studies are more comprehensive and allow the construction of more robust results.

The data gathering includes document review and archival analysis covering: records and other papers that describe the historical development of the NN, official website content, and other related reports generated within the context of each case studied here. We use this approach because there is a need to explore and describe the phenomenon of the NN process and the development of the principle in South America.

3 The General Context of Network Neutrality in South America

The NN is a current topic of discussion about Internet policies and is dealt with it in different ways by different countries. Many developed countries have implemented NN policies, usually in response to concerns over ISP potential abuse of market power, blocking or throttling of content. Until 2015 both the regulatory frameworks of the European Union and the Open Internet rules proposed by the Federal Communications Commission - FCC of the United States - US, indicated that NN is a subject of great importance and have a broad regulation scope to guarantee the Internet's open state [13, 14]. However, in the US case, in December 2017, the FCC rolled back the 2015 NN rules: throttling, banning, blocking, and paid prioritization. This resolution came into effect in June 2018. However, a federal court ruled that the FFC couldn't override state-level NN laws, and thus several states have passed local laws until 2020.

As is seen, NN laws change across authorities, but the dependence on transparency responsibilities to support user protection is a common denominator. Other controversial subjects are related to (1) the scope of laws, (2) the definition of what constitutes reasonable and permissible ITM, and (3) paid prioritization [15].

Table 1. The state on the Internet in five countries of South America.

Criteria	Countries				
	CH	CO	BR	EC	AR
Total population (millions)	19.03	50.61	211.8	17.51	44.99
Unique mobile phone connections (millions)	26.32	60.38	205.8	15.65	58.21
Internet users (millions)	15.67	35.00	150.4	12.00	35.09
Internet penetration	82%	69%	71%	69%	78%
Annual grow in the number of internet users	1%	2.9%	6%	0%	2.1%
Average speed of mobile internet connections (MBps)	21.10	18.71	24.79	21.46	25.13
Average speed of fixed internet connections (MBps)	91.46	28.32	48.75	24.79	34.22

Source: Digital 2020 January [3]

On the other hand, many developing countries in South America are considering implementing or already have implemented similar rules. However, these countries have a middle or middle-high percentage of Internet penetration, variable uses focused in big cities, and CAPs charged by ISPs in violation of NN could result in lower prices to users and higher adoption of Internet applications and services [16]. Thus, there could be incentives to apply diverse rules when increasing access is a priority. Table 1 shows the general context of the state of the Internet in five South American countries.

Even when NN is not clear in competitive markets, the market power in South America looks quite different. Two or three traditional operators are rolling out fiber to the premises, and lately, mobile Internet is predominant with two or more mobile operators (Movistar, Claro, and Mobile Virtual Operators). Also, it is relevant to consider that used to exist a semi-monopoly in the paid broadcasting market (e.g., satellite TV provider as DirecTv). However, this aspect has changed, and now there are some Internet services vertically integrated, with the same operators offering their content and that of third parties (e.g., Youtube, Movistar Play, ClaroTv, AmazonVideo, and Netflix). Those services have lower data prices and bring added competition for paid video/TV markets or others such as messaging and calls (Whatsapp, Line, Messenger), or music (Spotify, Deezer), especially for mobile access.

Based on this scenario, there are some means by which the ISPs can infringe the strict NN principles, such as 1) the discrimination (blocking, paid prioritization, zero-rating), and (2) vertically integrated ISPs that may have the ability and incentive to exclude rivals through prioritizing affiliated content over unaffiliated. Those aspects could have complex implications from an economic perspective for competition at ISP and CAP level, for incentives to innovate and invest in networks and content respectively, and ultimately, for consumers. Thus the NN regulations in South America should consider these aspects.

4 Net Neutrality Policies in South America

Given the multilayered scenarios and markets, policymakers in South America have reassessed their policy and regulatory frameworks to adjust them to current and future developments, including principles to ensure NN. Based on the rationales given by the initial countries to do so, they see NN as essential to stimulate competition, promote innovation and ensure that consumers can access any legal service or content provided over the Internet. On NN issues in South America, we concentrate in five countries. Notably, Argentina, Chile, Brazil, Colombia, and Ecuador, have taken decisions to prohibit blocking, throttling, and paid prioritization by ISPs while others are carrying out consultations or still debating on the topic [17].

4.1 Chile

In 2010, Chile was among the first countries to enact a specific law to protect NN. The Network Neutrality Act, Law No. 20.453/2010 in Article 24H, paragraph A, specifies the NN principle. It promotes (1) transparency by asking for the publication of the characteristic of Internet access and quality; (2) guarantees of the Internet service; and (3) prohibits blocking, interfering, discriminating, or disrupting any content, application, or legal service [8, 18]. In this regard, ISPs must offer connectivity and access to the

Internet without distinguishing arbitrarily content or applications by source, type, or ownership, considering the distinct configurations in the access contracts of all users [9].

The Act adopts a flexible view of traffic discrimination, letting providers take actions for network and traffic management, but without affect free competition. In 2013 was promulgated the Decree No. 368 [19] by the Secretariat of the Ministry of Transport and Telecommunication - SUBTEL, which forbids the arbitrary blocking of the Internet's content. But according to article seven of the rule, traffic shaping is regulated, but not prohibited [9]. In this sense, providers only can alter the quality or execute actions for the traffic shaping and net management, as long as those procedures do not harm privacy, the free competition, and were informed through an explicit publication.

It is worth mentioning that, despite the presence of the NN Law and decree, since 2013, several telcos were accused by civil associations of slowing down applications and specific contents without justifying these practices [9]. Huerta states, there is evidence of the absence of control of the obligations of the law, either due to a lack of technical capacity or political will [20]. In a 2015 report in counterpart, statistics about the level of compliance of the law were presented by [21]. It showed around 40 charges between 2011 and 2015 but did not indicate to what type of noncompliance it refers [18].

In recent years non-observance issues are maintained [22], especially with mobile operators that offer access to specific applications and contents with zero rating - ZR, as a sales strategy. For content providers, these strategies undermine free competition and NN. However, regulatory agents mention that commercial aspects such as ZR offers do not arbitrarily distinguish contents based on source or owner, and therefore does not apply discrimination since users can still access content and the Internet. This interpretation seems to favor the idea that technical aspects must meet the Law, and on the contrary, commercial aspects would be outside the scope of the obligations [17]. So, the SUBTEL only requires to telecom companies and ISPs to exclude from their plans the so-called no-charge access to applications and content [22].

Since 2019, based on the rollback of the NN rules in the USA, the debate continues. The SUBTEL has set up working groups with the multi-stakeholders to analyze the ZR, OTT services, and implications of the USA decision considering is a power in terms of content generation, as well as hosting 80% of the world information and data [23].

4.2 Colombia

The legislation that addresses the NN is contained in Article 56 of the Law No. 1450, Chapter II. It established the ISPs cannot block, interfere with, discriminate against, or limit the right of any Internet user to use, send, receive, or offer any legal content, application, or service on the Internet [8, 9] regardless on the origin or ownership. The law also sets the way to make offers according to the market segments' needs, based on the consumption and user profiles, and this could not be assumed as discrimination. Other aspects specify the users right to (1) use any class of legal devices; (2) be offered parental control services for content that violates the law; (3) the transparency of the information about characteristics and guarantees of the service; (4) have tools to protect the privacy of users against network security and viruses; and only block access to specific data upon express request of the user [24].

In addition to the Law, Resolution 3502 of the Colombian regulator, passed in 2011 generating criticisms from users, experts, and civic organizations who found in the text of the rule, inaccuracies, gaps, and ambiguities that could give operators a way of avoiding certain aspects of the NN. For some critics [25], the controversies in that process were due to the absence of the participation from civil society and academia, and therefore the vision of the operators influenced its development. The resolution itself allows providers to offer different prices for access according to the needs of the market, that is, to offer plans based on contents and applications the ISPs want to provide, through third parties, ZR, or vertical integration. Moreover, allow the implementation of reasonable and non-discriminatory traffic management by establishing QoS levels and prioritize traffic types according to latency and delay for access to the Internet. Also, clarify the requirements for content blocking due to security purposes that meet technical requirements or standards adopted [26]. Aspects that are banned or controlled in other NN laws [27].

The exception on Internet plans limited by content makes the resolution not faithful, ambiguous, and directly contradicts the Law since to limit content, applications, or services according to market deals or consumption profiles is to discriminate or prioritize the data, and make available a "reduced Internet" [18, 25]. Finally, Resolution 3502 regulates the relationship of consumers with ISPs, but not with content or application providers [8].

The Colombian Ministry of ICT started in 2016, a multi-stakeholder consultation to develop a new policy and regulatory framework for telecommunications and broadcasting to deal with convergence in communications markets [9]. Nowadays, Colombia has formulated the project "The role of OTT services" within the initiatives for the Regulatory Agenda 2018–2019. The objective is to review the regulatory trends for the digital economy, where analyzed and complement the regulatory framework on NN that allows offers and market prices oriented to costs, protection of consumers' interests, as well as updating the operating license conditions of existing operators and service providers [28].

4.3 Brazil

In 2014, after almost five years of discussion, the Congress passed the Marco Civil da Internet - MCI, Law No. 12.965/2014 which, among consolidating rights, duties, and principles for the use and development of the Internet, enshrined the principle of NN and privacy of Internet users [9]. Its importance lies not only in its principles but also in the way in which it was drafted, based on an open, public, and collaborative consultation process, implemented across the country that helped to construct the text of the Law.

The law sets (1) the obligations of transparency of information, by asking providers for clear and complete information on service contracts, including details of the data protection, mitigation, and network management; (2) non-blocking, by affirming that the providers must treat equally any data packages, despite origin and destination, content, service, or application; and (3) for anticompetitive practices, by offering services in non-discriminatory commercial conditions [8].

The law goes on to prohibit traffic discrimination or degradation, which can only be implemented as a result of essential technical requirements, and prioritization of emergency services. The Decree No. 8.771/2016 defines the "essential technical requirements" as network security issues, and exceptional situations of congestion or interruption; while "emergency services" as communications destined to the emergency services providers, or communications to inform the population in a disaster situation, emergency risk or state of public calamity [29]. The Decree also sought to regulate that commercial relations between the infrastructure operator - IO and the actors of the logical layer cannot "compromise the public and unrestricted character of Internet access", "prioritize data packages due to commercial arrangements" or benefit applications offered by the IO [18].

The MCI is considered one of the most advanced laws in the world and shows a significant effort to adapt fundamental rights and freedoms to the digital world, incorporating aspects to eliminate ambiguities and violations towards the NN that have the potential to stop innovation and competition. As a novelty, the decree prohibits ZR strategies, includes a tripartite system to supervise, and determines that content companies must have legal headquarters in Brazil [10]. However, when in 2018, consumer associations and other civil society organizations evaluated ZR practices as negative and claimed that violate NN. Private operators and regulatory agents claimed that NN was particularly restricted to the logical level and infrastructure of the network, so the market methods do not infringe NN [30].

4.4 Ecuador

The new Telecommunications Act in Ecuador - LOT, enacted in 2015, embraces the opportunities of convergence and stipulates in its Article 12 that the state will "propel the establishment and exploitation of telecommunication networks and services that promote the convergence of services, in conformity with public interests and with the dispositions of the Act and it's normative" [8, 17]. Also, included clauses specifically addressing the NN and its promotion (Article 3 and 4); and the rights that users and subscribers of Internet services have, indicating that they can access to any legal application or permitted content available (Article 22).

The Act prohibits blocking, throttling, prioritizing, or restricting the user's right to send, receive, use or offer any legal data, applications, service, or content through the Internet, their networks, or other ICT technologies. Also forbids to restrict the right to include or use any class of equipment or device on the network, whenever legal, except for cases established under the legal framework and those in which the competent authority decides, or when the client, subscriber, or user expressly demands the limitation or blocking of content. Providers are allowed to carry technical actions to manage their networks when considered necessary, and within the exclusive scope of its activities to guarantee the provision of service [31].

According to the new Act, the regulator body ARCOTEL will be responsible for regulations and norms that allow the provision of multiple services on the same network to drive, in an effective manner, the convergence of services and assist with the technological development in the country, following the principle of NN [17]. However, the LOT has omissions such as: not defining what NN is, using the concept issued by the ITU, which is ambiguous and not very applicable. It also ignores the meaning of the principle and allows the Art. 66 of the Law, in which service providers establish "tariff plans constituted by one or several services, or products of service." Besides, unlike the policy that previously defined the NN (TEL-477-16), which included the concepts of non-discrimination, the Law has not included them. All those aspects violate the NN and leave users defenseless against the potential abuses from providers [32]. As an example, there was a lawsuit against mobile operators for offering unlimited and free WhatsApp service, but these offers did not include the voice call functionality that normally is integrated. Regarding the situation, the Minister of Telecommunications stated that this practice of not including that function in WhatsApp's ZR packages went against the principle of NN and will investigate the facts and take actions [33]. However, even today mobile operators still offer unlimited use of some services and certain applications along with their limited data plans.

4.5 Argentine

In 2014, the Law 27.078 "Argentine Digital Law" passed, this new law governs telecommunication in the country and replaces the National Law of Telecommunications 19.798/72 and Decree 764/00 [34]. The law was an important update and established the state must ensure quality, accessibility, and affordability to the Internet. It also must guarantee the ICT services and ensure their conditions of quality, affordability, and at fair and reasonable prices, encouraging competition between service providers so that consumers can choose the best for them Article 1 established and guarantee NN as an objective. While Article 56 states the rights of users to access, send, receive, use or offer any application, service, or content without any restriction, discrimination, distinction, interference, obstruction, or degradation [34]. Article 57 prohibits ICT operators to (1) block or discriminate access or use of any content, application, or service except by court order. (2) Set the price of Internet access under its content, services, or applications; and (3) arbitrarily limit the right to access to the Internet using any software or hardware, whereas those do not harm the network.

Although the law was modified in 2015 by the Decree of Necessity and Urgency 267/2015, the changes did not modify the provisions of articles 56 and 57 about NN. However, it generated controversies and criticisms, due to the Law itself ensures ICT services that in practice should be Internet, content, or application providers, but in its disaggregation, it seems to only refer to ISPs. Also, neither the zero-rating aspects are specified nor do procedures to ensure NN [10], issues that affect free competition.

Since 2017 mobile operators have been offering free traffic for some applications and services, within this strategy providers paid third-party services to position their offer, and even one of them, integrated vertically, favored its content. This practice is an obstacle for other existing applications it creates an entry barrier for new ones, affecting competition and innovation [35]. Finally, it is important to note that in 2018 there was a lawsuit between operators claiming that a video ZR application of one of them, during the Soccer World Cup, violated the NN. The regulatory agent concluded that the promotion of that application was tolerable as long as it is applied to other similar applications as well [35]. This aspect implies that the argentine regulator is not independent or impartial, being more permeable to the interests and competition of companies.

5 Results and Discussion

The NN is a case of the required policy response to digital development since it is a principle for regulating the relationship among providers and users related to Internet access. Although rules related to NN have been enacted in some countries in the region, their implementation may vary for fixed or mobile networks and even commercial developments. Table 2 shows the list of countries in this study and the differences between the regulations or laws.

Table 2. Differences between regulations.

Criteria	Countries				
	CH	CO	BR	EC	AR
Define NN and other concepts	•	—	•	—	—
Guarantees to access or use of services or content	•	•	•	•	•
Prohibits to block or discriminate access or use	•	•	•	•	•
Transparency of the Information	•	•	•	•	•
Use of HW to access	•	•	•	•	•
Set price by content, service, or market	—	•	▲	•	▲
Implement ITM techniques	•	•	▲	•	•
Exceptions (technical requirements, emergency)	—	—	•	—	—
Zero Rating	▲	—	▲	—	—
Control of the obligations	•	—	•	—	•

• Specified, — Not Specified, ▲ Prohibits

The review shows that the NN in South America focuses on transparency of information. However, most of the time the terms of the service are not clear for users. On the other hand, addressing these challenges in some countries is complex because NN laws are ambiguous and contribute to regulatory uncertainty. Thus, the three most significant challenges that NN rules have are:

1) To determine the reach of regulations. Because the objective of the NN is to protect the open Internet and does not refer to all IP-based traffic. Thus, it is necessary to precisely what type of data flow and market players apply the regulation.
2) The definition of what is acceptable network management. Even when the NN rules are similar across jurisdictions, they differ and do not define how to deal with the arbitrary traffic management or the exceptions to the principle.
3) Zero-Rating is considered as a commercial aspect, and that NN is limited to the network and transport layer of the Internet, so the market practices do not infringe NN, since ZR offers do not arbitrarily distinguish contents, based on source or owner.

On the other hand, the development of streaming (audio and video), Internet of Things, Artificial Intelligence applications in the network need new devices and quality of service - QoS. This integration is allowed in NN rules as long as they are legal devices and do not affect the QoS of the network. However, the laws seek to establish exceptions to allow ITM and the provision of QoS to users, applications, and services, but no pending between reasonableness criteria and exceptions explicitly placed in policies. Thus allowing ISPs to interfere and control Internet traffic arbitrarily, and could blocking new services for technical o commercial reasons.

The level of compliance with the laws has problems with repeated transgressions. A case concerning how this principle could be affected is related to the ZR plans in the mobile Internet service, through which providers offer access to certain applications and services with no charge for the data used. Since it is not possible to do so without the necessary discrimination of data by origin, destination, or content, this practice acts against the NN. However, there could be reasons to implement other rules when increasing digital access is a priority in different countries of the region like Colombia, Ecuador, and Argentine.

Finally, an imperative condition for the efficient implementation of NN laws is the availability of suitable mechanisms that permit the reliable detection and consequent sanction of anticompetitive practices on the Internet, a failing aspect in South America either due to a lack of technical capacity or political will.

6 Conclusions

This paper provides an overview of the NN rules in five countries in South America. We discovered the following facts about NN. (1) The Chilean Law was the first NN Law in the world, and the Brazilian is considered one of the most advanced for adapting fundamental rights and freedoms to the digital world, even when they are not perfect, are used as a base for others. (2) The NN rules in the region are ambiguous and contribute to regulatory uncertainty, which opens gaps that violate the NN principle. (3) Colombia, Ecuador, and Argentine permit commercial arrangements to privilege applications or services. (4) There is a lack of the definition of what constitutes reasonable network management and what constitutes acceptable ITM.

Commercial market strategies are becoming increasingly popular among operators in those countries, ZR issue is one of them and is currently being debated by some authorities. The exception framework based on the market, presented in the policies and laws, allow that despite establishing prohibitions of discriminatory behavior, instead of open gaps to offer ZR plans that in practice violate the strict principle of NN. This aspect affects free competition and innovation, and although could help to lower prices and to generate greater adoption of the Internet in the region, it needs further research.

Although in South America each nation has a specific market needs to deal with, they are adapting the laws to preserve an open Internet, guarantee access and quality for users, and facing a challenge to create frameworks able to adjust to the technology trends and dynamic market. A suitable regulatory regime may pave the way for innovation and technologies and is the key to a prosperous digital economy.

Finally, the growing processes to implement the NN and the number of nations adopting it confirms its importance. Thus the NN as a principle is likely to be a topic of continuous debates, agreements, and resolutions soon in all South America. And as in other countries, scholars should analyze the NN implications on local markets because neither this principle of nondiscrimination nor its discussions will disappear due to its nature and the significance of the Internet in contemporary society.

References

1. Internet Society: Internet Governance (2016). https://www.internetsociety.org/issues/internet-governance/. Accessed 24 July 2020
2. Castells, M.: The impact of the Internet in Society: A global perspective. BBVA (2013). https://www.bbvaopenmind.com/wp-content/uploads/2014/03/BBVA-Comunicaci%C3%B3n-Cultura-Manuel-Castells-El-impacto-de-internet-en-la-sociedad-una-perspectiva-global.pdf. Accessed 15 Aug 2020
3. Wearesocial, Digital 2020: Global Digital Yearbook. https://wearesocial.com/digital-2020. Accessed 23 Aug 2020
4. Wu, T.: Network neutrality, broadband discrimination. SSRN Scholarly Paper, Social Science Research Network, Rochester, NY (2013)
5. Stocker, V., Smaragdakis, G., Lehr, W.: The state of network neutrality regulation. ACM SIGCOMM Comput. Commun. Rev. **50**, 45–59 (2020)
6. Krämer, J., Wiewiorra, L., Weinhardt, C.: Net neutrality: a progress report. Telecommun. Policy **37**(9), 794–813 (2013)
7. Hahn, R., Wallsten, S.: The economics of net neutrality. Econ. Voice **3**(6), 1–7 (2006). https://www.bepress.com/ev/vol3/iss6/art8
8. Triviño, R., Franco, A., Ochoa, L.: Net neutrality regulation in South America: a progress review. Latin-Am. J. Comput. LAJC **6**(1), 17–26 (2019)
9. Vargas-Leon, P.: Net neutrality: an overview of enacted laws in South America. In: Belli, L., De Filippi, P. (eds.) Net Neutrality Compendium. Springer, Cham (2016). ISBN 978-3-319-26424-0
10. Carboni, O., Labate, C.: America Latina por uma rede neutral: o principio da neutralidade em Chile e Brasil. Revista Famecos - Midia Cultura e Tecnologia **25**(2), 1–22 (2018)
11. Hooton, C.: An empirical investigation of the impacts of net neutrality. Internet Association. https://internetassociation.org/reports/an-empirical-investigation-of-the-impacts-of-net-neutrality/. Accessed 29 Aug 2020
12. Yin, R.: Case Study Research: Design and Methods, 4th edn. SAGE, Los Angeles (2009)
13. EU – European Union: Official J. Eur. Union L310, **58** (2015). https://eurlex.europa.eu/legal-content/EN/TXT/PDF/?uri=OJ:L:2015:310:FULL&from=EN. Accessed 04 June 2020
14. FCC - Federal Communications Commission: Open Internet, Comisión Federal de Comunicaciones (2015). https://www.fcc.gov/general/open-internet. Accessed 04 Apr 2020
15. Stocker, V., Knieps, G.: Network neutrality through the lens of network economics. Rev. Netw. Econ. **17**(3), 115–150 (2019)
16. Robb, G., Hawthorne, R.: Net neutrality and market power: the case of South Africa. Telecommun. Policy **43**, 101814 (2019)
17. Organization for Economic Cooperation and Development, Inter-American Development Bank: Broadband Policies for Latin America and the Caribbean: A Digital Economy Toolkit (2016). https://publications.iadb.org/en/broadband-policies-latin-america-and-caribbean-digital-economy-toolkit. Accessed 12 Apr 2020
18. Derechos Digitales: Neutralidad de red en América Latina: Reglamentación, aplicación de la ley y perspectivas, 1era. Edición, Sao Paulo, Brasil (2017)
19. MTT – Ministerio de Transportes y Telecomunicaciones y SUBTEL – Subsecretaría de Telecomunicaciones. Reglamento que regula las características y condiciones de la neutralidad de la red en el servicio de acceso a Internet, Biblioteca del Congreso Nacional de Chile. Biblioteca del Congreso Nacional de Chile, Santiago de Chile, Chile. https://www.leychile.cl/Navegar?idNorma=1023845. Accessed 10 May 2020
20. Huerta, J.: La neutralidad de la red en Chile: ¿Mucha sofisticación para la SUBTEL?, Digital Rights - Latin America & The Caribbean, no. 30 (2013). https://www.digitalrightslac.net/es/la-neutralidad-de-la-red-en-chile-mucha-sofisticacion-para-la-SUBTEL/

21. Huichalaf, P.: La Neutralidad de la Red: El Caso Chileno (2015). https://www.regulatel.org/wordpress/wp-content/uploads/2015/07/4.Neutralidad_de_la_red_version%20final.pdf. Accessed 21 Jan 2020
22. Layton, R.: Which Open Internet Framework is Best for Mobile App Innovation an empirical inquiry of net neutrality rules around the world. Aalborg Universitetsforlag. Ph.D.-serien for Det Tekniske Fakultet for IT og Design, Aalborg Universitet (2017). https://doi.org/10.5278/vbn.phd.engsci.00181
23. SUBTEL - Secretaria de Telecomunicaciones de Chile, Subtel establece mesa de trabajo para analizar implicancias del fin de la Neutralidad de Red en EE.UU (2017). https://www.subtel.gob.cl/subtel-establece-mesa-de-trabajo-para-analizar-implicancias-del-fin-de-la-neutralidad-de-red-en-ee-uu/. Accessed 02 Feb 2020
24. Congreso de la República de Colombia, Plan Nacional de Desarrollo: 2010–2014 "Todos por un nuevo país" (2011). https://www.mincit.gov.co/loader.php?lServicio=Documentos&lFuncion=verPdf&id=78676&name=LEY_1753_DE_2015.pdf&prefijo=file
25. Fundación Karisma: Como se contrata en América Latina el acceso a Internet? que tiene que ver con la neutralidad de la red? Karisma, Bogotá (2016)
26. CRC – Comisión de Regulación de Comunicaciones, Neutralidad de Red (2011). https://www.crcom.gov.co/es/pagina/neutralidad-en-internet. Accessed 04 Mar 2020
27. Casasbuenas, J.: Neutralidad en Internet: Factor que potencia el cierre de la brecha social digital, COLNODO (2013). https://colnodo.apc.org/es/novedades/neutralidad-en-internet-factor-que-potencia-el-cierre-de-la-brecha-social-digital. Accessed 06 Mar 2020
28. CRC – Comisión de Regulación de Comunicaciones El rol de los servicios OTT en el sector de las comunicaciones en Colombia, impactos y perspectivas regulatorias (2018). https://www.crcom.gov.co/uploads/images/files/Documento%20de%20consulta%20publica%20definicion%20de%20metodologia%20OTT%20rev%20PUB.pdf
29. Cámara de Diputados de Brasil, Decreto No. 8.771 Reglamentación del Marco Civil de Internet (2016). https://www2.camara.leg.br/legin/fed/decret/2016/decreto-8771-11-maio-2016-783094-publicacaooriginal-150360-pe.html. Accessed 12 Apr 2020
30. UNESCO: Assessing Internet development in Brazil using UNESCO's Internet Universality ROAM-X Indicators (2019). https://cetic.br/media/docs/publicacoes/8/20200117094619/Assessing_Internet_Development_in_Brazil.pdf
31. MINTEL – Ministerio de Telecomunicaciones del Ecuador. Ley Orgánica de Telecomunicaciones, Registro Oficial 439 (2015). https://www.telecomunicaciones.gob.ec/wp-content/uploads/downloads/2016/05/Ley-Org%C3%A1nica-de-Telecomunicaciones.pdf
32. Delgado, A.: Ecuador: la ley de telecomunicaciones afecta la neutralidad de la red, FayerWayer (2015) https://www.fayerwayer.com/2015/02/ecuador-la-ley-de-telecomunicaciones-afecta-a-la-neutralidad-de-la-red/. Accessed 15 Jan 2020
33. Carrillo, A.: Protección a la neutralidad de la red en Ecuador, Regulación de Internet y derechos digitales en Ecuador, pp. 77–92. Editorial USFQ, Quito (2016)
34. Gendler, M.: La Neutralidad de la Red: Debates y acciones en Argentina. In: XXXI Congreso ALAS, Uruguay (2017)
35. Lufrano, L.: El primero te lo regalan: zero-rating de las operadoras móviles de la Argentina. Revista Question **1**(62), e163 (2019)

Design Techniques of FTTH-GPON Networks for Segmentation and Data Traffic Relief

Pablo Loayza-Valarezo[1], Javier Guaña-Moya[2](✉), and Patricia Acosta-Vargas[3]

[1] Telecommunications Consultant, AndinaGestión S.A., Quito 170503, Ecuador
pablo.loayza@andinagestion.com
[2] Faculty of Engineering, Pontifical Catholic University of Ecuador, Quito 170523, Ecuador
eguana953@puce.edu.ec
[3] Intelligent and Interactive Systems Laboratory, Universidad de Las Américas, Quito 170125, Ecuador
patricia.acosta@udla.edu.ec

Abstract. Today, the exponential increase in subscriber data usage, in short time frames, is causing Internet Service Providers (ISPs) to deploy or migrate their networks to Fiber to the Home (Gigabit) passive optical network technology meet growing subscriber demand. However, in Ecuador, due to a lack of technical knowledge to reduce implementation costs, many of these ISPs deploy their fiber networks without Quality of Service (QoS) designs or subscriber growth. In this document, we propose specific and sequential sizing techniques for the ISP's core equipment. This article can serve as a guide for future studies related to FTTH-GPON network design for segmentation and data traffic.

Keywords: Data traffic · FTTH · GPON · Subscriber's growth · QoS

1 Introduction

At present, the technological progress in the telecommunications field is advancing with significant leaps. The average day for many individuals, in their personal and professional lives, require the use of technological tools for communication and information. Moreover, for these tools to be more effective, they depend on faster and reliable interchange of data frames [1].

In response to this growth, the Telecommunications Control and Regulation Agency (ARCOTEL, by its Spanish acronym) has issued permits to many small ISPs nationwide to satisfy the increased demand.

However, many of these ISPs deploy their Fiber To The Home – Gigabit Passive Optical Network (FTTH-GPON) networks with designs focusing solely on revenue rather than QoS, or even worse, preceding any method whatsoever. For these types of ISPs, their primary goal is to add more subscribers, in the short term, with minimal capital investment. Due to the absence of QoS design, this model jeopardizes the availability and speed of the service, over time, because the data traffic increases with the growth of subscribers and associated bandwidth [1].

M. Botto-Tobar et al. (Eds.): CIT 2020, AISC 1326, pp. 162–176, 2021.
https://doi.org/10.1007/978-3-030-68080-0_12

The rest of this article is created as follows: in Sect. 2, we present a review of the related work to previous studies of FTTH-GPON's designing methods; in Sect. 3, we referred to a case study and the applied approach; in Sect. 4, we offer the results and discussion, and in Sect. 5 we present the conclusions and future work.

2 Related Work

We found several previous studies on the optimization of telecommunications networks that are related to fiber-optic network designs. Poon, K. et al. (2006) article: "*Designing optimal FTTH and PON networks using new automatic methods*"; described a hybrid method called: Expert System and Evolutionary Computing techniques; which was used to produce 'most fit' PON designs taking account the planning process, design factors and optimization goals.

Al-Quzwini, M. (2014) article: "*Design and implementation of a Fiber To The Home FTTH access network based on GPON*"; described a step-by-step design method of a protected GPON FTTH access network for 1,000 subscribers. Lokhande, M. & Singh, A. (2017) article: "*Design and implementation of FTTH*"; described a field implementation design for a GPON Network using computing tools such as GIS and AutoCAD.

And lastly, we used the core resulting data of a case study of one of our previous articles: "*Data gathering methodological guide for designing FTTH-GPON networks with QoS*" [7]; into the equipment sizing methodology of this article.

2.1 Architecture, Access Mode, and Encapsulation

The most used type of architecture in FTTH networks is Point to Multipoint (P2MP) using a tree topology and, within the network, passive optic elements (with no need of electrical source) to help distribute data frames to its subscribers [3].

FTTH-GPON adopts Time Division Multiplex (TDM) and Time Division Multiple Access (TDMA) technology for access mode to download and upload data, respectively [4].

Data encapsulation is done in two stages: the first one via GPON Encapsulation Method (GEM mode) to encapsulate Ethernet data frames, and the second by GPON Transmission Convergence (GTC) mode to encapsulate GEM, ATM, and TDM traffic [5].

2.2 Data Frame Download Operation

The Optical Line Terminal (OLT) encapsulates Ethernet, ATM, and TDM data, through GEM and GTC, to later on broadcast them in a chain of data frames to all Optical Network Terminal (ONT) units connected to the Optical Distribution Network (ODN) [3]. The chain of data frames is transmitted through fiber-optic threads, of the feeder cable, within the correspondent download wavelength following the ITU-T G.984 standard.

Once the chain of data frames reaches the first optic passive element (i.e., coupler or splitter), they are all distributed, alike, to all ONTs; meaning that the same information will receive the ONT1 like the ONTn. Downloaded data frames are structured by Physical

Control Block downstream (PCBd) and Payload (GEM Header and GEM/GTC data) [4]. In the chain of data frames, PCBd is used by each ONT to ex-tract its corresponding Payload by TDM access mode.

The OLT sets the data period transmission value at 125µs per frame. If the OLT does not need to send downstream data, the chain of frames will still be transmitted to all ONTs for synchronization purposes [4].

2.3 Data Frame Upload Operation

For upstream data, the OLT generates a bandwidth map (BM), which establishes variable slots of time for each ONT to receive data frames. The length and position of these slots, within the BM, depends on the type of data traffic, classified by Class and Quality of Service (CoS, QoS) [5]. Once the BM is established, the ONTs send bursts of data frames, within their assigned time slots, to the OLT, which are queued and processed by the TDMA access mode according to CoS and T- CONTs frame [5].

Upstream data frames are structured by (Physical Layer Overhead upstream (PLOu), Alloc-IDs, Dynamic Bandwidth Report upstream (DBRu), and Payload [5]. Additionally, it can contain Physical Layer Operation, Administration, and Management upstream (PLOAMu) and Power Level Sequence upstream (PLSu) [4].

All of the ONTs, within the ODN, are installed in different physical locations; hence there may be cases where data framing collisions may occur caused by a delay time gap [5]. To prevent this from happening, the OLT executes a scoping process; once the ONT is registered in the ODN, it calculates a delay time based on the physical distance of each unit and then transmits this information to the other units. This process is called Equalization Delay (ED) [6].

3 Case Study and Method

3.1 Case Study

In Ecuador, a country located in Latin America, the number of fiber broadband subscribers has increased considerably from 2015 to 2019 (see Fig. 1) [2].

Fig. 1. Ecuador's fiber broadband subscribers' growth.

To show real case dimensioning, we had to refer to a case study of one of our previous articles: "*Data gathering methodological guide for designing FTTH-GPON networks with QoS*" [7]. Therefore, all calculations shown in this section are based on real and reliable data.

3.2 Methodology

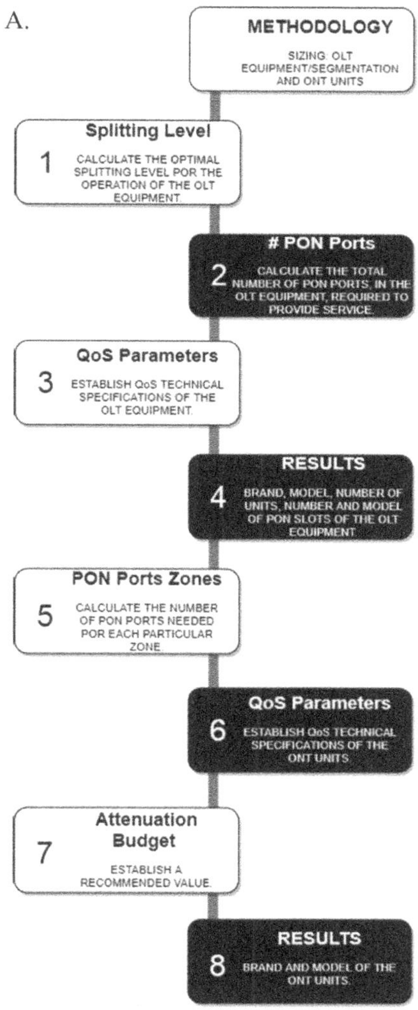

Fig. 2. Sizing methodology of OLT and ONT parameters.

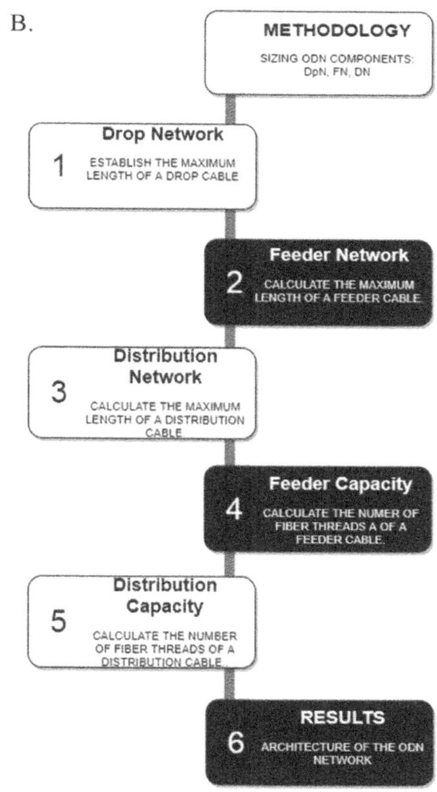

Fig. 3. Sizing methodology of ODN components.

The methodology of sizing the FTTH-GPON's equipment is executed in two processes: A. OLT and ONT technical specifications; B. ODN technical specifications; as shown in Fig. 2 and Fig. 3.

4 Results and Discussion

A. OLT and ONT Technical Specifications

OLT Technical Parameters

1. First, calculate the optimal level of division of the OLT's PON (Passive Optical Network) ports. The GPON Optical Line Terminal equipment can be configured in three splitting levels: 1:32, 1:64, and 1:128 (each PON port can provide service to 32, 64 and 128 subscribers). This process is where most ISPs, in the absence of QoS criteria, choose to provide service to more subscribers, with a single port, without paying attention to bandwidth and subscriber's growth.

The optimal splitting level is calculated below with the following equation:

$$BW_{proj} \leq \frac{DS_{GPON}}{SR} \tag{1}$$

Where,

BW_{proj} = Home Projected Bandwidth
DS_{GPON} = GPON Downstream Rate
SR = Split Ratio

Thus,

$$BW_{proj} = BW_{actual} \cdot Co_{BW} \tag{2}$$

$BW_{proj} = 20.00\,\text{Mbps} \cdot 1.43$; Section 4.3 [7].
$BW_{proj} = 28.60\,\text{Mbps}$

1:128 splitting level verification:

$$BW_{proj} \leq \frac{DS_{GPON}}{SR} \tag{3}$$

$28.60\,\text{Mbps} \leq \frac{2500\,\text{Mbps}}{128}$
$28.60\,\text{Mbps} \leq 19.53$ Mbps; It does not comply.

Hence, if the splitting level is configured to 1:128, each port could not provide an average BW_{proj} of 28.60 Mbps to each of the 128 subscribers.

1:64 splitting level verification:

$$BW_{proj} \leq \frac{DS_{GPON}}{SR} \tag{4}$$

$28.60\,\text{Mbps} \leq \frac{2500\,\text{Mbps}}{64}$
$28.60\,\text{Mbps} \leq 39.06\,\text{Mbps}$; it does comply.

For the case study [7], the 1:64 splitting level is optimal because it can provide an average BW_{proj} of almost 40 Mbps to each of the 64 subscribers with a single PON port.

The calculations shown above are an effective technique of data traffic congestion relief because choosing a splitting level of 1:64 over a 1:128 lessens: GTC's Payload and the period data transmission of the chain of frames. See Fig. 4.

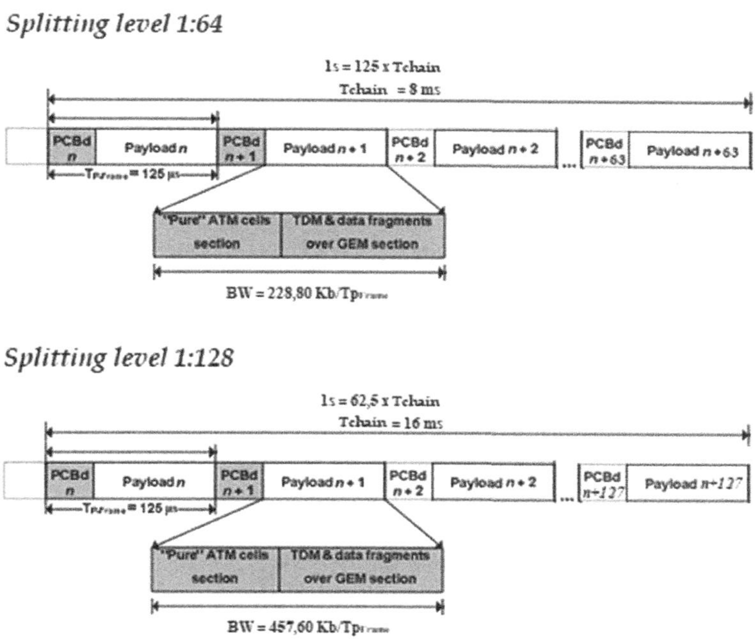

Fig. 4. Downstream data traffic relief.

2. Now, the number of PON ports required to provide service to the projected population has to be calculated. An example of which is shown below:

$$Pop_{proj} = Pop_{actual} \cdot Co_{FTTH} \tag{5}$$

$Pop_{proj} = 21,517 \cdot 1.82$; Sections 4.3 and 4.6 [7].
$Pop_{proj} = 39,160$ subscribers

Thus,

$$\#PON = \frac{Pop_{proj}}{SR} \tag{6}$$

$\#\text{PON} \leq \frac{39,160}{64}$
$\#\text{PON} = 611.87 \approx 612$

3. Once the number of PON ports is set, the QoS technical specifications must also be established [8, 9]:

   - QoS: Bandwidth management, latency, jitter, and packet loss rate;
   - QoS Model: DiffServ Model;
   - DiffServ Components: Traffic classification, traffic policing and shaping, congestion management, congestion avoidance;
   - ACL (Access Control List);
   - Multi-level scheduling mechanism: HQoS (Hierarchical Quality of Service);
   - Number of T-CONT: 16 per subscriber;
   - Number of GEM ports: 3872 per PON port;
   - Scalability: XG-PON1, XG-PON2, and NG-PON.

   The specifications 1 to 5, detailed above, have to be software configured in the OLT, within the time range, with the most data traffic congestion occurs: 6 PM–9 PM, according to the case study [7].
   GPON Technology allows a maximum physical distance of 20 km between OLT and ONT per PON port with the following ranges of link attenuation budgets [10]:

   - Class A: 5 dB–20 dB (Splitting level 1:32).
   - Class B: 10 dB–25 dB (Splitting level 1:64).
   - Class B+: 13 dB–28 dB (Splitting level 1:64).
   - Class C: 15 dB–30 dB (Splitting level 1:64).
   - Class C+: 17 dB–32 dB (Splitting level 1:128).
   - Class C++: 20 dB–35 dB (Splitting level 1:128).

   For the case study [7], the following specifications are established [11]:

   - Link Optical Budget: Class B+;
   - Reach Method: RTD (Round Trip Delay) and EqD (Equalization Delay);
   - The maximum distance between to ONTs in the same PON port: 40 km;
   - Configuration of a quiet zone for RTD and EqD in the time zone: 6 PM–9 PM.

4. Results

Once the QoS technical specifications are set, the next step is to determine the brand, model, and the number of units needed to provide service. For the case study [7], the following specifications are set: Huawei, Model MA5800-X17, 3 OLTs, 41 slots Model H902GPHF with 16 PON ports [12].

Identification and Segmentation OLT Data Traffic

To easily identify a corresponding data traffic zone, the segmentation of PON ports must be executed. This process means that each port must provide service to only one area. However, it is expected, among several ISP's FTTH designs, to provide service to many zones, regardless of proximity, to use the PON port's full capacity.

This process can create complications, operation wise because it will make it difficult to accurately identify an affected port during service problems due to data congestion. On the other hand, the one-port-to-zone approach maximizes the efficiency of subscriber growth as it allows for proper planning for the next PON port. Calculations and results are shown below and in Fig. 5.

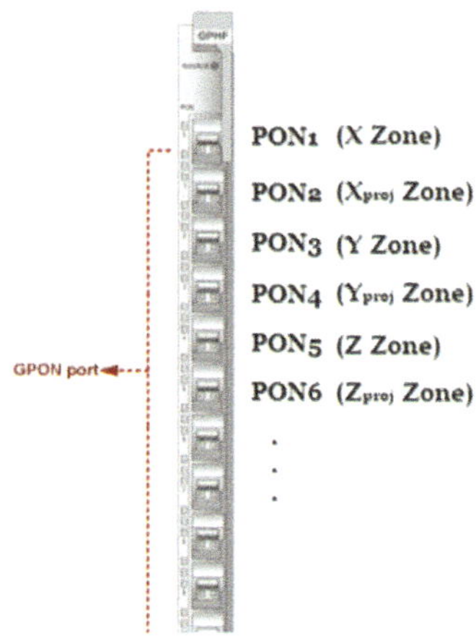

Fig. 5. Data traffic identification and segmentation.

$\#\text{SubscribersPON1} = 64(\text{X Zone})$

5. Then,

$$\#Subscribers_{proj}(XZone) = \#SubscribersPON_1 \cdot Co_{FTTH} \tag{7}$$

$\#\text{Subscribers}_{\text{proj}}(\text{X Zone}) = 64 \cdot 1.82$
$\#\text{Subscribers}_{\text{proj}}(\text{X Zone}) = 116.48 \approx 117$

Hence,
$\#SubscribersPON_2 = 53$(Xproj Zone); 11 subscribers left for CoBW.

ONT Technical Parameters

6. The QoS technical specifications for data traffic relief for the case study [7] are shown below [13]:

- Standard: ITU-T G.984.2 Class B+;
- Maximum reach: 20 km;
- Reception sensibility: −27 dBm;
- The flexible mapping between GEM and T-CONT ports;
- BW map: SR-DBA (Status Reporting - Dynamic Bandwidth Allocation) y NSR-DBA (Non-Status Reporting – Dynamic Bandwidth Allocation);
- Broadcast packet rate limitation;
- Queue data traffic relieves SP (Strict Priority) and WRR (Weighted Random Robin).

In the QoS technical specifications, shown above, the reception sensibility and reach were determined with the ONT's default parameters: −27 dBm and 20 km for maximum distance, respectively. However, for purposes of design efficiency, it is not recommended to use these values because they are geared towards maximum limits. Using alternative values is an excellent way to avoid future deficiencies in the service, especially during peak data traffic, because of collisions and loss of data frames.

7. The recommended value for the link attenuation budget is 25 dB, and the recommended length is shown in the next section [14].
8. Results

Once the QoS technical specifications are set, the next step is to determine the brand and model needed to provide service. For the case study [7], the following specifications are Huawei, Model HG8045H [13].

B. **ODN Sizing and Technical Specifications**

ODN Technical Parameters
This section shows the ODN's link attenuation budget and the calculations to determine maximum distances: the feeder network, distribution, and drop networks; for a 1:64 splitting level.

1. At first hand, we have established the following:
$DpN(DropNetwork)length = 0.25$ km; For maintenance purposes [14].

2. For the Feeder Network (FN), the maximum length of a fiber cable reel is 5 km, and in the case study [7], the Central Office (CO) is located at ¾ of a distance from the service area. Therefore, the Feeder Network's (FN1) length is set at 4 km. For the extension of the Feeder Network (FN2), a maximum distance of 1 km has been determined.

Then,

$$FN = FN_1 + FN_2 \tag{8}$$

FN = 4.00 km + 1km
FN = 5.00 km

Table 1. Link attenuation budget for a splitting level 1:64 [13]

Name	Type	Loss (dB)	Quantity	Total loss (dB)
Connection point	Pigtail/Patchcord	0.50	4.00	2.00
	Melt splicing	0.10	6.00	0.60
	Quick connector	0.60	1.00	0.60
Optical splitter	1:2	3.25	2.00	19.50
	1:4	6.50		
	1:8	9.75		
	1:16	13.00		
	1:32	16.25		
	1:64	19.50		
Optical cable (G652.D)	1310 nm (dB/Km)	0.35	FN + DN + DpN	
	1490 nm (dB/Km)	0.30		
	1550 nm (dB/Km)	0.25		
TOTAL				**22.70**

The ODN link attenuation budget, shown in Table 1, is broken down by element. The total attenuation loss, not including cables' length, is 22.70 dB. The recommended total attenuation loss is at 25 dB, which leaves 2.30 dB left to calculate the physical cables' lengths.

3. Given the values above, the following calculations must be used:

Loss for length of cable by km (1310 nm) = 0.35 dB.
dB left to recommended link attenuation budget = 2.30 dB.

Therefore,

$$Km\ Cable\ ODN = {}^{dB\ left}/_{Loss}\ \text{km } 1310\,\text{nm} \tag{9}$$

Km Cable ODN $= {}^{2.30\ \text{dB}}/_{0.35\ \text{dB}}$
Km Cable ODN $= 6.57$

Where,
DpN $= 0.25$ km
Hence,

$$KmODN = FN + DN + DpN \tag{10}$$

DN $=$ Km ODN $-$ FN $-$ DpN
DN $= 6.57$ km $- 5.00$ km $- 0.25$ km
DN $= 1.32$ km

Feeder Network Technical Capacity (FN)
To calculate the capacity of the fiber power cables, the total number of PON ports for the OLTs is needed. In this case study [7], it is determined using the criterion:
$\#\text{Fibers}_{\text{proj}}(FN) = \#PON_{ports}$; from (4)
$\#\text{Fibers}_{\text{proj}}(FN) = 612$
To calculate the number of feeder cables needed, segmentation by zones of the area of service is required. In the case study [7], based on the location of the CO, the area was segmented in 5 zones: Miravalle, Jacaranda, Cumbaya Central, San Juan de Cumbaya, and La Primavera.

4. Thus,

$$\#FiberCapacity_{proj}(FN) = \frac{\#Fibers_{proj}(FN)}{Zones} \tag{11}$$

$\#\text{FiberCapacity}_{\text{proj}}(\text{FN}) = \frac{612}{5}$
$\#\text{FiberCapacity}_{\text{proj}}(\text{FN}) = 122.40$ fibers (This value is not standardized).

Then,

$$\#\text{FiberCapacity}_{\text{proj}}(\text{FN}) = 144\text{fibers}$$

The FN will need five cables with a capacity of 144 fibers for each line. Note, although in the calculations above contemplate the feeder cable only requiring 123 fibers, it is highly recommended to have redundant threads. For the zone "Sector Central," the results

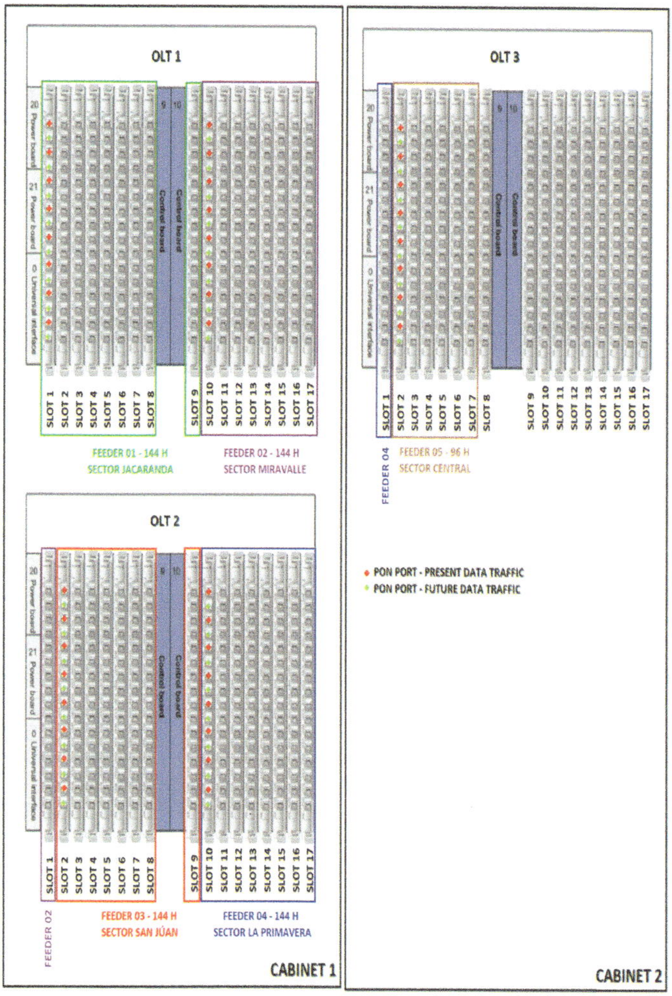

Fig. 6. FN segmentation and identification of data traffic.

of the surveys showed less interest in having an FTTH-GPON service [7], so for this sector, the feeder cable will only need 96 threads.

The segmentation and identification of data traffic of the FN is shown below in Fig. 6.

Distribution Network Technical Capacity (DN)

5. To calculate the capacity of the cables of the DN, for the case study [7] Sect. 4.3, it is necessary to use the number of projected subscribers and the optimal splitting level. Additionally, in the subsection "Identification and segmentation of OLT data traffic" of this article, it has been stated that one PON port will be used for present data traffic and another PON port for future data traffic, for one specific zone. Therefore, each PON port

will provide service to eight NAPs, and each one of these will provide service to eight subscribers.

However, because of CoFTTH (subscriber's growth), the SSC and each NAP will have space to install one more 1:8 splitter (next PON port), as shown in Fig. 7.

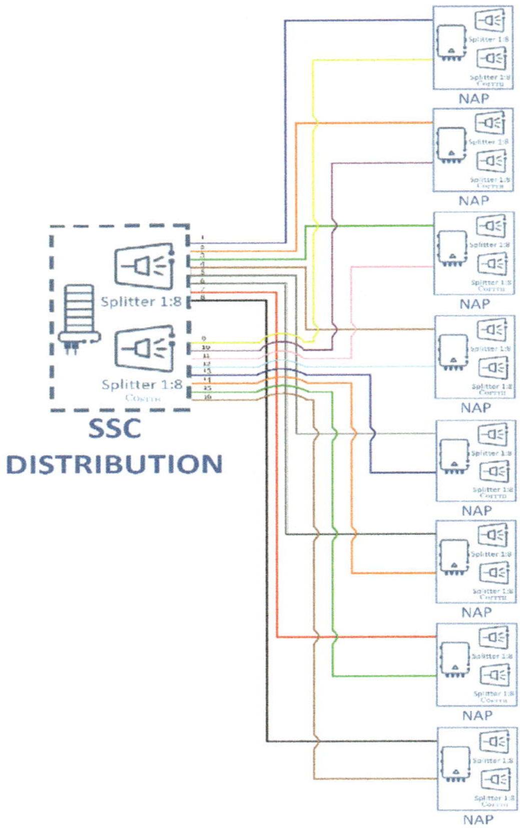

Fig. 7. DN fiber cable capacity.

Consequently, the cable's capacity for the DN is 24 fibers: 8 fibers to provide service to present subscribers, 8 for future subscribers, and 8 for redundancy purposes. The redundancy component is recommended by Ecuador's National Corporation of Telecommunications (CNT EP, by its Spanish acronym) [14].

$\#FiberCapacity_{proj}(DN) = 24$ fibers

6. Results

The ODN architecture will have two splitting levels of 1:8 (1:64 in total). The first splitting level is located at the Splitting and Splicing Closure (SSC) and the second at the Network Access Point (NAP). This process is shown in Fig. 8.

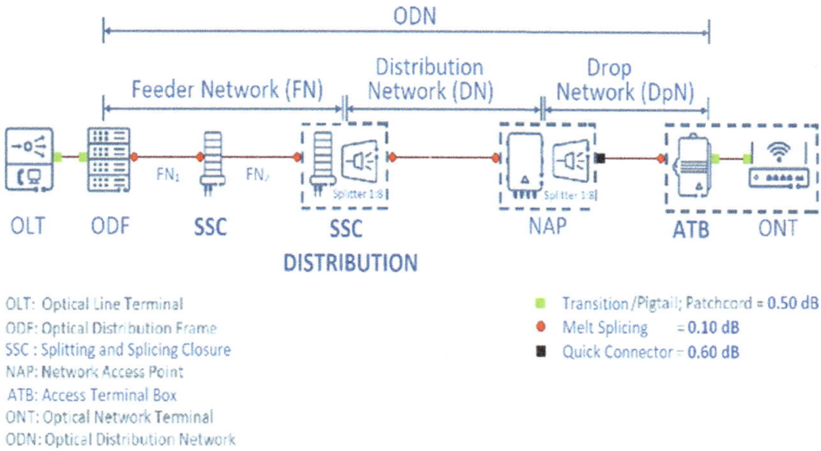

Fig. 8. ODN's architecture and components.

5 Conclusions and Future Works

This article helps demonstrate that the determination of the technical specifications of the core equipment of the FTTH-GPON network has to comply with adequate sizing for the handling and growth of data traffic, with calculations with real and reliable data from accredited sources. No technical parameters were calculated empirically.

Determining the make and model of an OLT requires meeting all technical specifications for current and future data traffic management rather than selecting by the number of PON ports and slots, which is a common error in most designs.

The identification and segmentation of data traffic in the OLT and ODN is of vital importance to solve, in a timely and efficient manner, problems such as congestion, collision, or packet loss.

As future work, we want aim at: 1) conduct a research related to the pandemic caused by the SARS-COV2 virus, which has generated the growth of FTTH-GPON subscribers and Internet bandwidth in Quito, Ecuador; 2) study how this subscriber growth affects the sizing of the GPON core equipment.

References

1. Loayza, P.: Diseño de Redes FTTH-GPON con enfoque QoS, pp. 1–36. Universidad Tecnológica Israel (2019)
2. Ecuador's Telecommunications Control and Regulation Agency (ARCOTEL): Internet Access Service - Statistical Bulletin, pp. 4–14, 19th May 2019
3. Lam, C.: Passive Optical Networks: Principles and Practice. Elsevier Inc., Burlington (2007)
4. Cale, I., Salihovic, A., Ivekovic, M.: Gigabit passive optical network – GPON. In: 29th International Conference on Information Technology Interfaces, ITI, Cavtat, p. 6 (2007)
5. Gregory, M.A., Abbas, H.S.: The next generation of passive optical networks: a review. J. Netw. Comput. Appl. **67**, 53–74 (2016)

6. Green, P.E.: Fiber to the Home: The New Empowerment. Wiley, Hoboken (2006)
7. Loayza, P., Guaña, E., Pumares, A.: Data gathering methodological guide for design-ing FTTH-GPON networks with QoS. Iberian J. Inf. Syst. Technol. RISTI 528–539 (2020). 26th Edition, ISSN 1646-9895
8. Huawei Technologies: GPON Tech. GPON Magazine, pp. 1–58 (2019)
9. Huawei Technologies: QoS Feature Guide. Huawei QoS, p. 139 (2019)
10. Huawei Technologies: Gigaband era. Access Solution Galaxy, pp. 18–23 (2017)
11. Smith, T.G., Hinton, K., Tucker, R.S.: Delay and jitter in long-reach GPON. In: 16th Opto-Electronics and Communications Conference, OECC, Kaohsiung, pp. 184–185 (2011). https://ieeexplore.ieee.org/6015074
12. Huawei Technologies: MA5800 Product Description. OLT Catalog, p. 15 (2018)
13. Huawei Technologies: HG8045H Product Description. ONT Catalog, pp. 1–2 (2018)
14. Corporación Nacional de Telecomunicaciones CNT EP. Normativa Técnica para el dimensionado de una ODN, Quito (2013)

Processing of Voice Signals in Telecommunications Systems Using MATLAB

Daniel Icaza[1,2,3,4](✉), Juan-Carlos Cobos-Torres[1,2,3], Geovanny Genaro Reivan-Ortiz[1,4], and Federico Córdova Gonzalez[1,3]

[1] Universidad Católica de Cuenca, Cuenca, Ecuador
{dicazaa,juan.cobos,greivano,ncordovag}@ucacue.edu.ec
[2] Proyecto de Investigación de Optimización Energética del Sistema de recaudo de Unidades de Transporte Urbano, Cuenca, Ecuador
[3] Grupo de Investigación en Radiación Visible y Prototipado GIRVyP, Cuenca, Ecuador
[4] RAT iN BOX: Análisis Experimental de la Conducta Animal, Laboratorio de Psicometría CIITT-Ricaurte, Cuenca, Ecuador

Abstract. In the article that is presented, we analyze the voice signals in the domain of time and frequency by using the matlab which we can perform the different commands to obtain the graphs, and analyze each of them taking into account the different types of signals that we will record with our own voice.

The analysis of voice systems can be decisive for making comparisons between the voice in a normal state and its voice caused by some adverse situation. In the case of communication systems they can be very useful to know if there are problems between the transmit and receive ends.

Keywords: Telecommunications · Signals · Coding · Processing · Voice

1 Introduction

For telecommunications one of the most important signals is voice, in this way telephony provides us with communication from one point to another, initially this was done through the analog channel based on the copper conductor, from here it evolves to different means of communication from the optical fiber that we obtain today [1–3]. In this way, as the means of communication evolve, it is necessary to study voice signals, know their characteristics and the behavior of frequencies.

In this document we will observe the characteristics of this type of signal and the frequency domain will be analyzed over time, in this way it will be analyzed with the study of the nature of voice and acoustic signals. In this way, exercises will be carried out in which we will record our voice through the use of matlab functions.

2 General Analysis

2.1 Acoustic of the Voice

These acoustic signals are formed by the movement of air particles by an impulse that generates pressure on them. At the moment of clapping, pressure is exerted on the

M. Botto-Tobar et al. (Eds.): CIT 2020, AISC 1326, pp. 177–190, 2021.
https://doi.org/10.1007/978-3-030-68080-0_13

particles found between the palms of the hands and these simultaneously transmit said pressure to the adjacent particles, causing the acoustic wave to propagate until it is received by the human ear in Fig. 1 [4].

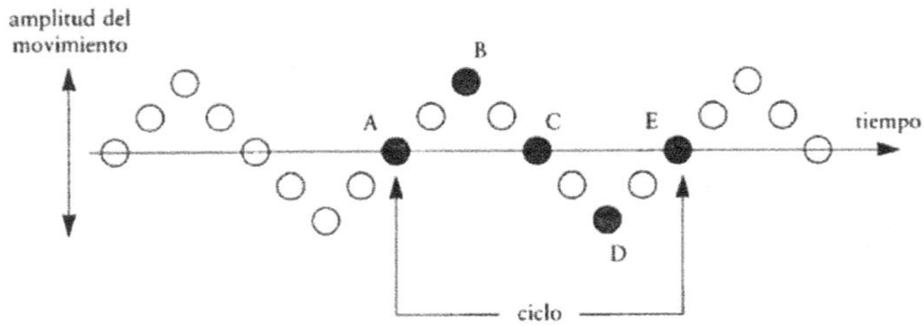

Fig. 1. Acoustic effect

The characteristics of the signals are:

- Periodicals that have a certain amplitude that determines the intensity with wich they are heard.
- The voice signals are produced in a range from approximately 50 Hz to 5 kHz, meanwhile the sound and music could reach a range of up to 20 kHz, if it is not within this range, the human ear no longer serves the channels of telephony up to 4 kHz. In the musical notes they are integrated by pure sinewaves as shown in Table 1. The fingerboard corresponding to the note when tuning the instruments of an orchestra that produces a frequency wave of 440 Hz.

Table 1. Frequency of musical notes.

Note	Frequency (Hz)
Do	264
Re	297
Mi	330
Fa	352
Sol	396
La	440
Si	495

2.2 Characteristics of the Organs that Make up the Voice

The voice signals that are heard are formed with the vibration of the vocal cords. The organs in charge of producing the voice integrate the vocal tract, which can be observed by means of a linear and time invariant system (SLIT) in this way an input signal is generated and the output then represents the frequency response of the same system [2–5].

The voice signal is the one that presents a fundamental and harmonic frequency that are of less amplitude. The fundamental frequency provides information on the speed at which the oral chords vibrate to produce a sound, the opening and closing of the oral chords is produced by small puffs of air [12–14].

The supra glottic cavity is made up of cavities that act as acoustic resonators. When performing a spectral analysis of sound after going through these activities, the resonance effect places an emphasis on certain frequencies of the spectrum, which are called voice formants [4, 12].

There are as many different forms as formants as resonators that the vocal tract has, in such a way it is considered the first three associated with the oral, buccal and nasal cavity that provide enough information to be able to differentiate the types of sound. The spectral envelope is a hallmark of the sound that is emitted, consisting of one main frequency and up to two voice-forming frequencies. An adequate analysis of the envelope allows obtaining information on the type of phoneme pronounced.

2.3 Voice Capture in Matlab

To start the voice analysis in matlab, as a first point we must capture a voice signal, either through an audio card and microphone, then through the example that we are going to highlight, we observe how we can capture a voice signal. [3–5].

Example 1

By using the computer's microphone or by means of a microphone, we capture a phrase or word for 6 s, and we will observe the graph within the program as shown in Fig. 2.

Matlab programming.

```
recObj=audiorecorder;
disp('Comienza grabacion');
recordblocking(recObj,6)
disp('Fin de grabacion')
play(recObj)
y=getaudiodata(recObj)
plot(y)
N=512;
y1=y(1:512);
w=hamming(N);
yw=y1.*w;
plot(y1)
```

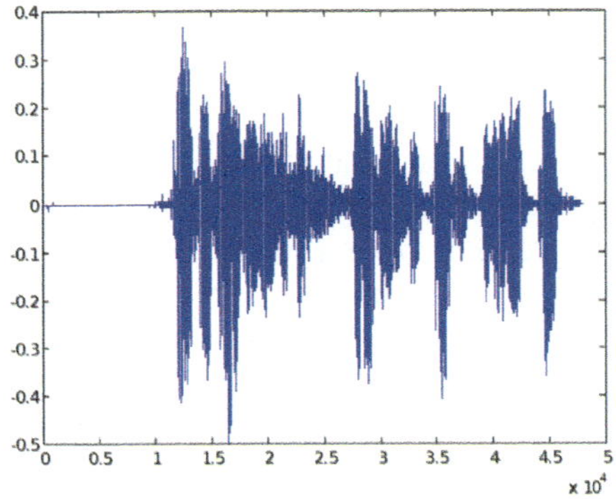

Fig. 2. Graph of the phrase expressed in MATLAB

Solution

By means of the commands applied in matlab it allows us to make the voice recording and at the same time make the graph of the frequency of the same which we can see in Fig. 2 shown below, in this way we can see the programming in Annex 1 in matlab which gives us the voice signal.

2.4 Analysis in the Time Domain

For voice analysis it is necessary to divide it into segments to carry out the individual study of each vowel or consonant.

The truncation caused by the rectangular window produces the so-called gibbs effect, in this way triangular, haming and hanning windows are used, in this way we can observe through example 2 that we observe below [4, 9–11].

Using example 1 we can see the voice segments obtained in said example. In this way we will observe in Fig. 3 the procedure performed in matlab.

Matlab programming.

```
recObj=audiorecorder;
disp('Comienza grabacion');
recordblocking(recObj,1)
disp('Fin de grabacion')
play(recObj)
y=getaudiodata(recObj)
N=512;
y1=y(1:512);
w=hamming(N);
yw=y1.*w;
subplot(2,1,1);plot(y1);plot(y);
```

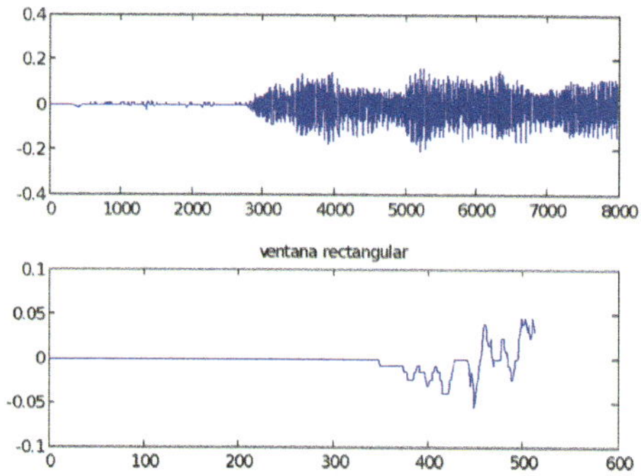

Fig. 3. Example of the segmentation and window.

2.5 Analysis in the Frequency Domain

The importance of the voice in the frequency domain lies in the determination of the voice formant frequencies. As mentioned above, the formants are the most important frequency peaks of each sound and are generally three the most representative. These form that allow us to differentiate vowels and recognize a vowel even when it has been pronounced by different speakers. They are also useful in identifying sounds together with certain consonants, that is, each sound in particular has certain specific formants, which allows the corresponding syllables to be recognized by identifying the maximum frequencies, a situation that gives rise to voice recognition systems [1, 4]. In this way we can see in the following example.

Example

Record in five different variables a, e, i, o. As seen in Figs. 4, 5, 6, 7, graph the oscillogram of each one and finally determine the frequency spectrum of each signal and graph it. Do the same exercise with different speakers and compare the respective spectra.

Matlab programming.

```
recObj=audiorecorder;
disp('Comienza grabacion');
recordblocking(recObj,1)
disp('Fin de grabacion')
play(recObj)
t=getaudiodata(recObj)

N=512;
y1=y(1:512);
w=hamming(N);
yw=y1.*w;

Fs=1000
t=0:1/Fs:(length(X)-1)/Fs;
f=(0:length(t)-1/length(t)*Fs);
f=f-Fs/2;
xf=fftshift(fft(X)/length(t));
figure
plot(t,X)
figure
plot(f,abs(xf))
```

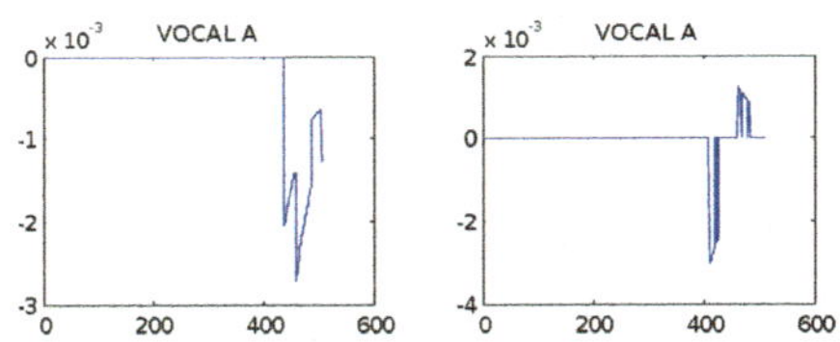

Fig. 4. Vowel A

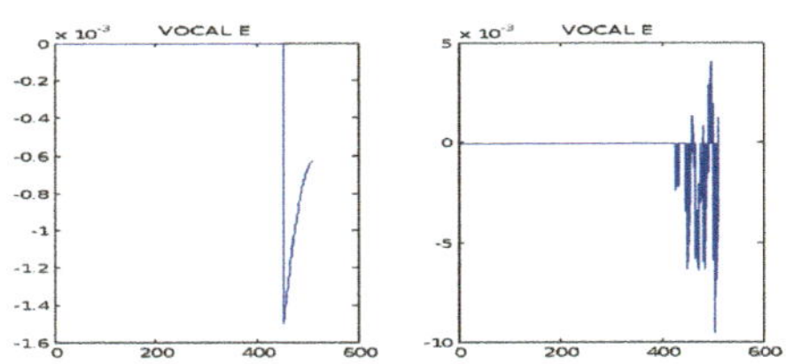

Fig. 5. Vowel E

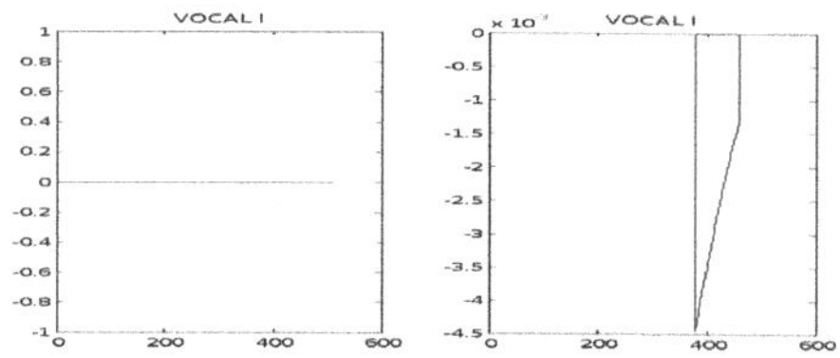

Fig. 6. Vowel I

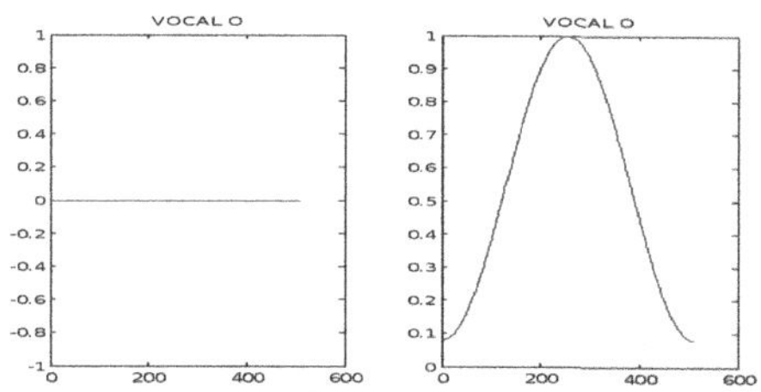

Fig. 7. Vowel O

In the exercise shown, it allows us to check the frequency spectrum of different speakers, whether male or female or child or adult. In this way, similar frequencies are seen in the figures, so it can be seen that there are variations in the peaks of the figures, taking into account the different frequencies of each vowel as shown in Table 2.

Table 2. Vowel formants.

Vowel	Main region
u	200 a 400 Hz
o	400 a 600 Hz
a	800 a 1200 Hz
e	400 a 600 y 2200 a2600 Hz
i	200 a 400 Hz y 3000 a 3500 Hz

2.6 Spectrum of the Voice Signal

The spectrogram is used to make voice signals much longer than a single vowel, in this way it can be seen in a 2-D figure where the frequency content of a signal is shown and how the time varies. This allows us to observe the voice forms as they are pronounced, this spectrogram can be generated in MATLAB using the spectrogram function which we will see later. This function generates a 2-D image where the warm colors are observed to indicate a greater amplitude, which a greater amplitude will be red and yellow, and the colors blue, violet to indicate smaller amplitudes as shown in the Fig. 8. In the following example we can see the 2-D image, which will make a long sentence to observe the colors already mentioned above.

Matlab programming.

```
rec=audiorecorder;
To=8;
disp('Hablar');
recordblocking(rec,To);
disp('Dejar de hablar...');
graba=getaudiodata(rec);
figure
specgram(graba)
'espectograma de la frase'
```

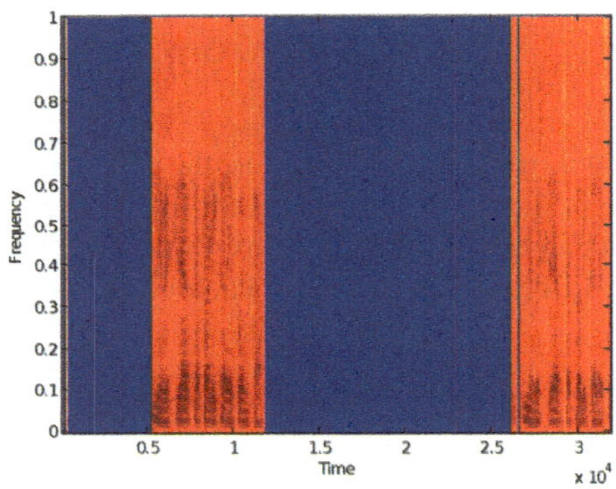

Fig. 8. Spectrum of the voice signal

3 Lineal Prediction of the Voice

In this section we will analyze the technique for voice analysis called linear prediction, so linear prediction is intended to estimate the parameters of a model that represents the voice signal being heard. In this way it is possible to know the behavior of the signal and determine the sample that is being processed from the previous samples, in this way the output of a digital filter is calculated.

3.1 FIR Filter

The linear phase property is that the response in the phase of a filter depends linearly on the frequency. This property is very important in voice applications where time alignment is essential to simplify the precision problem. The downside of FIR filters is that it is not easy to get them to cut sharply at the cutoff frequency compared to IIR filters [2, 8–10].

The model obtained is represented with a FIR filter and the frequency indicates the values of the representative frequencies of voice or voice formants. The problem is to determine the coefficients of the system that characterizes the voice signal, as will be seen in the coding stage called CODEX in GSM cellular telephony. It performs voice compression based on the transmission of the parameters that represent the system rather than the transmission of the parameters representing the system rather than the transmission of the minimum voice [3–5].

The linear prediction problem is represented by Eq. 1:

$$\underline{x} = \underline{H}\theta + \underline{\omega} \tag{1}$$

Where:

$\underline{H}$ is the observation matrix

$\underline{x} = x[n] = x[0]$.

$\underline{w} = w[0], w[1], \ldots, w[N-1]$.

3.2 Stimation Theory

There are two approaches to solving the problem of determining the coefficients, which are:

- Classical theory of where the value θ assumes is a deterministic variable.
- Bayesian theory of where θ is considered in a random variable

In this way θ represents the voice of a communication channel, in such a way that the message that is going to be transmitted is unknown, therefore it is considered θ a random variable [4, 5].

The properties that an estimator must have are:

- The mean of the estimator is the value of the parameter, $E\left(\underline{\theta}\right) = \theta$ where θ is the real value of the parameter, θ is the estimator and E represents the mathematical expectation.
- The variance of the estimator must be minimal

Thus, it indicates that the variation by $E\left(\underline{\theta} - \theta\right)^2$ must be minimal, an estimator is efficient if the variance is limited by a lower value given by Eq. 2:

$$Var\left(\hat{\theta}\right) \geq \frac{1}{-E\frac{\delta^2 P_x(X,\theta)}{\delta\theta^2}} \tag{2}$$

Where:

$P_x(X, \theta)$ It is the probability function between $xy\theta$

The estimator of minimum variance is given by Eq. 3:

$$\hat{\theta} = \left(H^T H\right)^{-1} H^T \underline{x} \tag{3}$$

3.3 Identification of a System

We will estimate the coefficients of a model of a system H (z) given in Fig. 9.

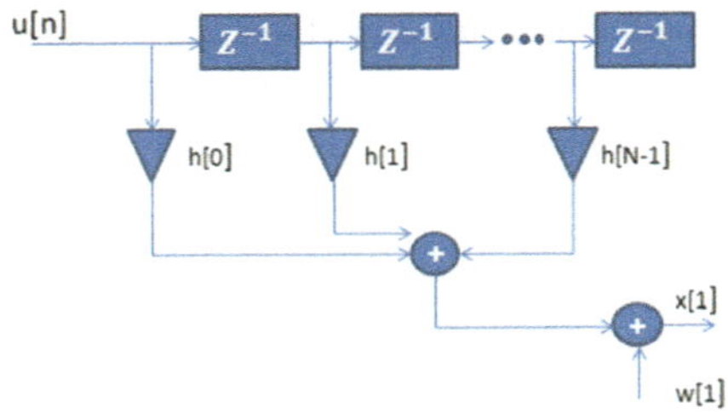

Fig. 9. Model representing voice signal.

From where *x (n)* you want to obtain the coefficients *h (n)* of a FIR type filter (it is an acronym in English for Finite Impulse Response or Finite Impulse Response. It is a type of digital filters whose response to an impulse signal as input it will have a finite number of non-null terms) [4–7].

The parameter to estimate is the vector of coefficients $\hat{h}$ and can be estimated as Eq. 4:

$$\hat{\theta} = \underline{h}^t \underline{x} = \underline{x}^t \underline{h} \tag{4}$$

The error is given by Eq. 5:

$$e = \hat{\theta} - \theta = \underline{x}^t \underline{h} - \theta \tag{5}$$

To obtain the least squared error and applied the perpendicular criterion, that is, Eq. 7:

$$Eerror.data = 0$$

$$Ee\,\underline{x} = 0$$

$$E\underline{x}\left(\underline{x}^t \underline{h} - \theta\right) = 0 \tag{7}$$

Where can you define

$$E\underline{x}\underline{x}^t = R_{xx} \tag{8}$$

Called the autocorrect matrix that is given by Eq. 9:

$$E\underline{x}\theta = r_{x\theta} \tag{9}$$

Called cros correlation vector. From where h is given by Eq. 10:

$$h = R_{xx}^{-1} r_{x\theta} \tag{10}$$

This is the so-called Wiener-Hopf equation

In such a way the solution expressed in matrix form is solved by means of a system of N equations by means of the Levinson-Durbin recursión

$$(r_{xx}[0]r_{xx}[1]\ldots r_{xx}[N-1]r_{xx}[1]r_{xx}[0]\cdots r_{xx}[N-2]r_{xx}[N-1]r_{xx}[N-2]\cdots r_{xx}[0])$$
$$(h[1]h[2]N-1) = (r_{x\theta}[1]r_{x\theta}[2]r_{xx}[N-1])$$

3.4 Lineal Prediction in MATLAB

Using the function **[a,g] = lpc(x,p)** which is the linear Prediction Coefficients which is the linear algorithm in Matlab, from which this function determines the coefficients of the linear predictor by minimizing the least squares applied error. The function outputs the vector **a** with the FIR filter coefficients representing the system model and the error vector **g**. In the case of speech comprehension, the sound or the fragment of the voice is represented by the coefficients that characterize the FIR filter.

Example

Determine the coefficients of a system from the signal obtained at the filter output, implementing with these coefficients for a random data entry, let the vector of coefficients: A = [1, 86.67, 55 0.68]; as shown in Fig. 10.

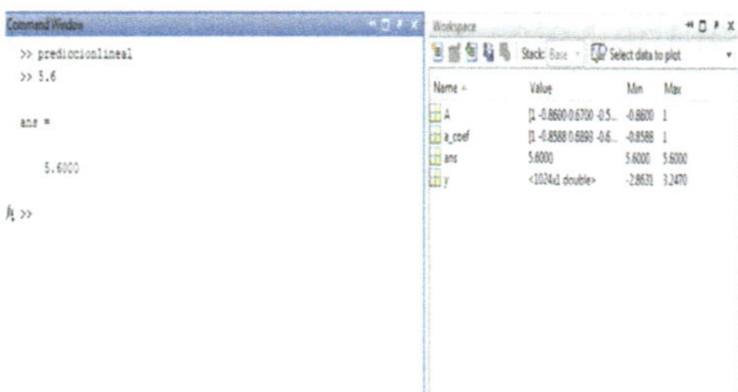

Fig. 10. Lineal prediction in MATLAB

The estimated coefficients that closely approximate the original coefficients are observed, which means that the estimator is a good approximation of the original model.

Apply a linear prediction algorithm to a voice fragment recorded in matlab and compare the signal obtained from the prediction model with the original signal as shown in Figs. 11 and 12.

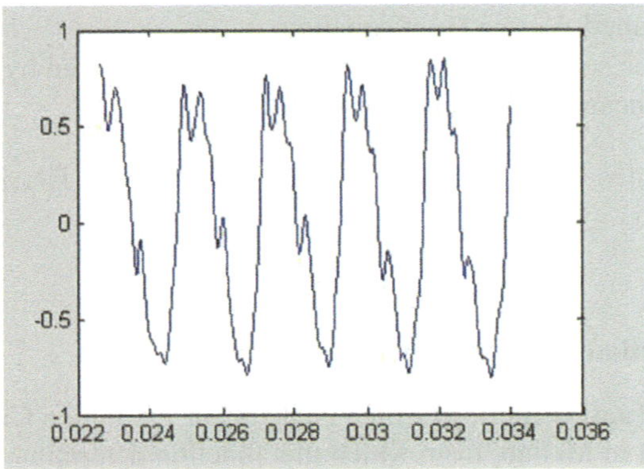

Fig. 11. Prediction model voice signals

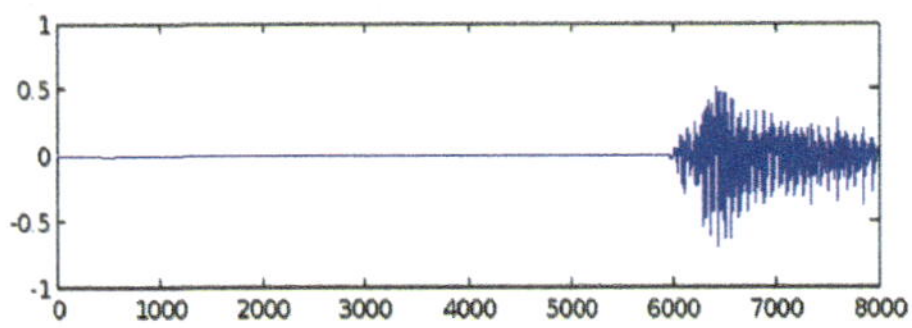

Fig. 12. Original voice signals

4 Compression of the Voice and Vocoders in Cellular Telephony

Vocoder is the term that represents the words 'voice coder' that maintains the voice quality within the bandwidth. A telephony channel covers 270 kbps, in 8 64 kbps slots it requires 512 kbps therefore it requires a process to compress the voice signal, this task is developed by DPS circuits so that the success of the implementation of the systems is allowed from GMS. Coding is based on linear prediction to find the parameters of a filter that characterizes the voice.

The voice signal is divided into 20 ms segments for which the estimation of the parameters of an 8 coefficient filter as well as an excitation signal is performed, this information is sufficient to reconstruct the original signal.

A significant aspect in the understanding of the information is that there are 20 ms segments similar to each other, in this way there are a large number of samples with redundancy in the information and correlations are sought between these data called Long Termtrecdiction or LTP. With this technique it has been achieved that the VOCODER compress the voice with a factor of eight while maintaining the sound of the natural voice, this technique has been used for optimization for the human voice therefore they do not work properly for music.

5 Case Analysis and Discussion

In the time domain of voice signals, a division into segments is made to carry out an individual study of each vowel or consonant, causing the effect of gibb.

In the frequency domain it is analyzed by formants that are the most relevant peaks of each sound, thus differentiating the most relevant words, vowels or consonants, giving rise to voice recognition.

In this way, with the help of matlab and each graph, we can easily observe each word, syllable or vowel that gets its corresponding frequency. On the other hand, to analyze in the time domain, the division by segments of each word must be carried out to be analyzed one by one, obtaining a more delayed analysis, by which the voice study by frequency is made.

In the spectrogram a voice signal is analyzed in which the different types of color are observed, the warm colors are greater when a red or yellow is observed and when they are blue or violet they are of smaller amplitudes in such a way that we can identify the amplitude of each of the words phrases or vowels that have a greater or less amplitude gives the mentioned phrase.

6 Conclusions

By mentioning voice recognition in matlab, it allows us to make a recording and thus be recognized through the matlab database according to its programming in which we will obtain the different types of analysis that we want to perform, either by time or frequency.

When performing the analysis of voice signals with MATLAB we observe how the different voices, words or syllables have a different sample of frequency and time signal since it dissipates every time you finish making a phrase or word, taking into account that it does not all the words get the same frequency.

In the spectrogram we can observe that the voice signals have their different colors depending on their amplitude, and frequency warmth since each of the words has their difference.

All these analyzes carried out can be very useful so that based on certain patterns we could identify voices and therefore detect people, there is a fairly wide field to study, such as criminalistics, the opening and closing of floodgates by voice (automation), etc.

On the other hand, it is also important to indicate that in most countries, legislation prohibits arbitrary incursion into telecommunications systems for violating the privacy of individuals, the analyzes carried out must be in accordance with the law. In this article we approach it from the technical side, but regarding studies and checks in specific fields, you must have the corresponding permits.

In addition, it is possible to have a variety of sounds to carry out experimentation with rats or mice with the intention of adapting a desired behavior, what is indicated is a purpose that we will be studying in depth in a new article.

References

1. Petrushin, V.A.: Detecting emotions using voice signal analysis. U.S. Patent No 7, 222,075, 22 Mayo 2007
2. Genoveva, V.R.: Guatemala, Sistema de reconocimiento de voz en matlab, Guatemala, abril de (2008)

3. José, A.C.I., Síntesis y Reconocimiento de Voz, Universidad Michoacana de San Nicolas de Hidalgo, Facultad de Ingeniería Eléctrica (2015)
4. Mauricio, O.R.: Matlab aplicado a telecomunicaciones (2015)
5. Pertrushin, V.A.: System method and article of manufacture for detecting emotion in voice signals by utilizing statistics for voice signal parameters. U.S. Patent No. 6,275,806. 14 Aug. 2001
6. Kwon, B., Farndale, E.: Employee voice viewed through a cross-cultural lens. Hum. Resour. Manage. Rev. **30**(1), 100653 (2020)
7. Lin, K.W.E., et al.: "Singing voice separation using a deep convolutional neural network trained by ideal binary mask and cross entropy. Neural Comput. Appl. **32**(4), 1037–1050 (2020)
8. Van Os, M., Gregory, N., Scott, H.: "Contextual voice commands." U.S. Patent No. 10,540,976. 21 Jan. 2020
9. Drugman, T., Thomas, D., Thierry D.: On the mutual information between source and filter contributions for voice pathology detection (2020). *arXiv preprint* arXiv:2001.00583
10. Liu, H., et al.: An epidermal sEMG tattoo-like patch as a new human–machine interface for patients with loss of voice. Microsystems & Nanoengineering **6**(1), 1–13 (2020)
11. Sugawara, T., et al.: Light commands: laser-based audio injection attacks on voice-controllable systems (2020). *arXiv preprint* arXiv:2006.11946
12. Gross, S.: Linguistic intuitions: Error signals and the voice of competence (2020)
13. Jeon, J.H., Kyoung, G.W.: Apparatus and method for processing control command based on voice agent, and agent device. U.S. Patent No. 10,607,605. 31 Mar. 2020
14. Miramont, J.M., et al.: Voice signal typing using a pattern recognition approach. J. Voice (2020)

Prototype of a Centralized Alert and Emergency System for Digital Terrestrial Television in Ecuador

Gonzalo Olmedo[1(✉)], Freddy Acosta[1], Diego Villamarín[1], Fabián Santander[2], Rosario Achig[2], and Villie Morocho[2]

[1] Ecuador WiCOM-Energy Research Group, Department of Electrical, Electronics and Telecommunications, Universidad de las Fuerzas Armadas ESPE, Sanqolquí, Ecuador
{gfolmedo,fracosta,dfvillmarin}@espe.edu.ec

[2] Department of Computer Science, Universidad de Cuenca, Cuenca, Ecuador
{fabian.santander,rosario.achig,villie.morocho}@ucuenca.edu.ec

Abstract. Latin America is living the migration towards the digital terrestrial television (DTT) with the ISDB-Tb standard. Some countries have even had the analog blackout. In relation to the Emergency Warning Broadcast System (EWBS), the process of transmission and reception is defined. In recent years, tests have been coordinated for the process of implementing EWBS for DTT. The Common Alert Protocol (CAP) allows emergency alerts to be exchanged between different communication systems. This article presents the implementation of a CAP module in a Spatial Data Infrastructure for risk management and integration of the EWBS for DTT in a single platform for emergency alerts. A web service that contains the codes representing the cantons of Ecuador was obtained, selected through the use of a map viewer and integrated with an application that processes this information and generates the alert to be transmitted via DTT. The main objective is the storage of the information in a database, and a WEB RSS service that contains the information in conformance with CAP format. A series of experiments show successful outcomes.

Keywords: EWBS · Spatial Data Infrastructure · Common Alert Protocol · Risk management

1 Introduction

The Ecuadorian public security law establishes the Risk Management Secretariat as an emergency technical body. Based on the law: "the functions of the technical governing body of the decentralized national risk management system are: to articulate the institutions so that they coordinate actions in order to prevent and mitigate risks, as well as to stop, recover and improve the above conditions to the occurrence of an emergency disaster and coordinate the actions necessary to

M. Botto-Tobar et al. (Eds.): CIT 2020, AISC 1326, pp. 191–201, 2021.
https://doi.org/10.1007/978-3-030-68080-0_14

reduce vulnerabilities and prevent, mitigate, attend to and recover eventual negative effects derived from disasters or emergencies in the national territory". In this way, it is essential to integrate different media and communication technologies that help generate early warnings of possible catastrophes or emergencies, which are also massive.

Ecuador and the other countries in the region are coordinating the migration of open terrestrial television from analog to digital. In Ecuador the analogical blackout is scheduled to start on the year 2020. The ISDTB-Tb standard, adopted by Ecuador, has the capacity to transmit data for interactive applications, programming guide, emergency alert, among others.

Field tests of the Emergency Warning Broadcast System (EWBS) in Quito were presented, integrated into the international ISDB-T digital terrestrial television system. This field tests were made through the broadcasting signal of a commercial channel on 635,143 MHz frequency, in UHF band channel 41 [1]. An EWBS server was implemented in the transmitter, where the physical locations that will be alerted were configured, through 12-bit codes defined by cantons for Ecuador in the Harmonization Document Part 3 "EWBS" [2], as well as also the edition of the alert message that will be displayed on the television superimposed on the video and audio signal of the programming, in accordance with the ARIB STD-B14 standard [3]. The server reconfigures the PSI/SI tables and generates a Transport Stream (TS) that multiplexed with the content of the television channel is transmitted as Broadcast Transport Stream (BTS) together with the emergency bit of the physical layer through a microwave link to the modulator located on Pichincha hill and by broadcast transmission distributed throughout the city. The EWBS server configuration was performed using a remote desktop to the EWBS server located in the television station, activating the emergency alert signal in the receivers located in the city with a delay of less than one second.

The successful experience of testing the EWBS system generated the requirement for a single platform led by the National Service for Risk and Emergency Management that configures the codes of the country's cantons and the editing of the emergency message on the servers that will be installed on television channels, in an agile and dynamic way.

This article presents the first results obtained from the integration of the emergency alert system platform that centralizes for EWBS servers. Structures the implementation of a Common Alert Protocol (CAP) module in a Spatial Data Infrastructure (SDI) for risk management and integration of the EWBS System for digital terrestrial television into a single platform for emergency alerts.

1.1 Core of the Early Warning System for Digital Terrestrial Television

The management of events produced by natural disasters is highly in need of information that provides agility to the institutions that will provide relief in case of emergencies. On the other hand, in the event of early alerts, the amount of information available to a user, be it the citizen as well as the relief agencies, must also maintain the greatest possible understanding and be flooded with validated and available information. In both cases it is totally desirable that the information to be used be of a geographical nature, that is, that it can maintain the geolocation of the details presented. The more description of these elements on a map and the less precision error is possible, the better any help, action or planning will be better structured, more effective and totally redundant for the benefit of those affected.

Therefore, since the projects related to Spatial Data Infrastructure, we have been working with the philosophy of geospatial information management that can contribute to this possible benefit. In this search, several projects have been produced whose progress and results have been the subject of innumerable presentations and publications [4–6,14]. In [7] the use of the UCuenca SDI is presented as a fundamental tool for risk management in the detection of mass movements. It even shows how this information can be supported with remote sensors and then serve as the basis for eventual risk management.

In [8] the authors describe the latest advances that allowed the use of SDI as the core of an early warning system for digital terrestrial television. Its structure is presented in Fig. 1. This core allows the integration of new functionalities that mainly complies two expectations, one for management by the body that issues the alerts and that can use a centralized platform in such work and another as an integrator of information to be presented to potential users (citizens, organizations, governments, etc.). This work presents the possibility of connecting said platform as a generator of information to be captured by the CAP, allowing the availability of these alerts to integrate them with other elements designed for risk management.

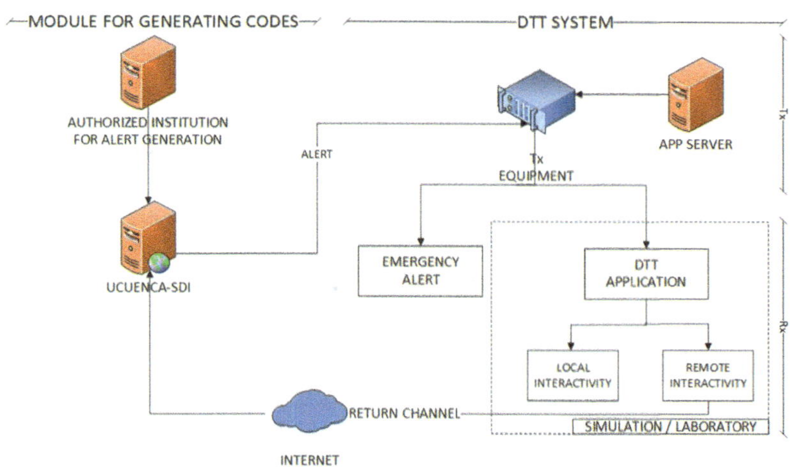

Fig. 1. SDI scheme as core of the early warning system for digital television

2 Common Alert Protocol

The CAP is a standard for emergency alerts, defined by Recommendation X.1303 of the International Telecommunication Union [9]. It is widely recognized internationally as the key standard for societies to overcome dangers, through all means of public alert communication for emergencies, according to the structure presented in Table 1.

Table 1. CAP implementation possibilities

Implemented in this project		
All hazard		All media
Landslide Public safety	Common Alerting Protocol	DTT Cell phone Internet
Future implementation		
All hazard		All media
Fire Flood Earthquake Volcano Tsunami Typhoon/Hurricane Disease Others	Common Alerting Protocol	Sirens Phone E-mail Radio Fax Others

Large emergencies involve large numbers of people, who obtain a lot of information in different ways. However, everyone should obtain key emergency data in the same way. For example: What is the event? What should people do? Where's the alert area? How soon is it coming? How bad will it be? How safe are the experts? These key facts are included in the elements of a CAP message format [10].

CAP uses the XML standard digital format to exchange emergency alerts that allows a consistent alert message to be broadcast simultaneously on many different communication systems. The XML base data format allows software and networks around the world to process it. When key messages are transmitted in the CAP message format, the emergency alert is much more accessible.

The CAP must be issued by an official alert authority, such as:

- National Service of Meteorology and Hydrology.
- Emergency Management Agency.
- Any other organization authorized to take alert functions.

Because alerts are now broadcast through powerful public networks, it is impossible to know the sources directly, as it could be in a small city. Different countries have their own policies on what it means to be "officially licensed." However, there is consensus that official alert authorities should be known internationally [11].

For this, it is necessary to make a record of the entity authorized to alert, in this case it is done in the WMO (Word Meterological Organization) [12], which is responsible for maintaining a record of official entities to issue an alert. In this registry, mainly data that characterizes the entity is entered, such as: the hazard categories for which this authority routinely issues alerts, the common alert zone for this alert authority, an alert news channel URL in CAP format.

3 CAP Implementation

The Spatial Data Infrastructure of the University of Cuenca in Ecuador has been working mainly on issues related to the mitigation of natural hazards, adapting to the technologies that are mainly developed in order to help the population, such as the case of CAP.

This section presents the work that has been done, relating the existing tools offered by the SDI with the development of a module that allows the entry of information to obtain an alert in the CAP format. We began by knowing the structure of the CAP and each of its parameters, in this case the indications proposed by OASIS were taken as a reference. OASIS is a non-profit consortium that drives the development, convergence, and adoption of open standards for the global information society [10].

After knowing the structure that makes up the CAP format, an analysis was made of each of the parameters, so that the form in charge of entering the information validates each of the data required by the CAP standard.

3.1 Development Tools

Various technologies were used to develop a module that adopts the characteristics offered by the CAP format. The main tools were: Angular, Java and PostgreSQL.

The Angular framework is maintained by Google and is used to create Single Page Application (SPA) websites. When a user enters a SPA website, everything is loaded at the same time on the same website and Angular internally changes the view when browsing the website so that it appears like a normal website [13].

The Angular framework allows the development of a dynamic WEB form, which manages to modify the browsers of desktop computers, tablets or mobile phones, in addition to having a user-friendly design.

PostgreSQL is an object-relational database management system. Its license and development is open source. In this case, this tool was used for information management and organization, through the generation of a database structure.

Java is an object-oriented programming language developed by Sun Microsystems that works at the server level. In this case, a library called "Cap-library" was used, which consists of a collection of code and tools to work with alert messages in the CAP.

Using these tools together, it was possible to carry out a process of entering, validating, storing, publishing and exporting the information.

3.2 CAP Module Integration to SDI

By having a module that allows information management to generate an alert in a CAP format, there is the possibility of adding functionalities that allow information to be entered dynamically in the form, as shown in Fig. 2.

Fig. 2. CAP Module Form - The CAP standard describes several essential fields to identify the alert, especially the alert classification, location, responsible for the issue. To that is added personalized coded information to communicate with the rest of the platform.

In this case, a functionality was implemented that allows interpreting data from other systems that contain information related to danger alerts. Such is the case of the SDI Ucuenca code generating system [8], that although, as a result of an alert report process, it has its own information format for EWBS, it is possible to read, interpret and adapt it to the CAP module.

The SDI Ucuenca code generator system has a format that contains a title, description, type of event and codes that represent the geometries of the cantons of Ecuador. To consult the information of the reported events of the SDI UCuenca code generator system, a function was added to the CAP module that allows listing all the events that have been reported through a WEB service provided by the SDI. In this way, it is possible to list and adapt the information of any entity that issues this type of report, using a process or geo-processes that allow this information to be adapted to the CAP format, according to the structure of Fig. 3.

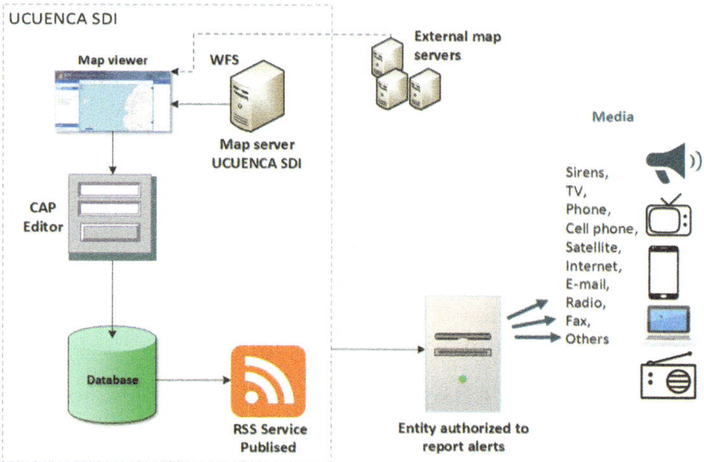

Fig. 3. Functional diagram of the CAP module with the SDI

3.3 Advantages and Disadvantages of the CAP System

Advantage

- CAP provides consistency across multiple channels, allowing accurate corroboration of alert information.
- CAP defines a digital message format compatible with all types of existing and emerging systems, data networks, as well as radio and television broadcasting.
- CAP is useful for multilingual and special needs populations.
- A sender of the CAP message can activate multiple alert systems with a single entry.
- Standardized alerts from many sources can be compiled for "situational awareness".
- Administrators can monitor the entire image in all types of local, regional, and national alerts (public alerts and messages among emergency personnel).
- By having a module that works together with a complete geographic information management system, it is possible to generate geometries dynamically from the information contained in the local server or other WEB map services.

Disadvantages

- For the integration of information from other entities to the CAP module, it is necessary to develop functions that interpret the format they have, this because each entity handles its own format.
- The information that is intended to be integrated into the CAP module from another entity, may contain characters that the CAP format does not allow, therefore, this information must be modified to adapt it to the format.
- If the information of other entities contains geographic information, it is necessary to know what the type of format is, or to have a URL from a WFS service to perform geoprocessing and convert it to the format that the CAP module accepts.

4 Results

4.1 Generation of Codes Through the Use of SDI for the Transmission of Information Through DTT

As a result of this work, a WEB service show in Fig. 4 was obtained, that contains the EWBS codes representing the cantons of Ecuador, selected through the use of a map viewer of the SDI UCuenca code generator system.

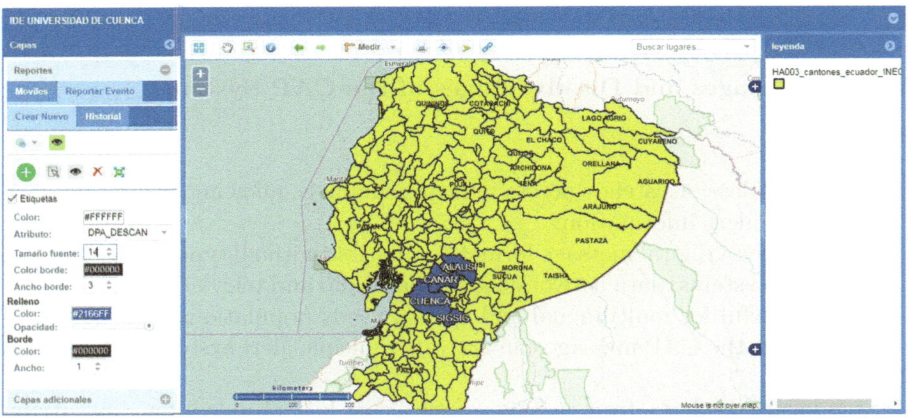

Fig. 4. EWBS code generator system of the SDI UCuenca

The WEB service is consumed by the EWBS server developed by the ESPE University presented in [1], which is in charge of processing the information necessary to transmit an alert via DTT to the cantons that have been selected in the map viewer from SDI UCuenca. Figure 5 shows the structure of events generated by the SDI and sent to the EWBS server on the television channel.

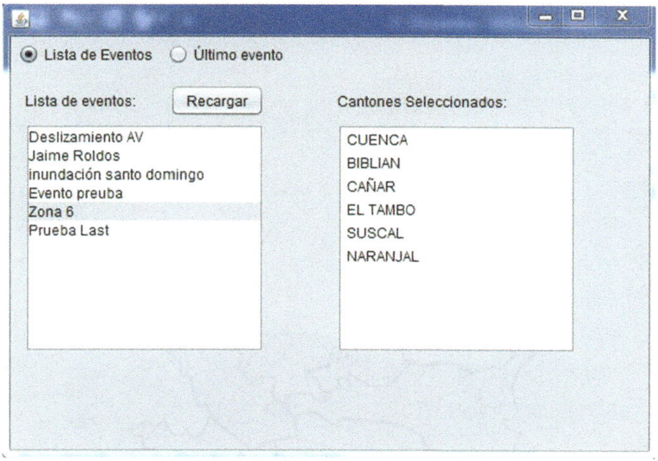

Fig. 5. Event history - App consults EWBS codes of the SDI platform

4.2 Generating an Alert Using the CAP Module

After entering information through the CAP Module form, integrating information generated by the SDI UCuenca code generator system, the result is the storage of the information in the database, and from there a WEB RSS service that contains the information entered in CAP format. An example of the data entered in the SDI can be found at the following link: https://ide.ucuenca.edu.ec/rest/feedcap.

To verify that this WEB service meets the requirements, a validation was carried out on the WEB site https://cap-validator.appspot.com/. This page performs a validation of all the parameters and data contained in the CAP format, and in this case, by validating the information contained in the RSS service, it gave a satisfactory result. An example of this validation is shown in Fig. 6.

5 Discussion

In the search for solutions to alert the citizen in ingenious ways and with the possibility of reaching the largest possible population, different types of alert elements have been designed. In this evolution, CAP is established as the last of the phenomena to follow and has been implemented internationally. Among the main advantages, the possibility of serving as an interface for both generating alerts and reading is described here.

This article presents the way of implementing a CAP service capable of integrating the early warning generation platform based on a Spatial Data Infrastructure with the possibility of serving as a CAP generator for other additional services. This opens the doors for several other organizations to make use of this encoder, which was also designed with an internal translator that allows

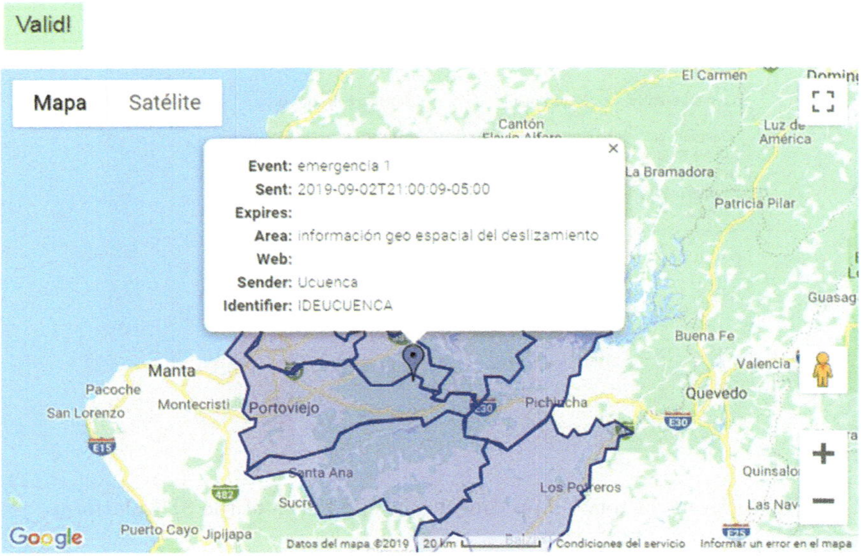

Fig. 6. Validation of the RSS service of the CAP module

any encoding sent under an established format to be translated into English. In such a way that the service is accessible by any possible international CAP reader and that it allows the appropriate interpretation. This allows the issuance of alerts to be used even by other international and national systems, such as sirens designed to read CAP services.

This proposal aims to generalize the generation of early alerts for the population.

6 Conclusions

A CAP module in a Spatial Data Infrastructure for risk management and integration of the EWBS for DTT in a single platform for emergency alerts have been implemented, and tests were carried out on television broadcasts.

A generator of an alert to be transmitted via DTT system trough a web service that contains the codes representing the cantons of Ecuador was created and implemented.

An graphic interface was developed to selected the cantons codes through the use of a map viewer and integrated with an application to processes this information.

Related to the main objective, the storage of the information in a database, and a WEB RSS service that contains the information in conformance with CAP format, was achieved.

Acknowledgments. This project is co-financed by CEDIA through the CEPRA projects from 2018–2019. It also has internal financing from the Research Directorate of the University of Cuenca and the *Universidad de las Fuerzas Armadas* ESPE.

References

1. Olmedo, G., Acosta, F., Haro, R., Benavides, N.: Broadcast testing of emergency alert system for digital terrestrial television EWBS in Ecuador. In: Communications in Computer and Information Science, vol. 1004, pp. 176-187. Springer, Cham (2019). https://doi.org/10.1007/978-3-030-23862-9_13
2. DiBEG.: ISDB-T Harmonization Documents Part 3 EWBS (2015). https://www.dibeg.org/techp/aribstd/harmonization.html
3. ARIB.: Data Multiplex Broadcasting System for the Conventional Television Using the Sound Sub Carrier. ARIB STD-B14 (1998). https://www.arib.or.jp/english/std_tr/broadcasting/std-b14.html
4. Morocho, V.,Morales, A., Feijoo, A., Sarmiento, A., Garcia-Almirall, P., Queraltó, P.: Estudio del impacto social, económico y académico de la infraestructura de datos espaciales de la universidad de Cuenca (Ecuador). ACE: architecture, city and environment, vol. 7, no. 19, pp. 255–274. Centre de Política del Sòl i Valoracions-Universitat Politècnica de Catalunya, Spain (2012)
5. Morocho, V., Santander, F.: De los rígidos códigos internos de una IDE a un ambiente gráfico de administración y gestión: IDE Ucuenca V3.5 (2015)
6. Morocho, V., Santander, F.: Los PP GIS como geo-herramienta para el mapeo de zonas vulnerables, en el contexto de la planificación territorial: una revisión bibliográfica. vol. 14, no. 2, pp. 116–125 (2017)
7. Morocho, V., España, D., Serrano, A., Achig, R., Crompvoets, J.: Mass movements detection from UAV images analysis. Remote Sens. Spat. Inf. Sci. Int. Arch. Photogram. **42**, 461–467 (2019)
8. Morocho, V., Achig, R., Santander, F., Bautista, F.: Spatial data infrastructure as the core for activating early alerts using EWBS and interactive applications in digital terrestrial television. In: Advances in Intelligent Systems and Computing, vol. 928, pp. 346–355. Springer, Cham (2019)
9. Kelly, T., Adolph, M.: ITU-T initiatives on climate change. IEEE Commun. Mag., **46**(10), 108–114 (2008)
10. OASIS.: Common Alerting Protocol (2010). http://docs.oasis-open.org/emergency/cap/v1.2/CAP-v1.2.html
11. Eliot, C.: Protocolo Común de Alerta (CAP) (2018). https://etrp.wmo.int/pluginfile.php/16534/mod_resource/content/1/2018-MISC-WDS-CAP-Protocol-Comun-18856_es.pdf
12. WMO: Register of WMO Members Alerting Authorities (2020). https://alerting.worldweather.org
13. Codingpotion: Qué es Angular y sus ventajas. Cómo se instala (2019). https://codingpotions.com/introduccion-instalacion-angular
14. Alvarez, M., Morocho, V., Morales, A., Rosanigo, Z., Jones, G., Lopez, L.: Contributions to the SDI from Latin American Universities - Some Undertaken Initiatives. In: Proceedings of GSDI 13, Quebec, Canada (2012)

Implementation of a Basic Risk Guide for Interactive Digital Terrestrial Television Using Learning Objects

Villie Morocho[1](✉), Paola Cárdenas[1], Lourdes Illescas[2], Jorge Maza-Cordova[3], and Rosario Achig[1]

[1] Computer Science Department, University of Cuenca, Cuenca, Ecuador
villie.morocho@ucuenca.edu.ec
[2] Philosophy Faculty, University of Cuenca, Cuenca, Ecuador
[3] Research Center of UTMACH, Machala, Ecuador

Abstract. The application "Disaster Risk Management Guide" for interactive Digital Terrestrial Television (iDTT), has been implemented based on the design that considers Learning Objects (LO) adapted to iDTT, where the receiving medium is the television and the main element of interaction is the remote control. The contents were worked on in coordination with the National Service for Risk and Emergency Management (SNGRE by its acronym in Spanish) of Ecuador, an organization that was also in charge of the validation. The implementation process, tests and improvements of the application are detailed, which includes three Learning Objects (LO) and a section with geographic information on emergency alerts in real time, obtained from a map server. Each LO addresses informational and educational topics, as well as a self-assessment for the viewer. The methodology for software development has been incremental, based on constant deliveries and progressive advances. The application has been developed in the Ginga NCL (Nested Context Language)-LUA programming language, and the tests have been carried out in two environments: 1) simulation with Ginga-NCL emulators and 2) using the Set Top Box (STB) *EiTV Smartbox* receiver equipment. The tests mainly detail the processes carried out to identify the audio and video formats supported in the two environments, the most appropriate options for integrating text in the interfaces, and the solution of problems encountered during the process. Finally, a usability analysis of the application is included with tests carried out in different age groups.

Keywords: Ginga · NCL · LUA · Interactivity · Set Top Box · ISDBT-TB · Return channel · Usability · DTT · iDTT

1 Introduction

In the main Digital Terrestrial Television (DTT) systems such as the Brazilian middleware Ginga supports declarative applications (it is through its declarative environment Ginga NCL) and procedural applications (it is through its execution or processing environment, Ginga-J) [1]. Ginga-NCL is a logical subsystem, responsible for the processing

M. Botto-Tobar et al. (Eds.): CIT 2020, AISC 1326, pp. 202–215, 2021.
https://doi.org/10.1007/978-3-030-68080-0_15

and presentation of NCL documents. "The programming language NCL has been developed by PUC-Rio to facilitate the specifications of interactivity, space-time synchronization between media objects, adaptability, support of multiple devices, and support of interactive on-live programs" [2]. On the other hand, Ginga-J is the middleware subsystem in charge of defining the Java application program interfaces (API), content, data formats, as well as protocols at the application level [3].

Depending on the functionality required in the design of each application, one paradigm will be more appropriate than the other [4], in addition, a Ginga application does not need to be purely declarative or procedural. Declarative applications often make use of Lua scripts, which is a language of a procedural nature. Also, a declarative application can refer to a built-in JavaTV Xlet. Similarly, a procedural request can reference declarative content, such as graphic content, or it can construct and initiate the presentation of content declarations. Therefore, any type of Ginga application can make use of application environments [3].

1.1 Learning Objects

The support of technologies has allowed innovation in methodologies and resources that strengthen educational processes. One of them the Learning Objects, this term was named for the first time in 1992 by Wayne, who associated LEGO blocks with standardized learning blocks, for reuse purposes in educational processes [5]. The concept of LO or Virtual Learning Object (VLA), is not a unified concept because "there is no consensus in the definition of learning objects. The basic idea allows a wide variety of interpretations" [6], however at an international level and with a more structured concept, a Learning Object is defined as any digital or non-digital entity that can be used, re-used or referenced for learning supported by technology [7].

Learning objects must meet basic characteristics that qualify them as such, however [8] affirms that reusability and granularity represent "the two most important properties of learning objects". The concept of granularity refers to highlighting a conception of objects as small units, which can be coupled or added in various ways, while [9] establish that "reusability is largely a function of the degree of granularity of objects". The reusability of the learning object will largely depend on the degree of granularity of the resource.

1.2 Related Works

There are some studies that present applications developed under the Ginga-NCL architecture, many of these are oriented to Game-type applications. Such as the studies [10] which proposes a framework for the development of games in Ginga. The study [11] where the development process of a training application is presented with a survey, creating libraries of graphic components in Lua and using these components in a Ginga NCL-Lua application called QuizTV. In the last two years of research published in documents, there are some studies related to areas such as telemarketing. For instance, the study [12] focused on the advertising of prepared foods, with navigation through menus using the remote control. In the health area, the study [13] presents a medical care application and reception tests with end users, with the aim of measuring the impact and scope of the project in the region.

In t-learning, studies like [11] propose digital television as a learning tool at home, with a language learning application. However, the study [14] suggests that applications that have the objective of teaching their users, especially when using learning objects, require a multidisciplinary team, which includes experts in the contents of the learning object, software developers, and graphic designers. With the aim of obtaining an application or software tool that is easy to use and flexible in its content, these considerations have been taken into account in all stages of the development of the Ginga NCL-Lua application presented in this document.

On the other hand, in accordance with the study [13], the objective of many of these applications is to increase the access to this type of information by ordinary Latin American citizens, whose access to the Internet is still scarce. "Full Internet connection is currently a privilege of the developed world, and it appears that it will continue to be so for a long time" [13].

This article presents an implementation worked for interactive Digital Terrestrial Television iDTT with content creation, which was supervised by the National Service for Risk and Emergency Management. It describes improvements regarding the interactive application and the use of learning objects oriented and adapted to the digital television format, the result of which is the generation of a Basic Course on Risk Management, using a methodology of design and adaptation of learning objects. The interface design is improved by integrating personnel specialized in the interface design, and a development methodology is used that allows constant validation and optimization of the application. Finally, usability tests are carried out, which allow to analyze the ease with which the user can perform the tasks and apply corrections to the deficiencies found.

2 Materials and Methods

In [2] a comparative study is realized between the two implementation alternatives for iDTT applications supported by the Ginga middleware. This study mentions that Ginga-J is under development and provides little support for developers. It also mentions the advantages that Ginga-NCL offers in the usability of an application for television, relating the usability of an application to the success or failure of the iDTT technology. Table 1 shows a summary of the advantages and disadvantages of the Ginga-NCL vs Ginga-J architectures, in accordance with the group experience and information described in [4].

After this analysis, the Ginga-NCL and LUA implementation scheme was selected for the development of the application in this work.

2.1 Methodology of Application Development

The "Disaster Risk Management Guide" application, described in this document, which can be considered as t-learning adapted to Digital Television, was implemented following the guidelines of the "Agile iDTT" methodology [11]. Technically describes the process

Table 1. Properties of the architectures supported by the Ginga middleware.

Ginga-NCL	Ginga-J
Declarative language	Procedural language
It can make use of procedural languages like LUA for the implementation of business logic for example	It can make use of declarative elements, for textual content or graphic structures
Open source with enough documentation	It is in development
Being declarative follows a high level of abstraction. "An NCL application, it only defines how multimedia objects are structured and related in time and space" [4]	Low level of abstraction in coding. Procedures and screen displays must be implemented
The developer provides a set of tasks that must be performed, without worrying about the details of the interpreter or compiler because the virtual machine takes care of that processing [4]	"The developer has more power over the code and tells it at every step what to do" [4]. It thus provides a more versatile programming environment
NCL allows the use and manipulation of audios, videos, html content among other multimedia elements, by assigning values to the attributes of the various elements that are part of the declarative language	Composed of API (Application Programming Interface) that aims to provide the necessary functionalities for the implementation of applications used in iDTT. With it, it is possible to manipulate videos, audios, texts and even access protocols [15]

to be followed in the development of applications for iDTT and associates the software requirements to the iDTT environment. This considers characteristics such as availability, usability, and efficiency in the use of resources, considering the critical factors in an iDTT environment, such as storage space, visibility, and adequate navigation, considering the television screen and the using the remote control.

In the first place, the scope of the application was defined through an analysis of the functional and non-functional software requirements.

i) Type of service: The interactive application can be executed on the television when a natural disaster event occurs and a television channel I sent the application through a return channel, or it can be executed manually by the viewer,
ii) Use of resources: Images and videos, with low processing performance, allowing the application to be embedded in the Set-Top Box
iii) Accessibility: The application is aimed at the general public and its navigability is through the use of remote control. Subsequently, the hardware/software restrictions, assumptions, and dependencies were analyzed.

The next step was the description of specific requirements that began with an interface design screen. Included considerations of style, location of elements, and colors that helped to de-fine the prototype of the interface, in which non-functional requirements such as performance, usability, and reliability were considered.

After, the LO interfaces were implemented in NCL source code, with the NCL composer development tool. In this app was defined regions, descriptors, and multimedia elements and was provided of navigation functionalities through the *focusindex* defined in the media elements. *Focusindex* is a property of NCL that indicates in which graphic element it is selected and also establishes the order of navigation on the screen.

On the other hand, the adaptation of interfaces of a Learning Object to an application of this type, includes interface considerations in design and navigability. This is because the applications for iDTT have different requirements to those of the software of other traditional platforms [11]. The main differences are in the use of the remote control instead of the mouse and keyboard or touch screens in the case of mobile applications. An application for television must adapt its interfaces and functionalities to those offered by remote control, considering the reduction of the resolution on the screen. This makes the multimedia elements larger and the almost obligatory omission of elements such as radio buttons, scroll bars, or hierarchical menus. In addition, the mobility limitations of using a remote control, which cannot mimic the movement of a mouse. Likewise, the limitations in the use of hardware resources and size of the application to be transmitted, join to television content in a transport stream on iDTT. Due to these limitations, this kind of applications must be non-complex applications without greater content scalability and with a high degree of usability [11].

As mentioned above, the implementation process was based on incremental software development methodology. Thus, the software development life cycle was based on increments and continuous delivery of features that are assigned a priority based on a plan of development of the project. Most of the increment has its own design, coding, and testing phases.

The iDTT application titled "Disaster Risk Management Guide" was implemented, mainly in Ginga - NCL 3.0 under the EDTV (Enhanced Definition Television) profile. EDTV refers to digital TV applications that include video and sound with intermediate quality, so that it provides quality in images but allows the application not to be as heavy as HDTV. It was developed in declarative language the graphical interfaces corresponding to the three Learning Objects that are part of the application.

These LO are:

- LO1. Introduction to Disaster Risk Management
- LO2. Family Emergency Plan and Self-Protection Measures
- LO3. Map of risks, resources and capacities

In the last option, there is georeferenced information obtained in real-time from a map server, which complements the functionality of the application. This information is integrated into an SDI (Spatial Data Infrastructure) where created maps for geo-tools can contribute as presents in [16]. This option was developed in the LUA programming language, which allows the viewer to be a recipient of information through events TCP class. It makes http requests to the server to display the processed result on the TV screen.

On the other hand, for development in the NCL declarative language, the usefulness of the NCL Composer tool [17] stands out. This allows a better understanding of the application's structure through the graphical view (Fig. 1). Thus alleviates the inconvenience because NCL, being a declarative language based on XML, tends to generate many lines of code and become complex due to the references to elements that are located in different areas of the NCL document.

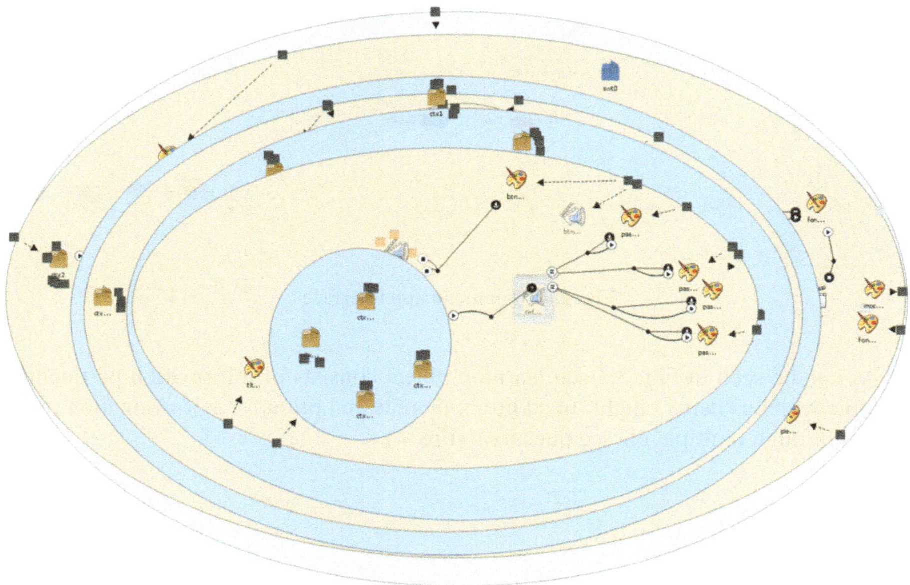

Fig. 1. Structural view of the application in NCL composer

Figure 1 the links that describe actions and events of the multimedia elements are displayed. Contexts containing media, contexts, switch, among others, are also displayed. A "switch" element allows Boolean logic rules to initiate one or another internal context. This function of the NCL language has been used within the application to choose options from the various menus that can be found in the application.

3 Results and Discussions

3.1 Results

The functionalities developed during the implementation process of the application can be seen in Fig. 2. In it, the main interface is presented, where the different elements of it can be seen: 1) main title, 2) video introduction, 3) graphic options of the main menu, which consists of access to the three learning objects and the visualization of live alerts, 4) interaction buttons on the remote control and 5) banner with application information.

Fig. 2. Main menu app interface

As can be seen in Fig. 3, each learning object consists of information pertinent to the topic which is being taught. In addition, there is an option for self-evaluation of the content through multiple-choice questions (Fig. 4).

Fig. 3. Content of a learning object

The general scheme of the "Disaster Risk Management" application is as follows:

Option 1: Introduction to Disaster Risk Management

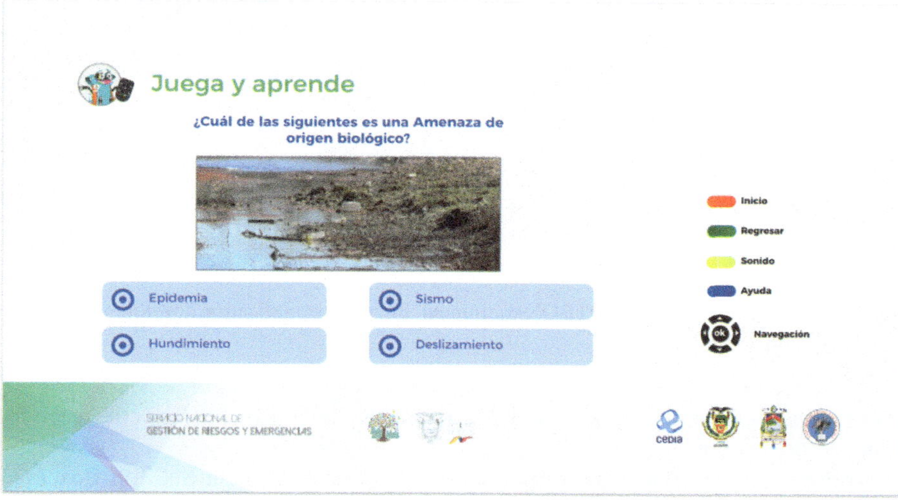

Fig. 4. Evaluation of a learning object

- Historical review
- Disaster risk factors
- Activities

Option 2: Family Emergency Plan and Self-Protection Measures

- Basic risk management terminology
- Family emergency plan
- Self-protection measures
- Play and learn

Option 3: Map of risks, resources and capacities

- Risk map
- Actions and responsible parties in the Community Plan
- Play and learn

Option 4: Live Alert Notifications

In this option there is a connection with server map on line.

Application Functionality Tests

In the implementation process, two kinds of tests were carried out on the different elements of the application. One of them in a simulation environment [18] and the other on the Set Top Box *EiTV Smartbox* receiver equipment (Fig. 5). During this process, the functionalities, loading times, and resolution of the multimedia content were evaluated. Also, tests yielded enough information for subsequent improvement or correction app. These activities were necessary because the Ginga middle-ware turns out to be highly variable in different contexts. In other words, some functionalities that are successful in the Ginga emulator on a PC are not always fully functional in a more real environment with the use of receiving equipment hardware. Table 2 describes some functional

Fig. 5. Set-Top Box EITV SMARTBOX equipment. Interactive application development receiver and decoder for ISDBT-Tb

results of the application "Disaster Risk Management Guide" executed over the receiving equipment hardware.

In Fig. 6 test interfaces performed on the receiver equipment Set Top Box *EiTV Smartbox* are presented.

Finally, in the last test, Ginga applications transmission is carried out through a TS (Transport Stream) frame in a digital television signal, obtaining preliminary results. But, the performance is affected by the size of the application. Therefore, this remains an open research line that will be studied in the future.

Usability Evaluation

To evaluate the usability of the application, a method based on the real experience of users was used. Empirical tests of the interface design to characteristic users under controlled conditions representative of reality [19–22] was applied.

Four aspects were considered:

- On-screen layout: related to the location of content and interactive applications on the TV screen.
- Applications: functional characteristics of interactive applications.
- Navigation: represents the aspects of the design of the navigation of content and interactive applications.
- Format: it has to do with the format that content and interactive applications must have (text, colors, etc.).

The results of the evaluation are presented in Fig. 7. In reference to the aspects evaluated, the results of "Applications" and "Format" exceed 80%, so it can be deduced that the application is more intuitive for the user, allowing better performance in the application's functionalities. Regarding "Design on-screen" and "Navigation", a percentage of less than 80% was obtained, therefore it can be interpreted that the position of certain elements was not the most appropriate and this makes navigation between the contents is not very intuitive, it should be noted that for certain users it was a new browsing experience that they had just tried. Therefore, as a general result of the evaluation, we can mention that users are satisfied with the interactivity functionalities, content navigation and with the appearance of the design in terms of color, typography and composition.

3.2 Discussions

In the tough task of implementing an iDTT application, several drawbacks had to be overcome. From a support point of view, there is very limited support nationally and internationally. There is also not enough achievable documentation, so many of the achievements were based on trial and error. From the point of view of national research

Table 2. *Functionality* tests on the receiving equipment Set Top Box *EiTV Smartbox*

Functionality	Set Top Box *EiTV Smartbox*	Implemented solutions
Remote control keys	When pressed, the "OK" key sets the variable service.currentFocus and assigns a value to the focusSrc property and the media element is selected The back, info and number keys (0–9) are recognized correctly	Se establece un *link "onSelectionStopStartSet_var"* para desactivar, activar y ubicar nuevamente el elemento seleccionado permitiendo la navegación sobre otras opciones del menú The back, info and number keys (0–9) perform functions of the television, but the actions programmed by the application are not recognized
Return to the previous screen	The functionality of returning to the previous screen ends the current context, and does not start the previous context because the attributes of the closed context are still loaded in memory	Additionally, stop type events are established to switch elements and internal contexts and the attributes of the elements are re-established to the values assigned before the start of the context to close
Audio Playback	The **mp3** audio format does not play in the application, no matter what attribute is assigned to the media element that contains the audio	The audios of the application were converted to mp2 format, since this and others such as wav format are reproduced correctly in the EiTV decoder
Playing Videos	Videos in mp4 format are played but the resizing attribute does not apply when the video is less than or equal to **four seconds long**	More than four seconds video duration was established for ever video that the application includes
Texts in flat files	Texts with special characters such as accents are displayed correctly, displaying legible text on the TV	By obtaining a favorable result on the STB not considered setting encoding in Ginga Ginga emulators

(*continued*)

Table 2. (*continued*)

Functionality	Set Top Box *EiTV Smartbox*	Implemented solutions
Html content	**HTML content** is displayed correctly after a considerable load time (5–10 s.). But when loading the content a second time the application tends to hang or close	Plain text files were used assigning color and font properties
Attributes of < media > elements in NCL	**Type attribute**: when a value is assigned to this field, for example in mp2 audio elements and png or jpg images, the elements are not shown on the screen	Attributes such as type do not need to have a value assigned for the presentation of multimedia elements
NCL element loading	Fast loading time, in a matter of milliseconds and no more than 3 s for content (text, audio, video)	Loading time does not cause annoyance to viewers (based on usability tests)
Attribute assignment to LUA *media* elements	The attributes and functionalities programmed in LUA are correctly reflected in the application as well as the passing parameters between languages	
Resizing images on televisions with different size screens	The images from both LUA and NCL are correctly adjusted to the screen size on which the application is running	The application correctly displays the interface in a real environment, such as an STB, so adjustments are not necessary

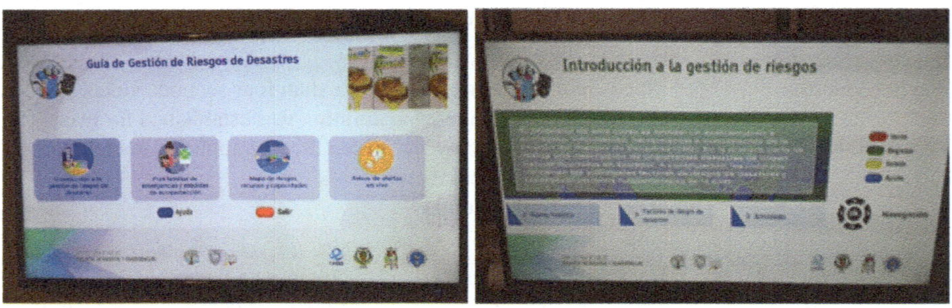

Fig. 6. Application tests in Set top box *EiTV Smartbox*

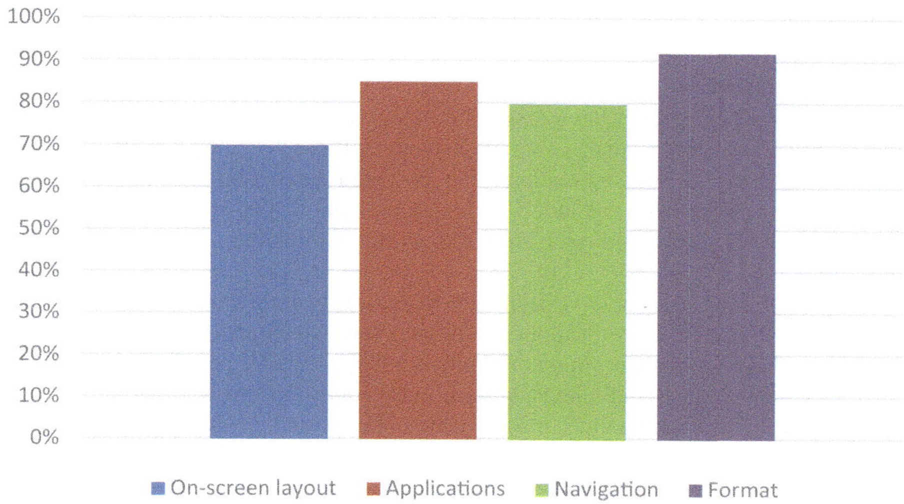

Fig. 7. Usability evaluation results

groups, there are also few who are working on the issue. Although it is highly developed internationally, countries like North America and others from Asia have complete risk and emergency management systems on digital television. And finally, from the point of view of manufacturers, few are interested in developing interactivity, since it is less costly for them in the production chain to introduce these elements in both hardware and software.

The contribution of the article with the table of solutions presented offers detailed steps to the researchers who are interested in solving implementation problems both in simulation and environment. This work demonstrates the possibilities of using technology, as well as the predisposition of some state agencies to develop it. However, a greater push is needed from the entire research community to open up new elements to use the iDTT, as well as from content production companies, the television channels themselves, and finally the television assembly industries.

The empowerment of the citizen to use the technological advantages present in interactivity is necessary to awaken. Despite the fact that technological obsolescence is 2 to 5 years depending on the equipment, it is necessary that in the new technological wave it is considered a real use of interactivity for the different public facilities, as well as other benefits such as natural disaster management.

4 Conclusions

The implementation of applications for iDTT has great limitations that could be seen as inconveniences when compared to other languages. However, the main advantage of this technology is to be able to reach the largest number of citizens who have a television team and who, with the planned transition from the national government, will have to evolve to DTT teams. Understanding that broadcasting applications over the air would allow all of these viewers to have access to an application without the need for the internet is a real advantage.

This article presents the advantages and disadvantages of implementing specialized languages for iDTT. By not having specific manuals or common details in other languages, as well as being a small group of developers. It has become an endless trial and error to ensure that the implementation is the most appropriate. On the other hand, there have been two philosophies included in this article, that of Learning Objects and that of Usability, each corresponding to its field but which allow the presented result to be complete.

The potential of an interactive digital television application would even allow communicating with the citizen with applications that allow understanding not only to people related to technology such as young millennials but also to older people. It is much easier to seek the use of digital television with older adults than to wait for an adequate response with screens as small as those of cell phones. This hypothesis is demonstrated with the usability tests described in this article.

Acknowledgments. This project received co-financing from CEDIA through the CEPRA projects for the year 2018–2019. It also had internal funding from the Research Department of the University of Cuenca. In addition to the important contribution of the research groups of the UTMACH and digital television of the ESPE.

References

1. Soares, L.F.G., Rodrigues, R.F., Moreno, M.F.: Ginga-NCL: the declarative environment of the Brazilian digital TV system. J. Braz. Comput. Soc. **13**(1), 37–46 (2007)
2. Muñoz, T., Siguenza, A.: TV digital fija utilizando middleware ginga-ncl aplicado a un noticiero digital, Universidad de Cuenca (2012)
3. de Souza Filho, G.L., Leite, L.E.C., Batista, C.E.C.F.: Ginga-J: the procedural middleware for the Brazilian digital tv system. J. Braz. Comput. Soc. **13**(1), 47–56 (2007)
4. Zancanaro, A., Santos, P.M., Todesco, J.L.: Ginga-J ou Ginga-NCL: características das linguagens de desenvolvimento de recursos interativos para a TV Digital. In: I Simpósio Internacional de Televisão Digital, pp. 1084–1108 (2009)
5. Hodgins, H.W.: Into the future: A vision paper for American Society for Training and Development. National Governors' Association. Commission on Technology and Adult Learning (2000)
6. Downes, S.: Learning objects: resources for distance education worldwide. Int. Rev. Res. Open Distrib. Learn. **2**(1) (2001)
7. Barker, P.: What is IEEE learning object metadata/IMS learning resource metadata. CETIS Standards Briefing Series, JISC (Joint Information Systems Committee of the Universities' Funding Councils) (2005)
8. Wiley, D.A.: Connecting learning objects to instructional design theory: a definition, a metaphor, and a taxonomy. Instr. Use Learn. Objects **2830**(435), 1–35 (2000)
9. South, J.B., Monson, D.W.: A University-wide System for Creating. Capturing, and Delivering Learning Objects (2003). https://reusability.org/read/chapters/south.doc
10. Barboza, D.C., Clua, E.W.G.: Ginga game: a framework for game development for the interactive digital television. In: 2009 VIII Brazilian Symposium on Games and Digital Entertainment, pp. 162–167 (2009)

11. Quevedo, E., Cárdenas, P., Mejía, M., Muñoz, L.: ShimiTV, a T-learning system for Android TV. In: 2018 XIII Latin American Conference on Learning Technologies (LACLO), pp. 73–76 (2018)
12. Arroyo Arroyo, J.D.: Desarrollo de una aplicación interactiva de promoción de un producto y/o un servicio mediante la plataforma Ginga NCL/Lua para el estándar ISDB-TB de Televisión Digital y con canal de retorno. B.S. thesis, Quito (2020)
13. Kieling, A., et al.: Healthcare application: an analysis of reception studies in interactive digital television. Journalism **9**(5), 235–242 (2019)
14. Damasceno, A.L.B., Busson, A.J.G., Lima, T.S., Neto, C.S.S.: Authoring hypervideos learning objects. In: Roesler, V., Barrére, E., Willrich, R. (eds.) Special Topics in Multimedia, pp. 149–181. IoT and Web Technologies. Springer, Cham (2020)
15. Kulesza, R., Lima, J.F., Guedes, Á.L., Junior, L.L., Meira, S.R., Guido Filho, L.S. : Ginga-J-an open Java-based application environment for interactive digital television services. In: IFIP International Conference on Open Source Systems, pp. 34–49 (2011)
16. Morocho, V., España, A., Serrano, C., Achig, R., Crompvoets, J.: Mass movements detection from UAV images analysis. In: The International Archives of the Photogrammetry, Remote Sensing and Spatial Information Sciences, vol. XLII-2/W13, pp. 461–467 (2019)
17. Azevedo, R.G.A., Araújo, E.C., Lima, B., Soares, L.F.G., Moreno, M.F.: Composer: meeting non-functional aspects of hypermedia authoring environment. Multimedia Tools Appl. **70**(2), 1199–1228 (2014)
18. Morocho, V., Cárdenas, P., Illescas, L., Achig, R.: Proceso de implementación de un software de capacitación en Televisión Digital Interactiva para gestión de riesgos y eventos provocados por fenómenos naturales, Río de Janeiro, Brasil, pp. 3–14 (2019). https://sedici.unlp.edu.ar/handle/10915/90396
19. Febles Estrada, A., Pozo, D.: Evaluando la usabilidad en aplicaciones para la TVDi (2013)
20. Pantoja, W.L., Arciniegas, J.L., Collazos, C.A., Chica, L.G.: Método para evaluar la usabilidad de aplicaciones de televisión digital interactiva: RECOEXTDI. Revista UIS Ingenierías **15**(1), 15–26 (2016)
21. Rusu, C., Solano, A.F., Collazos, C.A., Arciniegas, J.L.: Evaluando la usabilidad de aplicaciones interactivas en entornos de Televisión Digital (2013)
22. Solano, A.F., Collazos, C.A.C., Arciniegas, J.L.: Evaluando la usabilidad de aplicaciones de televisión digital interactiva desde una perspectiva colaborativa. Entre Ciencia e Ingeniería **6**(12), 43–49 (2012)

Using H5P Services to Enhance the Student Evaluation Process in Programming Courses at the Universidad Politécnica Salesiana (Guayaquil, Ecuador)

Joe Llerena-Izquierdo(✉) and Jonathan Zamora-Galindo

Universidad Politécnica Salesiana, Guayaquil, Ecuador
jllerena@ups.edu.ec, jzamorag@est.ups.edu.ec

Abstract. This study analyzes the effectiveness of H5P services used to create interactive student evaluations on a Moodle platform. The content evaluated in this study comes from the Universidad Politécnica Salesiana (Guayaquil, Ecuador) Programing Course syllabus. Study participants are first year university students (18–20 years old). The students participate in a learning validation experience through Python Experiential Educational Platform (PEEP), designed for this purpose. A 75% improvement in retention of acquired knowledge is achieved by using active learning strategies during assessment. With H5P technology professors can reuse audio and video material previously recorded, together with the material they generate in each new class, allowing them to reuse resources not only for informative or didactic explanation, but also as an interactive evaluation resource made possible by interactive technologies designed for this purpose.

Keywords: Interactive assessments · Active learning · Technology enhanced learning · H5P

1 Introduction

1.1 Evaluation Strategies for an Active Learning Processes

The current generation of youth studying at the postsecondary level are known as the generation of "digital natives", as their daily activities involve the continuous use of technological instruments [1]. To ensure an effective teaching – learning processes their teachers must constantly incorporate new technologies and develop new methodologies to effectively achieve specific learning goals while also participating in networks of educational professionals to share knowledge through the exploration of didactic resources that facilitate an effective educational process [2, 3].

Since March, 2020 due to restrictions in place to combat the COVID 19 pandemic [4], Ecuador requires social isolation measures. Technology and technological services have become an increasingly necessary support for teachers that use them for online, virtual and distance education [5]. Professors must define instructional strategies that

M. Botto-Tobar et al. (Eds.): CIT 2020, AISC 1326, pp. 216–227, 2021.
https://doi.org/10.1007/978-3-030-68080-0_16

guarantee learning in virtual environments [6]. The pressing question for educational researchers in the coming years is how to create an active, effective, sustainable learning process that combines online learning with autonomous student work [7, 8].

An active learning process implies that a student is involved in teacher planned activities in an autonomous learning scenario [9, 10], such as study groups or work exchanges between students as well experiential stimulation processes such as the use of simulators [11, 12] or activities based on projects or challenges. The student learns at his or her own pace on his or her own terms and the role of the professor is that of a learning guide [13]. For professors the question arises "how can learning evaluations be optimized in this day and age?" The question opens the door to the exploration of interactive dynamic evaluation methods through known technological platforms [14, 15]. The use of new technologies or technological services within platforms that facilitate the continuous development of an active teaching-learning process poses an interesting challenge for professors as they must learn to use the new technology to potentiate student learning.

1.2 Technology Supported Learning and Interactive Evaluations

With the arrival of new technologies that offer educational services [16], learning experts worldwide agree that to ensure active and flexible learning, validation strategies must not only evaluate knowledge but must also motivate attention, assimilation and retention of knowledge through experiential approaches [17].

The application of immersive technologies [18–20] during a learning process requires that students have devices capable of producing an active experience whereas internet based technological services [21, 22] allow an experience with similar characteristics while hosting infrastructure in a virtual environment [23, 24].

H5P [25] services offer open source technologies for interactive content generation. Its philosophy of creating, sharing and reusing teaching resources with HyperText Markup Language and version 5 (HTML5), compatible with integration through platforms such as Blackboard, Brightspace, Canvas, Moodle among other learning management systems (LMS) and virtual learning environments (VLE), allow public and private institutions, educational institutions in particular, to create interactive learning strategies[1] while optimizing reusable evaluations [26, 27].

2 Materials and Methods

This study uses an empirical methodology with a quantitative approach in a quasi-experimental model. It applies the survey technique to four groups of first-year university students. Sixty-eight students out of a total of one hundred and seven participated in the experience of measuring their knowledge through two types of evaluations: the first a questionnaire tool through the VLE platform (in Moodle) [28] and the second through the interactive H5P tool (integrated in the Moodle platform). This study has a confidence level of 89% and a margin of error of 5.9%.

[1] Available at https://h5p.org/.

The student participants are enrolled at the Universidad Politécnica Salesiana in Ecuador. The University has three campuses located in the cities of Guayaquil, Quito and Cuenca (main campus). This study is conducted on the Guayaquil campus. The syllabus of the Programing course is modified to incorporate H5P technology, introducing interactive course videos, which can then be used during the evaluation process. A Moodle based learning platform hosted at https://academiadevcode.com/moodle30/ is used for the experiment. In this setup, the administrator configures the H5P integration, a professor prepares appropriate instructional content and students are assigned a username and password to participate (see Fig. 1).

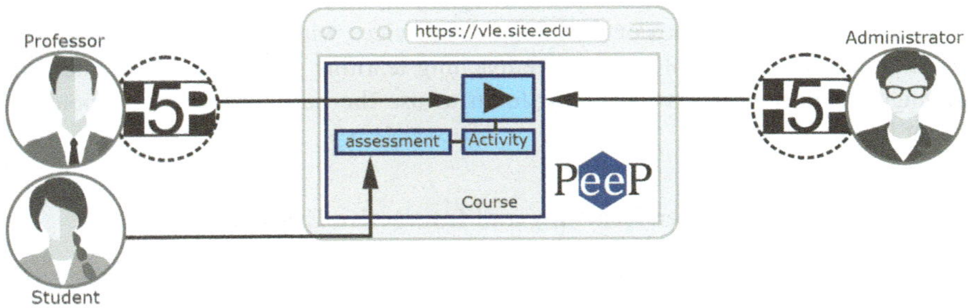

Fig. 1. H5P implementation diagram to prepare an activity within a virtual course

Today's society increasingly demands study methods that use communication technologies with adequate instructional design and technological tools, where a teacher can rapidly manage and development interactive activities for learning and evaluation [25]. Traditional teaching methods used in face-to-face classroom education have proven ineffective with the arrival of COVID-19 restrictions which have pushed schools to virtual environments. Teachers are now looking for new, creative, easy to use methods to promote the teaching-learning process in a virtual environment. H5P technology improves a variety of didactic resources by creating interactive experiences from traditional learning resources through its inter-active HTML5 package (see Fig. 2) [29].

Although many different digital learning tools are currently available, they are often expensive and difficult for teachers and students to manage [30]. H5P allows an appropriate and adaptable integration into learning systems with which teachers and students are already familiar (see Table 1).

H5P is a software as a service (SaaS) solution that, through technological resources, uses a standard integrated communication protocol to provide internet-based support. It distributes a software resource with technology available on its servers, known as Learning Tool Interoperability (LTI) that can be used in a transparent way by a variety of clients. Table 1 shows the different integrations available for H5P applications[2].

[2] Obtained from https://h5p.org/node/287135.

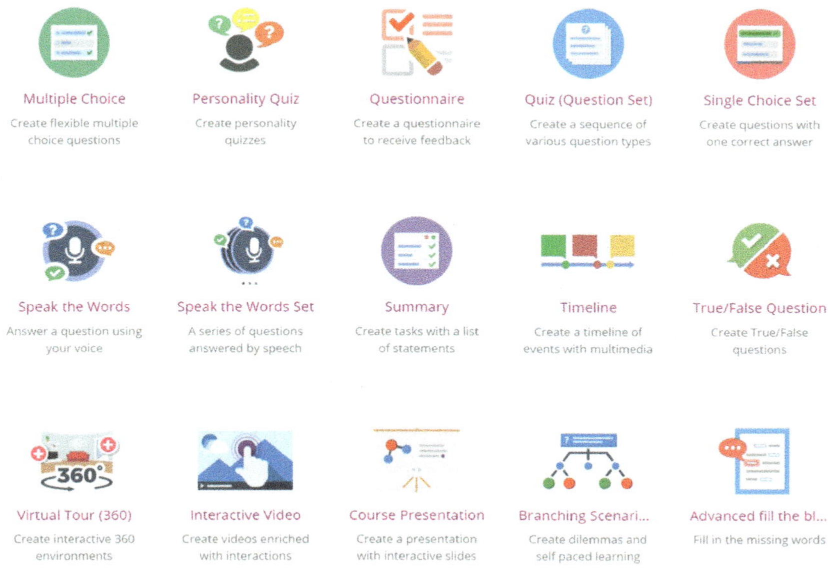

Fig. 2. H5P applications available at H5P.org

Table 1. Types of H5P integrations

Systems	Embed	Integrations via LTI	Plugins
Canvas	X	X	
Brightspace	X	X	
Blackboard	X	X	
Moodle	X	X	X
WordPress	X	X	X
Drupal	X		X
Sakai and others	X	* (systems that supports LTI)	

2.1 Python Experimental Educational Platforms

This study evaluates learning achievements of first-year engineering students enrolled in a programming course at the Universidad Politécnica Salesiana (Guayaquil, Ecuador) through a Python Experimental Educational Platform (PEEP) hosted in a virtual Moodle 3.9 environment integrated with H5P. PEEP[3] is used to create unique virtual instructional spaces including digital materials, collaborative work (discussion forums, chats, emails, etc.) and interactive evaluation material meeting the specific needs of users (both students and teachers) in a specific study program (Fig. 3).

[3] Available at https://academiadevcode.com/moodle30/login/index.php.

Fig. 3. Python experimental educational platform web portal

The structure of the platform on Moodle allows for the integration of H5P, adapting the platform's own classroom resources and turning them into interactive learning objects. By using the Learning Tool Interoperability (LTI) it incorporates objects with different formats such as video formats or links from YouTube to integrate them into VLE and enrich them with H5P technology (tool provider) for interaction (see Fig. 4).

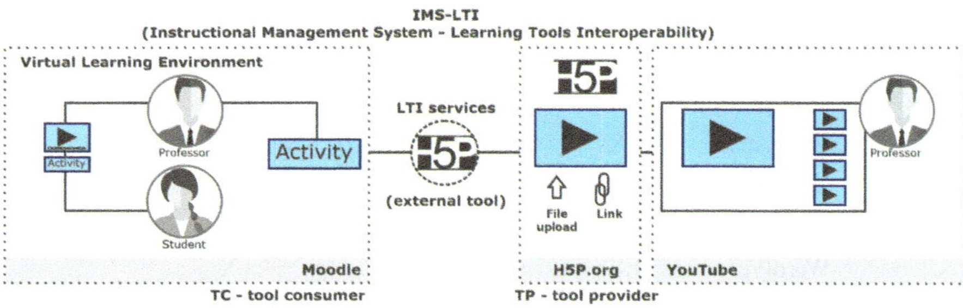

Fig. 4. Structure to integrate H5P in virtual learning

For example, the interactive videos of a teacher hosted on a YouTube channel, created for a class, can be reused in the VLE by automatically incorporating interactive questions (tool consumer) assigned at times, and specific to each activity. The student responds to the questions as he or she watches the prerecorded class, verifying acquired knowledge and testing comprehension by choosing an answer (in real time), and finally receiving the total score achieved at the end of the video. A traditional multiple-choice questionnaire is improved by using it as a dynamic interaction learning tool (see Fig. 4).

The study conducted two evaluations at different times with questions about the Python programming language and good practices for developing programs (see Fig. 5). Students were invited to participate in the evaluation experience using new technologies. Of a universe of one hundred and seven students (four courses of engineering students) sixty-eight students freely chose to participate in the study. Two evaluations are applied,

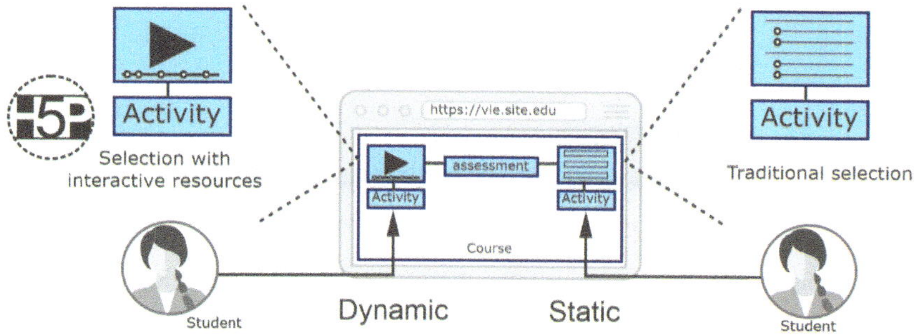

Fig. 5. Interactive technology from the H5P package applied to learning resources

one without H5P technology and one incorporating H5P technology in the teacher's previously developed evaluation resource. The first test used a VLE questionnaire whereas the second used an interactive resource integrated with H5P in the VLE. After completing the tests, a survey was applied to teachers and students to gather data about the design of the learning environment using H5P resources, learning motivation and its evaluation, and satisfaction levels while using the PEEP integrated with H5P.

3 Results and Discussion

Students participating in the PEEP testing were assigned a username and password to authenticate their identities. Sixty-eight students volunteered to participate in the study, 64% of the total universe. Participants had the following age distribution: 10% under 18 years old, 78% between 18 and 20 years old, and 7% older than 24 years old (see Fig. 6). The heterogeneity in students' ages allows us to understand generational differences in classroom interaction and how students benefit or understand differently the use of technology in an evaluation process. Students adaptation to new technologies is improved when the new technology is used consistently.

Two tests are given: the first with traditional content we will define as "static" and the second integrated with interactive H5P resources we will define as "dynamic".

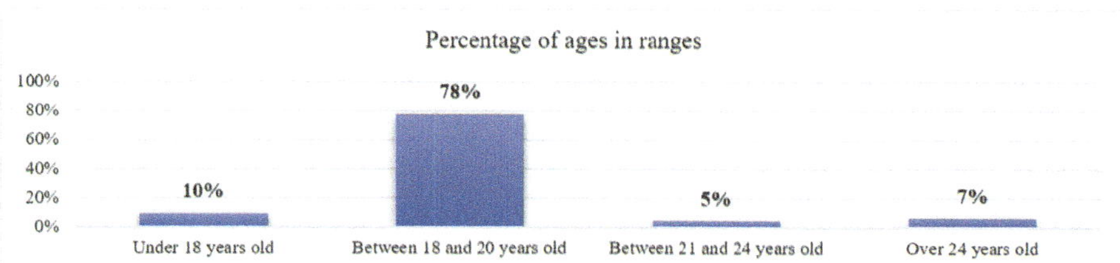

Fig. 6. Range of student age distribution, percentage

The results of the evaluations show a variation in the grade obtained by the students before and after using H5P. 6% receive a lower grade, 19% maintain the same grade and 75% improve their grade (see Fig. 7).

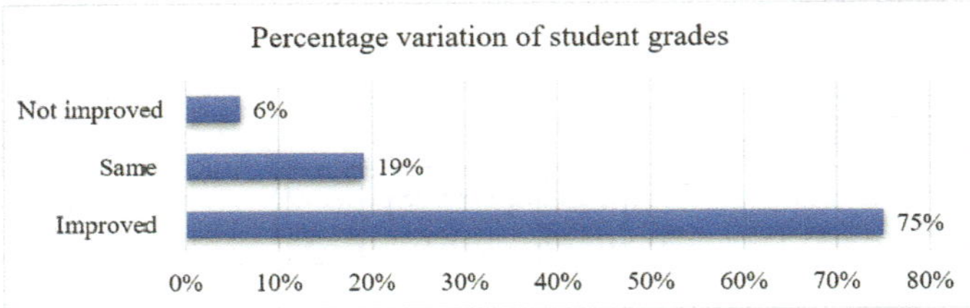

Fig. 7. Variation of students' grades after the application of the evaluations, percentage

The results show the percent that grades improved or reproved (see Fig. 8). The group that received a lower grade on the second test received generally high grades on the first test (7, 9, 10 and 9 out of 10 on the first test and 6, 8, 6 and 4 on the second test). This data shows that for the second evaluation to be effective participants must understand the use of technology and the way to select the correct response, and they must be more observant during the evaluation. Furthermore, these results indicate that the connectivity and proper use of the device used by the student should not affect the evaluation process, although this possibility remains open.

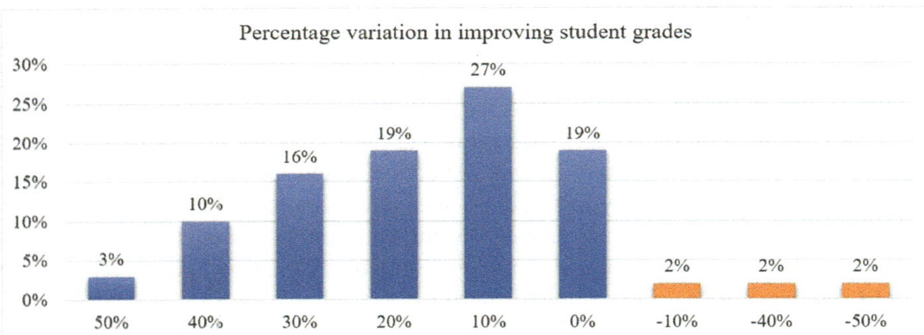

Fig. 8. Variation of student grades in the respective percentages of improvement, percentage

Participants were asked to evaluate on a scale of 1 to 5, 5 being the highest evaluation, the assessment potential of the H5P interaction tool applied in the evaluation process. 62% of the sample valued it as 5, 28% as 4, and 10% as 3, showing that 90% of participants accept the tool. No participants evaluated the tool's assessment potential as 1 or 2 (see Fig. 9).

To the question "By using videos with automatic questions from time to time, do you believe that student attention improves?" 97% responded that "Yes, attention improves" and only 3% replied that "No, attention does not improve", implying that attention improves when students must answer interactive questions such as those generated through the application of H5P resources (see Fig. 10).

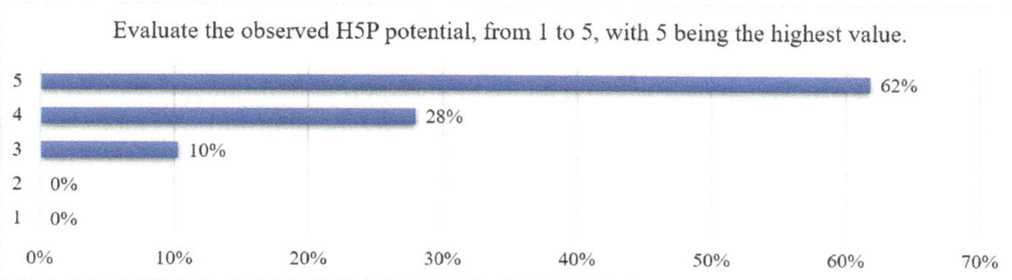

Fig. 9. Assessment of the potential of H5P tools for learning, percentage

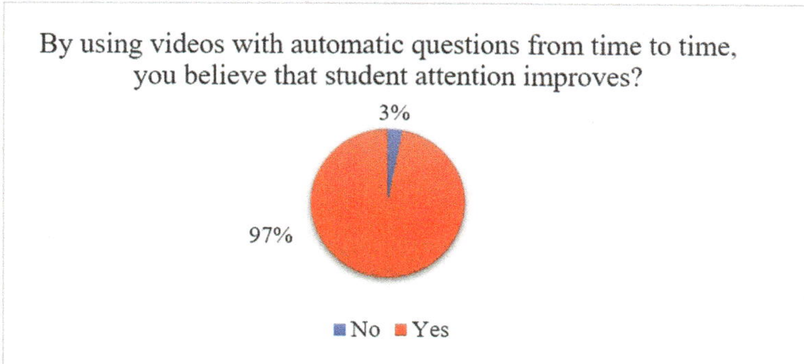

Fig. 10. Perception on whether the use of videos with interactive questions improves the attention of students, percentage

When asked if technology and technological services can improve learning, the results show that 59% believe that learning can be improved by 100%, 29% believe that learning can be improved by 75%, 9% believe that learning can be improved by 50% and 3% believe that learning can be improved by 25%. This means that 88% of participants believe that technology and technological services can improve learning by at least 75% (see Fig. 11).

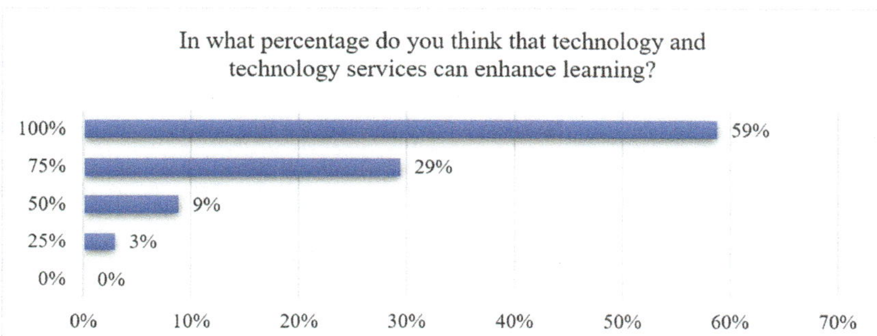

Fig. 11. Perception on whether technological technologies and services enhance learning, percentage

Lastly, participants were asked about their motivation and interest in the instructional content used for the development of the teaching and learning of the Programming subject through interactive resources housed in the Python Experimental Educational Platform, PEEP. 94% indicated that interest in the content used on the platform was motivated by the interactive resources used by the teacher during the Programming class (see Fig. 12).

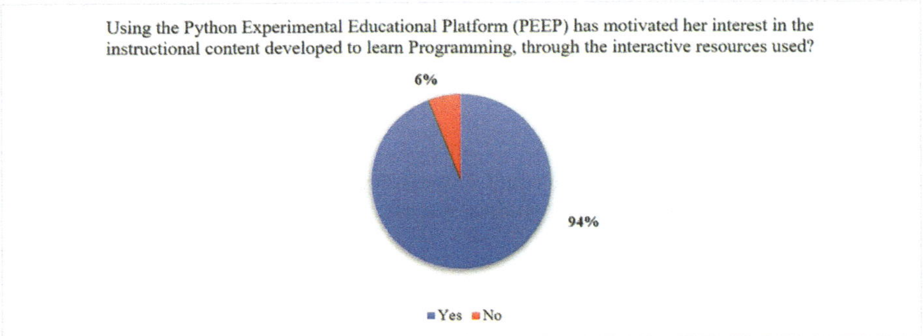

Fig. 12. Motivation generated while using the interactive resources of the Python Experimental Educational Platform (PEEP) to learn programming, percentage

Currently, choosing instructional content means choosing tools that not only permit the replication of a classroom activity in a virtual environment, but also empowering learning even more than the same activity would have in a traditional classroom. According to the research contained in the Bibliography and the author's experience, there exist technology and technological services that allow a teacher to develop a learning activity (for example, an interactive video) that can also enhance said learning. Teachers must assess the learning results that are expected to be obtained from the application of a new technological resource versus the time it will take for the teacher and students to master the technology to prioritize (or not) its use in a learning environment; this subject continues to be open to discussion.

4 Conclusions

Active learning allows a student to involve him or herself in a more dynamic way, stimulating abilities that allow detailed observation, acuity in attention and strength knowledge retention, increasing interest in the proposed activity and enhancing learning. Active learning evaluations involve techniques that make use of existing technologies or technological services chosen specifically for their applicability from the teacher's experience.

Students are motivated when they are involved in dynamic, interactive activities. The new generation of students accepts multimedia resources as they already have great familiarity with them from their daily lives. Teachers increasingly require technological tools that are easy to manage and effectively achieve learning goals.

Worldwide, communities of professionals, technicians and developers have services that can be applied to digital educational resources that can be reused to improve and augment learning. Teachers can validate technological educational tools through a constant exchange of experience between professionals and specialized entities examining changes in student achievement when using technological tools. This study also shows that traditional methodologies do not need to be replaced by digital resources, rather that technology can strengthen traditional methodologies making them more effective for virtual learning.

This study shows that 75% of participants improve their learning process when using interactive technology. More research can be conducted to further understand how different learning tools can be strengthen by technologies to further achieve the learning goals of university students. This study provides direction for others analyzing the feasibility of using interactive technologies to improve inclusive learning.

References

1. Li, Y., Wang, Q., Lei, J.: Exploring technology professional development needs of digital immigrant teachers and digital native teachers in China. Int. J. Inf. Commun. Technol. Educ. **16**, 15–29 (2020). https://doi.org/10.4018/ijicte.2020070102
2. Prenger, R., Poortman, C.L., Handelzalts, A.: Professional learning networks: from teacher learning to school improvement? J. Educ. Chang. (2020). https://doi.org/10.1007/s10833-020-09383-2
3. Wahyuni, D.: Improvement of teacher competence towards industrial revolution era 4.0. In: Pusat Penelitian Badan Keahlian DPR RI, pp. 350–352. Atlantis Press, Paris (2018). https://doi.org/10.2991/assehr.k.200620.068
4. University Johns Hopkins: COVID-19 dashboard by the Center for Systems Science and Engineering (CSSE)
5. Llerena-Izquierdo, J., Valverde-Macias, A.: Google Classroom as a Blended Learning and M-learning Strategy for Training Representatives of the Student Federation of the Salesian Polytechnic University (Guayaquil, Ecuador). Presented at the July 27 (2021). https://doi.org/10.1007/978-3-030-59194-6_32
6. Porczyńska-Ciszewska, A.: The use of e-learning in education for people with mild intellectual disability in the context of their mental well-being (2019)
7. Bao, W.: COVID-19 and online teaching in higher education: a case study of Peking university. Hum. Behav. Emerg. Technol. **2**, 113–115 (2020). https://doi.org/10.1002/hbe2.191
8. Ayala Carabajo, R.: Pedagogical relationship: max van manen's pedagogy in the sources of educational experience. Rev. Complut. Educ. **29**, 27–41 (2018). https://doi.org/10.5209/RCED.51925
9. Huguet, C., Pearse, J., Noè, L.F., Valencia, D.M., Ruiz, N.C., Heredia, A.J., Avedaño, M.A.P.: Improving the motivation of students in a large introductory geoscience course through active learning. J. Geosci. Educ. **68**, 20–32 (2020). https://doi.org/10.1080/10899995.2019.1588489
10. Díaz Redondo, R.P., Caeiro Rodríguez, M., López Escobar, J.J., Fernández Vilas, A.: Integrating micro-learning content in traditional e-learning platforms. Multimed. Tools Appl. **80**, 1–31 (2020). https://doi.org/10.1007/s11042-020-09523-z
11. Stranford, S.A., Owen, J.A., Mercer, F., Pollock, R.R.: Active learning and technology approaches for teaching immunology to undergraduate students. Front. Public Heal. **8**, 114 (2020). https://doi.org/10.3389/fpubh.2020.00114

12. Hodges, L.C.: Student engagement in active learning classes. In: Active Learning in College Science, pp. 27–41. Springer International Publishing, Cham (2020). https://doi.org/10.1007/978-3-030-33600-4_3
13. Ruiz, H.: ¿Cómo aprendemos?. Una aproximación científica al aprendizaje y la enseñanza. Graó. 3–23 (2020)
14. Mpungose, C.B.: Is Moodle or WhatsApp the preferred e-learning platform at a South African university? First-year students' experiences. Educ. Inf. Technol. **25**, 927–941 (2020). https://doi.org/10.1007/s10639-019-10005-5
15. Dahdal, S.: Using the WhatsApp social media application for active learning. J. Educ. Technol. Syst. 004723952092830 (2020). https://doi.org/10.1177/0047239520928307
16. Zayapragassarazan, Z.: COVID-19 : Strategies for online engagement of remote learners COVID-19 : strategies for online engagement of remote learners department of medical education jawaharlal institute of postgraduate medical. F1000 Research, **9**, 1–11 (2020). https://doi.org/10.7490/f1000research.1117835.1
17. Nerantzi, C.: The use of peer instruction and flipped learning to support flexible blended learning during and after the COVID-19 pandemic. Int. J. Manag. Appl. Res. **7**, 184–195 (2020)
18. Izquierdo, J.L., Alfonso, M.R., Zambrano, M.A., Segovia, J.G.: Aplicación móvil para fortalecer el aprendizaje de ajedrez en estudiantes de escuela utilizando realidad aumentada y m-learning. Rev. Ibérica Sist. e Tecnol. Informação, 120–133 (2019)
19. Llerena, J., Andina, M., Grijalva, J.: Mobile application to promote the Malecón 2000 tourism using augmented reality and geolocation. In: Proceedings - 3rd International Conference on Information Systems and Computer Science, INCISCOS 2018 (2018). https://doi.org/10.1109/INCISCOS.2018.00038
20. Llerena-Izquierdo, J., Cedeño-Gonzabay, L.: Photogrammetry and augmented reality to promote the religious cultural heritage of San Pedro Cathedral in Guayaquil, Ecuador (2020). https://doi.org/10.1007/978-3-030-42520-3_47
21. Daniela, L., Strods, R., Kalniņa, D.: Technology-enhanced learning (TEL) in higher education. In: IGI Global (ed.) Knowledge-Intensive Economies and Opportunities for Social, Organizational, and Technological Growth, pp. 12–24 (2018). https://doi.org/10.4018/978-1-5225-7347-0.ch002
22. Llerena-Izquierdo, J., Idrovo-Llaguno, J.: Introducing Gamification to Improve the Evaluation Process of Programing Courses at the Salesian Polytechnic University (Guayaquil, Ecuador). Presented at the July 27 (2021). https://doi.org/10.1007/978-3-030-59194-6_33
23. Daniela, L.: Smart pedagogy for technology-enhanced learning. In: Didactics of Smart Pedagogy, pp. 3–21. Springer International Publishing, Cham (2019). https://doi.org/10.1007/978-3-030-01551-0_1
24. Daniela, L., Visvizi, A., Gutiérrez-Braojos, C., Lytras, M.D.: Sustainable higher education and technology-enhanced learning (TEL). Sustain. **10**, (2018). https://doi.org/10.3390/su10113883
25. Singleton, R., Charlton, A.: Creating H5P content for active learning. Pacific J. Technol. Enhanc. Learn. **2**, 13–14 (2019). https://doi.org/10.24135/pjtel.v2i1.32
26. Kleftodimos, A., Lappas, G., Evangelidis, G.: Edutainment and practice in video-based learning: enriching educational videos with interactive activities and games. Int. J. Entertain. Technol. Manag. **1**, 5 (2020). https://doi.org/10.1504/ijenttm.2020.105689
27. Llerena Izquierdo, J.F., Ayala Carabajo, R.: El uso de Grabaciones por Video como Recurso de Evaluación de Conocimientos de Aprendizajes. 4to. Congr. Int. Ciencia, Tecnol. e Innovación para la Soc. (2018). https://pure.ups.edu.ec/es/publications/el-uso-de-grabaciones-por-video-como-recurso-de-evaluaci%C3%B3n-de-con-2

28. Kerimbayev, N., Nurym, N., Akramova, A., Abdykarimova, S.: Virtual educational environment: interactive communication using LMS moodle. Educ. Inf. Technol. **25**, 1965–1982 (2020). https://doi.org/10.1007/s10639-019-10067-5
29. Skvortsova, S., Onopriienko, O., Britskan, T.: Training for Future Primary School Teachers in Using Service H5p Teaching Mathematics (2019)
30. Reyna, J., Reyna, J., Hanham, J., Todd, B.: Flipping the classroom in first-year science students using H5P modules. EdMedia + Innov. Learn. **2020**, 1077–1083 (2020)

Embedded System for Fall Detection in Activities of Daily Living ADLs: A Time Window Analysis

Solange Bustos, María José Flores, David Pozo-Espín(✉), and Santiago Solórzano

Facultad de Ingeniería y Ciencias Aplicadas (FICA), Universidad de las Américas, 170122 Quito, Ecuador
{damaris.bustos,maria.flores.anchatuna,david.pozo, santiago.solorzano}@udla.edu.ec

Abstract. According to the world health organization (WHO), the falls are the second cause of death worldwide related with unintentional injuries, affecting mainly adults over 65 years old. In this article, a portable embedded system is developed through the use of a 32-bit microcontroller and an inertial sensor, that is used for three activities of daily living (ADLs) detection: falls, walking and stationary state. In embedded systems the amount of data to be processed (time window) affects directly the response time and define the features of the hardware to be selected and used. Within a different time windows, the accelerometer signals a_x, a_y, a_z are acquired in order to calculate the acceleration module over the time. By the use of descriptive statistics, this module will provide features as: mean, median, variance, 25th percentile and asymmetry in order to be used in the Naive Bayes classifier, which was selected based on a previous work. The system has wireless communication to send information from the detected ADLs to the PC for test purposes. The device was carried by 6 people on the waist, who carried out the tests of the different ADLs. System accuracy is between 60% and 82% with processing times between 0.5–1.35 ms depending of the time window size. For the offline training of the Naïve Bayes algorithm, the Data Set of the Institute of Communications and Navigations is used, corresponding to samples of 16 male and female subjects between 23 and 50 years old.

Keywords: Embedded system · Accelerometer · 32-bit microcontroller · Time window · Naive Bayes

1 Introduction

The WHO defines falls as "involuntary events that make you lose your balance and hit the body on the ground or another firm surface that stops you." According to the data compiled by the WHO, it is known that approximately 646,000 fatal falls occur per year; ranking falls as the world's second leading cause of death from unintentional injuries [1]. Every year there are 37.3 million falls that, despite not being fatal, require medical attention. The highest morbidity corresponds to people over 65 years old, young people between 15 and 29 years old and those under 15 years old [1].

M. Botto-Tobar et al. (Eds.): CIT 2020, AISC 1326, pp. 228–238, 2021.
https://doi.org/10.1007/978-3-030-68080-0_17

Currently, the use of smart devices and mobile applications are very common to provide security for children, youth and adults. These can be divided into four large groups: immediate detection devices, devices for unusual behavior, another method of detection of falls and environmental monitoring for unusual behavior [2]. Due to the technological advances in the microcontroller development related with its processing capabilities and affordable prices, this paper is focused on the immediate detection devices for fall detection in ADLs.

The aim of this article is to develop an embedded system based on a 32-bit microcontroller and an inertial sensor that could be capable of detect three different ADLs: fall, walk and stationary state. Moreover, in necessary to define the proper sampling range in order to minimize the processing time and increase the accuracy for the ADLs detection. In order to acquire and analyze de data form de accelerometer is used different time windows with 130, 150, 170, 190, 210 y 230 samples per window, which let calculate the acceleration module over the time. The embedded system will use the Naïve Bayes Classifier, which will be trained previously offline based on the Data Set of the Institute of Communications and Navigations, that corresponds to samples of 16 male and female subjects between 23 and 50 years old. The features to be used in the Naïve classifier will be extracted from the acceleration module in the sample windows. The features to be used are mean, median, variance, 25th percentile and asymmetry. The results are collected on a PC via Bluetooth communication.

2 Related Work

Immediate detection devices are those portable devices, which can detect a fall when it occurs and sending an immediate alarm. The technologies that are usually used are shock devices, position, inclination sensors and accelerometers together with suitable control algorithms in a microcontroller [2, 3].

Devices of unusual behavior are small devices carried by users, but they do not detect falls. These only monitor the activity of the person and detect when unusual behavior occurs compared to the typical pattern of a person. These de-vices alert when detecting abnormal actions, which does not differentiate if it is a fall [2].

Another method of detection of falls is immediate detection environmental monitoring, which consists of the installation of sensors around the person to detect a fall. The technologies used are: video recording [4], image analysis, sound analysis, installation of shock sensors on the floor and carpets [2].

In environmental monitoring for unusual behaviors, environments are equipped with sensors to monitor people such as doors, windows, infrared barriers, passive IR detectors. These devices require a large infrastructure and a good analysis system [2].

Portable fall detection systems have the facility that they can be used outside the home, as long as they are not uncomfortable for the user [5]. As an example, Angel4 can be mentioned, which is a device that detects a fall using a triaxial accelerometer. This device consists of a small sensor placed on the waist, this sensor is connected to a telecare system, the health center or the mobile phone via Bluetooth and is controlled through a mobile application [3, 6].

Vigi'Fall is a detection device which sends an alert in the event of detecting a fall followed by the inability to get up by itself. It is made up of a small biosensor in the shape of a triangular patch. This device is attached and can be used even in the shower. Its function is to emit signals that are received by wireless sensors placed on the walls of the house, allowing to send alerts in case of falls [7, 8]. Vibby, is another fall detection device, which is worn on the wrist, this device will emit an alarm in case of detecting a fall and call easily to be able to request help manually. This device works with a pressure sensor and an innovative algorithm that measures the loss of altitude and speed, in addition to being able to count on vibrations to confirm that alerts have been sent [9].

Several of the previously mentioned devices have been able to help in the detection of falls thanks to the application of Machine Learning, this is an analytical method that allows a system to learn certain patterns without the need for a human being. This is done through training with large volumes of data and algorithms. The latter can be classified into three large groups: supervised, semi-supervised and unsupervised algorithms [10, 11].

In addition, you can also find specific algorithms with different characteristics for data treatment, among these we have: Deep Learning, is a machine learning algorithm with multiple levels of information abstraction, this algorithm consists of a complex structure in large data sets. The purpose of which is to be able to create models capable of automatically creating a hierarchy from simple concepts to creating more complex concepts [10–12]. Active Learning is a semi-supervised algorithm which can achieve greater precision with fewer training instances, this algorithm interacts with a user in order to obtain the desired precision [10, 13].

One of the most used algorithms for its efficiency and effectiveness is Naïve Bayes, which infers that an action belongs to some class based on the assumption that all its attributes are independent of each other [14]. Naïve Bayes is a very effective algorithm as it has been proven in many applications including text classification, medical diagnostics and systems performance management [15]. The advantage of this algorithm is the independence of the features, which means that they are not related, resulting in the features agreeing on the most probable class. This allows each feature to contribute to the final decision [15, 16].

3 Methodology

The general scheme of the implementation of the embedded system can be seen in Fig. 1 and consists of three parts. The first is the offline training of the Naïve Bayes algorithm by means of a Data Set with information from ADLs, where data preparation and modules of the accelerations of two classes (fall and walk) are prepared for the subsequent obtaining of results. Training stands. The second part is the implementation of Naïve Bayes in the embedded system (32-bit microcontroller and accelerometer) for the detection of falls in real time, and the third is the sending of the predicted class via Bluetooth to the PC for the analysis of the results. The Naïve Bayes algorithm was selected based on a study carry out in a previous work, where is shown its performance in comparison with another techniques [3].

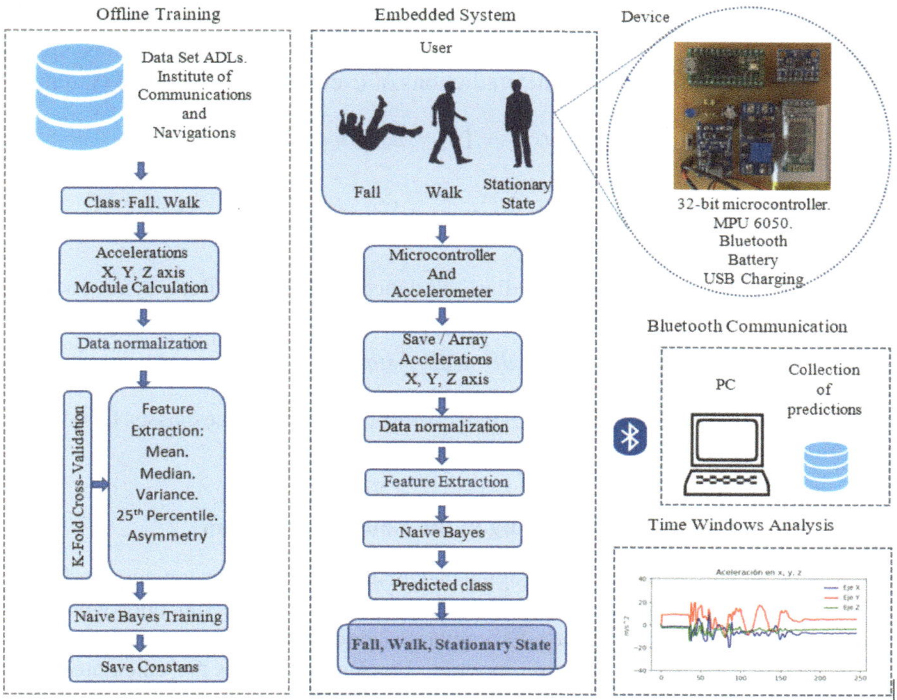

Fig. 1. General scheme.

3.1 Naïve Bayes Training (Offline)

In the offline part the "Institute of Communications and Navigations" database is used containing 17 classes of ADLs [17], of which it has been decided to use two classes in this work: falls and walking. Each of the classes contains the acceleration variables in the a_x, a_y, a_z axes. The falls class consists of 32 events, where each of the events have a certain number of samples that vary from 92 to 280. The walking class It consists of 42 events and the number of samples varies from 661 to 22084. In addition to the two classes obtained from the database, we worked with a third class named stationary state, which was created by taking the data from the sensor in real time. This class has 23 events with several samples that varies from 133 to 192.

From the accelerations of the two classes used, walking and falling, the calculation of the modules and extraction of characteristics is carried out according to [3]. Subsequently, the modules are processed by a K-Fold algorithm, which consists of dividing the data into k blocks, which are used to partition the information and divide 20% in test and 80% in training. In K-Fold, it was decided to use cross - validation as this allows taking values from different parts and doing tests to be able to choose the one with the best performance within the algorithm and in this way the algorithm has greater accuracy.

Once the division of the data is ready, 80% is taken and the characteristics of each of the events are calculated. Subsequently, the calculation of the constants is carried out, resulting in 10 constants for each class, given by the mean and standard deviation of each feature (mean, median, variance, 25th percentile and asymmetry), which will be applied in the Naive Bayes algorithm for the predictions of events in the embedded system.

$$Mean\ constants\ u = \{u_{1Cj}, u_{2Cj}, u_{3Cj} \ldots u_{mCj}\}$$

$$Standard\ deviation\ constants\ \sigma = \{\sigma_{1Cj}, \sigma_{1Cj}, \sigma_{1Cj} \ldots \sigma_{mCj}\}$$

$$Feature\ vector\ X = \{\text{mean},\ \text{median},\ \text{variance}, \text{25th percentile},\ \text{asymmetry}\}$$

$$Classes\ C = \{fall,\ walk,\ stationary\ state\}$$

Where m = 1,2,3….5 and j = 1,2,3 that corresponds to the three classes C: fall, walk and stationary state.

3.2 Acquisition and Data Processing in the Embedded System

In the Fig. 1, the different stages of signals acquisition and processing acquired by the device can be seen in the Embedded System section. The microcontroller works with the a_x, a_y, a_z signals from the accelerometer that are stored in arrays and processed in order to calculate the acceleration module. The module values are calculated in the microcontroller through Eq. (1):

$$\left|\vec{A}_i\right| = \sqrt{a_{xi}^2 + a_{yi}^2 + a_{zi}^2} \tag{1}$$

Where i = {1,2,3, …. n}, and "n" is the total number of samples of each time window of the ongoing activity.

With the modules already calculated, it is possible through descriptive statistics to calculate the feature vector X. These values are the features that the Naive Bayes algorithm will use to predict a class.

3.3 Naive Bayes in the Embedded System

The implementation of Naive Bayes in the microcontroller has two main stages: The first receives as parameters the features calculated in each time window and the constants obtained in the training stage of the offline algorithm. The probability density function to be implemented is presented in (2) [18]:

$$P(x_i|C_j) = \frac{1}{\sqrt{2\pi\sigma_{iC_j}^2}} * e^{-\frac{\left(x_i - u_{iC_j}\right)^2}{2\sigma_{iC_j}^2}} \tag{2}$$

Where:

δ_{iC_j} = standard deviation of i-th training offline feature of a class j.

u_{iC_j} = sample mean of i-th training offline feature of a class j.
x_i = features of X, $i \in \{1, 2, 3 \ldots m\}$.
C_j = class to be analyzed, j $\in \{1, 2, 3\}$.
$P(x_i|C_j)$ = conditional probability of each feature related to a class j.

The second stage is to implement the Bayes' theorem (3) with the results obtained from (2) in order to get the posterior probability of each class C_j related with the feature vector X [18].

$$P(C_j|X) = \frac{P(X|C_j) * P(C_j)}{P(X)} \tag{3}$$

Where:

$$P(X|C_j) = P(x_1|C_j) * P(x_2|C_j) * \ldots\ldots P(x_m|C_j) \tag{4}$$

$$P(X) = \sum_{j=1}^{3}\left(\prod_{i=1}^{m}(P(x_i|C_j)) * P(C_j)\right) \tag{5}$$

$P(C_j|X)$ = posterior probability.
$P(X)$ = evidence.
$P(X|C_j)$ = conditional probability.
$P(C_j)$ = prior probability of a class. Is assumed that is 0,33.

Once calculated the posterior probability for each class $P(C_j|X)$, in order to classify the time window data acquired is performed a comparison that evaluates which is higher and this result will give the final prediction for the activity ongoing.

4 Test and Results

The algorithm is implemented in the microcontroller development board Teensy 4.0, and the offline training is performed in Python. The next tests are carry out with the participation of a group of 6 person, which perform the requested activities or classes (walking, falling and stay in stationary state) while wearing the device on the waist. For the tests, the embedded system will be configured in six time windows (samples per event in a determined time) of 130, 150, 170,190, 210, 230 taking into account the following variables:

- Walk: Number of walk predictions
- Stationary state: Number of stationary state predictions
- Fall: Number of fall predictions

The analysis of the time windows sizes are relevant due to the amount of data to be acquired and processed, affect directly the processing time and the internal resources necessaries (program memory, clock speed, volatile memory, etc.) of the embedded system to be used.

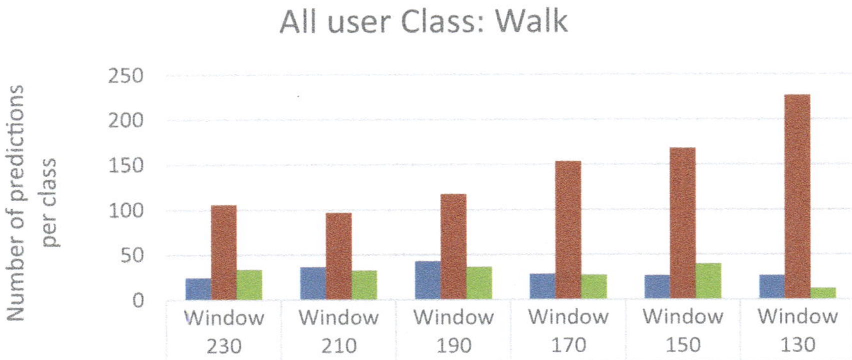

	Window 230	Window 210	Window 190	Window 170	Window 150	Window 130
Fall	25	37	43	29	27	27
Walk	106	97	118	154	168	227
Stationary S.	34	33	37	28	40	13

Fig. 2. Tests users walk activity

In order to analyze the performance of the algorithm for the Walk detection, the tests users proceed to accomplish this activity. In Fig. 2 is showed the results of the implemented algorithm in the embedded system, it is considered that the activity Walk is the correct and the activities fall, and stationary state are incorrect.

In Fig. 2 can be seen the Y axis, that represents the number of predictions per activity and on the X axis the time windows. For example, in the 230 samples per window when the person is walking, the system detected 106 times that the users are walking, while 34 are considered as stationary state and 25 are register as falling. In the Walk activity, the best window for all participants is 130.

In Fig. 3 is presented the average time that takes to the algorithm to process de data and detect the activity. It must consider that this time represent only the processing data algorithm and not the sampling time. Is clearly that less samples per window involves better time response, in this test corresponding to 130 samples.

The test for fall activity could be seen in Fig. 4, here the activities walk, and stationary state are considered incorrect. In this scenario the 150 sample window shown the best results.

The average time for the best performance window (150) is about 742,56 ms (See Fig. 5).

Finally, for the stationary state, the tests are carried out with 6 different time windows and in this case the stationary state activity is considered as a success and the walking and falling events will be taken as a failed event (See Fig. 6).

Based on Fig. 6 the stationary state activity prediction shows its better performance with 190 samples in the time window. Moreover, the prediction performance of this activity could be considered the best related with the last two except by the window with 130 samples.

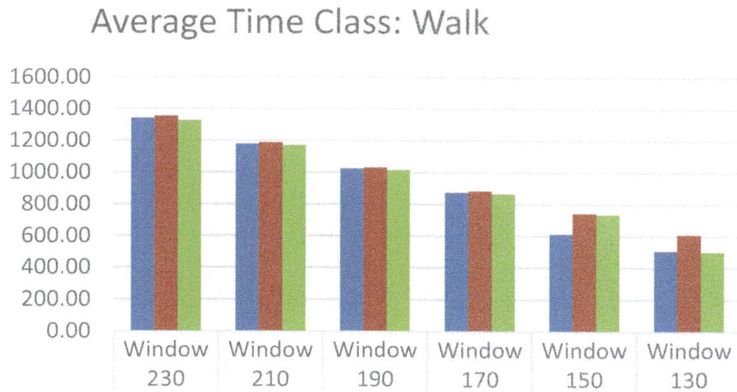

	Window 230	Window 210	Window 190	Window 170	Window 150	Window 130
Fall (microseconds)	1344.17	1181.67	1025.49	877.54	616.16	509.42
Walk (microseconds)	1357.51	1190.28	1032.97	884.59	743.43	611.54
Stationary S. (microseconds)	1330.01	1172.49	1017.90	869.65	738.32	503.77

Fig. 3. Processing time for Walk activity prediction

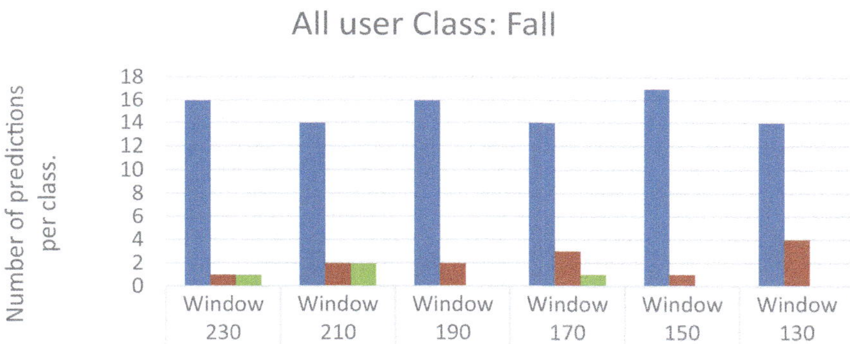

	Window 230	Window 210	Window 190	Window 170	Window 150	Window 130
Fall	16	14	16	14	17	14
Walk	1	2	2	3	1	4
Stationary S.	1	2	0	1	0	0

Fig. 4. Tests users Fall activity

The average time spent for the stationary activity prediction related with the 190 sample window is 1034.17 ms. In the Fig. 7 is not shown the time on 130 samples due to there is no predictions for stationary state at that time window.

In order to stablish the performance of the embedded system related with the time windows, in Fig. 8 is presented a comparative among the accuracies of the different samples in the activity prediction.

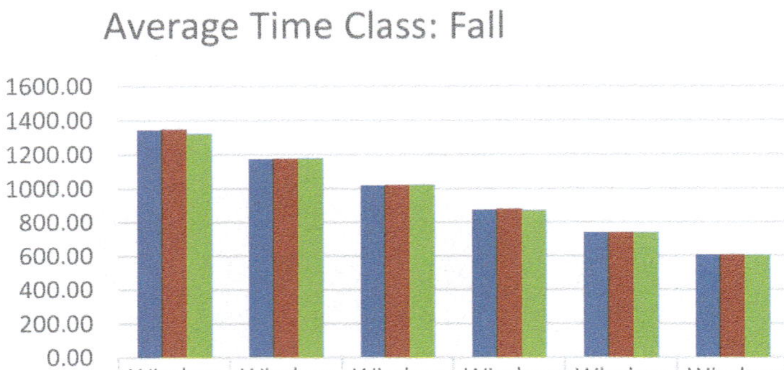

	Window 230	Window 210	Window 190	Window 170	Window 150	Window 130
Fall (microseconds)	1350.17	1178.54	1021.34	877.79	742.56	609.69
Walk (microseconds)	1352.65	1179.36	1023.71	882.73	742.37	611.15
Stationary S. (microseconds)	1331.20	1183.39	1027.42	876.30	743.64	610.01

Fig. 5. Processing time for fall activity prediction

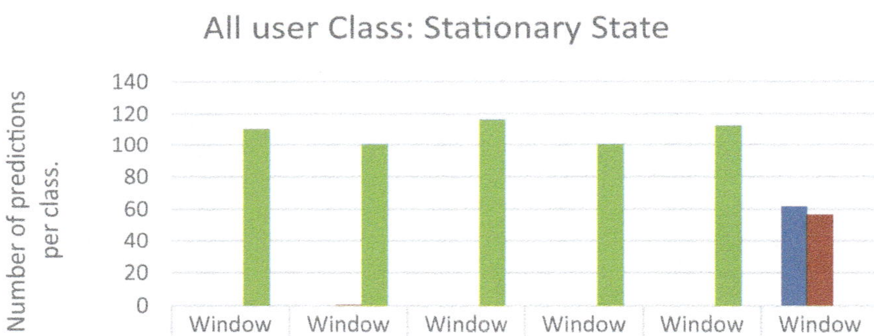

	Window 230	Window 210	Window 190	Window 170	Window 150	Window 130
Fall	0	0	0	0	0	62
Walk	0	1	0	0	0	57
Stationary S.	111	101	117	101	113	0

Fig. 6. Tests users stationary state activity

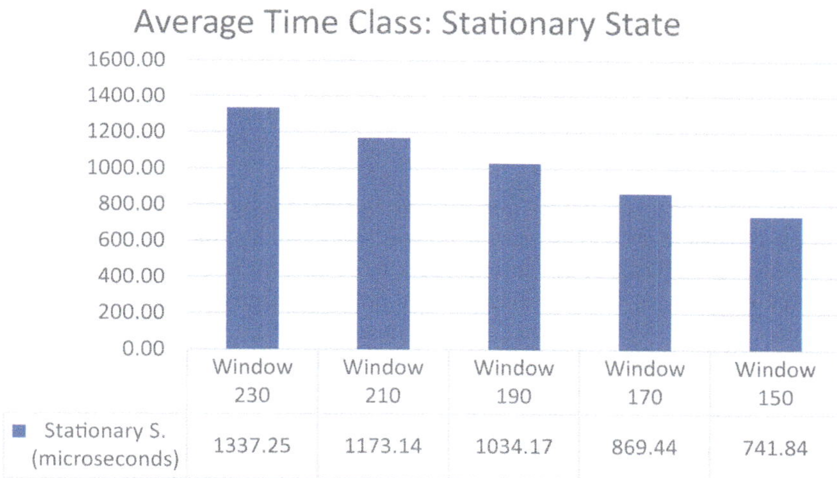

Fig. 7. Processing time for stationary state activity prediction

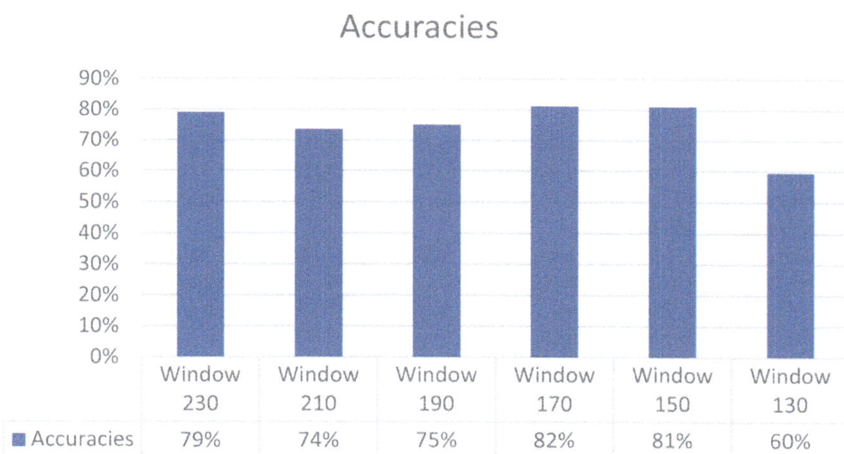

Fig. 8. Accuracy performance of the algorithm for each time window.

5 Conclusions

In this work, the development of an embedded system based on a 32-bit microcontroller for fall detection in ADLs was presented. It was proposed the use of different time windows in order to analyze the effect of it in the data processing time and the performance of the algorithm. This analysis let stablish a range for a window data size that could be the best for carry out ADLs recognition within an embedded system implementing a Naïve Bayes classifier. Could be concluded that a window with 170 samples will give the best results related with accuracy of the algorithm reaching the 82%. Nevertheless, the use of windows of 150 samples could be a good option too, with an accuracy of 81%. In terms of time, the best results in the windows size of 170 and 150 are 869,65 and 616,16 ms respectively.

In the other hand the worst results in terms of accuracy are related with windows of 130 samples, that only reach the 60%, however, as it was expected this window size has the best processing time.

Finally, based on this work, it is possible to stablish the appropriate range for sampling data in terms of accuracy and processing time for walk, fall and stationary state activities based on the Naïve Bayes classifier implemented in a 32-bit microcontroller.

For future works, is possible to implement new algorithms in the embedded system in order to compare its performance. Moreover, an important step is to extend the tests users in order to analyze in a better way the system, and to add new activities to be recognized by the system.

Acknowledgment. The authors thanks to the staff Unidad de Innovación Tecnológica (UITEC) of Universidad de Las Américas for their infrastructure, equipment and support.

References

1. Salud de la, O. M.: "Caídas" (2018)
2. González, R., et al: Desarrollo de un sistema de detección de caídas basado en acelerómetros (2015)
3. Solórzano, S., Pozo, D., Morales, L., Villalonga, C.: Análisis comparativo de algoritmos de aprendizaje supervisado para la detección de caídas (2018)
4. María, J., Plaza, C., Agüero, C.E., González, T.: CHAPTER X Detección visual de caídas para ambientes inteligentes
5. Persona-Ordenador, I.: Interacción 2018 XIX Congreso Internacional de (2018)
6. SENSE4CARE: The best fall detector in the market. Reliable, easy to use and economical
7. Vigi'Fall: Detector de caída a distancia - Vigi'Fall TM - Vigilio Telemedical
8. Barreras, V.S.: El detector de caídas 'Vigi'fall
9. Asistencias, T.: Detector de caídas
10. Russo, C., et al.: Tratamiento masivo de datos utilizando técnicas de Machine Learning. In: XVIII Workshop de Investigadores en Ciencias de la Computación (Entre Ríos, Argentina) (2016)
11. Perspective, A.A.: Machine Learning & Pattern Recognition Series. Chapman & Hall/CRC, Boca Raton (2015)
12. Casas Roma, J., Bosch Rué, A., Lozano Bagén, T.: Deep learning : principios y fundamentos. UOC (2019)
13. Settles, B.: Computer sciences department active learning literature survey (2009)
14. Chen, S., Webb, G.I., Liu, L., Ma, X.: A novel selective naïve Bayes algorithm. Knowl. Syst. **192**, 105361 (2020)
15. Rish, I.: An empirical study of the naive Bayes classifier. (2020)
16. Ting, S.L., Ip, W.H., Tsang, A.H.C.: Is naïve Bayes a good classifier for document classification? (2011)
17. "DLR - Institute of Communications and Navigation - Data Set."
18. Morales, L., Pozo, D.: An experimental comparative analysis among different classifiers applied to identify hand movements based on sEMG. In: 2017 IEEE Second Ecuador Technical Chapters Meeting (ETCM), pp. 1–6 (2017)

Parallel CPU-Based Processing for Automatic Crop Row Detection in Corn Fields

Marco Pusdá-Chulde[1,2](✉), Armando De Giusti[2], Erick Herrera-Granda[1], and Iván García-Santillán[1]

[1] Facultad de Ingeniería en Ciencias Aplicadas, Universidad Técnica del Norte, Ibarra, Ecuador
{mrpusda,epherrera,idgarcia}@utn.edu.ec
[2] Facultad de Informática, Universidad Nacional de La Plata, La Plata, Argentina
{marco.pusdac,degiusti}@info.unlp.edu.ar

Abstract. With technological advancement, current computer applications generate large volumes of heterogeneous data (text, images, video, audio), causing a considerable increase in processing execution times. The use of heterogeneous architectures in the systematization of processes is an alternative to obtain better processing times, optimizing computational resources, through the interaction between CPU (Central Processing Unit) and GPU (Graphics Processing Unit) architectures for parallel processing. Specifically, in precision agriculture (PA) Image analysis for the automatic detection of crop lines is essential for the use of autonomous systems applied to different agricultural processes in real-time. In the present work, the use of parallel programming is proposed for the implementation of an existing algorithm for the detection of curved and straight lines of crops, based on micro-ROI (Small regions of interest), with images captured in cornfields during the early stages of growth. Matlab (Parallel Computing Toolbox) libraries were used for parallel and sequential implementation in a multicore CPU to contrast execution times in a set of 300 images. The results obtained were statistically validated using the T-Student test in the R programming language. The evaluated times indicate accelerations of 20% on average in recognition of crop lines, demonstrating a notable improvement in the performance of the parallel algorithm over the sequential one.

Keywords: Image analysis · Heterogeneous architectures · GPU · Multicore · Crop row detection

1 Introduction

Computing architectures have evolved the way they perform computational operations, including multi-core CPU and GPU processors through vector instructions SIMD (Single Instruction stream, Multiple Data stream) [1, 2]; which facilitates the implementations of algorithms that require reduction of execution times to solve more complex computational problems [3, 4], as in precision agriculture, which requests automation in agricultural activities taking advantage of technological and electronic advances to improve the efficiency and quality of agricultural production. Recognition of crop lines

M. Botto-Tobar et al. (Eds.): CIT 2020, AISC 1326, pp. 239–251, 2021.
https://doi.org/10.1007/978-3-030-68080-0_18

in cornfields [5, 6] and other types of crops [7], it is an essential phase to discriminate weeds, detection of crop lines [8], identification of pests and other factors that affect crops using different techniques, processes and technologies through digital image processing [9–12].

The use of heterogeneous architectures together with artificial vision techniques are promising alternatives in offline and real-time agricultural image processing applications, reducing time, complexity, and computational costs using software tools to address parallel programming in heterogeneous architectures [12–16]. Matlab is a numerical computing system that offers an integrated development environment for technical-scientific programming, includes parallel programming tools and functions to accelerate computational calculations [17]. The Parallel Computing Toolbox, Fig. 1, provides various methods of parallel processing locally using labs or workers [18], to reduce execution time and programming effort in different computer applications in several application domains [19–21].

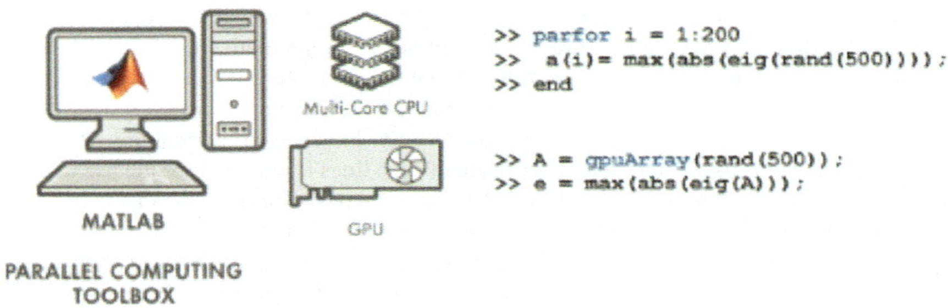

Fig. 1. Parallel computing toolbox architecture [22]

However, published research work on heterogeneous architectures applied to precision agriculture is still limited [23]. Therefore, in this work, the use of parallel programming is proposed in the implementation of an existing algorithm for the detection of curved and straight lines of crops, based on micro-ROI, with images captured in cornfields during initial stages of growth. [24], using Matlab Parallel Computing Toolbox for both parallel and sequential implementation in CPUs to improve the performance of the sequential something-rhythm (temporal and spatial limitations), which will contribute to its future implementation in vehicles or autonomous devices in real-time.

As background on the use of parallel programming in heterogeneous architectures for AP applications, three more current works were considered that served as the theoretical basis in this work. In [25] they developed a system for detection and localization of weeds among crops of a farm using parallel processing in GPU to apply in real-time. In [7] they developed a real-time crop detection, crop segmentation, and crop line detection system implemented in an FPGA (Field Programmable Logic Gate Array). In [34] presents an innovative procedure to calculate three-dimensional (3D) geometric characteristics in almond tree crops by generating photogrammetric point clouds with Unmanned Aerial Vehicle (UAV) technology, and analysis of point clouds using Object-based image analysis (OBIA). The rest of the document is structured as follows: In the materials and methods section, the parallelization of the sequential algorithm for detection of culture lines explained; the results section shows the main results obtained and discussion with similar works, and in the conclusions section, the main contribution of the study.

2 Materials and Methods

2.1 Image Dataset

The images evaluated for the present work took randomly from a dataset (300) used in the investigation for automatic detection of curved and straight lines of cultivation from images in cornfields in different growth stages [24]. The images were captured for 40 days, with 6 days of difference to obtain images with various lighting conditions and different growth stages. Figure 2 shows two examples of the images captured in the field with corn crops that have some characteristics, such as the number of leaves in the range [4, 6], plant heights between 100 and 300 mm and plant population around 40,000 plants per hectare.

The images used in the study are 24 bits with resolutions of 3000 x 2250 pixels (7Mpx) with RGB color space and JPG format. A region of interest (ROI) was considered in the process of detecting crop lines, it is selected four rows of crops with a width of 3.4 m and 5 m long, take into account a space between rows of 0.85 m to identify other plants (weeds) and also enough points in recognition of crop lines [26].

(a) (b)

Fig. 2. Examples of images obtained in the cornfield

2.2 Algorithm Parallelization

Parallelizing the CPU it is characterized by sharing the computational load among the different cores (physical) contained the processor. What it requires for this work is the Matlab R2017b programming environment and Parallel Computing Toolbox (PCT) taking advantage of the cores available in the machine [18, 27, 28]. The sequential algorithm proposed by [24], allows the automatic detection of curved and straight lines of crops, based on small regions of interest (micro-ROI) considering three phases: (i) segmentation, (ii) identification of starting points, and (iii) detection of crop lines. For the implementation in Matlab in parallel, sentences allowed by the programming tool (PCT) in the third phase of the algorithm (detection of crop lines) they were modified using vectorization to eliminate loops using other functions and equivalent operations, moreover parfor was used for cycles where all its interactions are distributed and executed among all the physical cores (workers) available within a workspace (MatLabpool)

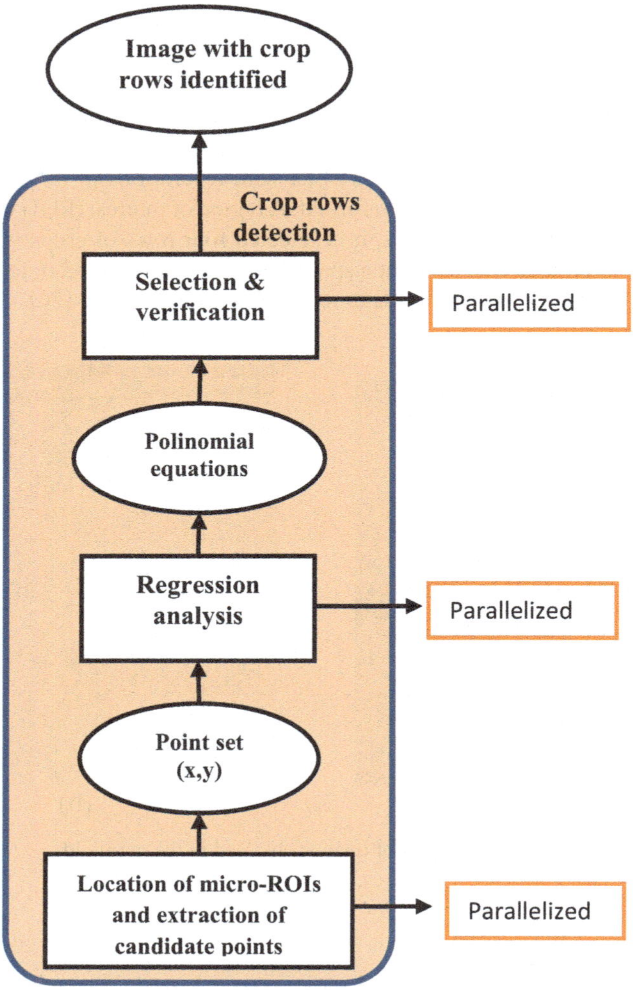

Fig. 3. Parallelized sections of the micro-ROI-based sequential algorithm [24]

to reduce the execution time of the calculations. The detection phase of straight and curved crop lines follows the next steps: (i) Extraction of candidate points; (ii) regression analysis; and (iii) selection and verification. These sections were parallelized because they include different operations based on micro-ROI to consider the maximum number of crop plants, leaving most of the weeds out of the micro-ROI in the spacing between the crop rows. Figure 3 details the parallelized section of the sequential algorithm.

Extraction of Candidate Points

In this section, vectorization and the parfor instruction were used to define the micro-ROIs for each crop line (blue rectangles - Fig. 4a), The initial micro-ROI has a width of 150 pixels, the rest of the micro-ROI are gradually reduced by 5% in width, based on the previous micro-ROI generated, until reaching a total of 12 micro-ROIs to form the curve of each crop row. The crops lines are extracted sequentially from left to right, numbered 1 to 4, considering the alignments of each micro-ROI from a starting point about the estimated straight-line slope used as a guide.

Regression Analysis

In this phase, vectorization and the parfor instruction were used to draw the micro-ROIs, taking the coordinates as a reference to identify the crop plants. Moreover, the generations of the lines of the slopes were parallelized to detect the trajectory of the crop plants.

Selection and Verification

Vectorization and parfor instruction were used in this phase for the selection of crop lines and later verification of the selected lines identifying the curves that best fit each crop line (Fig. 4b). In the verification of crop lines, the separation space be-tween the crop lines and the orientation is considered to accept or reject the image.

The result of the sequential algorithm that presents the division of the micro-ROI Fig. 4a and the detection of the crop lines in Fig. 4b.

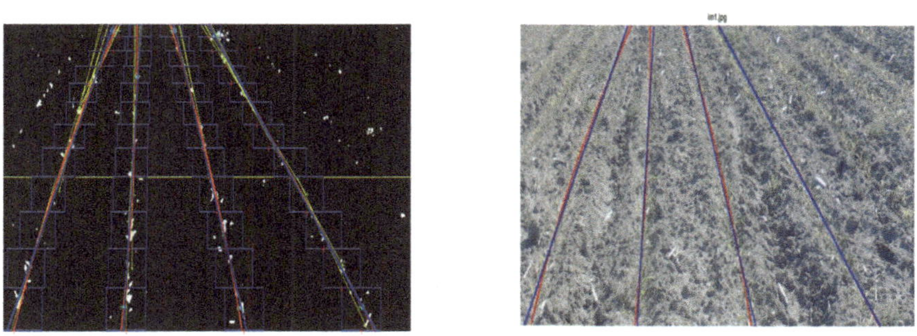

(a) (b)

Fig. 4. Line detection in corn crops based on micro-ROI

2.3 Acceleration of the Algorithm

The acceleration of an algorithm in parallel implementation concerning the sequential implementation is called Speed-UP (Amount of global improvement of the performance of a system) and is a dimensionless quantity. Equations 1, 2, 3 show the calculation of acceleration, where U are the units of computation, N the number of process elements per unit, R (N) the theoretical maximum performance, S (N) Speed-Up or acceleration, T sequential time (Time from the beginning until the last element of processing finishes its execution) TP (N) parallel time, and E (N) is the efficiency (Fraction of time in which computational resources used productively). The real acceleration of an algorithm is carried out by Amdahl's law [29], which explains how the parallel time of an algorithm is calculated and based on this predicts the real behavior of its acceleration [30, 31].

$$R(N) = U \times N \tag{1}$$

$$S(N) : \frac{T}{TP(N)} \tag{2}$$

$$E(N) = \frac{S(N)}{N} = \frac{T}{N \times TP(N)} \tag{3}$$

In Eq. 4 Amdahl's law is presented, where f is the sequential part of the code, replacing Eq. (4) in (2) we obtain Eq. (5) and making a limit on it, having process units an infinity get the (6). Equation 6 means that the real acceleration of an algorithm tends to a constant value, this is because as computation units increase, the processing time decreases, however, the transfer and communication times between the units increase.

$$TP(N) = f \times T + (1 - f) \times \frac{T}{N} \tag{4}$$

$$S(N) = \frac{N}{f(N-1)+1} \tag{5}$$

$$\lim_{n \to \infty} = \lim_{n \to \infty} \frac{N}{f(N-1)+1} = \frac{1}{f} \tag{6}$$

The Eq. 7 allows to calculate of the improvement value, and a rule of 3 was considered evaluating the time obtained in series and parallel.

$$\text{Improvement} = (TS - TP) * 100/TS \tag{7}$$

2.4 Processing Equipment

The tests of the serial and parallel algorithm were conducted on an HP Pavilion 14 Notebook Laptop with Windows 10 64-bit, Intel Core I7 processor, 16 MB of RAM. The technical characteristics of the computer and the CPU used for parallel processing are presented in Table 1.

Table 1. Technical specifications of Laptop with Window 10 and multicore-CPU processor

Windows 10 Features	Processor Features
Operating system: Microsoft Windows 10	Architecture: x86_64
RAM: 16	CPU op-mode (s): 32 bit, 64 bit
System type: 64-bit operating system, x64-based processor	Byte order: little Endian
Processor: Intel® Core ™ i7-5500U CPU @ 2.40 Ghz. 2 main processors, 4 logical processors	CPU(s): 4
Edition: Windows 10 Pro	On-line CPU(s) list: 0–3
Version: 1803	Threads(s) per core: 2
OS version: 17134.1130	Core(s) per socket: 2
Operating system manufacturer: Microsoft Corporation	Sockets(s): 1
System name: MARCK-PC	Vendor ID: GenuineIntel
System Manufacturer: Hewlett-Packard	Model name: Intel ® Core™ i7-5500U CPU @ 2.40 Ghz
System Model: HP Pavilion 14 Notebook PC	CPU MHz: 2401.000
System type: 64-bit based PC	CPU max MHz: 2401.0000

2.5 Statistical Tests

To evaluate the validity of the proposal using parallel programming, the deductive statistical technique T-Student [32], in the programming language R [33], was used to determine if there is a significant difference between the measures of two types of algorithms (Series, Parallel). The T-student test is a parametric test therefore, before its execution, it was verified that the sample complies with the assumptions of linearity, normality, homogeneity, and homoscedasticity. For this purpose, a false regression analysis was performed, where the standardized and studentized quantiles were obtained for each element of the sample, the same ones that were compared with the theoretical quantiles of the χ ^ 2 distribution.

To detect and eliminate possible atypical observations [33], based on the Z score obtained for each element of the database using a statistic of 3, where a cut-off score of 1,069 was obtained. In this way, it was concluded that atypical observations were not detected and the database was made up of the same 300 observations. To con-firm the distribution of the evaluated data, a boxplot was used as described in Fig. 5, which shows the quartiles for the series and parallel configurations.

The Fig. 6 shows the results of the linearity test employing the false regression analysis with the execution times obtained for the two types of algorithms In the graph, Q-Q Plot is observed that the quantiles included in a range of −2 to 2 for statistical they are oriented mostly in a linear trend with a strong positive correlation, therefore, the assumption of linearity was accepted.

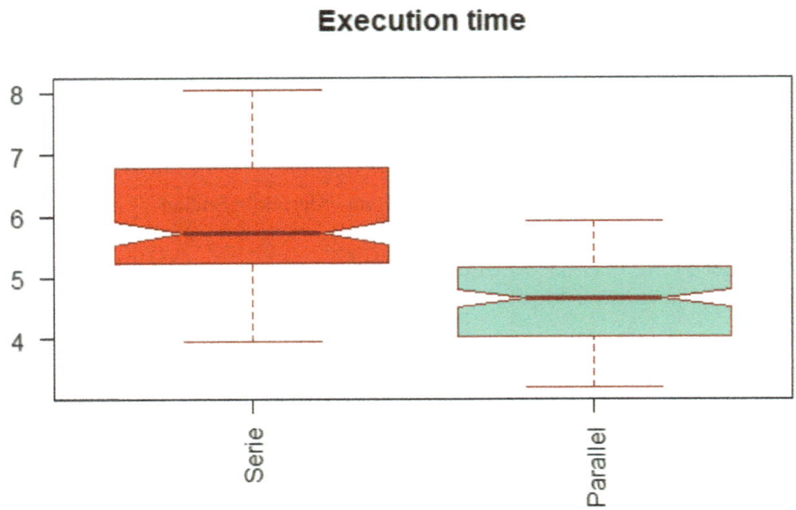

Fig. 5. Quartile distribution for Series and Parallel configurations

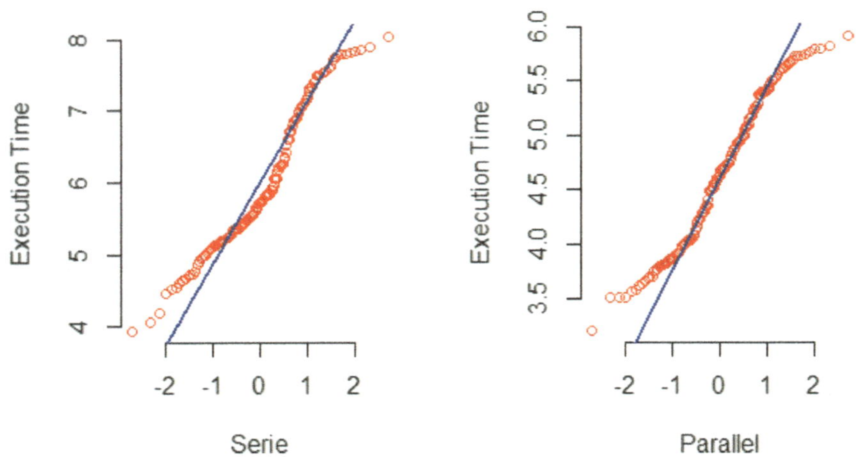

Fig. 6. Linearity test series and parallel configurations

In Fig. 7, the histogram of the standardized quantiles is presented, where a distribution centred on zero and very similar to a normal distribution is observed, for which the normality assumption was also accepted.

Finally, in Fig. 8, the scatter plot of the adjusted quantiles concerning the standardized ones is presented. Almost homogeneous distribution of the quantiles is observed in the four quadrants, and there are no patterns or trends in the distribution of the points, hence, the assumptions of homogeneity and homoscedasticity are accepted.

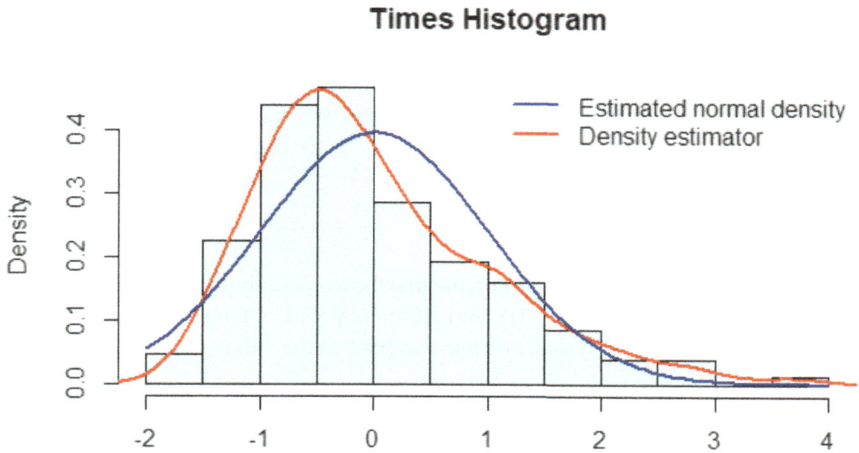

Fig. 7. Normality test of standardized times of serial and parallel algorithms

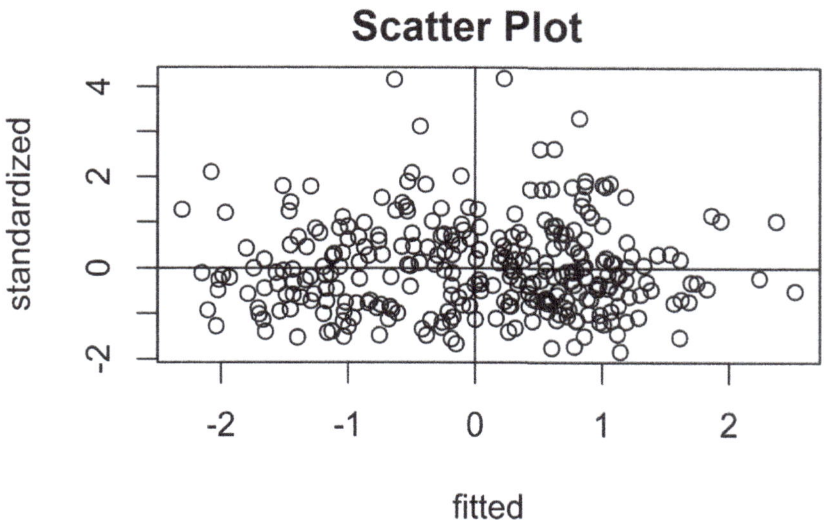

Fig. 8. Homogeneity and homoscedasticity test with standardized times of the series and parallel configurations

3 Results y Discussion

As indicated above, we worked with 300 images of corn crops in growth stages (40 days), with different densities and lighting used in the research proposed by [24]. The measurements were made with 300 images with the two algorithms (sequential-parallel), the first 300 measurements corresponded to the execution times with a serial configuration, and the final 300 measurements were concerned with the parallel configuration. Table 2 shows the average of the times (seconds) obtained from the 300 images.

Table 2. Average time values (seconds) of the 300 images.

Variables evaluated (time seconds)				
Time Series	Parallel Time	Improvement (Ec. 7)	Speed-UP (Ec. 2)	Efficiency (Ec. 3)
5.9683	4.6388	20.5805	0.6545	0.3273

The Improvement values (Eq. 7) indicate that the average time is favorable by approximately 20%. The value of Speed-UP and Efficiency was obtained through Eqs. 2 and 3, considering two cores (physical) of the computer, after having carried out the tests, positive values.

Once the assumptions are verified, the T-test is carried out using the statistical programming language R [33], or the two configurations in series and parallel as independent samples, accepting the equality of variance between them. In Table 3 the results are shown.

Table 3. Results of the T-test for two independent samples

Dataset: Time ~ Configuration					
t:	-13.729	**df:**	298	**p-value:**	2.2×10^{-16}
Alternative hypothesis:			True difference in means is not equal to 0		
95% Confidence interval:			$-1.5201 \leq x \leq -1.1389$		
Sample estimates:					
$\bar{x}$ **in group paralelo:**	4.64		$\bar{x}$ **in group serie:**	5.97	

As can be seen in Table 3 and Fig. 9, the results obtained show that there is a significant difference in the processing times of the images in the detection of crop lines when using the two configurations in series and parallel, with a p-value of 2.2 × 10 ^ (−16). When estimating the mean of each configuration from the samples, a time of 4.64 s was obtained for the parallel configuration and 5.97 for the series configuration, therefore, it concludes that the parallel configuration presents advantages in the task of detecting crop lines since it shows a significant decrease in processing time.

The present work demonstrates a performance improvement of the serial algorithm based on micro-ROI, using parallel programming tools in multicore architectures for the detection of crop lines (four) in cornfields considering the algorithm execution times. in series and parallel. Unlike the work done on FPGA architectures [7] that recognizes two lines of cultivation of other types of plants and includes the evaluation of execution time, memory consumption, and operating speed of the FPGA device. The limitations of the present study (memory use, disk use, load balance) will be solved in future works considering multicore architectures with many cores, GPU, and FPGA.

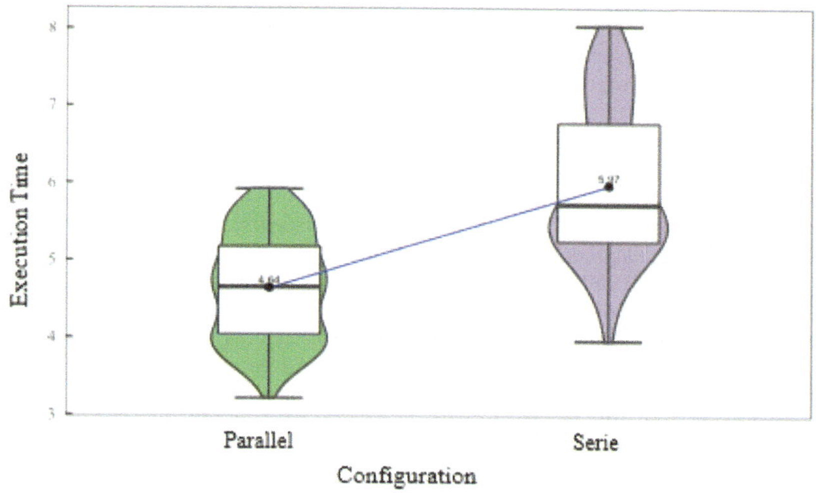

Fig. 9. Box-plot and density plots of processing times for parallel and serial configurations

4 Conclusions

This research used the Parallel Computing ToolBox tool included in Matlab R2017b for the implementation of sequential and parallel algorithms for the detection of crop lines in images of corn crops, to reduce processing times, taking advantage of hard-ware resources available in the current computers (Fig. 9). With the use of CPU parallelism, it is possible to improve the performance of the algorithms, obtaining lower execution times, speed-up, efficiency, and an average improvement of 20% in the processing of the 300 images (Table 2).

The use of the T-Student distribution (Table 3) confirms that the proposal presented in the present work is viable because the times evaluated, on average, both in parallel (4.64) are lower than the mean of the times in series (5.97) from a sample of 300 measurements for each type of configuration.

An important aspect to consider in the design of algorithm is the availability of several physical cores on current PCs, enabling the reduction in execution times by parallel instructions, without changing the previously parallelized algorithm, without the need to change the previously parallelized algorithm.; because Matlab tools distribute the processing among the available physical cores; therefore, the algorithms adapt automatically to each computer.

As future work, based on this work, algorithms with parallelization of all phases (segmentation, detection of crop lines, and weeds) will be implemented in corn crops or other crops using processors with more cores, GPU, and FPGAs.

References

1. Frati, F.E., De Giusti, A.: Software para arquitecturas basadas en procesadores de múltiples núcleos. In: Detección automática de errores de, UNLP (2015)

2. Dávila-Guzmán, M.A., Alfonso-Morales, W., Caicedo-Bravo, E.F.: Arquitectura heterogénea para el procesamiento de los algoritmos de enjambres. TecnoLógicas **17**(32), 11 (2014)
3. Tabares Soto, R.: Programación paralela sobre arquitecturas heterogéneas. Facultad de Ingeniería y Arquitectura, p. 80 (2016)
4. Rucci, E.: Evaluación de rendimiento y e ciencia energética de sistemas heterogéneos para bioinformática (2015)
5. García-Santillán, I., Guerrero, J.M., Montalvo, M., Pajares, G.: Curved and straight crop row detection by accumulation of green pixels from images in maize fields. Precision Agric. **19**(1), 18–41 (2017). https://doi.org/10.1007/s11119-016-9494-1
6. García-Santillán, I., Peluffo-Ordoñez, D., Caranqui, V., Pusdá, M., Garrido, F., Granda, P.: Computer vision-based method for automatic detection of crop rows in potato fields. In: Rocha, Á., Guarda, T. (eds.) Proceedings of the International Conference on Information Technology & Systems (ICITS 2018). ICITS 2018. Advances in Intelligent Systems and Computing, vol 721. Springer, Cham (2018). https://doi.org/10.1007/978-3-319-73450-7_34
7. Li, S., Zhang, Z., Du, F., He, Y.: A new automatic real-time crop row recognition based on SoC-FPGA. IEEE Access **8**, 37440–37452 (2020)
8. García-Santillán, I., Pajares, G.: On-line crop/weed discrimination through the Mahalanobis distance from images in maize fields. Biosyst. Eng. **166**, 28–43 (2018)
9. Zhao, S., Zhang, Z.: A new recognition of crop row based on its structural parameter model. IFAC-PapersOnLine **49**(16), 431–438 (2016)
10. García-Santillán, I., Pajares, G.: Detección automática de líneas de cultivo: estado del arte y futuras perspectivas. Av. y Apl. Sist. Intel. y nuevas Tecnol. **54**, 381–398 (2016)
11. García-Santillán, I., Pusdá-Chulde, M., Pajares, G.: Identificación automática de vegetación utilizando imágenes agrícolas: Una revisión de métodos. In: Imbaquingo, D. (ed.) Tecnologías Aplicadas a la Ingeniería, pp. 155–162. Universidad Técnica del Norte, Ibarra (2017)
12. Partel, V., Charan Kakarla, S., Ampatzidis, Y.: Development and evaluation of a low-cost and smart technology for precision weed management utilizing artificial intelligence. Comput. Electron. Agric. **157**, 339–350 (2019)
13. Demski, A.J., Di Donato, A.L., Maudet, S.F., Furfaro, A.: Algoritmo de visión estereo en tiempo real implementado en GPGPU. Simp. Argentino Intel. Artif. (2015)
14. Miranda, N.: Cálculo en tiempo real de identificadores robustos para objetos multimedia mediante una arquitectura paralela CPU-GPU (2016)
15. Osio, J.R., et al.: Análisis de Eficiencia en Arquitecturas Multiprocesador para Aplicaciones de Transmisión y Procesamiento de Datos (2017)
16. Giusti Armando, D., et al.: Arquitecturas multiprocesador en computación de alto desempeño: software, métricas, modelos y aplicaciones. La Plata (2017)
17. MathWorks, "MATLAB - El lenguaje del cálculo técnico - MATLAB & Simulink," 2020. https://la.mathworks.com/products/matlab.html?s_tid=hp_products_matlab. Accessed 29 Jun 2020
18. Suh, J.W., Kim, Y.: MATLAB and parallel computing Toolbox. Accel. Matlab with GPUs, pp. 99–125 (2014)
19. Ruiz de Miras, J., Martínez-Lledó, G., Orwig, W., Sepulcre, J.: A MATLAB tool for computing the spherical harmonic fractal dimension of the cerebral cortex. Comput. Phys. Commun. **254**, 107381 (2020)
20. Ploskas, N., Samaras, N.: Parallel computing toolbox. (2016)
21. Wei, P., Wang, B.: Multi-sensor detection and control network technology based on parallel computing model in robot target detection and recognition. Comput. Commun. **159**(January), 215–221 (2020)
22. MathWorks, "Parallel Computing Toolbox," 2020. https://la.mathworks.com/products/parallel-computing.html. Accessed 10 Jun 2020

23. Pusdá-Chulde, M.R., Salazar-Fierro, F.A., Sandoval-Pillajo, L., Herrera-Granda, E.P., García-Santillán, I.D., De Giusti, A.: Image analysis based on heterogeneous architectures for precision agriculture: a systematic literature review. In: Nummenmaa, J., Pérez-González, F., Domenech-Lega, B., Vaunat, J., Oscar Fernández-Peña, F. (eds.) Advances and Applications in Computer Science, Electronics and Industrial Engineering. Advances in Intelligent Systems and Computing AISC, vol. 1078, pp. 51–70. Springer, Cham (2020). https://doi.org/10.1007/978-3-030-33614-1_4
24. García-Santillán, I., Montalvo, M., Guerrero, J., Pajares, G.: Automatic detection of curved and straight crop rows from images in maize fields. Biosyst. Eng. **156**, 61–79 (2017)
25. Umamaheswari, S., Arjun, R., Meganathan, D.: Weed detection in farm crops using parallel image processing. In: Conference on Information and Communication Technology (CICT), pp. 1-4 (2018)
26. Pajares, G., et al.: Machine-vision systems selection for agricultural vehicles: a guide. J. Imaging **2**(4), 34 (2016)
27. Pascual, O.D.: Uso de Herramientas de Paralelizado para Optimización de Algoritmos de Procesado de Iimágenes. Universidad Politécnica de Valencia (2017)
28. Azzini, I., Muresano, R., Ratto, M.: Dragonfly: a multi-platform parallel toolbox for MATLAB/Octave. Comput. Lang. Syst. Struct. **52**, 21–42 (2018)
29. Universidad Europea de Madrid. Introducción y Medidas de Rendimiento (2017)
30. Zaripov, D.I., Li, R., Mikheev, N.I., Dushin, N.S.: Speed-up algorithm based on parallel projection correlation technique for planar PIV: accuracy and limitation. Flow Meas. Instrum. **60**, 88–94 (2018)
31. Salza, P., Ferrucci, F.: Speed up genetic algorithms in the cloud using software containers. Futur. Gener. Comput. Syst. **92**, 276–289 (2019)
32. Amat, J.: T-test: Comparación de medias poblaciones. RPubs, (2016). https://rpubs.com/Joaquin_AR/218467
33. Conesa, D.: Computación y programación en R : tema 4 Tema 4 : Análisis de datos con R. Valencia – España (2018)
34. Torres-Sánchez, J., et al.: Mapping the 3D structure of almond trees using UAV acquired photogrammetric point clouds and object-based image analysis. Biosyst. Eng. **176**, 172–184 (2018)

Detection of Suspected of Breast Cancer Using Thermal Images

Brian Meneses-Claudio(✉), Witman Alvarado-Díaz, and Avid Roman-Gonzalez

Image Processing Research Laboratory (INTI-Lab), Universidad de Ciencias y Humanidades, Lima, Peru
{bmeneses,walvarado}@uch.edu.pe, avid.roman-gonzalez@ieee.org

Abstract. Breast cancer is a type of cancer that forms in breast cells, this type of cancer is very common, diagnosed in most cases in women. Approximately 5% to 10% of breast cancers are estimated to be associated with inherited genetic mutations. It is important to know that most breast lumps are benign and not cancerous (malignant), it means that there are tumors that are not life threatening because they spread outside the breast, but if it should be examined by a specialist to obtain a clearer diagnosis of what it could be. That is why in this research work the detection of suspicions of breast cancer using thermal images is proposed, because some discomfort presented in the breast could be identified in time and determinate if it could possibly be a benign or malignant cancer, in addition keeping in mind that the analysis does not require physical contact with the patient and, furthermore, the infrared camera does not produce a congenital defect, being ideal for quick analyzes and to rule out illnesses. It was obtained as a result that the patients with suspicions of breast cancer, were a little reluctant to the diagnosis of the image processing, but after checking it with a specialist they were surprised, it was also identified that the different temperature between the body and breast cancer is 3 °C to 5 °C.

Keywords: Breast cancer · Thermal images · Image processing · Temperature difference · Heat map

1 Introduction

Breast Cancer is a disease in which breast cells grow and multiply uncontrollably. Breast cancer can start in different parts of the breast, because the position or the area where it can occur is not exact. Generally, they usually originate in the lobules, these are the glands that produce milk. In addition, breast cancer can spread outside the breast, through the blood or lymphatic vessels, which means that it cannot necessarily stay in the breast, in other words, the cancer is considered to have metastasized [1].

Breast Cancer is classified by stages, these being by levels from 0 to 4, in addition, some levels consist of variables, these levels differ by the size of the cancer tumor and in addition to its proximity to the axillary lymph nodes, in other words, under the arm;

M. Botto-Tobar et al. (Eds.): CIT 2020, AISC 1326, pp. 252–262, 2021.
https://doi.org/10.1007/978-3-030-68080-0_19

the higher the level, it is because the cancer has spread to the chest wall or also to the lymph nodes near the sternum. Furthermore, it is known that the more the cancer tumor spreads, the riskier the treatment, as well as its removal [2].

According to the WHO and PAHO, each year in the Americas, more than 462,000 women are diagnosed with breast cancer, and almost 100,000 die from this disease. Furthermore, if these statistics are maintained, it is provided that by 2030, the number of women diagnosed with breast cancer will increase by 34% in the Americas. Also, in Latin America and the Caribbean, the highest percentage of deaths from breast cancer occurs in women younger than 65 years (56%) compared to the United States and Canada (37%). Early diagnosis of this disease and advances in treatment have resulted in increased survival for women with breast cancer. Unfortunately, many countries in Latin America and the Caribbean continue to have limited access to these interventions [3].

The United States Food and Drug Administration (FDA) indicates that mammography, which is a radiological image of the breast with a low radiation dose, is the most effective screening test for breast cancer. A breast cancer screening test allows health personnel to find the cancer even before they show any signs and symptoms of the disease, which means that it also works as a prevention method [4].

In [5], it mentions that currently there is no effective method to prevent or know the reasons for the growth of cancer cells, however, if detected early it could save many lives in a higher percentage of women. This research work proposes to extract characteristic features of the breasts of the region of interest that is segmented using an approach through image processing, then classified as normal or abnormal using a neural network classifier. Obtaining how resulting a success rate of 96.51% effectiveness in the classification of breast cancer.

In [6], it reviewed various methods of Artificial Intelligence using image processing to diagnose breast cancer, making a comparison to obtain the most relevant methods for early detection of breast cancer. They found that the Super Vector Machine had the highest percentage of precision for different types of images (ultrasound = 95.85%, mammography = 93.069%, thermography = 100%), means that it is the most effective at the time of processing and detection of breast cancer, it is also known that computer-assisted methods increase the precision of diagnosis by reducing false positives.

In [7], they present a numerical and experimental research on the parameters of the breast cancer tumor by the reverse heat transfer method using a Genetic Algorithm (GA) and Image Processing (IP), in order to determine the depth and the rate of heat generation produced by the breast cancer tumor because it is known in theory that the tumor differs from the body temperature being more intense. They applied thermal heat to a point with a cylinder depth at which the depth and heat generation rate of the breast tumor were determined. Furthermore, it shows that the average computational error between the numerical and experimental results in this method to determine the depth of breast cancer is approximately 8–10% and to determine the rate of tumor heat generation is approximately 0.01–1%.

The main objective of the research work is the detection of suspicious breast cancer using thermal images to prevent the spread of cancer and also to prevent the proliferation of cancer with prevention because it can counter the future consequences of this disease that the majority of the cases is women. Detection of suspicions is through image processing where the points with the highest temperature other than the body temperature of 3 °C to 5 °C [7], will be identified, which is considered an abnormal variation between body temperature.

Thermography is a technique that allows temperatures to be determined at a distance without the need of physical contact with the object to be studied. It allows the infrared radiation of the electromagnetic spectrum to be captured using thermal cameras, each pixel that makes up the image corresponds to the measurement of radiation, so each pixel is a temperature value. This technology is being widely used in the field of medicine, industry and climatology. Variation in temperature provides relevant data on surface anomalies and in addition to overheating in areas where further analysis is required. There are various thermal cameras such as the FLUKE or the FLIR ONE Pro [8], which is currently on the market and they are presented as thermal cameras capable of obtaining the real-time temperature of an object to which the camera is pointed and also download the image with information in each pixel for future image processing.

The thermal images show the thermal composition of the body and objective, it also identifies the temperature difference that exists in each point or pixel of the image, in addition the thermal cameras have software where it can edit and process the images, these being limited due to that it can only change the range and the heat map, for this reason other software is required that is capable of processing the image through matrix variation and pixel management as MATLAB. These thermal images are used in many applications, one of the best known is in the identification of hot spots and in the case of this research work, the identification of anomalies through the variation of body temperature.

The following research work is structured as follows: In Sect. 2, the development of thermal image processing of suspicious breast cancer will be shown, as well as the sectorization process of the breast cancer abnormality. In Sect. 3, the results of thermal image processing and the identification of suspected breast cancer will be shown, in addition to the parameters that were followed to capture the images. In Sect. 4, the discussions of the research work are presented and finally in Sect. 5, the conclusions, as well as the recommendations for future work to be achieved with the research work.

2 Methodology

In this section, each part of the thermal image segmentation is developed for the detection of suspicions of breast cancer, which consists of image acquisition, thermal image processing using different processing techniques and finally the obtaining of the resulting image, all the mentioned steps are shown in the flow chart in Fig. 1.

2.1 Image Acquisition

A camera capable of capturing thermal images in other words that has an infrared lens was used, for that reason the FLIR ONE Pro thermal camera was used, which is compatible with Android, IOS and Windows mobile devices. Also, all of the products FLIR have their own application so there is one for mobile devices and software to process the

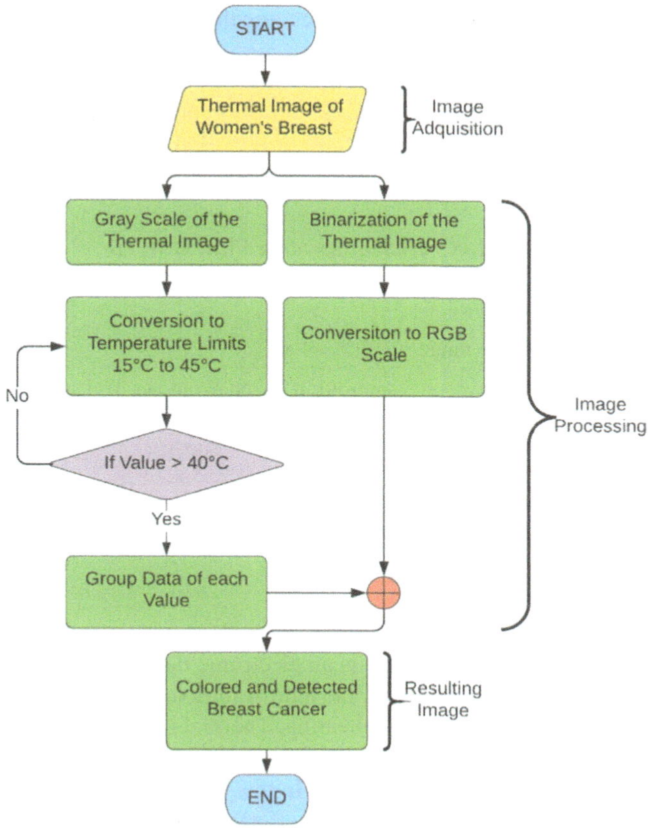

Fig. 1. Diagram of flow for the detection of suspicious of breast cancer

Fig. 2. FLIR ONE Pro thermal camera

thermal images saved by the mobile device. The FLIR ONE Pro have 2 lenses as can be seen in Fig. 2, the upper lens captures images in RGB, as well as a normal camera, the downer lens captures the thermal images that will be saved in the connected mobile device. For this research work, an Android device with a C connector was used, so the camera was used without any inconvenience because the thermal camera has C connector a default [9].

The thermal camera has characteristics in which the images and their operation are followed, these characteristics are shown in Table 1, in addition it should be noted that the data presented is based on the information provided by the same manufacturing company because they are very rigorous in their tests.

Table 1. Characteristics of the FLIR ONE Pro

FLIR ONE Pro	
Temperature range	−20 °C–400 °C
Compatibility	Android, IOS and Windows devices
Allowable distance	1.8 m
Weight	36.5 g
dimensions	68 × 34 × 14 mm
Thermal resolution	160 × 120
Operating time	1 h

The FLIR ONE Pro thermal camera is an electronic device whose main function is to capture images in temperature ranges, the temperature difference could help to verify places where there are anomalies or overheating. Thermal images are important because different diseases and tumults that show on the surface of the skin can be identified, such as varicose veins, diabetic foot, bruising, etc. In this case, it was confirmed that the use for the detection of suspected breast cancer is effective and it is possible to prevent this disease if detected early [8, 10, 11].

2.2 Image Processing

The image obtained is processed to identify the points where the maximum temperature variation is found, as shown in Fig. 3, where the thermal image is observed, then the gray-scale image, then the binarized image and finally the Superposition of the detection of 3 to 5 °C identifying a breast cancer due to the area in which it is found.

The MATLAB software reads the image previously saved in the computer, each thermal image is made up of 3-dimensional pixels because it is in the RGB format, so as the first step in processing the image, it will convert to grayscale due to that it will be in 2 dimensions and will be easier to identify the area where the temperature variation is greatest. To carry out the conversation in RGB to gray scale format, the software internally uses the following formula:

$$0.2989 * R + 0.5870 * G + 0.114 * B \tag{1}$$

After its grayscale conversion, it is known that the pixel value is in a range of 0–255 in two dimensions, so, it will indicate that each pixel is within the range of the usual body temperature [8].

To convert the grayscale image to the temperature scale, a technique called image mapping will be used, the following formula will be used to map the grayscale image:

$$\begin{aligned}\text{Mapping} =&((\text{Image\_Pixel} - \text{BeforeLow}) * (\text{AfterHigh} - \text{AfterLow})/\\ &(\text{BeforeHigh} - \text{BeforeLow}) + \text{AfterLow})\end{aligned} \quad (2)$$

Where:

- BeforeLow and BeforeHigh = 0 and 255 (Gray Scale)
- AfterLow and AfterHigh = Temperature Range that is from 15 °C to 40 °C.

Replacing formula 2, the values will be placed in the correct variables, as shown in the next formula (3):

$$\text{Resulting\_Image} = ((\text{Image\_Pixel}-0) * (40 - 15)/(255 - 0)) + 15 \quad (3)$$

When it has the mapped image, it can identify the highest temperature values in addition to identifying where points are, so the authors will first have to apply to the thermal image an image processing technique called binarization that consists mainly of each data that has a low threshold level will be taken as a value of 0 and all data that has a high threshold will be taken as 1, this is to obtain the area where a possible breast cancer is occurring or proliferating [9].

According to [5], it explains that body temperature is always maintained between the ranges of 15 °C to 40 °C, which means that ranges higher or lower than them could mean some anomaly, pathology or disease depending on the place and the level of variation of temperature that is detected, they also indicate that breast cancer always generates additional heat to the body, being 3 °C, but in this research work, temperature differences of 3 °C to 5 °C were found, which is why, rethinking what indicating in the research work, whether it is possible to obtain breast cancer by thermal imaging.

So when a pixel is different from the body temperature and it is also in the breast area, it is indicative of suspicions of breast cancer, also how it can be seen in Fig. 1, the data obtained through the temperature filter will be grouped to thus identify the size and depth of breast cancer.

After grouping the pixels with the highest temperature, it superimpose on the binarized image in order to identify the suspected breast cancer, also to locate in the area where it is located and also the level of heat it has, in order to classify whether it is already harmful or is just proliferating.

Finally, all the thermal images obtained were processed with the same algorithm in order to obtain the hottest points other than body heat and to identify if there is any suspicion of breast cancer. Furthermore, these images were analyzed by an oncologist, who approved the detection algorithm and also the use of the thermal camera in intimate areas because it did not generate harmful ills due to its superficial and non-contact analysis.

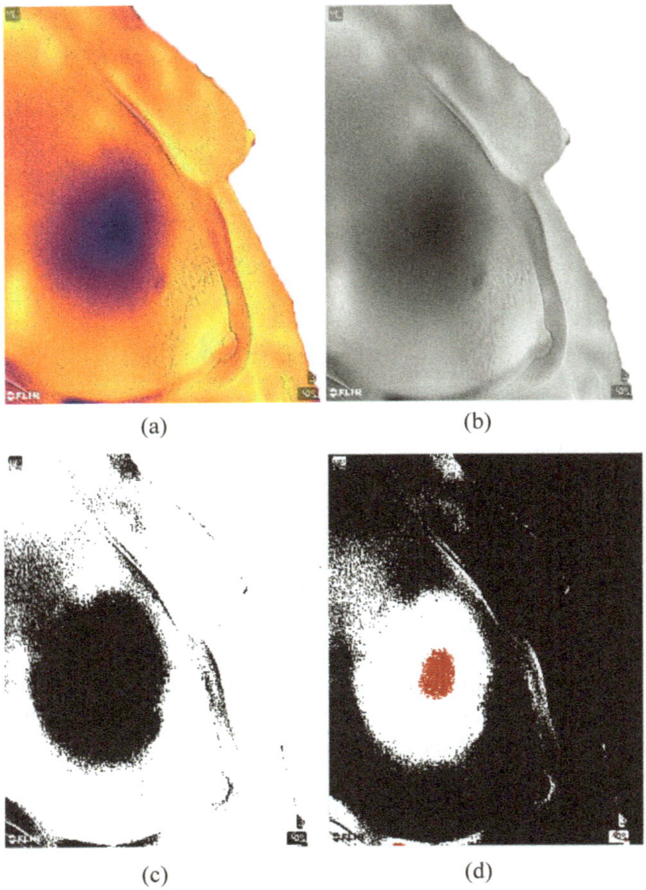

Fig. 3. (a) Thermal image. (b) Grayscale image. (c) Binarized image. (d) Superimpose of breast cancer detection and binarized image.

3 Results

The thermal images were acquired in an oncology center and the image capture protocols were followed because all the images have the same characteristics and also the same processing to obtain the detection of the suspicion of breast cancer. Table 2 will show the protocols followed to obtain the thermal images.

In Fig. 4, the reader can see the results of different patients who had suspicions of breast cancer where in some pictures, they are small spots and in other cases larger dimensions are shown. This research work aimed at detection and thus also propose the use of thermal images for the rapid diagnosis of female patients with suspected breast cancer, in addition to adding that the use of the thermal camera does not produce any congenital malformation to the patient and also it is contactless. This research work had 25 patients of which 10 (40%) had detected advanced breast cancer, 6 (24%) had early

Table 2. Characteristics of thermal images

Thermal images	
Distance	60 cm and 40 cm (depending of the breast size)
Temperature range	15 °C–40 °C
Connected device	Redmi Note 7
Place where the image was taken	• In front of the breasts • From both sides of the breasts • Individually to easch breast
Rest time before study	10 min

stage breast cancer, 7 (28%) thought they did not have breast cancer, but it applied the analysis to detect suspicions of breast cancer using a thermal camera and it was identified that they had signs, and finally 2 (8%) did not have breast cancer and did not show any signs.

4 Discussion

The research work confirms the use of thermal imaging for the detection of suspicious of breast cancer due to the variation of temperature in the breasts from the front, the sides.

For image processing, the thermal image must have good resolution, therefore the protocols shown in Table 2 must be followed, in order to homogenize all the images and obtain the expected results.

It disagrees with the research work [5], because they indicate that the temperature difference of 3 °C is enough but, in this research work, higher temperatures up to 5 °C were identified, which means that it depends on the stay of the breast cancer, the variation temperatures will also be different.

The patients were alarmed when verifying that, if small spots of different parts were verified, where a large percentage realized that they did have a proliferation of breast cancer due to the study of our research work.

This research work is for the prevention of proliferations of breast cancer, in addition to detection in early stages so that treatment is easier for patients, we also do not consider the clinical history of patients, so this study has more variables that can be taken into account.

(a) (b)

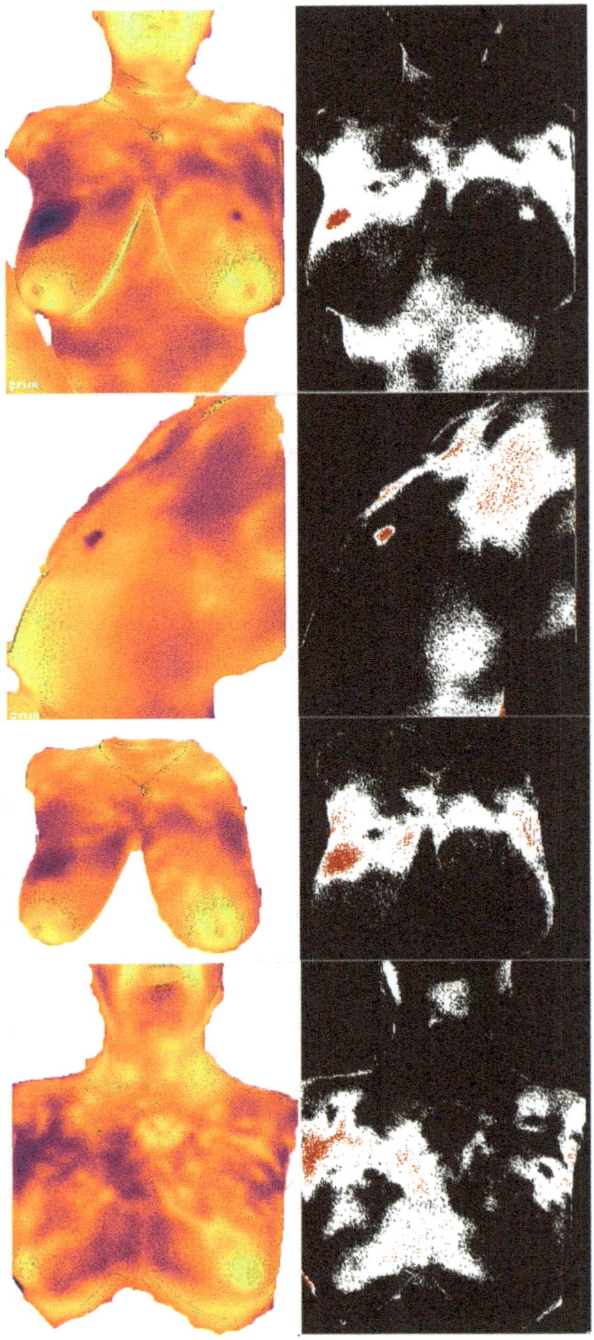

Fig. 4. (a) Thermal image. (b) Detection of suspicious of breast cancer

5 Conclusions

It is concluded that it is possible to detect suspicious of breast cancer through the processing of a thermal image in an efficient and fast way, because there is software that can segment the area of interest, identify temperature variation points and finally automatically obtain the tumor detection.

It is concluded in many people, mostly women, they can touch their breasts with the palm of their thumb to identify any lump, being an indication of any abnormality. In addition to having check-ups at the doctor about breast cancer because it is a very common disease in Latin America and treatable when it is detected early and actions are taken immediately.

It is concluded that an oncology center provides a lot of relevant information about breast cancer, treatments and home diagnoses that each person can do, but it was identified that people are very afraid that cancer is detected outside the established time and can be fatal to the lives of the patients.

As a future work, it is necessary to detect breast cancer in a short time so that an established treatment can be followed and this disease can be cured, which is why we want to propose a portable equipment validated by a group of doctors specialized in cancer and thus to diagnose immediately if it is a cancer or another similar anomaly. It should be noted that this research work was validated by oncologists, unfortunately it is not a portable electronic device.

References

1. Yi-Sheng, S., et al.: Risk factors and preventions of breast cancer. Int. J. Biol. Sci. **13**, 1387–1397 (2017). https://doi.org/10.7150/ijbs.21635
2. Wielsøe, M., Kern, P., Bonefeld-Jørgensen, E.C.: Serum levels of environmental pollutants is a risk factor for breast cancer in Inuit: A case control study. Environ. Heal. A Glob. Access Sci. Source. **16**, 1–16 (2017). https://doi.org/10.1186/s12940-017-0269-6
3. Doltra, A., et al.: Magnetic resonance imaging of cardiovascular fibrosis and inflammation: from clinical practice to animal studies and back cardiovascular MRI view project magnetic resonance imaging of cardiovascular fibrosis and inflammation: from clinical practice to ani. Biomed. Res. Int. **676489**, 1–2 (2013). https://doi.org/10.1155/2013
4. Walker, A.J., et al.: FDA approval of palbociclib in combination with fulvestrant for the treatment of hormone receptor-positive, HER2-negative metastatic breast cancer. Am. Assoc. Cancer Res. **22**, 4968–4972 (2016). https://doi.org/10.1158/1078-0432.CCR-16-0493
5. Soliman, O.O., Sweilam, N.H., Shawky, D.M.: Automatic breast cancer detection using digital thermal images. In: 2018 9th Cairo International Biomedical Engineering Conference, CIBEC 2018 – Proceedings, Institute of Electrical and Electronics Engineers Inc., Cairo, Egypt, pp. 110–113 (2019)
6. Sadoughi, F., Kazemy, Z., Hamedan, F., Owji, L., Rahmanikatigari, M., Azadboni, T.T.: Artificial intelligence methods for the diagnosis of breast cancer by image processing: a review, Breast Cancer: Targets Therapy, **10**, 219 (2018)
7. Bahador, M., Keshtkar, M.M., Zariee, A.: Numerical and experimental investigation on the breast cancer tumour parameters by inverse heat transfer method using genetic algorithm and image processing. Sadhana - Acad. Proc. Eng. Sci. **43**, 1–10 (2018). https://doi.org/10.1007/s12046-018-0900-4

8. Meneses-Claudio, B., Alvarado-Díaz, W., Flores-Medina, F., Vargas-Cuentas, N.I., Roman-Gonzalez, A.: Muscles heating analysis in sportspeople to prevent muscle injuries using thermal images. Int. J. Adv. Comput. Sci. Appl. **10**, 40–44 (2019). https://doi.org/10.14569/ijacsa.2019.0100606
9. Meneses-Claudio, B., Alvarado-Díaz, W., Flores-Medina, F., Vargas-Cuentas, N.I., Roman-Gonzalez, A.: Detection of suspicious of diabetic feet using thermal image. Int. J. Adv. Comput. Sci. Appl. **10**, 379–383 (2019). https://doi.org/10.14569/IJACSA.2019.0100648
10. Kirimtat, A., Krejcar, O., Selamat, A., Herrera-Viedma, E.: FLIR vs SEEK thermal cameras in biomedicine: comparative diagnosis through infrared thermography. BMC Bioinformatics **21**, 1–10 (2020). https://doi.org/10.1186/s12859-020-3355-7
11. Meneses-Claudio, B., Alvarado-Díaz, W., Roman-Gonzalez, A.: Detection of suspicions of varicose veins in the legs using thermal imaging. Int. J. Adv. Comput. Sci. Appl. **10**, 431–435 (2019)

IoT-Based Sensor Nodes for Structural Health Monitoring of Bridges

José Luis Molina[1], Darwin Zhagñay[1], Milton Muñoz[2], Remigio Guevara[2], Santiago González[1](✉), and Jaime Enrique García[3]

[1] Department of Electric, Electronic and Telecommunication Engineering, Cuenca, Ecuador
{jluis.molinal,darwin.zhagnayl, santiago.gonzalezm}@ucuenca.edu.ec

[2] Red Sísmica del Austro, Cuenca, Ecuador
{milton.munozc,remigio.guevara}@ucuenca.edu.ec

[3] Department of Civil Engineering, University of Cuenca, Av.12 de Abril, CP. 010203 Cuenca, Ecuador
enrique.garcia@ucuenca.edu.ec

Abstract. The road infrastructure represents a strategic area for the planification and development of urban and rural environments. In this context, essential facilities such as bridges are subject to damages mainly due to the daily load that represents the vehicular traffic, the environmental conditions and also due to the natural degradation of the structural elements. Therefore, it is fundamental performs a continuous evaluation about the state of an infrastructure. Following, this purpose, this paper presents an architecture for the Structural Health Monitoring (SHM) focused on bridges. The proposal solution involves the application of emerging technologies such as MEMS accelerometers, IoT devices and the deployment of a Wireless Sensor Network topology. Particularly, the architecture implemented consists of a set of ten sensor nodes which are management and synchronized by means of a central station. Furthermore, an additional node with capacity of capture and video recording was implemented Therefore, this last node provides to the system with a visual option for identifying the connection between the accelerations values and the vehicular traffic. In order to evaluated the system, the architecture was deployed along a bridge characterized by a heavy traffic load. Results on the scenario reveals the system allows for determining the main parameters related to state of the structure (i.e., acceleration values, load distribution and the fundamental frequencies of vibration). Consequently, the proposal architecture contributes significantly in the development of solutions focused in the analysis and early detection of damages on essentials structures or facilities.

Keywords: Wireless Sensor Networks · MEMS sensor · Internet of Things · Structural Health Monitoring

M. Botto-Tobar et al. (Eds.): CIT 2020, AISC 1326, pp. 263–277, 2021.
https://doi.org/10.1007/978-3-030-68080-0_20

1 Introduction

Currently, the monitoring of operational conditions in civil engineering projects involves the application of diverse emerging technologies. For example, solutions based on Internet of Things (IoT) platforms, Wireless Sensor Networks (WSN), Digital Signal Processing (DSP), among others. In particular, this research area focused in locating and identifying damages in infrastructures or facilities is named Structural Health. It that sense, a damage refers to some internal or external alteration detected in structural elements as well as changes in the geometrical properties of a structure which can affect its functionality [1, 2].Therefore, the Structural Health applications are various. For example, in [3] is described a system for identifying seismic signals caused by landslides along roads. In particular, the proposal employs a total of seven sensor stations which allows for detecting events a few hundred of meters from the epicenter. In this context, it is worth mentioning that seismic events represent one of the most critical conditions with a significant impact on infrastructures. In this sense, the north region of South American (i.e., Venezuela, Colombia, Ecuador and Peru) exhibit an important earthquake record along several centuries. This condition is mainly due to the complex structure of the northern Andes consisting of three continental plates (South American, Caribe and Nazca plate), which has generated the volcanic features and the high seismicity along this region [4]. Consequently, the analysis and identification of risks related to seismic events is a priority task in order to achieve a development based on the resiliency and efficiency in infrastructures.

On the other hand, it is important to note that seismic events can also be generated from human activities. In [5], is presented a proposal for monitoring a metallic bridge with more than 100 years old whose structure is subjected to daily passenger and heavy vehicles. In particular, the evaluation of bridges presents a great potential for structural health applications. For example, in [6] is described a proposal for deploying an ITS system (Intelligent Transport System), for real-time traffic management and monitoring, considering the stress conditions on bridges (i.e., load applied on structural elements due to the vehicular traffic). Regarding, the implementation of a monitoring architecture, Wireless Sensor Networks (WSN) represent a proper solution. This technology is based on the communication paradigm known as Ad hoc Networks [7]. The main advantage of a WSN is that the network topologies are more robust with regard to route breakages and packet losses. Furthermore, the WSN applications are suitable for both urban and rural environments as is highlighted in [8] and [9], respectively.

This paper presents the design and implementation of a structural health monitoring system applied to bridges. The architecture proposed consists of a set of ten stations or sensor nodes. Each node includes an accelerometer based on MEMS technology (Microelectromechanical Systems) as well as a microcontroller device dsPIC. The sensor nodes are synchronized by means of a central station implemented with a SBC platform (Single Board Computer). Finally, the system includes an additional sensor node with capacity of capture and video recording. Therefore, this last node provides to the system with a visual option for identifying the connection between the accelerations values and the vehicular traffic. Results from experiments describe the proper operation of the system implemented which represents the contribution of this proposal for deploying structural health monitoring applications using emerging and low-cost technology. The paper is organized as follows. Section 2, presents related works. Section 3, describes the architecture and the components employed for the monitoring system. Section 4, presents the experimental evaluation and results. Finally, Sect. 5, presents conclusions.

2 Related Works

The management and maintenance of infrastructures and facilities represent one of the essential areas for the development of a country. For example, the road infrastructure, the telecommunication and energy infrastructure, the water and gas distribution networks, among others, are particularly susceptible to damages caused by environmental conditions and natural phenomena. In this context, the support of emerging technologies (e.g. IoT devices, wireless sensor networks), is crucial for deploying monitoring systems and collect data about the integrity and capacity of structures, continuously. The following are some of the most relevant and representative studies in the literature. In [10] and [11], are described studies which evaluate the potential impact of a seismic event in distribution networks of water and gas, respectively. Regarding tests, supervised detonations are performed with the aim of determining acceleration, strain and displacement effects on the pipes. In this context, the preventive maintenance focused on strategic infrastructures, presents a great interest, especially for those elements or facilities subjected to drastic environmental conditions such as seismic activity. For example, in [12] and [13], are presented proposals which aim to monitor the normal operation of a set of wind turbines by means of deploying a sensor network including accelerometers devices.

Additionally, in [14], is highlighted the relevance of performing seismic surveys in those locations where wind turbines are deployed. Particularly, with respect to the vibration behavior and frequency response of the structures as well as regarding the seismic register of the zone (i.e. microzoning studies in order to identify seismic risk). In other hand, in [15], is described a project which aim to monitor the Pier Luigi Nervi Exhibition Centre (Turin-Italy) continually affected by the seismicity of that region. Therefore, one the main task consists in determining the optimal placement of the sensors considering the complex architecture. In that sense, a similar proposal is described in [16], in this case in order to evaluate the operational status of the Hospital Cardarelli (Campobasso-Italy) and identifying risks due to structural damages caused by recent earthquakes.

On the other hand, in [17], is proposed a monitoring system for a bridge using accelerometers based on MEMS technology (Microelectromechanical Systems). In particular, the data collected by the sensors are useful in order to characterize the dynamic response of the bridge, methodology known as Experimental Modal Analysis (EMA). Moreover, in [18, 19], the reliability of MEMS sensors for structural health applications is emphasized due to features such as the low-noise density, high resolution, low cost and reduce size. It is worth mentioning that the operational status of a structure can be analyzed from additional parameters. For example, in [20], is detailed a study where both the environmental conditions and vehicle traffic levels are analyzed. Thus, these parameters are used as indicators of the load and stress levels which are subjected essential infrastructures (e.g. highways, bridges among others).

Regarding, implementation and deploying of monitoring stations, a key aspect to take into account is the proper synchronization of nodes. Specially, it is fundamental that the timestamp differences between the nodes are in the range of the microseconds, particularly for applications focused on early warning systems. In [21] and [22], are presents studies that assess the challenges of implementing synchronization mechanisms in wireless sensor networks. Finally, in [23], are evaluated some synchronization algorithms usually employed in sensor networks. Following, in Sect. 3, the architecture proposed for structural health monitoring in bridges is presented.

3 System Architecture and Methodology

In this section, the architecture proposed for the monitoring system applied to bridges is detailed. Especially, the main components employed for implementing the stations are described as well as the solution designed in order to enable the communication and synchronization among the nodes, Finally, the monitoring software is presented. Figure 1, presents a functional diagram of the proposed system. As can be observed, the architecture consists of different types of stations or nodes. In first place, a set of ten sensor nodes, which can be placed along the structure where monitoring is required (e.g. bridges, roads, buildings among others). Each sensor node is provided mainly with an accelerometer for detecting events (i.e. vibrations on structural elements) as well as a wireless module in order to enable the connectivity with a central station. Moreover, the architecture includes an additional type of sensor node with video recording capabilities. This sort of node, allows the system to contrast the event detection with the presence of elements that cause different load levels on the structure under study (e.g. vehicular traffic along a bridge). Finally, the system includes a central station which is in charge of managing nodes (i.e. initialization and synchronization processes). Regarding, nodes components, the main elements used as well as the characteristics are summarized in Table 1. Following, in next subsections the implementation of each type of node, is detailed.

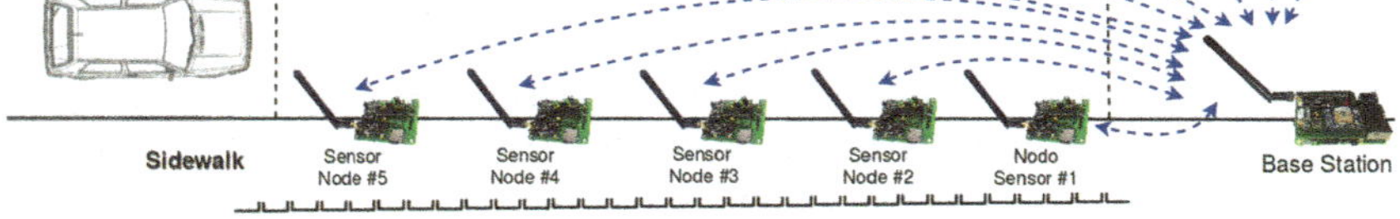

Fig. 1. Functional diagram of the monitoring architecture

Table 1. Main stations components and specifications

Component	Specifications
Single Board Computer RPi 3B +	Broadcom BCM2837B0; cortex-A53 (ARMv8) 64-bit SoC
Wireless module NRF24L01	2.4 GHz ISM bandwidth; supply range 1.9 to 3.6 V
Microcontroller Dspic-33EP256MC202	Input power 3.0 V to 3.6 V; 16-Bit CPU; 2 SPI modules; 5 16-Bit / 32-Bit timers; 28 Pins
Accelerometer ADXL355	Analog to Digital conversion (ADC); Supports measurement range of ±2,048 g, ±4,096 g, ±8,192 g
Current sensor ACS722	Supply voltage 3 1 3.6 V; typical sensitivity of 132 mV/A
RTC module DS3234	Maximum SPI communication 4 MHz; packed 20 pins SO; backup battery

3.1 Base Station

The main component used for implementing the central node is the SBC Platform Raspberry Pi 3B + which incorporates diverse interfaces for the connection of peripherals devices (e.g. SPI, I2C, USB among other interfaces).

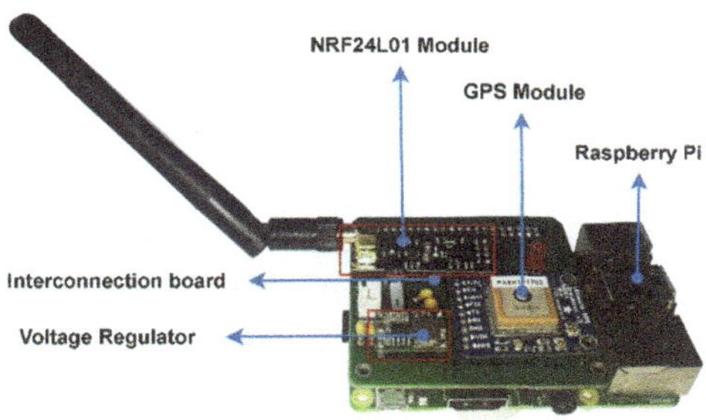

Fig. 2. Base station and components

Figure 2, presents a picture of the implementation performed, additionally the main components of the stations are highlighted. In particular, the prototype includes a GPS module [24] for both geolocation and synchronization purposes. In regard to network connectivity, the module NRF24L01 [25] was employed, which allows for setup a wireless link in the 2.4 GHz band and with a data rate values between 250 kbps and 2 Mbps. Moreover, as can be noted an electronic board was designed with the aim to integrate all the peripherals properly. Regarding, energy autonomy, the prototype includes a battery of 4 Ah that facilitate both the station setup and for developing experiments on a scenario under study. Figure 3, presents a functional diagram of the base station. As can be seen, the central node has in charge generating a signal in order to synchronize and initialize the sensor nodes. In regard to the event detection, it is carried out through a request

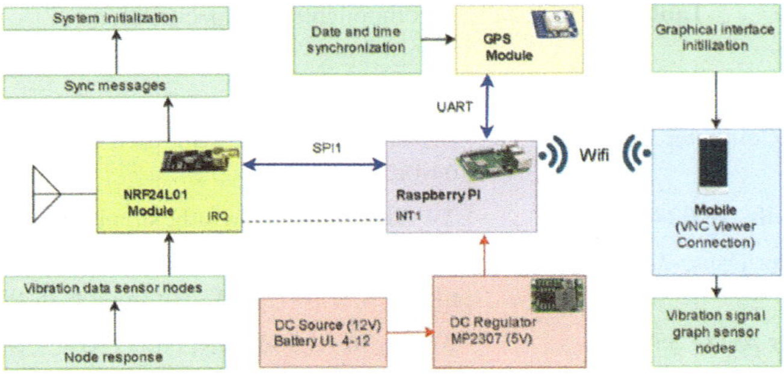

Fig. 3. Functional diagram of the base station

message that is sent from the base station with the aim to collect the acceleration values captured by the sensors. Finally, the information is analyzed and depicted by means of monitoring software implemented on the Raspberry Pi platform.

3.2 Video Monitoring System

The sensor node for video recording, presents a similar design to the base station, but including in this case a web camera attached to the USB port of the Raspberry Pi. Figure 4, depicts a picture of the implementation, additionally the main components are remarked.

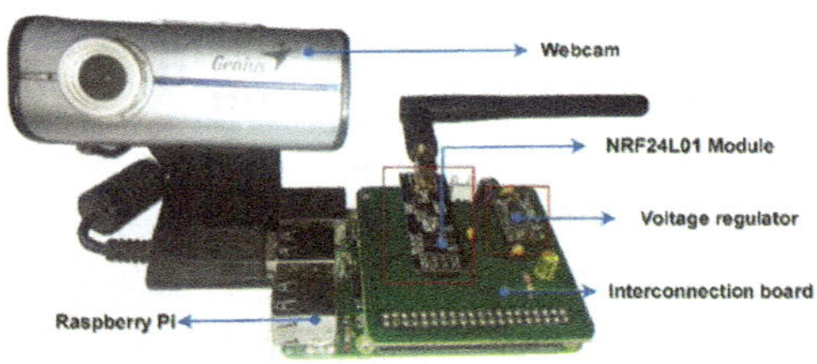

Fig. 4. Video station implementation and components

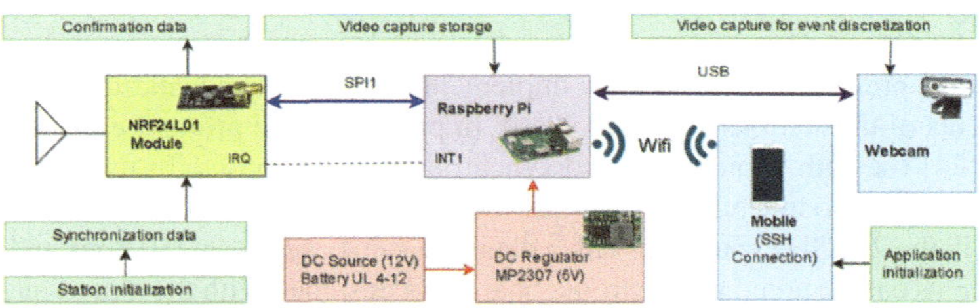

Fig. 5. Functional diagram: sensor node for video monitoring

On the other hand, Fig. 5, present the functional diagram. As can be seen, the video node is initialized independently of the others sensor nodes (i.e. a manual starting is required) but the synchronization is carried out through the signal sent from the central node. Following, the video is captured during a lapse of two hours that represent a proper time interval for deploying experiments. It is worth to mentioning, that in order to manage efficiently the storage resources, the video is captured and is compressed using de H.264 codec. With this end, the FFMPEG [26] software was used, that is an open source tool for video processing and coding. Finally, the video captured is useful in order to verify the events occurrence and to synchronize this information with the acceleration values captures by the sensor nodes.

3.3 Sensor Node

The monitoring architecture includes a set of ten sensor nodes for capturing acceleration values along a bridge while events take place (i.e. vehicular traffic along a bridge).

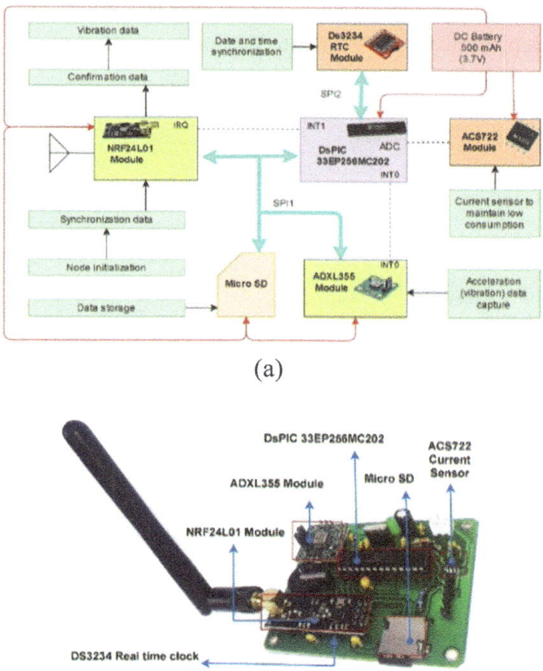

(a)

(b)

Fig. 6. Sensor node architecture (a) schematic diagram (b) implementation

In particular, the central component of the sensor node is the microcontroller DSPIC 33EP256MC202 [27]. In that sense, it is worth to indicating that the use of a microcontroller allows for reducing the time of data acquisition, while the task of processing and managing are executed by the central node (i.e. by means of the Raspberry Pi platform). Furthermore, each sensor node includes a set of peripheral devices which are attached to the microcontroller through the SPI bus (Serial Peripheral Interface). Specifically, the design incorporates the RTC module DS3234 (Real Time Clock) [28] for synchronization purposes, a digital accelerometer ADXL355 [29], which consists in a 3-axis sensor based on MEMS technology widely used in structural health applications. Additionally, the node includes a micro SD card for storing and backup the data collected by de sensors. Also, the wireless module NRF24L01 was used in order to enable the connectivity with the base station. Regarding, the power supply, each sensor node incorporates a battery of 500 mAh as well as the ACS722 current sensor with the purpose of characterizing the energy consumption. In Fig. 6 (a) a schematic diagram of the node architecture is depicted, additionally Fig. 6 (b) presents a picture of the implementation performed.

3.4 Synchronization

The synchronization between the nodes with respect to the base station was carried out by means of the standard PTP (Precision Time Protocol) [30] which allows getting an accuracy at around microseconds and is based on the master-slave hierarchy. In this case the sensor nodes represent the subordinate devices and the base station is the master device.

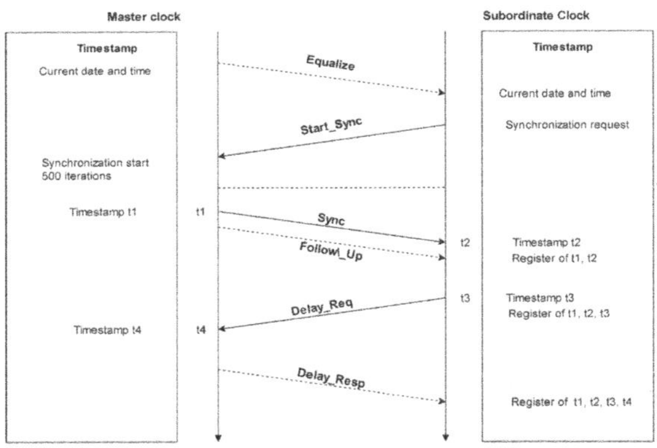

Fig. 7. Synchronization process on the nodes by means of the PTP protocol

The synchronization of the slave devices is performed one at a time. In particular, the standard PTP messages were implemented using software libraries (i.e. the time library as well as the timer functionalities available at the microcontroller). In order to start the synchronization, process the master sends to the slave the current date and time this is carried out by the message named adjust. Then, the slave generates a request for a timestamp (Init_Syc message). This timestamp (t1) is stored by the slave with along an internal timestamp named t2 (Sync and Follow_Up phases). Subsequenty, the slave generates a new timestamp (t3) and send this information to the master (Delay_Req message). Similarly, the master store the t3 timestamp, generates a new timestamp named t4 and sent both timestamps (t3 and t4) to the slave (Delay_Resp message). Finally, the slave uses all the timestamp (t1, t2, t3 and t4) in order to compute the offset and time delay. The process described is depicted in Fig. 7.

3.5 Monitoring Software

The monitoring software implemented allows the system for both control the nodes and also in order to display the acceleration values in real time. Following this purpose, the user interface has three main sections. In the GPS data section, the geolocation information is provided (i.e. latitude and longitude), additionally the current time corresponding to the internal clock of the Raspberry Pi is presented. On the other hand, the Sampling section, allow the users to define the time for capturing the acceleration data from a sensor node as well as the time for capturing the video in the case of the node that incorporates such functionality. Regarding, the Devices section, it incorporates a set of options in order to select a node (i.e. video station or sensor nodes). Finally, in the bottom on the interface is located a function that present three states and allow the system to perform the synchronization process, start an experiment or stop a test. In Fig. 8, a capture of the user interface is depicted.

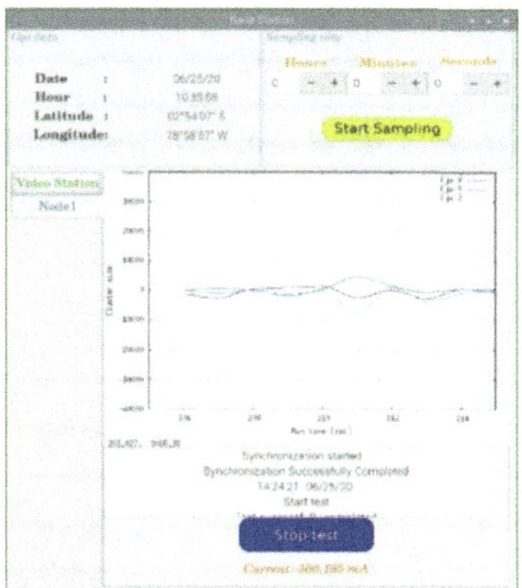

Fig. 8. User interface for monitoring nodes and displaying acceleration information

4 Experimental Evaluation and Results

This section details the experiments designed in order to evaluate the operation of the proposed architecture. Particularly, the proper data acquisition was verified on a real scenario consists in a bridge characterized by a heavy traffic load. Finally, the system performance (i.e. synchronization and energy consumption) is discussed.

4.1 Structural Health Data Acquisition

The experimental test was carried out on the Nulti bridge (Lat. 2°53′18"S, Long. 78°56′3"W), that is located in the highway between the cities of Cuenca and Azogues (Ecuador). Specifically, the system was deployed following the scheme depicted in Fig. 9, which is based on the recommendations defined in a previous structural analysis performed on the bridge and detailed in [31]. Regarding the orientation of each node, the

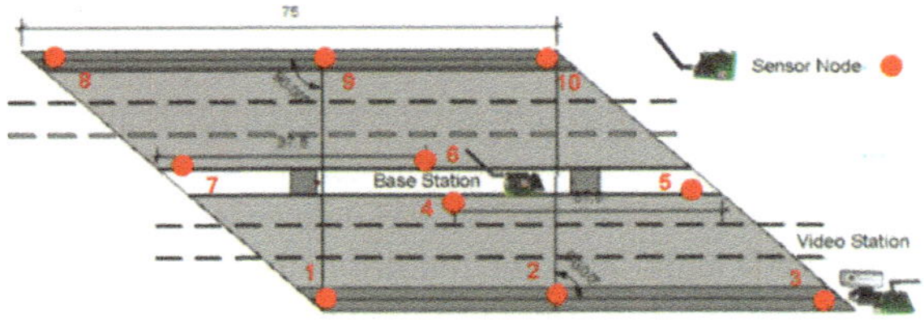

Fig. 9. Scheme implemented for the experimental test on the bridge

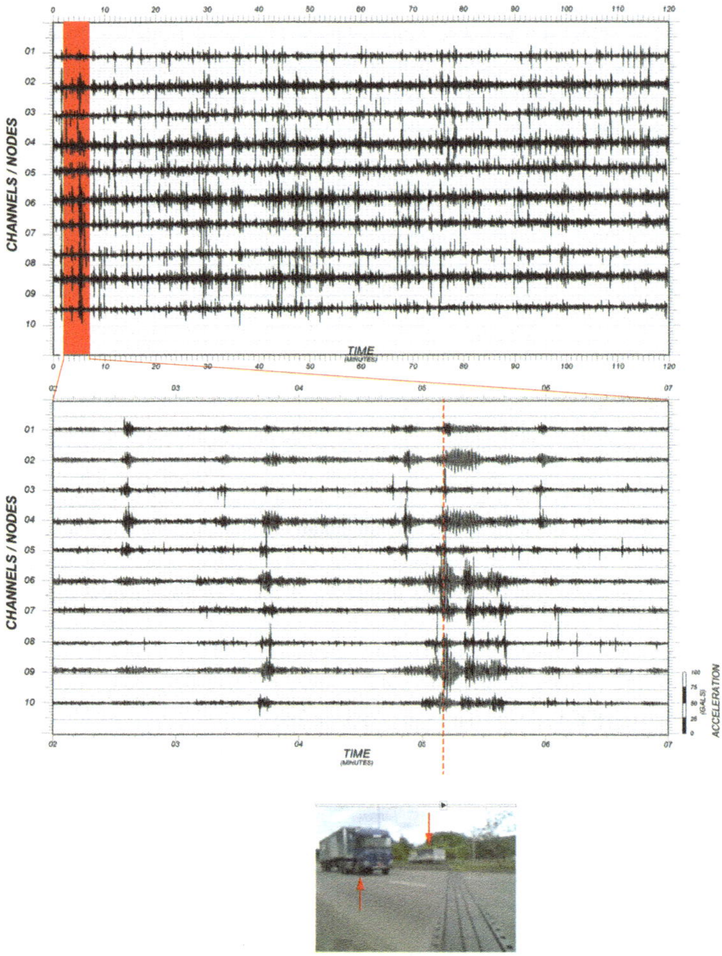

Fig. 10. Register of the acceleration values of each node (channels) corresponding to heavy vehicle traffic

x-axe was defined as the transversal component, the y-axe as the longitudinal component (i.e. parallel to the bridge) and the z-axe as the vertical component.

The data acquisition was collected during rush hours (from 15:59:21 to 17:59:20 on July 28, 2020). Particularly, Fig. 10, presents the acceleration values (z-axe) for a time interval where a significant event due to heavy vehicle traffic was register by the monitoring system. Also, the event was captured by the video sensor node as is highlighted in the top section of the Fig. 10. As can be seen, the event took place around of the minute 3 for the selected interval. It is worth pointing out that the higher signal amplitudes were registered by the nodes located in the central section of the bridge, especially on the node 6 and node 9.

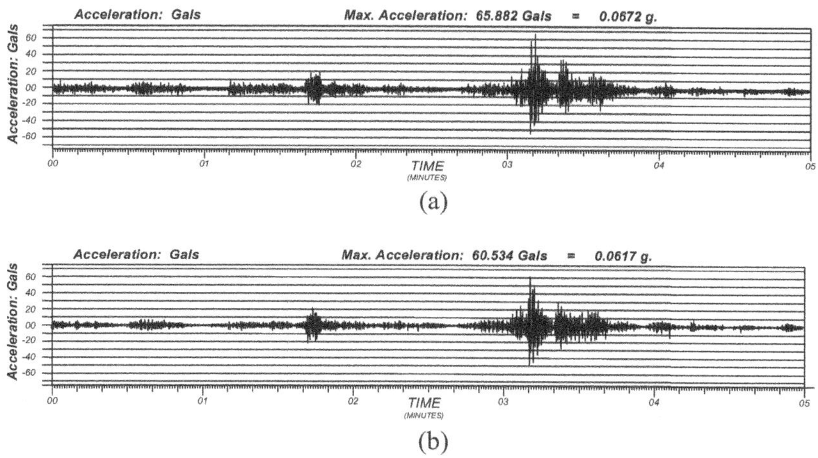

Fig. 11. Accelerograms for the central nodes (a) Node 6 (b) Node 9

In this sense, in Fig. 11, the information provides for these nodes is detailed. As can be appreciate, the maximum acceleration value detected on node 6 was of 65.8 gals and 60.5 gals in the case of the node 9. These results reveal an unequal load distribution of the acceleration values along the structure, that can be useful in order to evaluate the actual state of the structure and for identifying requirements of a structural intervention. Furthermore, regarding the analysis in the frequency domain, results are presented in Fig. 12, as can be seen, the main components are in the range of 1 to 10 Hz for the nodes analyzed.

4.2 System Performance Evaluation

Figure 13, presents results concerning to the synchronization of the nodes during the experimental test. As can be seen, the time deviations among the sensor nodes are in the order of the micro seconds, which represents a suitable range in order to ensure the coherence between the samples collected by the sensors. Particularly, the maximum time difference was obtained between the node 7 and the node 6, specifically 17.7 μs.

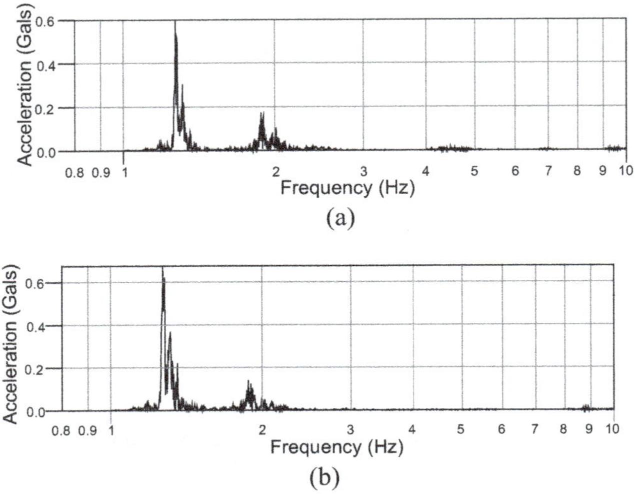

Fig. 12. Frequency spectrum for the central nodes (a) Node 6 (b) Node 9

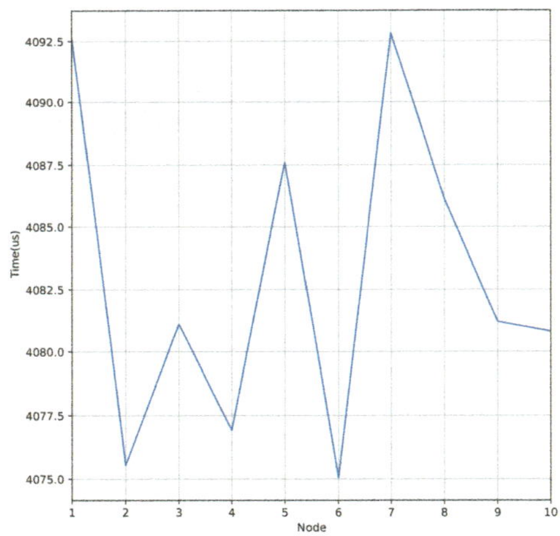

Fig. 13. Synchronization differences among the sensor nodes

Finally, Fig. 14, shows the energy consumption on each sensor node during the experimental test, where it can be noted that in most of the cases the average energy consumption is in the order of the 150 mA. On the other hand, the nodes 1 and 6 present an additional energy expenditure with a maximum value of 228 mA. Consequently, considering the capacity of the batteries used (i.e. 500 mAh), the autonomy of the system is of around two hours or more, which facilitates the development of outdoor experiments.

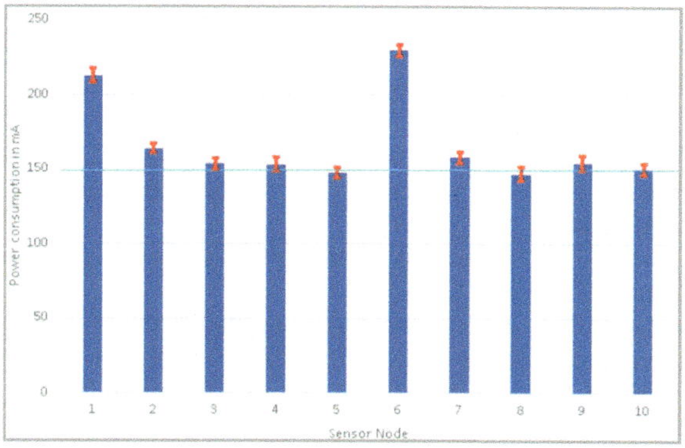

Fig. 14. Energy consumption on each sensor node

5 Conclusions

In this paper we presented an architecture focused in applications of structural health monitoring. The proposal incorporates the use of recent technology such as IoT devices and Wireless Sensor Networks topologies. Furthermore, a monitoring software was implemented which provides users with tools in order to develop outdoor experiments as well as for analyzing data in real time. Regarding, the energy evaluation, the sensor nodes presents a maximum consumption of 228 mA which ensures an energy autonomy of around two hours or more. The system operation was evaluated on a real scenario, specifically the sensor nodes were located along a bridge characterized by a heavy traffic load. Results reveals an unequal load distribution of the acceleration values along the structure. Particularly, the highest values were detected in the center of the bridge on the z-component of the node 6 (65.882 gals) and the node 9 (60.534 gals). Respecting, the frequency domain analysis, results show that the main components are in the range of 1 to 10 Hz. Consequently, the information obtained is useful in order to evaluate the actual state of the structure and for identifying requirements of a structural intervention. Additionally, the proper system synchronization was verified during the experiments. In particular, the maximum delay determined between the sensor nodes and the base station was around of 17.7 μs, which is an admissible value in order to performs an adequate analysis of the data captured by the sensors. Finally, the maximum transmission range determined between the sensor nodes and base station was of around 90 m. Consequently, the proposal architecture contributes significantly in the development of solutions focused in the analysis and early detection of damages on essentials structures or facilities.

Acknowledgment. This work is part of the research project "IoT Technologies and Wireless Sensor Networks for Structural Health Monitoring of Essential Facilities of the City of Cuenca". The authors gratefully acknowledge to the Research Management Department of the University of Cuenca- Ecuador (DIUC), the RSA Department (Red Sísmica del Austro) for the support and resources provided during the development of this research work.

References

1. Wenzel, H.: "Bridge Rating and Risk Assessment", in *Health Monitoring of Bridges*, pp. 39–61. Wiley, First edit. (2009)
2. Noel, A.B., Abdaoui, A., Elfouly, T., Ahmed, M.H., Badawy, A., Shehata, M.S.: Structural health monitoring using wireless sensor networks: a comprehensive survey. IEEE Commun. Surv. Tutorials **19**(3), 1403–1423 (2017)
3. Zimmer, V.L., Sitar, N.: Detection and location of rock falls using seismic and infrasound sensors. Eng. Geol. **193**, 49–60 (2015)
4. Bilek, S.L.: Invited review paper: seismicity along the South American subduction zone: review of large earthquakes, tsunamis, and subduction zone complexity. Tectonophysics **495**(1–2), 2–14 (2010)
5. Ghafoori, E., Hosseini, A., Al-Mahaidi, R., Zhao, X.L., Motavalli, M.: Prestressed CFRP-strengthening and long-term wireless monitoring of an old roadway metallic bridge. Eng. Struct. **176**(June), 585–605 (2018)
6. Malik, H., Zatar, W.: Software agents to support structural health monitoring (SHM)-informed intelligent transportation system (ITS) for bridge condition assessment. Procedia Comput. Sci. **130**, 675–682 (2018)
7. Conti, M., Giordano, S.: Mobile ad hoc networking: milestones, challenges, and new research directions. IEEE Commun. Mag. **52**(1), 85–96 (2014)
8. Mukherjee, S., Biswas, G.P.: Networking for IoT and applications using existing communication technology. Egypt. Inform. J. **19**(2), 107–127 (2018)
9. González-Briones, A., Mezquita, Y., Castellanos-Garzón, J.A., Prieto, J., Corchado, J.M.: Intelligent multi-agent system for water reduction in automotive irrigation processes. Procedia Comput. Sci. **151**(2018), 971–976 (2019)
10. Wang, C., Liu, W., Li, J.: Artificial earthquake test of buried water distribution network. Soil Dyn. Earthq. Eng. **79**, 171–185 (2015)
11. Miao, H., Liu, W., Wang, C., Li, J.: Artificial earthquake test of gas supply networks. Soil Dyn. Earthq. Eng. **90**(February), 510–520 (2016)
12. Herrasti, Z., Val, I., Gabilondo, I., Berganzo, J., Arriola, A., Martínez, F.: Wireless sensor nodes for generic signal conditioning: application to Structural Health Monitoring of wind turbines. Sens. Actuators, Phys. **247**, 604–613 (2016)
13. Kilic, G., Unluturk, M.S.: Testing of wind turbine towers using wireless sensor network and accelerometer. Renew. Energy **75**, 318–325 (2015)
14. Dai, K., Gao, K., Huang, Z.: Environmental and Structural Safety Issues Related to Wind Energy. Elsevier Inc., (2017)
15. Lenticchia, E., Ceravolo, R., Chiorino, C.: Damage scenario-driven strategies for the seismic monitoring of XX century spatial structures with application to Pier Luigi Nervi's Turin Exhibition Centre. Eng. Struct. **137**, 256–267 (2017)
16. Gargaro, D., Rainieri, C., Fabbrocino, G.: Structural and seismic monitoring of the 'cardarelli' hospital in Campobasso. Procedia Eng. **199**, 936–941 (2017)
17. Bedon, C., Bergamo, E., Izzi, M., Noè, S.: Prototyping and validation of MEMS accelerometers for structural health monitoring—the case study of the Pietratagliata cable-stayed bridge. J. Sens. Actuator Netw. **7**(3), 30 (2018)
18. Madhukumar, N., Baiju, M.R.: "MEMS based Wireless Sensor Network for Structural Health Monitoring, **6**(8), 865–870 (2017)
19. Sabato, A., Niezrecki, C., Fortino, G.: Wireless MEMS-based accelerometer sensor boards for structural vibration monitoring: a review. IEEE Sens. J. **17**(2), 226–235 (2017)
20. Rekha, K.S., Sreenivas, T.H., Kulkarni, A.D.: Remote monitoring and reconfiguration of environment and structural health using wireless sensor networks. Mater. Today Proc. **5**(1), 1169–1175 (2018)

21. Avci, O., Abdeljaber, O., Kiranyaz, S., Hussein, M., Inman, D.J.: Wireless and real-time structural damage detection: a novel decentralized method for wireless sensor networks. J. Sound Vib. **424**, 158–172 (2018)
22. Huang, G., Zomaya, A.Y., Delicato, F.C., Pires, P.F.: Long term and large scale time synchronization in wireless sensor networks. Comput. Commun. **37**, 77–91 (2014)
23. Abdaoui, A., El Fouly, T.M., Ahmed, M.H.: Impact of time synchronization error on the mode-shape identification and damage detection/localization in WSNs for structural health monitoring. J. Netw. Comput. Appl. **83**, 181–189 (2017)
24. Adafruit: Adafruit Ultimate GPS, technical specifications. https://cdn-learn.adafruit.com/downloads/pdf/adafruit-ultimate-gps.pdf. Accessed 15 Aug 2020
25. Nordic Semiconductor: nRF24L01, Data sheets. https://www.sparkfun.com/datasheets/Components/SMD/nRF24L01Pluss_Preliminary_Product_Specification_v1_0.pdf. Accessed: 15 Aug 2020
26. ffmpeg.org: FFmpeg a complete, cross-platform solution to record, convert and stream audio and video. https://ffmpeg.org/. Accessed: 15 Aug 2020
27. Microchip: Microchip dsPic data sheets. https://www.microchip.com/wwwproducts/en/dsPIC33EP256MC502. Accessed: 15 Aug 2020
28. Dallas Semiconductor: Extremely Accurate SPI Bus RTC. https://www.sparkfun.com/datasheets/BreakoutBoards/DS3234.pdf.
29. Analog Devices: Low Noise, Low Drift, Low Power, 3-Axis MEMS Accelerometers. https://www.analog.com/media/en/technical-documentation/data-sheets/adxl354_355.pdf. Accessed 15 Aug 2020
30. IEEE: 1588–2008 - IEEE Standard for a Precision Clock Synchronization Protocol for Networked Measurement and Control Systems. https://standards.ieee.org/standard/1588-2008.html. Accessed 15 Aug 2020
31. Narvaez, M., Zapata, J., Garcia, J.: Evaluación experimental y verificación teórica de la respuesta dinámica de uno de los puentes de Nulti, ubicados en la abscisa 7+670 de la autopista Cuenca-Azogues, Universidad de Cuenca (2019)

Software Architecture for IoT-Based Solutions

Carlos Jones-Ortiz[1] and Félix Fernández-Peña[2](✉)

[1] ESIT, Universidad Nacional de La Rioja,
Avda. de la Paz, 137, 26006 Logroño, La Rioja, Spain
carlosvicente.jones401@comunidadunir.net
[2] ESAI, Universidad Espíritu Santo, Km. 2,5 vía Samborondón,
092301 Guayaquil, Ecuador
fofernandez@uees.edu.ec

Abstract. Multiple specific solutions reinventing the wheel on how to integrate the IoT technology characterise the actual scenario of Industry 4.0. This paper proposes a software architecture for integrating IoT-based solutions in a generic scenario. Our main contribution is the definition of formal processes for capturing, processing and presenting data to the user. The proposed architecture was evaluated in the construction of a system for controlling access to a parking lot in *Universidad Nacional de Educación*. The experimental evaluation of the architecture shows that the design of a solution based on the proposed architecture increases the extensibility of the IoT solution and increases the easiness to integrate the solution with a legacy system. As a result, an open source, easy to extend software architecture is presented.

Keywords: IoT solutions · Software architecture · Extensibility

1 Introduction

Internet of Things (IoT) refers to the connectivity of devices that can send and receive data through Internet [9]. The automation of processes based on the Internet of Things (IoT) has given rise to the definition of multiple concepts around the intelligent behaviour of software in homes [13], cities [13,15], universities [1]. Boundaries are been shifted due to generation of information in IoT environments that will change markets and exchanges [11]. Without any doubt, artificial intelligence solutions are more ubiquitous than ever thanks to the data that an exponentially growing amount of sensors collect around the world [4,13].

In the same vein, specific IoT-based systems have successfully scaled to the cloud. This way, it is possible to talk about IoT-based big data solutions [5]. Nevertheless, scaling IoT systems is a challenging task. This is because, unlike the traditional Internet, the IoT systems integrate both, physical and logic components into its programming [18]. Homogeneous interoperability is the key and the challenge is the interoperation of dispared systems [5,6,18].

M. Botto-Tobar et al. (Eds.): CIT 2020, AISC 1326, pp. 278–289, 2021.
https://doi.org/10.1007/978-3-030-68080-0_21

Meanwhile, a software architecture defines methods for the efficient segmentation of a computer system and establishes how its components interact among them [7]. Having identified the disparity of IoT devices as the main issue for the interoperability of IoT-based solutions, this paper proposes a software architecture for scaling IoT solutions on behalf of their users.

The hypothesis of our work states that *if IoT devices are encapsulated as services then it is possible to define a software architecture for building IoT-based systems easy to extend.* For the evaluation of this hypothesis, the development of an IoT-based solution for controlling the access to a parking lot in the *Universidad Nacional de Educación* was taken as case study. The experimental validation of the results included the calculation of the extensibility of a software architecture using the Extensibility Metric for Software Architecture (EMSA), proposed by Kim et al. [8].

This paper is structured as follows. Section 2 depicts related work. The proposed architecture, and its main components, are described in Sect. 3. Section 4 presents the results of the experimental evaluation of the tool through the construction of the application of the case study. Finally, in Sect. 5, we arrive to conclusions and we present some ideas for future work.

2 Related Work

There is a growing interest in the integration of IoT solutions for smarter services in modern society [1,2,4] and, in this scenario, the quality of service is recognized as a main concern [4]. Kumar et al., compared the performance of four IoT hardware architectures using a smart city benchmark [9]. The results of their experiments show that the Edge architecture achieves the least latency in scenarios characterized by the proximity to query devices and sensors. However, the Cloud architecture obtained a minimum latency for compute-intense cases.

Silva et al. proposed a RESTful interface for a smart gateway system controlling an energy provision management system to homes in a smart city [13]. In their study, they concluded that a Restful interface facilitates IoT platform compatibility and ubiquitous access to services [13]. In fact, Silva et al. consider the REST architectural style as a reference paradigm to integrate smart things into the web.

On the other hand, Lee considers that enterprise sector has become the largest sector among various IoT sectors. His intention is to stimulate further interest in enterprise IoT with the definition of an ecosystem, architecture, and IoT service business model for innovating hotel rooms [10]. This architecture is structured in five layers: service management layer, application layer, processing layer, network layer and perception layer. Something missing in his proposal is a detailed definition of the responsibilities of each of his architecture's layers.

In the area of building situation-aware IoT applications, Teixeira et al. recently proposed the LAURA architecture [16]. In their paper, a service orientation is proposed for the development of decoupled IoT applications. Furthermore, their contribution is focused in the generation of low-level code of generic IoT devices. This way, the need for specialized IoT low-level knowledge is reduced while showing an acceptable performance. In the literature review that we carried out, there is a clear trend to the use of a RESTful API for the integration of IoT as a service. Meanwhile, the primary sources review of Alreshidi and Ahmad shows that recent works focuses mainly in the integration of artificial intelligence in IoT software architectures [2]. However, to the best of our knowledge, none of the proposals takes into account the insights of the management of data and interfaces of IoT devices from the perspective of a generic software architecture. We consider this approach as fundamental for the development of extensible IoT systems.

3 Description of the Proposed Software Architecture

The proposed software architecture integrates the Lambda architecture and the Reference architecture for IoT (IoT ARM). It is structured in four layers. Those layers are: perception, storage, processing and presentation. Figure 1 shows the functional model of the proposed software architecture. These layers are structured in a way that maximizes the extensibility of an IoT solution based on the proposed software architecture.

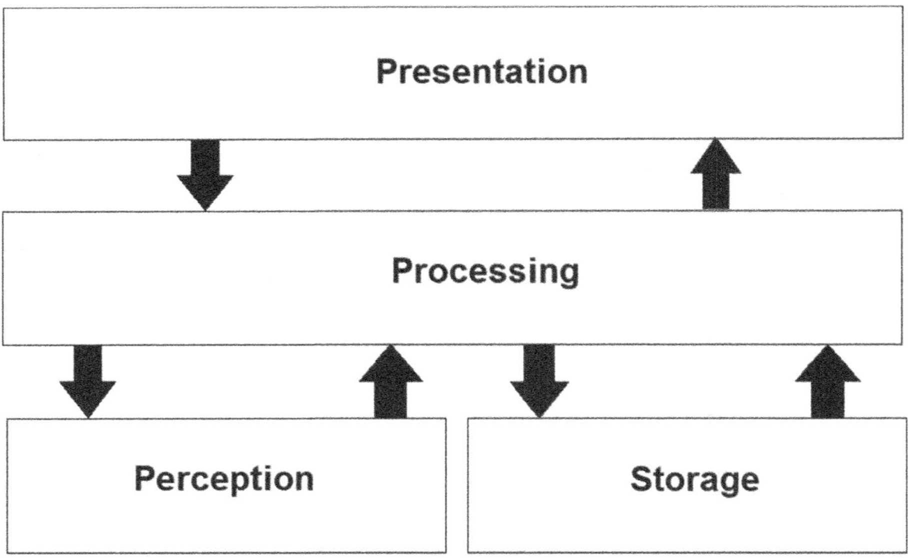

Fig. 1. Functional model of the software architecture.

Functionalities of actuators and sensors at the perception layer are abstracted as services exposed in the processing layer. High level processes, implemented in the presentation layer, interacts with the RESTful API that the processing layer exposes. Processing layer read data from the perception layer and save relevant data in the storage layer. The processing layer analyses stored data, interacts with external systems, and take decisions on behalf of the end user. Those decisions are notified through corresponding interfaces defined in the presentation layer.

Furthermore, users interact with the presentation layer for entering configuration parameters. Those parameters are sent by the processing layer into the storage layer and into the perception layer. Storage layer has specific interfaces used for registering actions that support auditing the system. Meanwhile, the perception layer implements specific interfaces taking into account the kind of devices to be configured.

The proposed software architecture is focused on integrating IoT-based solutions. Python is chosen as the programming language for implementing the processing layer in the study case but the architecture is not tied to a specific language; using json for the exchange of data among architecture's layers allows to encapsulate the complexity of each layer regardless of the programming language used in each software architecture's layer.

3.1 Perception Layer

The perception layer encapsulates the interaction with IoT devices. A RESTful API hides the complexity of the hardware interfaces and supports the auditing of the quality of the service in a standard way. This support for auditing services is inspired in the work of Bhandari and Cho [4]. Figure 2 shows the composition of the interface of this layer (class Infrastructure) of as many instances of IoT devices (class Device) as needed.

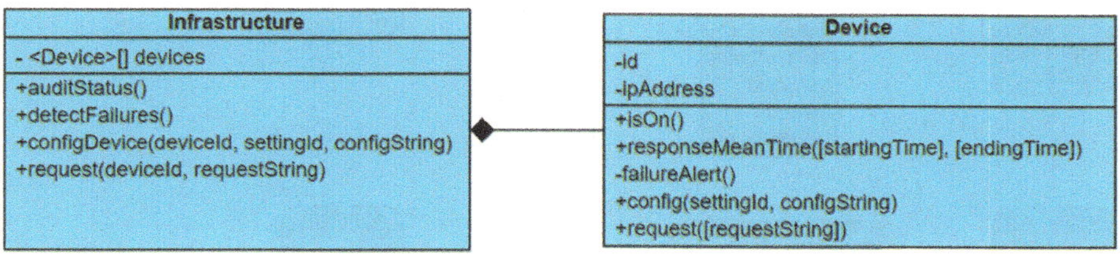

Fig. 2. Abstraction of IoT hardware for supporting standard communication and auditing.

3.2 Storage Layer

A data model was defined for supporting the auditing of the quality of service of IoT devices. Figure 3 presents the design of this database which determine the autonomy of this layer against the data model of any particular scenario.

In this regard, those entities included in Fig. 3 are generic for any case study. The management and operational auditing of the IoT devices is independent of specific business rules, as proposed by Bhandari and Cho [4]. The entity by which this data model is connected to the data model of an specific scenario is the entity *User*. The key field in this entity is the only property required for the integration of data into this data model.

This proposal considers the general access policies for users upon business rules non-dependant from access policies to IoT devices. That is why this data model includes entities for setting access permissions in the configuration of IoT devices (*allowedConfiguration*) as well as which authorization items are enough for changing any configuration (property *authItemId* in the relation *allowedConfiguration*). A hierarchy of authorization items (through the *authItemChild* relation) supports any level of complexity in the definition of permissions for managing IoT devices; even bizRules can be used for this purpose.

In addition, other entities were defined for storing actual configuration settings (*configuration*) and observations of their readings to be processed (*reading*). Furthermore, this layer is responsible for adding and removing logic instances of IoT devices and device types from the infrastructure (entity *device* and *deviceType*).

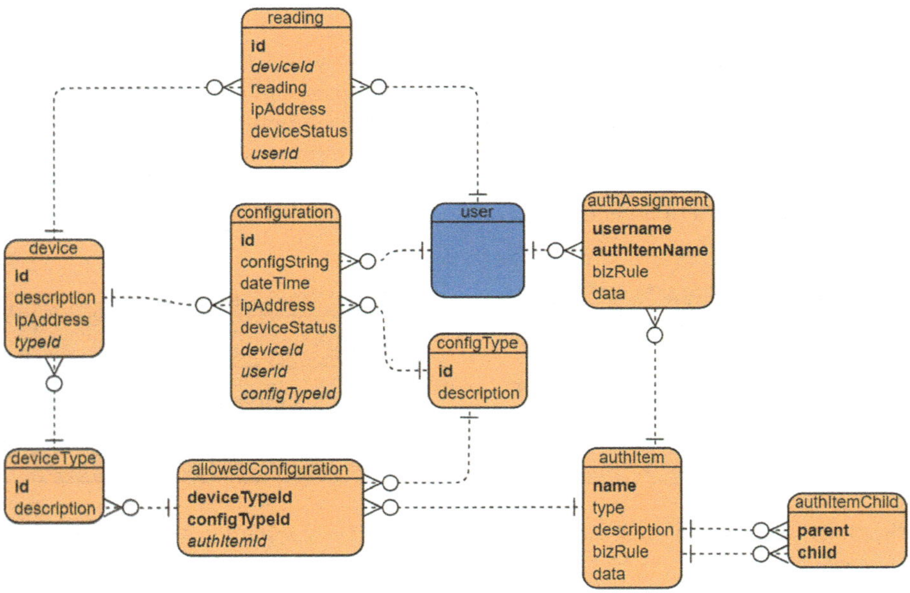

Fig. 3. Defined data model for supporting the auditing of the operation of IoT devices.

3.3 Processing Layer

The processing layer entirely depends on the business logic of each scenario. The services of this layer are supported by the RESTful API of the perception layer and the storage layer. This means that this layer exposes business-logic specific endpoints for retrieving the results of the analysis of data read by sensors. Other endpoints would be defined for triggering high level orchestration of services provided by IoT devices. Figure 4 shows the class diagram resulting of the design of the processing layer. By extending one of these classes, a high level web service would be defined. The whole arsenal of the RESTful API of the perception and storing layers get available through these classes. The result of extending classes of this level would be the implementation of high level IoT-based web services dependant of an specific business logic.

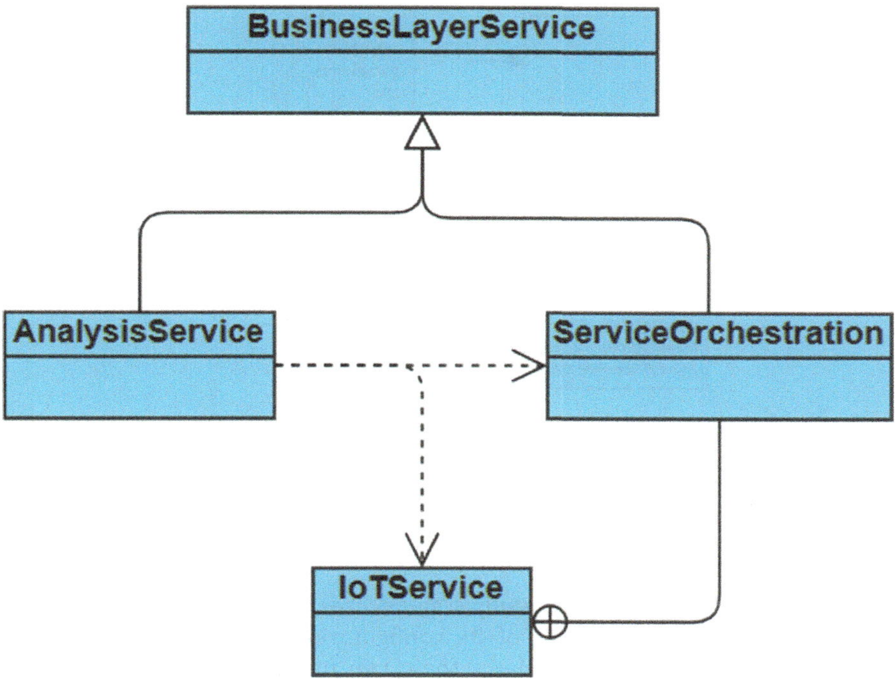

Fig. 4. Web services of the proposed architecture that encapsulate the processing of data coming from IoT devices and external systems' functionalities too.

It is in this respect that we recommend to deploy services of this layer using a RESTful API. In terms of efficiency, sending data in JSON format and using the traditional communication protocols of Internet remains as the most accepted mechanism for integrating dispared systems [1,7]. This way, the integration and extensibility of solutions gets easier [7].

3.4 Presentation Layer

The user interaction with the IoT systems takes place through many different kind of interfaces. Devices like PCs, tablets and phones, and even more specific hardware, are used.

We define an standard interface for the user to interact with IoT systems. No matter of the scenario, the interface can be automatically generated based on the data model behind the classes of Fig. 5.

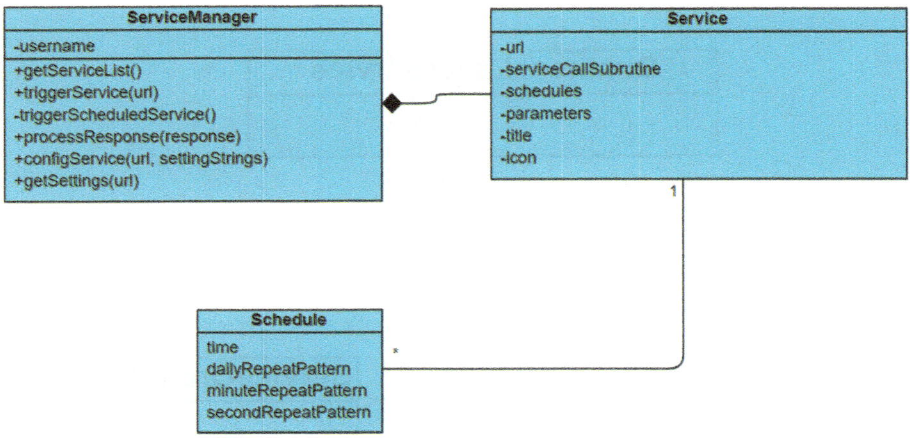

Fig. 5. Classes for managing the presentation layer.

4 Analysis of Results

The proposed architecture successfully standardizes the mechanism for managing the IoT infrastructure in terms of its configuration and functioning (Fig. 2). The class *Infrastructure* encapsulates these two functionalities and provides a standard interface for configuring and triggering devices' functionalities. New devices are added to the infrastructure without changes in the interface of the RESTfull interface of this layer but with internal add-ons and the insertion of a new instance of the class *Device*.

For its part, the storage layer standardizes a mechanism for auditing the access to any device and its configuration by any authorized user. Regardless any other specific business rules, the data model of Fig. 3 restricts access to the IoT infrastructure. The access to the infrastructure is then encapsulated by the processing layer (Fig. 4). This layer is responsible for the integration of the IoT infrastructure with third-party services, reducing the impact on the extensibility of any IoT solution. The abstraction of business services in this layer also makes easier the integration of an IoT-based solution with legacy systems.

Finally, the classes defined in the presentation layer (Fig. 5) allow to schedule calls to the services of the business layer. This layer is responsible of the interaction with the end user, making it possible to interact with the rest of the functional layers of the software arquitecture.

Controlling access to a parking lot of *Universidad Nacional de Educación* was used as a case study for determining the extensibility of the proposed software architecture. A software implemented in Python 3.7 uses Open CV 3 and Numpy 1.17 for transforming and analyzing the lastly taken picture in order to recognize the plate of the vehicle. A solution designed using the proposed architecture was compared with an ad hoc solution.

Internally, KNN algorithm was used for interpreting the patterns and determining the actual plate of any car trying to get access to the parking lot. This solution is based in the work of Varma et al. [17]. Other implementations are based on neural networks [3, 12, 14] and decision tree [15]. KNN is a less resource-consuming algorithm; nevertheless, the decision of which algorithm to use is actually out of the scope of this research. Replacing the implemented algorithm with a different one represents an internal change that does not affect the overall behaviour of the system. The functionalities of the business logic are exposed using a RESTful API implemented using Requests 2.2. The data of plates of vehicles that can get into the parking lot are stored in a PostgreSQL 11 server and Psycopg2 2.8 was used as Pythons connector to this database.

Five software engineers participated in the evaluation of this architecture. Each engineer was asked to design new functionalities for the parking lot case study. The new functionalities were the following:

- F1 Identifying car's plates with two cameras instead of one.
- F2 Opening the gate of the parking lot without any human interaction.
- F3 Notifying when a specific user is actually using the service of the parking lot.
- F4 Reporting data of users once they are permitted to enter to the parking lot.

In this study, the metric EMSA [8] was measured for quantifying the extensibility factors of understanding, designing, implementing and adapting new functionalities to a software architecture. This way, it was possible to compare the extensibility of an ad-hoc IoT software architecture against the extensibility of the software architecture that we proposed (see Table 1).

As defined by Kim et al. [8], $IE(SA)$ quantifies the complexity of the software architecture. Its value is calculated as the entropy of the data flow event to occur on the links of a module in the architecture. For its part, $IE(SA^{new})$ quantifies the complexity of the software architecture when new functionalities are included.

Meanwhile, $NM(SA, SA^{new})$ refers to the number of new modules in the changed software architecture and $IR(SA, SA^{new})$ refers to the number of connections between new modules and the modules of preexisting software architecture.

As defined by Kim et al., $EMSA(SA, SA^{new})$ is calculated as a value of the understandability of the software architecture. Other criteria that $EMSA$

Table 1. Measuring EMSA metrics.

Metrics	SA_{ad-hoc}	$SA_{proposed}$
$IE(SA)$	8.299	1.287
$IE(SA^{new})$	9.178	1.358
$NM(SA, SA^{new})$	6	3
$IR(SA, SA^{new})$	8	6
$EMSA(SA, SA^{new})$	88.609	50.531

takes into account are the ability to design and to implement new functionalities. Lastly, $EMSA$ is a measure of the ability to interconnect old and new functionalities in a software architecture [8]. That is why, the Eq. 1 allows to calculate the metric $EMSA$. In this equation (equivalent to that one proposed by Kim et al. in [8]), $SIA(SA, SA^{new})$ takes the value specified in Eq. 2 and, $V_N(SA, SA^{new}) = 65.5$, $V_I(SA, SA^{new}) = 4.5$, $C_{UD} = 0.295$, $Y_{UD} = 6.711$, $C_{IA} = 0.164$, and $Y_{IA} = 6.386$. These factors we use are the same that Kim et al. validated in an experimental correlation analysis [8].

$$\begin{aligned} EMSA(SA, SA^{new}) = {} & (C_{UD} \cdot (IE(SA) + IE(SA^{new})) + Y_{UD}) \\ & + (C_{IA} \cdot SIA(SA, SA^{new})) + Y_{IA}) \end{aligned} \tag{1}$$

$$\begin{aligned} SIA(SA, SA^{new}) = {} & V_N(SA, SA^{new}) \cdot NM(SA, SA^{new}) \\ & + V_I(SA, SA^{new}) \cdot IR(SA, SA^{new}) \end{aligned} \tag{2}$$

It is our opinion that the differences between the IE metrics respond to the high *cohesion* and loose *coupling* of modules in our approach. The values of $IE(SA)$ and $IE(SA^{new})$ show that understanding the proposed approach is easier. Furthermore, when analyzing the *extensibility* of the software architecture, it is evident that our approach requires the creation of less modules with less interdependent relationships among new modules and preexisting modules. These features determine the values of $NM(SA, SA^{new}$ and $IR(SA, SA^{new})$.

Table 2 sums up the amount of modules required for the previously listed functionalities (F1 - F4). In this table, $IoT - D$ refers to modules required for working with a new IoT device to the system, whilst IM refers to modules required for the integration of functionalities and RM refers to modules for generating reports in the user interface.

In Table 2, we can observe that for the first of the functionalities in the experiment ($F1$), our approach only requires a module for integrating more than one IoT device. Adding the device requires its instantiation in the architecture and not any additional software module. Something similar happens in the case of the second functionality ($F2$). For the 3rd and 4th functionalities ($F3$ and $F4$), the software architecture that we propose does not require any module programming but modifying the programming of a module previously registered in the set of services of the presentation layer.

Table 2. Modules to be created because of new functionalities.

Functionality	SA_{ad-hoc}	$SA_{proposed}$
F1	1 $(IoT - D)$, 1 (IM)	1 (IM)
F2	1 $(IoT - D)$, 1 (IM)	1 (IM)
F3	1 (RM)	-
F4	-	-

Making an analysis of the results, we can say that a solution based on the proposed architecture would have lower costs when extending its functionalities. In opinion of the experts that participated in the study, this architecture actually impacts the development of a generic IoT-based solution. Nevertheless, these experimental results are not determinant but indicative of the validity of this proposal. Because of that, further research will include more than one case study in the experimental evaluation of this software architecture.

5 Conclusions

The proposed software architecture guarantees the extensibility of IoT solutions. The definition of the layers of this architecture guarantees the separation of concerns when new functionalities are to be added to the system. It is easier to maintain the code because features are more isolated and therefore less likely to introduce bugs. Software maintenance is also safer since auditing procedures are inherent to the data models defined in the layers *storage* and *presentation*.

We carry out an experimental validation of the results of this research. Using the proposed architecture, the extensibility of the software architecture was improved in 38.078PH when compared with the software architecture of a system for controlling the access to a parking lot in the *Universidad Nacional de Educación*. This result allows us to deduce that using the software architecture that we proposed would be easier to extend the use of IoT devices in substantial processes in this institution.

As future work, we propose to define the semantics of the RESTful API of the software architecture in order to make easier the integration of functionalities as semantic web services. This way, we expect to increase the autonomy of the software architecture in the service orchestration that takes place in the processing layer.

References

1. Adeyemi, O.J., Popoola, S.I., Atayero, A.A., Afolayan, D.G., Ariyo, M., Adetiba, E.: Exploration of daily internet data traffic generated in a smart university campus. Data Brief **20**, 30–52 (2018). https://doi.org/10.1016/j.dib.2018.07.039
2. Alreshidi, A., Ahmad, A.: Architecting software for the internet of thing based systems. Future Internet **11**(7) (2019). https://doi.org/10.3390/fi11070153

3. Asif, M.R., Qi, C., Wang, T., Fareed, M.S., Raza, S.A.: License plate detection for multi-national vehicles: an illumination invariant approach in multi-lane environment. Comput. Electr. Eng. **78**, 132–147 (2019). https://doi.org/10.1016/j.compeleceng.2019.07.012
4. Bhandari, K., Cho, G.: Resource oriented topology construction to ensure high reliability in IoT based smart city networks. Int. J. Syst. Assur. Eng. Manage. **11**(4), 798–805 (2020). https://doi.org/10.1007/s13198-019-00861-2
5. Bruneo, D., Distefano, S., Giacobbe, M., Longo Minnolo, A., Longo, F., Merlino, G., Mulfari, D., Panarello, A., Patanè, G., Puliafito, A., Puliafito, C., Tapas, N.: An IoT service ecosystem for smart cities: the #smartme project. Internet Things **5**, 12–33 (2019). https://doi.org/10.1016/j.iot.2018.11.004
6. Chen, G., Jiang, T., Wang, M., Tang, X., Ji, W.: Modeling and reasoning of IoT architecture in semantic ontology dimension. Comput. Commun. **153**, 580–594 (2020). https://doi.org/10.1016/j.comcom.2020.02.006
7. Cortellessa, V., Eramo, R., Tucci, M.: From software architecture to analysis models and back: model-driven refactoring aimed at availability improvement. Inf. Softw. Technol. **127**, 106362 (2020)
8. Kim, J., Kang, S., Ahn, J., Lee, S.: Emsa: extensibility metric for software architecture. Int. J. Software Eng. Knowl. Eng. **28**(03), 371–405 (2018). https://doi.org/10.1142/S0218194018500134
9. Kumar, H.A., Rakshith, J., Shetty, R., Roy, S., Sitaram, D.: Comparison of IoT architectures using a smart city benchmark. Procedia Comput. Sci. **171**, 1507–1516 (2020), third International Conference on Computing and Network Communications (CoCoNet'19)
10. Lee, I.: The internet of things for enterprises: an ecosystem, architecture, an IoT service business model. Internet Things **7**, 100078 (2019)
11. Ng, I.C., Wakenshaw, S.Y.: The internet-of-things: Review and research directions. Int. J. Res. Mark. **34**(1), 3–21 (2017). https://doi.org/10.1016/j.ijresmar.2016.11.003
12. Omar, N., Abdulazeez, A., Sengur, A., Al-Ali, S.: Fused faster RCNNs for efficient detection of the license plates. Indonesian J. Electr. Eng. Comput. Sci.**19**(2), 974–982 (2020). https://doi.org/10.11591/ijeecs.v19.i2.pp974-982
13. Silva, B.N., Khan, M., Han, K.: Integration of big data analytics embedded smart city architecture with restful web of things for efficient service provision and energy management. Futur. Gener. Comput. Syst. **107**, 975–987 (2020)
14. Silva, S.M., Jung, C.R.: Real-time license plate detection and recognition using deep convolutional neural networks. J. Vis. Commun. Image Represent. **71**, 102773 (2020). https://doi.org/10.1016/j.jvcir.2020.102773
15. Suresan, A., Venkatraman, M., Suresh, S., Supriya, P.: Comparison of machine learning algorithms for smart license number plate detection system. In: Advances in Intelligent Systems and Computing, vol. 1200. AISC, pp. 63–75 (2021). https://doi.org/10.1007/978-3-030-51859-2_7, cited By 0
16. Teixeira, S., Agrizzi, B.A., Filho, J.G.P., Rossetto, S., Pereira, I.S.A., Costa, P.D., Branco, A.F., Martinelli, R.R.: Laura architecture: towards a simpler way of building situation-aware and business-aware IoT applications. J. Syst. Softw. **161**, 110494 (2020)

17. Varma, R.K., Ganta, S., Krishna B, H., Svsrk, P.: A novel method for Indian vehicle registration number plate detection and recognition using image processing techniques. Procedia Comput. Sci. **167**, 2623–2633 (2020). https://doi.org/10.1016/j.procs.2020.03.324, International Conference on Computational Intelligence and Data Science
18. Verdouw, C., Sundmaeker, H., Tekinerdogan, B., Conzon, D., Montanaro, T.: Architecture framework of IoT-based food and farm systems: a multiple case study. Comput. Electron. Agric. **165**, 104939 (2019). https://doi.org/10.1016/j.compag.2019.104939

Bootstrap as a Tool for Web Development and Graphic Optimization on Mobile Devices

Oscar López-Gorozabel(✉), Emilio Cedeño-Palma, Jenmer Pinargote-Ortega, Walter Zambrano-Romero, and Marcos Pazmiño-Campuzano

Facultad de Ciencias Informáticas, Universidad Técnica de Manabí, Portoviejo, Ecuador
{olopez7574,emilio.cedeno,maricela.pinargote,walter.zambrano, marcos.pazmino}@utm.edu.ec

Abstract. The advance of technology requires that people need publicity and administrative advice of their business through the Web. It is there where the role of the web developer appears since this topic must make use of the appropriate tools that enhance their work. The style pages have been facilitating the elaboration of web pages for some years now. However, programmers have not created any framework to optimize the size of the interfaces automatically.

By doing so, navigation from any device without any anomaly. Currently, the massive use of mobile devices requires the optimization of the design of the graphic interfaces developed for computers to the resolution of mobile devices. One could say that this motivated the company Twitter in 2011 to create Bootstrap, a framework with the function of adjusting the resolution of websites that can be developed from a computer and then be displayed correctly in the different screen resolutions of mobile devices.

This article tries to orient the evaluation of the Bootstrap tool for the re-engineering of the web system "EducArte Comunidad" interacting with the following design languages: HTML5, CSS, and JQuery. Specifically, this research is carried out based on the observation in the development of computer applications under the multiplatform modality.

Keywords: Websites · Bootstrap · Graphic languages · Adaptive frame · Graphic interfaces

1 Introduction

1.1 Evolution of Online Computer Systems

Web systems have evolved rapidly, and then desktop systems have been left behind. Some companies have migrated their obsolete systems to the web, as online services provide users with faster data processing and improved information availability.

According to [10], in his article on the development of the educational platform Zera (Cuba), he expresses that the use of the Bootstrap framework facilitated the programming of interfaces and improved the user experience in navigation from any mobile device.

M. Botto-Tobar et al. (Eds.): CIT 2020, AISC 1326, pp. 290–302, 2021.
https://doi.org/10.1007/978-3-030-68080-0_22

On the other hand, [15] state in their article on M-learning, that educational systems must currently focus on pocket technology, that is, mobile devices and tablets that facilitate the learning process at an economic level. For this reason, we intend to perform the re-engineering of the EducArte Comunidad system since the system only has a presentation for computers, but it will be taken to the mobile part through adaptability.

Following the line of development of educational systems, [7] considers that the use of languages that provide adaptability, especially Bootstrap allowed the library web system of the University of Maryland to be adaptable. This adaptability allowed access through cell phones and tablets. The results of this implementation were beneficial by improving the number of virtual visits and users experiences.

There is no doubt that the adaptable aspect in the development of applications is an advantage when it comes to creating systems. Therefore, [9] states that predecessor websites were created with design aspects that were only relevant to PCs. This type of development posed a large number of problems, having to create a different design for each device where the application was running. To solve this problem, Bootstrap was implemented in the graphic interfaces of the Juang Restaurant System, finally generating an optimal presentation in the content displayed on mobile devices.

This situation, according to [3], is conducive to the emergence of receptive websites, such as the SANEC Portal, which was developed with the methodology of adaptability, seeking to adjust the desktop system to all mobile devices, increasing the number of visits exponentially. Likewise, the aim is to reach the demand of users in the EducArte Community system. Currently, frameworks dedicated to web interfaces are so important as to improve the visual interaction between the user and a computer system. According to [11], the most used framework nowadays is Bootstrap, due to the ease and resources it offers to create web interfaces.

The development of this adaptive web system has social importance for the benefit of the community to be implemented in vulnerable sectors, where children and young people have the facility to access the services offered by the system through a mobile device. The reason is that almost every home has a mobile device and not a computer due to economic difficulties. Therefore, with the development of this web system, we intend to reduce the limitations in the design and architecture of the "EducArte Comunidad" system, opening the way to the m-learning paradigm, conditioning the availability of its contents and its correct functioning in mobile devices. All this, to present one of the main results achieved in the research related to the adaptive web, highlight the theoretical references that support the solution.

2 Materials and Methods

The methodology of this research is of a bibliographic and experimental nature. Applied analysis and synthesis is made to summarize the theories on which the research is based, citing the criteria of web development experts on the Bootstrap framework and its characteristics, to conclude the importance and guarantee of its use for the graphic optimization of web applications. The following aspects will be studied in depth: definition of Bootstrap, its main characteristics, the compatible browsers, the types of language it integrates. Attached to the above, the following items will also be detailed:

- Definition of Bootstrap principles and usability guidelines.
- Experimentation and evaluation of the "EducArte Community" web system, which will be re-engineered with Bootstrap.

2.1 Bootstrap for the Production of Web Systems

Nowadays, you can find tools for web design, which Twitter developed for those users who do not have much knowledge in HTML and CSS programming. It is worth emphasizing that there is currently a fine line between designers and programmers in this area of development, making possible the creation of ideal tools for creative processes, which contribute to the development of software without previous knowledge of engineering [13].

[21] defines that: The user regularly changes electronic devices. Therefore, programmers must develop web systems to adapt to a variety of resolutions, specifically image size and command sequences. In short, the web system must automatically respond to different devices to satisfy user preferences. The adaptive design of Bootstrap is based on the use of third level style sheets with CSS Media Queries, which allows styles to be applied depending on the size of the screen on which the web system is displayed. According to [17], Boostrap has surprising features for front-end development, being visually coupled to the needs and experiences of the user, the features presented are:

- It allows the insertion of several elements online: icons, drop-down menus, combining HTML5, CSS and Javascript languages.
- The design will be adaptable, not affected by the scale or resolution of the mobile device.
- The design of columns is faster and easier to configure.
- It integrates with the main Javascript and Media Queries libraries.

Responsive Web System Against Mobile Applications

For [20], the application can read user contacts, calendars, photo galleries, and other resources. But in the case of a web system, it does not allow access to system resources. A difference in compatibility that exists in web development is that the development code of the mobile application depends on the phone manufacturer; at the same time, any mobile browser can read the HTML-based web system. On the other hand, the user must download the application and its updates separately, which will not be necessary for receptive systems.

2.2 Screen Adaptation on Mobile Devices

For [1], small screen mobile devices are gaining popularity as time goes by. Video game consoles and smart TVs are also becoming more common among consumers, as they allow them to browse the Internet on high-resolution devices in comfort.

That is why web developers have begun to bet on the graphic adaptability of systems in a wide variety of screen resolutions, with the primary objective of giving the user maximum usability and presentation of content in a clean and appropriate. Below, Table 1 shows the visualization preference that users have on a website, a survey conducted by Google in 2014 [14].

Table 1. Preference in screen accuracy according to Google users.

Range accuracy (px)	Popularity (%)
1366 × 768	20.74
1024 × 768	9.61
1280 × 800	9,35
1920 × 1080	6.63
1440 × 900	5,44
768 × 1024	5.26
1280 × 1024	5.22
1600 × 900	4.14
320 × 480	2.90

Source. Data taken from a Google survey published by W3C.

Composition of the Adaptive System

According to [9], when the Internet became widespread in the 1990s. It was usual for web systems to be compiled into a spreadsheet structure. Tables were the only effective way to link text and graphics. These issues occurred when sites became more sophisticated and extensive, and table structures were transformed into more complex nested tables.

File Download Formats

For [6], Bootstrap is free to download and also for unlimited commercial use, allowing web designers to use it as a tool to create web systems for different purposes. The source code contains the files Bootstrap CSS, Less, which is a CSS extension, Javascript, and other original fonts, as well as documentation. These files can be found in GitHub.

2.3 Preparing Bootstrap for Web Development

[3] defines that: Bootstrap contains compiled and compressed source code, as well as versions of CSS, JavaScript, and font files. Also, when downloading it, the files will be extracted to the root of the web system to link locally in the document header. The Bootstrap file structure consists of three folders: css, js, and fonts. We can recognize compressed files by the min file name. These files contain the same code as the translated files but are compressed, which takes less time to download from the server.

Bootstrap file structure (see Fig. 1).

```
bootstrap/
├── css/
│   ├── bootstrap.css
│   ├── bootstrap.min.css
│   ├── bootstrap-theme.css
│   └── bootstrap-theme.min.css
├── js/
│   ├── bootstrap.js
│   └── bootstrap.min.js
└── fonts/
├── glyphicons-halflings-regular.eot
├── glyphicons-halflings-regular.svg
├── glyphicons-halflings-regular.ttf
└── glyphicons-halflings-regular.woff
```

Fig. 1. Compressed files can be recognized by the. min file name. To access or read Bootstrap files it is recommended that they are compressed [14].

Bootstrap Compatibility and Support

Bootstrap is compatible with the latest versions of desktop and mobile browsers. When running on older browsers, it differs in style, functionality, and performance of some components. In Windows, Bootstrap runs in Chromium, Chrome, and Firefox browsers, which are not officially compatible. However, the network version allows them to automatically link to the repositories where the Bootstrap libraries are located. On the other hand, it unofficial sources states to work quite well with Linux [12] (Fig. 2).

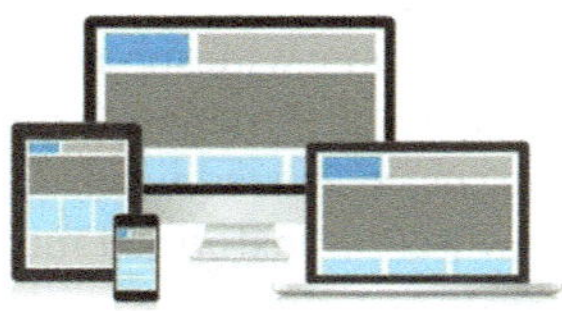

Fig. 2. Multiplatform web design with Bootstrap

Grid System

[14] claims that:

Bootstrap always works with a sensitive and predefined grid system that is distributed in the twelfth column. The grid system creates a design with content, on the site divided into sets of columns and rows. The grid system is surrounded by a div.container (fixed width).container-fluid (adaptive width) element. A line is then inserted into the container. (Row).

<div class = “container”>
<div class = “row”> </div>
</div>

The site columns are integrated into the rows. The purpose of the rows is to divide the columns into horizontal groups. The site content is placed within individual columns.

<div class = “container”>

```
<div class = “row”>
<div class = “col-md-6”> Hello Word </div>
<div class = “col-md-6”> My name Oscar López </div>
</div>
</div>
```

You can use the HTML table containing the line (tr) to help outline the structure of the cell (td). The content of the table, such as text, graphics, and so on, is written inside the cell. The height of cells determines the height of the row around them. In a grid system, columns are similar to cells in a table, since the site content is embedded. The row always begins with a new group of columns. Unlike a grid, rows can contain a different number of columns [14].

```
<div class = “container”>
<div class = “row”>
<div class = “col-md-12”> EducArte Comunidad </div>
</div>
<div class = “row”>
<div class = “col-md-4”> Home </div>
<div class = “col-md-4”> Gallery </div>
<div class = “col-md-4”> Notes </div>
</div>
</div>
```

The spacing of the columns is 15 pixels by default on the left and right sides created by the fill. Because the column fill is at both edges, the content area of the first and last columns in the row is not aligned with the container. The left and right borders were created for columns with negative margins at-15 pixels. This will change the position of the first and last columns in the row so that their content area is aligned with the container [14]. Bootstrap has a total of four predefined breakpoint definitions for columns: lg, md, sm, and xs (Table 2).

Table 2. Acceptance of breakpoints in pixels.

Category	Width in pixels	Devices
.col – lg- *	>1200	Large desks
.col – md- *	>992	Small desks
.col – sm- *	>768	Tablets and small screens.
.col – xs- *	<768	Smart phones

Source. Own Elaboration

Breakage Points

For [14], a website can be divided into four columns when viewed on a desktop screen. In tablet view, the columns are divided into two rows and there are only two columns next to each other. In a cell, columns are always 100% wide, unless you specify a separate column. xs- * category.

```
<div class = "container">
<div class = "row">
<div class = "col-sm-8 col-md-3"> Part 1 </div>
<div class = "col-sm-6 col-md-3"> Part 2 </div>
<div class = "col-sm-6 col-md-3"> Part 3 </div>
<div class = "col-sm-4 col-md-3"> Part 4 </div>
</div>
</div>
```

Graphics Window Configuration

According to [8], is the visual area in which documents are displayed in HTML, is where primarily applies a scalability to the interface, this is done because you must set a value for mobile devices, then detaches some attributes that has the graphical window:

Width: device width or number of pixels.
Height: height of the device or value in pixels.
Initial scale: 0–1 (1 indicates that it is not scaled).
User scale: yes/no (indicates if the user can change the scale).
Minimum scale: 0–1 (minimum scale of the graphic window).
Maximum scale: 0–1 (maximum scale of the graphic window).
To apply a receptive design:
<meta name = "viewport" content = "width = device - width, initial - scale = 1.0"
For mobile devices, it is preferable to disable the zoom:
<meta name = "viewport" content = "width = device - width, initial - scale = 1.0, maximum - scale = 1.0, user - scalable = no">

3 Implementation of Bootstrap in the EAC EducArte Community Web System

See (Fig. 3)

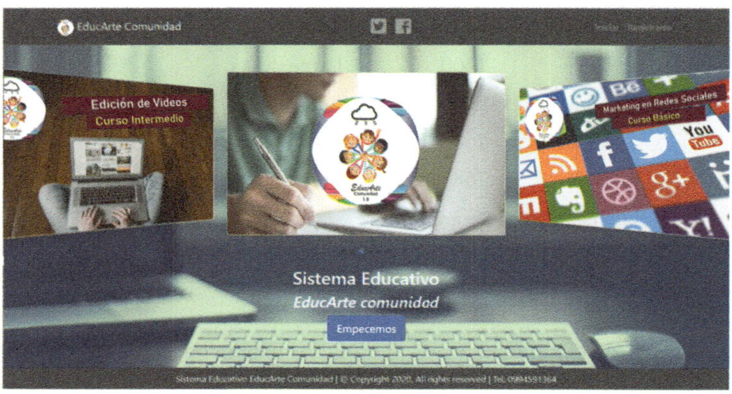

Fig. 3. Web system "EducArte Community".

The EducArte Comunidad system is an educational web platform dedicated to offering diverse knowledge courses to the communities of the city of Portoviejo, Ecuador. This system is accessible without registration costs. It was developed by the student author of this article, belonging to Faculty of Computer Science at Technical University of Manabí. The software has efficient functions in the backend, but in the frontend aspect, only CSS style sheets are utilized. They were not optimal due to the exuberant coding, and therefore the lack of graphic adaptability in mobile devices. In the next section, we will present some code fragments where Bootstrap was implemented to give the educational system an automatic resolution adjustment. In the reengineering of the system, the linked Bootstrap libraries will be used online (Fig. 4).

```
<meta name="viewport" content="width=device-width, user-scalable=no, initial-scale=1.0, maximun-scale=1.0,
<meta http-equiv="X-UA-Compatible" content="ie=edge">
<title>{{datas.datasI.abreviatura}}</title>
<link rel="icon" type="image/png" href="/img/{{datas.datasI.logo}}"/>
<link rel="stylesheet" href="https://use.fontawesome.com/releases/v5.4.2/css/all.css" integrity="sha384-/rX
<link href="https://fonts.googleapis.com/css?family=Oswald" rel="stylesheet">
<link rel="stylesheet" href="https://stackpath.bootstrapcdn.com/bootstrap/4.3.1/css/bootstrap.min.css" inte
<link rel="stylesheet" href="/css/styles.css">
<link rel="stylesheet" href="https://cdnjs.cloudflare.com/ajax/libs/Swiper/4.5.1/css/swiper.min.css">
<link rel="stylesheet" href="https://stackpath.bootstrapcdn.com/bootstrap/4.3.1/css/bootstrap.min.css" inte
```

Fig. 4. Bootstrap libraries taken online.

The following code details the creation of the login with Bootstrap classes, using the div constant for the separation of text entry: email and password (Fig. 5).

```
<div class="container p-2 ">
    <div class="row">
        <div class="col-md-4 mx-auto">
            <div class="card text-center shadow">
                <div class="card-header bg-dark text-white">
                    <h5>Iniciar Sesión</h5>
                </div>
                <div class="card-body">
                    <img src="/img/user.png" alt="Logo" class="card-img-top mx-auto m-2 rounded-circle w-50">
                    <form action="/signin" method="POST">
                      <div class="form-group">
                        <input type="email" name="email" placeholder="Ingrese su correo electronico" class="for
                      </div>
                      <div class="form-group">
                        <input type="password" name="password" placeholder="Ingrese su clave" class="form-contr
                      </div>
                      <button class="btn btn-primary btn-block"> Ingresar </button>
                    </form>
                </div>
            </div>
```

Fig. 5. Form (login) with Bootstrap.

Using the nav_link class, a vertical drop-down bar is being created, which is located on the left side of the interface (Fig. 6).

Sidebar with Bootstrap. With these lines of code you could create a horizontal navigation bar, with multiple options, commonly called nabvar (Fig. 7 and Fig. 8).

```
<nav class="container p-2" >
    <ul class="nav flex-column">
        <li class="nav-item"><a class="nav-link text-dark quita" href="#">{{user.asignatura}}</a></li>
        <div class="dropdown-divider"></div>
        <li class="nav-item"><a class="nav-link text-dark quita" href="/users/edit"><span class="fa fa-user ico
```

Fig. 6. Sidebar with Bootstrap.

```
            <div class="dropdown-divider"></div>
            <a class="dropdown-item" href="/users/edit">Mis datos</a>
            <div class="dropdown-divider"></div>
            <a class="dropdown-item" href="/logout">Cerrar sesión</a>
        </div>
    </li>
else}}
    <li class="nav-item">
        <a class="nav-link" href="/signin">Iniciar</a>
    </li>
    <li class="nav-item">
        <a class="nav-link" href="/signup">Registrarse</a>
    </li>
/if}}
```

Fig. 7. Navbar created with Bootstrap classes.

Fig. 8. The following images were captured from a Samsung J5 cell phone, and it can be seen that the adaptability is 100% fulfilled.

4 Results and Discussions

During the Covid-19 pandemic, it is clear that not every family has a laptop or a desktop computer for online activities, but they generally do have a smart cell phone. According to [4], he states that the process of incorporating ICT into education is a fundamental component in the development of education in the 21st century. In this sense, the indicators of acceptability in cellular networks and the use of mobile devices to access the Internet in Latin America show a growing and highly favorable scenario for educational processes mediated by mobile devices. This issue is due to their adaptation to the system in which we live, where there is a tendency to different anomalies (pandemics, civil wars, among others). Time is money, and an adaptative application would avoid increases in its creation. According to [20], several tools integrate Bootstrap for the development of web systems and save development time. They are Pingendo, Brix, Jetstrap, Pinegrow, Bootstrap Studio, Bootply, Bootmetro, JSfiddle with Bootstrap, among others.

Emphasizing a tool that could allow the accelerated development of interfaces would be Bootstrap Studio, as it is considered a Web 2.0 tool due to its synchronization of graphic libraries and database integration. As mentioned [2], at present, the potential of the cell phone as an educational tool would improve the presentation and availability in time and resource of computer applications, such as the case of (DpWcE) "Web page design in educational contexts". This design was a Spanish educational system of the UNED for the doctoral process. This web portal was developed under the W3C standard by using layout languages and CSS dynamics. It is worth mentioning that the results of this article indicate the importance of migrating the desktop aspect to the mobile one to detect user satisfaction and an increase in the number of visits.

Following the same line of research, for [18], the application of Bootstrap in the Web System of the Faculty of Computing and Informatics (University of Malaysia) improved performance on mobile devices. According to the experimental results, the Page Speed Insights tool indicated that mobile speed scores increased from 27/100 to 91/100 after the Bootstrap framework was implemented. On the other hand, speed scores for the desktop also increased from 28/100 to 97/100. Besides, the Web Page Test tool revealed that the loading time decreased from 10,175 s to 3,099 s after optimizing the responsiveness of the website. The following paragraphs will detail the values corresponding to navigation speed, loading time, and the number of weekly visits, both on mobile and desktop of the EAC (EducArte Comunidad) platform. The data could be estimated thanks to the software Page Speed Insights of Google, which allows analyzing the performance of web portals. Likewise, the Google Analytics API was used to measure visits to the portal (Table 3).

Table 3. Assessment of browsing speed, loading time, number of visits and adaptability of the EAC Web System.

Measured aspects	Tools	Portal with bs	Portal without bs
Navigation speed	Page speed insights	1, 7 s	2, 0 s
Charging time	Page speed insights	1, 7 s	2, 0 s
Number of visits to the portal	Google analytics	75 users	45 users
Adaptability according to Woorank analysis	Woorank analytics	100% adaptive	No BS Compressed css files

Source. Own Elaboration

After analyzing the web system EAC in its two versions, it was possible to obtain a positive result in terms of browsing performance: The implementation of BS manages to minimize with 3 ms, the loading time, and browsing speed concerning EAC without BS. On the other hand, it was possible to increase the audience of the portal through the Google Analytics tool. The access to the portal has increased from 45 to 75 users. In summary, EAC users feel satisfied because they can navigate from their mobile devices (Table 4).

Table 4. User experience through surveys in Google forms

User experience		
Questions	Answers	
How do you consider the navigation quality of EAC with Bootstrap on your mobile device?	High (40)	Medium (10)
How do you consider the navigation quality of EAC without Bootstrap on your mobile device?	Under (45)	Medium (5)
In which presentation do you like the EAC system best?	With Bootstrap (50)	Without Bootstrap (0)
Regarding the loading speed of the system, which one do you think is slower?	EAC without BS (35)	EAC with BS (15)

Source. Own Elaboration

The data obtained through the experience questionnaire shows that for users, Bootstrap improves the presentation and adaptability of interfaces, including in aspects of load and runtime. According to [19], they state that the use of responsive tools potentially improves the presentation of web portals, bypassing mobile development, and allowing obsolete systems (desktop) to apply re-engineering. Also, the web education service "Labyrinth" that, after being a desktop system, transformed it into an adaptive web system supported by a cloud infrastructure to update it. We concluded that the re-engineering done was the optimal solution to improve the availability and performance of the system.

Also, the EAC portal improved its availability, speed of navigation, and loading time. Also, its amount of code decreased because Bootstrap contains embedded code that allowed through a sentence to design essential components, such as login, navbars, jumbotrons, footers, tables, warning messages, among others.

5 Conclusions

The use of Bootstrap allows the adaptation of web environments to different devices, without the need for operating systems, so it is an advantage in time and money, which is what currently would be convenient for companies. There is no doubt that this "responsive" movement has come to have an impact on user access to different websites. The company Google has incorporated into its servers the use of these graphic libraries that allow users to upload sites to the Internet and make them adaptable. Finally, the objective of this article is to present the web system "EducArte Comunidad" to the population of "El Florón" in the future. Children from this community will have the facility to access the services offered by the system through a mobile device since it is evident that almost every family has a smart device and not a computer due to financial difficulties. The supported data collected by a survey conducted last year by the author of this article, where it is evident that the majority of the population has smart mobile devices and not computers.

Finally, positive results were achieved with the reengineering of the EAC Web System, improving graphic and navigation aspects, such as 100% adaptability to various mobile devices, loading time, and system speed. In the analysis, it was also possible to diagnose that Bootstrap increased the weight of the system because it creates strings of code through its graphic libraries. This disadvantage was diagnosed but ignoring that small excess, it improved the appearance and speed of the system and, therefore, satisfied users. The advantages provided by Bootstrap had an impact on the increase in the audience of the EAC web system. Currently, the system is available at the address: eacomunidad.xyz, where the system will appear in its two versions: with BS and without BS. As future improvements, they will try to create modules such as online chat, library, academic activities, games, among other interactive modules. It is worth noting that it seeks that this system in the future is free software, becoming a significant contribution to distance education or online, for schools, colleges, and inclusive universities. There is no doubt that this "responsive" movement has come to have an impact on user access to different websites. The company Google has incorporated into its servers the use of these graphic libraries that allow users to upload sites to the Internet and make them adaptable. Finally, the objective of this article is to present the web system "EducArte Community" to the population of "The Florón" in the future. Children from this community will have the facility to access the services offered by the system through a mobile device since it is evident that almost every family has a smart device and not a computer due to financial difficulties. The supported data collected by a survey conducted last year by the author of this article, where it is evident that the majority of the population has smart mobile devices and not computers.

References

1. Afia, Z., Syed, J., Khurram, S., Usman, A.: The users experience quality of responsive web design on multiple devices. In: ICFNDS 18: Proceedings of the 2nd International Conference on Future Networks and Distributed Systems (2018). https://dl.acm.org/doi/abs/10.1145/3231053.3234632
2. Brazuelo, F., Cacheiro, M.: Design of Educational web pages for mobile phones. Electronic magazine of educational technology (2010). https://www.edutec.es/revista/index.php/edutec-e/article/view/437/172
3. Carranco, J., Maya, E.: Design and development of the adaptive web portal of SANEC (Ecuadorian sanctuaries) in the diocese of Ibarra. Magazine UTN (2016). http://repositorio.utn.edu.ec/bitstream/123456789/5611/3/ARTICULO.pdf
4. Chiappe, A.: Trends in digital educational content in Latin. Cuaderno SITEAL UNESCO (2016). https://unesdoc.unesco.org/ark:/48223/pf0000245673

5. Danlin, O.: Responsive web design and php e-commerce. Conferences THESEUS (2018). https://www.theseus.fi/handle/10024/153778
6. Git Hub. Git HUB Wbs/bootstrap: The most popular HTML, CSS, and JavaScript framework for developing responsive, mobile first projects on the web (2016). https://github.com/twbs/bootstrap
7. Glassman, N., Shen, P.: One site fits all: responsive web design. J. Electron. Res. Med. Lib (2015). https://www.tandfonline.com/doi/abs/10.1080/15424065.2014.908347
8. Jiang, W., Zhang, M., Jiang, Y., Zhang, Y.: Responsive web design mode and application. In: Workshop on Advanced Research and Technology in Industry Applications (WARTIA) (2015). https://ieeexplore.ieee.org/document/6976522/
9. Lee, J., Jung, M., Kim, H.: Responsive web design according to the resolution. IEEE Explore (2016). https://ieeexplore.ieee.org/document/7434345
10. Manso, Y., Cañizares, R., Febles, J.: Responsive web design for educative platform ZERA. Cuban J. Comput. Sci (2016). https://rcci.uci.cu/?journal=rcci&page=article&op=view&path%5B%5D=1048&path%5B%5D=422
11. Marcotte, E.: Responsive Web Design. A List Apart (2011). https://alistapart.com/article/responsive-web-design/
12. Nilesh, J.: Review of different responsive css front-end frameworks. J. Glob. Res. Comput. Sci (2014). http://www.rroij.com/open-access/review-of-different-responsive-css-frontend-frameworks.pdf
13. Pekka, J., Solanen, T.: On the design of a responsive user interface for a multi-device web service. In: 2nd ACM International Conference on Mobile Software Engineering and Systems (2015). https://ieeexplore.ieee.org/abstract/document/7283029
14. Peltomäki, V.: Bootstrap Framework as a web design tool (2014). https://www.theseus.fi/bitstream/handle/10024/74557/Peltomaki_Veera.pdf?sequence=1&isAllowed=y
15. Pisanty, A., Enríquez, L., Chaos, L., García, M.: "M-learning en ciencia" - Introducción de aprendizaje móvil en física. RIED. Revista Iberoamericana de Educación a Distancia, **13**(1), 129–155 (2010). http://dx.doi.org/10.5944/ried.1.13.892
16. Powers, D.: Introduction to media queries. Adobe Developer Connection (2011). https://www.adobe.com/devnet/archive/dreamweaver/articles/introducing-media-queries.html
17. Rodríguez, R., Vera, P., Marko, I., Zain, G.:Adaptive web frameworks analysis based on html5. XX Workshop of Researchers in Computer Science (2018). https://pdfs.semanticscholar.org/69a6/6e9bdb643c94c561f1c1895cf7c4ce199a4b.pdf
18. Shan, T., Obit, J., Rayner, A., Tahir, A.: Enhancing the Performance of University's Website for Mobile Devices Based on Responsive Web Design Approach. American Scientific Publishers (2017). https://www.researchgate.net/publication/322098135_Enhancing_the_Performance_of_University's_Website_for_Mobile_Devices_Based_on_Responsive_Web_Design_Approach
19. Stočes, M., Masner, J., Jarolímek, J., Šimek, P., Vaněk, J., Ulman, M.: Cross-Platform user interface of E-learning applications. 11th International Conference Mobile Learning (2015). https://files.eric.ed.gov/fulltext/ED562430.pdf
20. Suraj, G., Adkar, P.: A review paper on bootstrap framework. IRE J. (2019). https://irejournals.com/formatedpaper/1701173.pdf
21. Thomas, A., Walsh, P., Kapfhammer, M.: Automatic detection of potential layout faults following changes to responsive web pages. IEEE Explore (2015). https://ieeexplore.ieee.org/abstract/document/7372059

Remote Variable Monitoring App for Mechanical Ventilators Used in COVID-19

Carlos Miño[1], Omar Flor[2](✉), Josué Quiroga[3], and Andrés Cuaycal[4]

[1] Universidad Tecnológica Israel, Facultad de Ingeniería en Electrónica y Telecomunicaciones, Quito, Ecuador
calolomino@gmail.com
[2] Universidad de las Américas, Facultad de Ingeniería y Ciencias Aplicadas, Ingeniería Industrial, Quito, Ecuador
omar.flor@udla.edu.ec
[3] Universidad Central del Ecuador, Facultad de Ciencias Médicas, Quito, Ecuador
josueesteban_12@hotmail.com
[4] Escuela Politécnica Nacional, Facultad de Ingeniería Eléctrica y Electrónica, Quito, Ecuador
andres.cuaycal@epn.edu.ec

Abstract. In this article, the development and testing of an App that can be used to control and monitor parameters in mechanical ventilators in intensive care units is presented. Mechanical ventilation variables of a pressure-controlled ventilator are monitored and configured by the App with the use of a cellular device, wireless communication, and remote communication. The presented alternative improves the time response of intensivist specialists who are very demanded due to the COVID-19 pandemic. The use and performance of the developed App is demonstrated in a high-fidelity emerging mechanical ventilator.

Keywords: Remote monitoring · Emerging mechanical ventilator · COVID-19

1 Mechanical Ventilator

A mechanical ventilator is a medical equipment that assists a patient in case of Acute Respiratory Deficiency (ARD). It permits the proper support in inhalation and expiration in the most severe cases that occur by COVID-19 pandemic. The collapse of the respiratory system is a major complication that can be aided by a mechanical ventilator which helps in transporting air and oxygen to the patient's lungs [1, 2].

Invasive-type mechanical medical ventilation equipment require that an endotracheal tube is inserted through a sedated patient oral cavity [3]. The inspiration is completely controlled by the ventilation mode, pressure, or volume. In order to maintain stable vital signs in a patient, the intensive care specialist determines the inspiration and expiration time. Therefore, the most adequate respiratory frequency (RR) is established considering the patient condition.

Mechanical ventilators allow the user to configure various parameters such as pressure and volume. Moreover, the user interface presents reference values which help

M. Botto-Tobar et al. (Eds.): CIT 2020, AISC 1326, pp. 303–315, 2021.
https://doi.org/10.1007/978-3-030-68080-0_23

monitoring the patient. Also, there are configurable alarms settings that are triggered when variations in reference values are detected [4, 5].

There are multiple mechanical ventilation modes. The most common modes are Controlled Ventilation (VC), Assisted Ventilation (AC), Pressure Support Ventilation (PSV) [6].

Controlled Ventilation (VC), is a method in which the ventilator controls the number of respiratory cycles by means of pressure or volume as a control parameter [7]. Given that the respiratory system is not able to perform the inspiration and expiration process. The ventilator regulates these processes and the intensivist specialist is the one who decides the respiratory rate, tidal volume or pressure according to the ventilation method [8, 9].

Figure 1 shows the internal design of the emergent mechanical ventilator SURKAN which was designed and fabricated in Ecuador. This ventilator is connected to an oxygen and medicinal air source to provide gas to the patient lungs using an Assisted/Controlled Pressure ventilation mode [10].

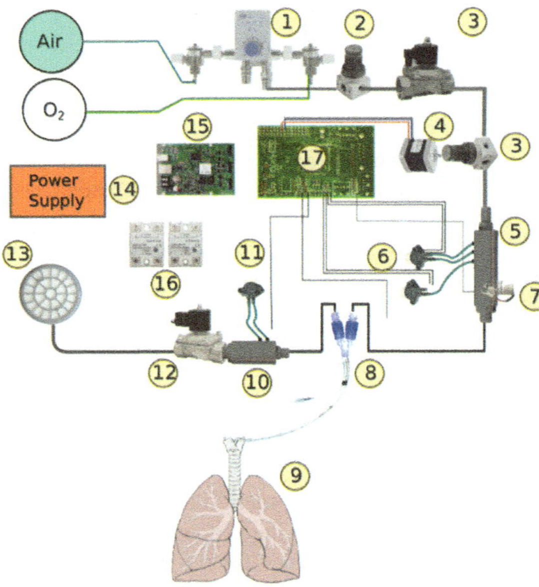

Fig. 1. SURKAN mechanical ventilator schematic [10]

Components enumerated in Fig. 1 are: (1) Air and Oxygen Mixer, (2) Precision pressure regulator, (3) Pressure regulated valve, (4) Stepper motor NEMA 17, (5) inspiration manifold, (6) pressure and flow sensors, (7) oxygen sensor, (8) patient circuit, (9) patient respiratory system, (10) expiration manifold, (11) flow sensor, (12) solenoid valve, (13) HEPA Filter, (14) 12VDC Power supply, (15) main controller unit, (16) electric relays, and (17) controller board. Additionally, the implemented system requires a user interface that shows relevant information and should be ease of use.

2 Ventilation Parameters in Pressure Control Ventilation

There are two fundamental parameters that must be provided in mechanical ventilation: maximum pressure and plateau pressure. The first is reached when the ventilator provides air to the patient's lungs, while the second is reached at the end of inspiration. In the operation, parameters that support the control and monitoring must also be included, such as Tidal Volume (TV), Respiratory Frecuency (RR), Minute Volume (MV), Inspiratory Time (Ti), Inspiration/Expiration Ratio (I/E) [11].

The Tidal Volume (TV), is the amount of gas that the ventilator sends to the patient in each inspiration. Typically, a TV of 7–10 ml/kg is programmed (Except in neonates and patients with hypoxemic lung disease, in whom lower volumes are programmed). It must be considered that some respirators automatically compensate the volume of the tubing (Compression volume) and others do not, therefore, with the same programmed volume, the effective volume may be different [12].

The Respiratory Frequency (RR), or number of breaths per minute (bpm), administered by the ventilator, depends on the age and pathology of the patient. RR of 40–60 bpm are used in neonates, 30–40 bpm in infants, 20–30 bpm in children, 12–15 bpm in adolescents, and 8–14 bpm in adults.

The Minute Volume (MV) is the volume of gas that the respirator sends to the patient in each minute of ventilation. It is the product of the TD by the RR. The MV is the parameter that is most directly related to ventilation and, therefore, to the arterial pressure of carbon dioxide (PaCO2). In some ventilators the MV is programmed instead of the TV [12]. Therefore, to improve ventilation, it can be modified, depending on the patient's condition, the parameters TV, RR or both.

The period of gas entry into the airway (tubulators, endotracheal tube, trachea, and bronchi) and lungs is known as Inspiratory Time (Ti). This parameter is programmed in both volume and pressure modes. In volume ventilation, inspiration is divided into 2 phases: In the first phase, gas enters (Ti) and in the second phase, the Inspiratory Pause Time (Tp), in which the air is distributed by the lung. In this phase the flow becomes 0. The inspiratory pause confirms that the ventilation is more homogeneous since it allows a redistribution of the gas through all the alveoli, even though they have different time constants (resistance and elasticity). In pressure ventilation, the Tp is not programmed [13].

The Inspiration/Expiration (I/E) ratio is the ratio of time spent on inspiration and expiration in each respiratory cycle. Usually an I/E ratio of 1/2 to 1/3 is used. The programming of the I/E ratio depends on the ventilator model. In some ventilators, the RR and the I/E ratio are programmed; in others, Ti and expiratory time in seconds; in others, the RR, Ti and Tp in percentage; and in others, RR and Ti in seconds [13].

3 Monitoring and Control Apps

Table 1 shows apps and software developed to monitor and control respiratory variables in mechanical medical equipment utilized in ICUs. The app name, developers and compatibility with cell phones, tablets, or computers [14–21].

The latter apps are developed to be used in commercial specialized equipment and allow the control of respiratory variables in invasive and non-invasive medical equipment.

However, these apps can not be implemented in a emerging mechanical ventilator due to the exclusivity and privacy of commercial apps. Many of these specialized apps were

Table 1. Apps utilized in invasive and non-invasive mechanical ventilators

Aplicative	Autor	Characteristics	Cell phone	Tablet	PC
Guide to NIV	Phillips	Free clinical application for professional training in Non- Invasive Mechanical ventilation	X	X	
My VENT	Felip Miralles (Oxigen, Eurecat e IRB Lleida)	Controls pressure and volume in invasive mechanical ventilation and non-invasive	X	X	
Aventho	Google play	Control and calculation of PEEP, pulmonary compliance, transairway pressure, CO2 correction in non-invasive ventilation	X	X	
Ventilator mode map	Google play	Gas exchange, pressure and lung volume monitoring in mechanical ventilators	X	X	
Hamilton-C6	Hamilton medical	Monitoring of patient oxygenation and ventilation parameters and simulated blood gas	X	X	X
VentilO	L'Institut universitaire de cardiologie et de pneumologie de Québec (IUCPQ)	Control of pressure, tidal volume and breathing speed	X	X	
Mechanical Ventilation Advanced	Google play	High frequency control, basic and advanced ventilation modes, pressure and volume control	X	X	

(*continued*)

Table 1. *(continued)*

Aplicative	Autor	Characteristics	Cell phone	Tablet	PC
Ventilator simulator	Google play	Pressure, flow and volume control for positive pressure mechanical ventilation	X	X	

developed by Google and others. Many of them with the support of brands and world renown institutions.

4 User Interface Design

In order to design the user interface various aspects were considered such as: ease of use and parameter visualization in a single screen. Respiratory variables were taken based on intensivist experts and commercial ventilators. A dark color palette is considered to prevent visual fatigue for the user and discomfort for the patient.

The Human Machine Interface (HMI) utilized in the ventilator was developed using web technology by means of HTML5 and CSS3 and Javascript to give dynamism.

The implemented dynamic interface allows the real time screen refreshment without updating the web page. Also, AJAX in Javascript allows the communication between the backend and control to send and get data.

Web technology is very useful because the user interface can be used in multiple devices. The ventilator will work as a web server, that is, a hotspot can be created or a router can create a local network so multiples devices such as smartphones, tablets, or laptops can be connected and get remote control of the mechanical ventilator

Table 2. User interface design used in the emerging mechanical ventilator SURKAN

Patient identification	Alarms	Ventilatory mode indetification
Parameter control: - PIP - PEEP - FR - Ti - Inspiration pause - Trigger	Graphs: Pressure vs time Flow vs time Sensed parameter viewer	Sensed parameter viewer: - PIP - Tidal volume - Minute volume - FR - FiO_2 - Trigger
Configurate equipment buttons		

Table 2 shows the user screen divided in sections such as: control parameters input, graphs, monitoring of main and secondary parameters, a section with multiple buttons to configure the equipment. Through configuring options, the user can use the equipment in automatic mode and could also set variables.

Raspbian operating systems offers the possibility to implement a web server very easily by using the command *apt-get install apache2*. By executing the last command all necessary libraries are installed and a web service is created usually in the 80 port.

Once the installation is finished the user has access to the web server via its IP and can be remotely accessed within the same local network where the equipment is connected.

5 Control Algorithms

The implemented code was programmed in Python using multiprocessing.

In Fig. 2, there are 4 processes. The main process that ran all the processes that control the equipment. As sub-processes we have one that is in charge of acquiring data from the pressure, flow sensors incorporated in component (6), (11) and FiO2 (7) to be used in all processes, the second is in charge net of performing pressure control and the last one calculates references to the air volume in each breath.

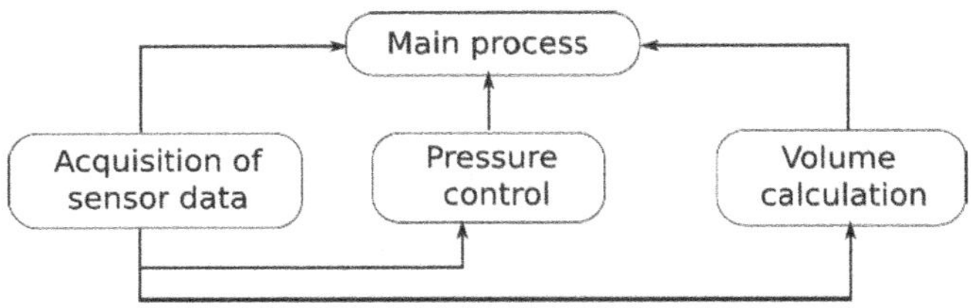

Fig. 2. Block diagram of principal processes

PIP is determined using Eq. 1, which provides a relation between the sensor in voltage (6) and the actual value of the measurement. Equation 1 has been obtained from a tabulation of sampled and compared data with a reference value.

$$PIP = (21.743 * \text{ voltage}) - 3.8927 \tag{1}$$

The algorithms presented in this section were of our own development and to provide adequate control of the mechanical ventilator variables.

As a first calculation the equipment gets from the control panel the frequency and inspiration time then calculate the numbers of cycles by minute must to execute including the expiration time as shown in Fig. 5.

Figure 3 presents the pseudo-code of the program that controls the PIP pressure. By using components (4) and (3) the equipment must provide in each ventilation a value previously set by the user to the patient (9). It consists of a conditional in which it is verified that there is a ventilation or cycle, then in a loop it is verified that everything runs while there is an inhalation and also the pressure (6) that is delivered to the patient is measured. Moreover, through a PID control the values for the proportional valve (3) are obtained until meeting the objective that is the set PIP. In addition, minimum pressure

```
PIP PRESSURE CONTROL ALGORITHM
read cycle state
read maximum_pressure
read minimum_pressure
if (cycle=1)
    while
        read pressure sensor voltage
        convert to pressure units (cmh2o)
        if (read_pressure <= minimum_pressure) or
           (read_pressure >= maximum_pressure)
              pressure_alarm=1
        else
              pressure_alarm=0
        end if
        if (read_pressure >= 0) and (read_pressure <= 5)
              patient_circuit_alarm=1
        else
              patient_circuit_alarm=0
        end if
        error calculation between sensed and configured pressure
        control pid
        proportional valve on/off
      end while
end if
```

Fig. 3. PIP pressure control algorithm

```
PEEP PRESSURE CONTROL ALGORITHM
read cycle state
if (cycle=0)
   while
      read pressure sensor voltage
      convert to pressure units (cmH2O)
      error calculation between sensed and configured pressure

      if (read_pressure <= configured_pressure)
         expiratory valve off
      end if
   end while
end if
```

Fig. 4. PEEP pressure control algorithm

values are verified to determine a minimum pressure alarm. Furthermore, it is established whether the patient circuit is connected or not, all with component (6).

Figure 4 presents the pseudo code of the program that controls the PEEP pressure that must be provided to the patient (9) and that is configured by the user. The algorithm consists of a conditional that verifies that the equipment is in an operating state. Through a loop it is checked that it is in an expiration period and by means of the measured value of pressure (6) it is compared with that configured in the interface. When the condition that the measured value is less than or equal to the set value is fulfilled, then the expiration valve (12) will close. Therefore, the desired residual pressure will be left in the patient circuit (8).

```
RESPIRATORY FREQUENCY CONTROL ALGORITHM
read respiratory_frequency
read inspiratory_time
read inspiratory_pause
cycle_time = respiratory_frequency / 60
expiratory_time = cycle_time – inspiratory_time
if (inspiratory_pause>0)
|  expiratory_time = inspiratory_time – inspiratory_pause
end if
```

Fig. 5. Respiratory frequency control algorithm

Figure 6 presents the pseudo-code of the control for the start of a ventilation cycle by means of the trigger or trigger sensitivity that is configured by the user. By checking a cycle, the pressure obtained or measured (6) is compared with the Trigger value. If this value is less than or equal to the Trigger, it stops the expiratory cycle (12) and starts a new inspiration (3). In addition, breathing apnea and patient alarms are activated and will be visible and audible in the equipment. The time is taken to compare it with the apnea time set in the equipment and when this is fulfilled the ventilation cycle is restarted by the configured ventilation parameters, that is, frequency, and normal inspiration and expiration times. This is done until a new respiratory effort is obtained from the patient (9).

```
TRIGGER CONTROL ALGORITHM
read cycle state
read trigger
read apnea_time
if (cycle=0)
|  read pressure_sensor voltage
|  convert to pressure units (cmH20)
|  compare sensed and trigger pressure values
|  if (sensed_pressure<=trigger_pressure)
|  |  stop expiration cycle
|  |  start inspiration cycle
|  |  apnea=1
|  |  patient_breathing_alarm = 1
|  |  counting_time = 0
|  else
|  |  counting_time
|  |  if (counting_time=apnea)
|  |  |  patient_breathing_alarm = 0
|  |  |  apnea = 0
|  |  end if
|  end if
end if
```

Fig. 6. Trigger control algorithm

6 Commutation System

A centralized communication system was developed with the Ethernet protocol or the IEEE 802.11 standard, the assignment of the equipment IP addresses is performed dynamically using DHCP (Dynamic Host Configuration Protocol). The main system or BackOffice is assigned by a fixed IP on the same network.

In each equipment, the mentioned IP is configured to send and receive data, with a transmission speed of 150 Mbps which is convenient to avoid delays in the visualization and control of each equipment.

Figure 7 presents a diagram of the communication network used for monitoring and control of various mechanical ventilation equipment using a PC and cell phone app. For the user to unequivocally control the desired computer, assigned IP addresses are used on each computer. All patient information is reflected on the interface.

Fig. 7. Equipment distribution and connectivity in a local network with centralized system

7 User Interface

The user interface was developed and shown in a 14-in. touch screen with touch capacitive technology of the "ASUS" brand. The developed ventilator graphical interface consists of 4 blocks (Fig. 9): parameter settings and buttons (a), real-time monitoring (b), function configuration buttons (c) and graphs display (d). The interface allows setting and monitoring of parameters, while the electronic system performs the control and communication process with the screen.

Ventilator operation variables must be entered by the specialist who has assessed the patient's condition and determined its optimal parameters.

Parameters settings are detailed as follows: PIP parameter is entered in the interface through the "PIP" option that allows the entry of values in a range 0 to 40 cmH_20. The pressure variable at the end of expiration is entered into the interface using the "PEEP" option and the parameter can be varied from 0 to 15 cm H_20. The respiratory frequency parameter is entered into the interface using the "Frequency" option that allows values to be entered in a range from 0 to 30 bpm. The inspiratory time variable is entered into the interface using the "Ti" option and the parameter can be varied from 0 to 5 s in decimal steps. The inspiratory pause parameter is entered using the "Ti" option and the user can change the parameter in a range of 0 to 5 s. The trigger parameter is entered through the option in the interface "Trigger" and the user can vary the value in cm H_20.

The following buttons are used to control the mechanical ventilator settings. Pressing the "Start" button will cause the mechanical ventilator to begin cycling by sending the pre-programmed airflow. When the user presses the "Manual" button the ventilator will send an inspiration cycle to the patient. By pressing "events" button, the user will be able to access a record of all the ventilator actions. By pressing the "Ventilation Modes" button the user can select the "Pressure Controlled" and "Patient Assisted" modes.

Patient monitoring can be viewed on the developed interface using the following indicators: "Exp Tidal V" the expiratory tidal volume in ml, "I:E" indicates the inspiration/expiration ratio, "Exp V min" monitors the expiratory volume per minute in LPM, "PEEP" indicates the real-time value of expiratory pressure on exhalation in cm H_20, "Ti" shows the inspiratory time in seconds, "Compliance" is displayed on the interface in ml/cm H20, "Te" displays the expiratory time in seconds, "Insp R" the inspiratory rate is displayed in cmH_20/LPM and finally "Vt/PCI" in units of ml/Kg.

In the last section of the screen the Pressure vs. Time graph is shown. Figure 8 shows the resulting interface design and the remote application for controlling the SURKAN mechanical ventilator.

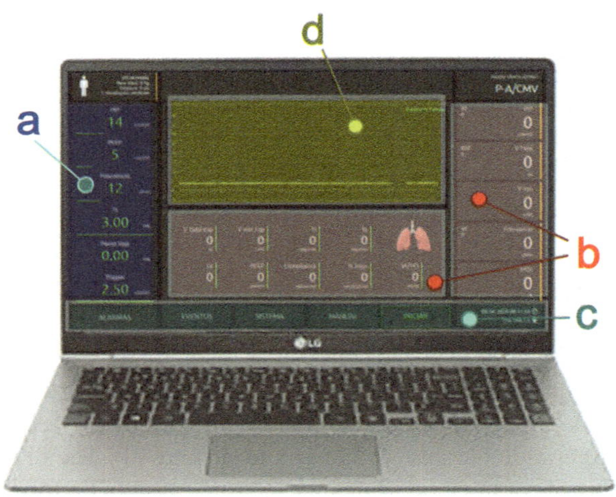

Fig. 8. Graphic interface design to remotely control the SURKAN mechanical ventilator, the 4 sections that correspond to the distribution proposed in Fig. 2 are displayed.

8 Results

The app was developed and implemented in the SURKAN mechanical ventilator which was designed and fabricated in Ecuador. The ventilator offers most respiration parameters found in a commercial mechanical ventilator.

Multiple tests were performed including, parameter changing with response analyzing, delays, communication, ease of use. The last test was verified by an intensive care specialist.

The following steps need to be performed to properly control the app and its remote controlling feature: (1) Enter the local network (2) Pairing with the communication network and identification of the desired equipment, (3) Set the following variables: PIP, PEEP, Frequency, Ti, Insp Pause, Trigger, depending on the professional's criteria; (4) Alarms setting. (5) Reading and monitoring of ventilatory parameters.

Remote monitoring is possible since the mechanical ventilator and the app are connected to the same network. Due to the high speed of data transmission and low latency in the network, the programmed and monitored changes in the mechanical ventilator will be shown in real time with an imperceptible delay.

The implemented app and interface provide advantages for the equipment user by avoiding physical presence and using a smartphone, tablet or computer. In addition, since the alarms have a limited physical range, this alternative allows to extend these signals and facilitate the timely action by the intensivist. In the case of a lack of personnel in health institutions, this technology multiplies the range of action of professionals and offers the possibility of digitizing information with the work and implementation of databases.

Users of the app can monitor the patient who is connected to the emergent mechanical ventilator, in this way the patient will have a better treatment and possible complications can be avoided due to the lack of intensive care physicians present. In addition, remote monitoring will save several resources and time. A timely action can save the life of the patient. Figure 9 shows the application working on the SURKAN emergent mechanical ventilator.

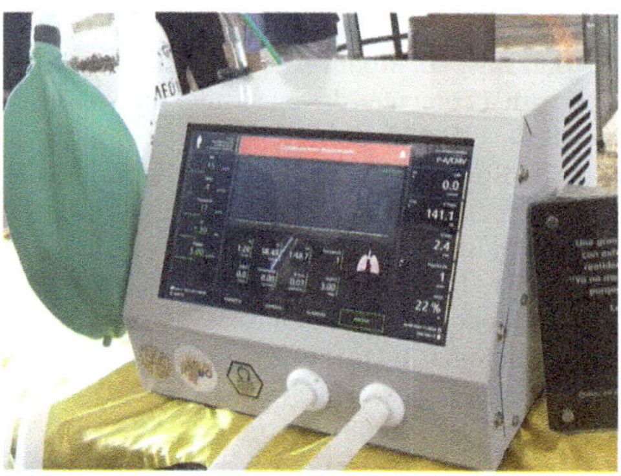

Fig. 9. Ecuadorian emergent mechanical ventilator running the designed control and monitoring app.

9 Conclusions

The app was developed for the control of the SURKAN emergent mechanical ventilator, of Ecuadorian origin and can be replicated in equipment that has a controller board with network access and where a web server can be configured. The characteristics and functionality of this application allow the use and configuration of equipment to be implemented in intensive care units as an alternative to support in the COVID19 pandemic.

According to specialized personnel in the use of these devices, the layout and variables presented on the 15-in. screen are friendly, easy to interpret and visualize, allowing intuitive use.

The possibility of remote management that this application allows, reduces the response time of specialized personnel given the high demand for ventilators and specifically in the cases of patients who totally depend on the use of this equipment and require greater attention and care by intensive care doctors.

References

1. Gutierrez, F.: Ventilación mecánica. Acta Médica Peruana **28**(2), 87–104 (2011)
2. Gutiérrez, F.: Diagnóstico, Monitoreo y Soporte Inicial del Paciente con Insuficiencia Respiratoria Aguda. In: Simposio: "Atención Inicial Del Paciente Crítico Para No Especialistas" (Parte 1). Revista Acta Médica Peruana (2011)
3. Rodríguez, Y., Rodriguez, D., Monsalve, A., López, K., Arévalo, A.: Recomendacones clínicas para la atención del paciente agudo y crítico con COVID-19 (2020)
4. Grace, K.: The Ventilador: selection of mechanical ventilators. Crit. Care Clin. **14**(4), 563–580 (1988)
5. Farias, J.A., Alía, I., Retta, A., Olazarri, A., Fernández, A., Esteban, A.: An evaluation of extubation failure predictors in mechanically ventilated infants and children. Intensive Care Med. **28**(6), 752–757 (2002)
6. Chavez, A., De la Cruz, R., Zaritsky, A.: Spontaneous breathing trial predicts successful extubation in infants and children. Pediatr Crit Care Med. 7(4), 324–328 (2006)
7. Reina, C.M.: Ventilación mecánica controlada y asistida y controlada. Sci. Direct **59**, 82–85 (2003)
8. Jouvet, P., Farges, C., Hatzakis, G., Monir, A., Lesage, F., Dupic, L: Weaning children from mechanical ventilation with a computer-driven system (closed-loop protocol): a pilot study. Pediatr Crit Care Med. **8**(5), 425–432 (2007)
9. Scopesi, F., Calevo, M.G., Rolfe, P., Arioni, C., Traggiai, C., Risso, F.M.: Volume targeted ventilation (volume guarantee) in the weaning phase of premature newborn infants. Pediatr Pulmonol. **42**(10), 864–870 (2007)
10. Naranjo, C., Flor, O., Tapia, J., Flores, E., Coba, A., Chango, E.: Diseño de Ventilador Mecánico Emergente en Modo Asistido/Controlado y Espointáneo por Presión, Revista Universidad Ciencia y Tecnología, **1**(1), 130–137 (2020)
11. Lopez, J., Carrillo, A.: Ventilación mecánica: indicaciones, modalidades y programación y controles., Elsevier, **6**(6), 321–329 (2008)
12. Shults, J., Martsma, J., Slutsky, S., Gajic, O.: What tidal volumes should be used in pattents without acute lung Injury? ASA Publ. Anesthesiol. **106**(6), 1226–1231 (2007)
13. Carrillo, A., Lopez, J.: Parameters of mechanical ventilation, National Library of medicine (2003)

14. World Diagnostic News. https://www.diagnosticsnews.com/empresas/28176-guide-to-niv-philips-estrena-app-para-formar-profesionales-en-ventilacion-mecanica. Accessed 28 Nov 2017
15. Centre Tecnologic de Catalunya. https://eurecat.org/es/aplicacion-telemedicina-controlar-terapias-ventilacion-mecanica-pacientes/. Accessed 7 Jun 2019
16. Google Play. https://play.google.com/store/apps/details?id=com.icc.ortientacionmedica&hl=es_419. Accessed 18 Jun 2020
17. Google Play. https://play.google.com/store/apps/details?id=mandupress.ventilatormodemap&hl=es. Accessed 2 Aug 2016
18. Google Play. https://play.google.com/store/apps/details?id=com.HamiltonMedical.c6simulation&hl=es_EC. Accessed 17 Dec 2019
19. Google Play. https://play.google.com/store/apps/details?id=ca.qc.iucpq.ventilo&hl=es. Accessed 14 Apr 2020
20. Google Play. https://play.google.com/store/apps/details?id=soooooonandroid.mechanicalventilationadvanced&hl=es_EC. Accessed 4 Oct 2018
21. Google Play. https://play.google.com/store/apps/details?id=com.ssafadi.vent_sim&hl=es_EC. Accessed 23 Sept 2019

Wearable Devices Information Security: An Overview

Angelo Brichetto[1], Santiago Criollo-C[1(✉)], Ángel Jaramillo-Alcázar[1], and Sergio Luján-Mora[2]

[1] Facultad de Ingeniería Y Ciencias Aplicadas, Universidad de Las Américas, Quito 170125, Ecuador
luis.criollo@springer.com
[2] Department of Software and Computing Systems, University of Alicante, 03690 Alicante, Spain

Abstract. Wearable technology has become a subject of research and analysis. However, the importance of the information carried across these devices incentivizes malicious individuals to look for ways to attack them and intercept such information for various illicit purposes, it makes information security a major concern for users and manufacturers. The purpose of this research is to analyze wearable devices security through different parameters centered on two different models found after a thorough research based on scientific articles, where the authors demonstrate their importance and effectiveness in the analysis of computer security. The first model selected was the DREAD model, which analyzes the Damage, Reproducibility, Exploitability, Affectability, and Discoverability of each attack examined. The second model chosen is the STRIDE model, this one allows to classify attacks within the main categories of threats, which are: denial of services, elevation of privileges, repudiation, among others. The analysis performed shows the injection of code attack as the one that represents a higher level of risk for wearable devices. On the other hand, the threat of information disclosure ends up being to which these devices are most susceptible because it houses the most attacks analyzed.

Keywords: Wearable technology · Information security

1 Introduction

Wearable technology provides the user with friendly and personalized services with multiple information sensors. However, they handle personal and sensitive information. If this data is used by malicious users, it could cause a lot of social problems. Within the security aspect, previous analysis has been conducted on real examples of wearable devices such as Google Glasses, Fitbit devices and Samsung smartwatches. The Google Glasses were the first wearable devices to start the growth of wearable technology. However, there is concern about the vulnerability to privacy of the person who uses them [1]. About Fitbit devices, whose best-known device is the Smart fitness band used on the wrist, it provides measurements of human activity such as number of steps

M. Botto-Tobar et al. (Eds.): CIT 2020, AISC 1326, pp. 316–327, 2021.
https://doi.org/10.1007/978-3-030-68080-0_24

walked, sleep quality, pulse, body temperature, among others. However, one of the most important issues encountered on this device has been the lack of authentication in the data monitoring aspect, which allows any attacker to get the data easily without the user noticing [1]. Samsung Smartwatch devices also contain vulnerabilities because a study conducted by HP claims that 100% of the devices evaluated have different attack doors such as weak authentication, do not contain encryption, among others [2].

2 Analyzed Parameters

2.1 DREAD Model

Its name represents Damage, Reproducibility, Exploitability, Affectability and Discovery as described in Table 1. This model allows you to calculate the risk for each of the threats. Threats have a level of severity assigned as follows [3]: Low: 5 to 7; Medium: 8 to 11; High: 12 to 15.

Table 1. DREAD model description.

Threat	High (3)	Medium (2)	Low (1)
Damage	The attacker can ignore the protection mechanism of the security system and obtain authorization from the administrator to execute programs arbitrarily	The attacker can obtain sensitive information from the user	The attacker can obtain only general identifying information
Reproducibility	The attack can be carried out several times and executed at any time without restriction	The attack can only be repeated at a specific situation or point in time	Despite known security vulnerabilities, it is difficult to repeat the attack
Exploitability	A novice attacker can perform the attack directly using scripts or simple tools	An attacker with some experience can carry out the attack but relevant technical knowledge is required	Requires the experience of a computer security professional with a high understanding of the attack
Affectability	All users are affected	Only some users are affected	It affects only a small percentage of users and can be quickly recovered
Discoverability	Vulnerability information is defined by severity levels, is publicly exposed, and is easy to implement repeatedly	The vulnerability exists in a service that is rarely used, and its impact is very low, requiring specific information to identify it	The attack on vulnerability is slight. Its implementation requires specific conditions and is extremely difficult to detect

Kappa coefficient. To analyze the results obtained from the DREAD model through surveys of two experts. These were chosen according to their experience in security matters. In addition, they are researchers on the subject of cybersecurity and data analysis from the university sponsoring the research. Only two teachers were chosen since the other researchers stand out in other areas that are not part of the central theme of this research. It is considered necessary to evaluate the consistency between their evaluations, for which the Kappa coefficient will be used. The Kappa coefficient proposed by Jacob Cohen in 1960 refers to a statistical measure that allows comparing the agreement observed by two evaluators in a group of data, with what could happen by chance [4]. The equation used to obtain the Kappa coefficient (K) is the following:

$$\kappa = \frac{\Pr(\alpha) - \Pr(\mathfrak{e})}{1 - \Pr(\mathfrak{e})}$$

Where, Pr(a) refers to the agreement observed between the evaluators and $\Pr(\mathfrak{e})$ is the probability of agreement by chance. In case there is a complete agreement between the two evaluators, the K coefficient will be equal to 1, otherwise it will approach 0 or even negative values in a 2×2 matrix. The Table 2 shows the strength of agreement proposed by Landis and Koch in 1977 based on the kappa coefficient.

Table 2. Strength of agreement Kappa.

Kappa coefficient	Strength of agreement
<0	Null
0.0–0.2	Poor
0.2–0.4	Weak
0.4–0.6	Moderated
0.6–0.8	Good
0.8–1.0	Very Good

2.2 STRIDE Model

To identify threats, it is necessary to categorize them in the classification in Table 3. For the identification of each threat, the STRIDE model is used, this identifies and examines its security aspects. This model categorizes threats according to the intention of the attack, this allows developing certain security policies [3].

2.3 Analyzed Attacks Classification

The different types of attacks on wearable devices are the main point of analysis of this document, therefore, it is important to understand their behavior. Each of the types of attacks to which these devices are exposed will be described below. Bluetooth Pairing Attacks. Bluetooth pairing techniques have significant security flaws which lead to certain vulnerabilities. These devices can be attacked using techniques such as: man-in-the-middle attacks, packet injection, and spying. Through these attackers can actively spy on portable devices and misuse information [5]. Regarding surveillance attacks, they have been classified as follows [5]:

- Blue Printing: Refers to spoofing MAC addresses to execute a man in the middle attack.

Table 3. STRIDE model description.

Threat	Security property	Description	Action
Spoofing	Authentication	Unauthorized access to the system using an illegal identity or wrong information to deceive the system	This threat uses the credentials of other legitimate users to authenticate themselves to the system and access information
Tampering	Integrity	Unauthorized malicious modification of the information to mislead the recipient	It allows to modify the information to be stored in databases, file systems or network traffic
Repudiation	Non-repudiation	The user declares not having and not being able to perform any action	The system does not have the ability to track user activities
Information disclosure	Confidentiality	The information is disclosed to unauthorized users	Allows access to unauthorized files or data in transit
Denial of service	Availability	Deny or interrupt user access to applications or services	Denies valid users' access to temporarily unavailable databases and applications
Elevation of privilege	Authorization	Users with limited permissions can use the app to obtain unauthorized permissions	Obtain access permissions to resources to access or damage information

- Blue Stumbling: This is a forced re-match attack.
- Blue Tracking: It allows determining the encryption key through brute force.

Code Injection. It is defined as the introduction of malicious code into a system that contains programming errors. This form of attack is used for various purposes, such as stealing information, taking control of the system, and spreading worms. Examples of these attacks are: scripts html and shell injection. If the attack is successful, it causes the loss of control of the application and compromises the privacy of the user [6].

Buffer Overflow. This attack exploits the vulnerabilities that an application has to transcend the limits of the code or the data buffer. The attack consists of writing a sequence of data in a specific area, this produces an overflow of the sequence beyond its predefined region of residence. The result can be the modification of other data, the execution of malicious code and the destruction of the control flow of the application [7].

Handling of Sensitive Information. This attack makes an illegal access to the information handled by portable devices, or can manipulate it without restrictions. The wearable device sends sensitive information to the application through events, which are used by the application to monitor the wearable device. However, this delivery of the event is done without privacy, this can cause a complete filtration and great damage to the user [8].

Phishing. Here the attacker tries to impersonate a legitimate entity that offers non-public services in order to obtain sensitive information from them. It is generally done using email, where the attacker acquires sensitive information when the user opens the sent email [6].

Authentication and Authorization. Authentication is a fundamental part of device security and privacy. Current authentication mechanisms do not provide adequate verification. For example, when updating applications, malicious code can be downloaded, at which point attackers can use it to control the terminal remotely. Furthermore, keeping the default setting is also part of the root of the permittivity problem [9].

Cloud Flooding. It works as an attack that denies services hosted in the cloud. Attackers usually send requests to a cloud service constantly, this causes resources to be exhausted and the quality of service is affected. Also, when the cloud-based service realizes that the current instance does not meet the requirements, it transfers it to other servers. This increases your workload and the denial becomes apparent [10].

Cloud Malware Injection. The attacker has permission to: modify information, take control and execute malicious code on a service or virtual machine in the cloud [11]. Attackers copy and load a victim's service instance, this instance only responds to the request when some service requests the victim's instance. In this way, the attacker obtains information that is confidential [10].

Signature Wrap. The cloud-based system uses XML signing to ensure that no unauthorized person can change something about a service. The attacker changes the exchanged data without invalidating the signature [12]. In this way, the attacker has executed, edit permissions to arbitrarily perform operations as a legitimate user.

SQL Injection. If an application is poorly designed, incorporating SQL statements into the input data makes it a vulnerable target to attack [13]. Attackers use SQL statements to perform read and write operations. These attacks allow to obtain private data of the user; in addition, it constitutes a threat to the enterprise database. When web applications are attacked by SQL injection, the browser displays various results compared to the true information [14].

Denial of Service. His type of attack is achieved by sending multiple requests to the victim, which produce a very large amount of traffic compared to the usual [15]. Denial of service exhausts all available resources, rendering users unable to access them. Furthermore, a large amount of unencrypted information can be leaked [6]. In addition, there is an attack that denies services in a distributed manner (DDoS), this can transfer many devices into a controlled attack platform to execute denial attacks on one or more targets. Other types of attacks could be: Sybil Attack, Replay Attack and Sleep Deprivation Attack, etc.

Attack by Bluetooth Pairing. The matchmaking techniques have significant security holes and potential vulnerabilities including man-in-the-middle attacks, spying, and packet injection. Through these types of attacks, attackers have the ability to actively spy on portable devices and misuse information [16].

3 Analysis

We will proceed to create comparative tables using the DREAD and STRIDE models previously described to give the reader a clearer view of the different threats and attacks to which they may be exposed when using wearable devices, so that they are aware of these aspects when acquiring them. Finally, we will proceed to create a table showing some security recommendations against the analyzed threats.

3.1 DREAD Model Based Analysis

The following tables show the results of the survey conducted by two evaluators with extensive experience in the topic of information security, these will allow to calculate the risk of each attack through parameters such as Damage (D), Reproducibility (R), Exploitability (E), Affectability (A) and Discoverability (D), where each aspect will be graded on a scale of 1 to 3 to finally obtain a total on a scale of 5 to 15 that will allow the level of risk to be decided, being this low, medium or high.

Tables 4 and 5 show that the most dangerous attack for wearable devices is the Phishing attack with the highest risk rating, 15. Through this attack, you can ignore the system protection mechanism, execute malicious code, modify data, and even cause loss of control of the application and compromise the user's privacy. In addition, it can be performed several times at any time, even by a novice using simple and easy-to-implement scripts or tools that are publicly exposed on the Internet, thus affecting all users of the victim device. On the other hand, the least risky attack is the signature wrapping attack, this one received a rating of 5 by both evaluators, which places it at the Low Risk level. Through this attack, you can obtain only general identification information, its technical complexity is high, and it is difficult to repeat the attack. In addition, it requires the experience of a professional with a high knowledge of the attack to execute it correctly. Also, the affectation would be only a small percentage of users and can be recovered quickly. Kappa Coefficient Agreement Analysis. Once the results of the expert surveys have been obtained, the concordance between their evaluations is assessed using Cohen's Kappa coefficient. Table 6 shows the tabulation of the matching data between the evaluators.

Subsequently, the observed and expected probabilities have been calculated as shown below, obtaining the results shown in Table 7.

$$\Pr(a) = \frac{3+5+4}{18} = 0.67$$

$$\Pr(e) = \frac{\frac{(4*4)+(10*6)+(4*8)}{18}}{18} = 0.33$$

Table 4. DREAD model based analysis Evaluator 1

Attack	D	R	E	A	D	Total	Risk
Bluetooth pairing attacks	2	3	2	2	2	11	Medium
Code injection	2	2	2	3	3	12	High
Buffer overflow	1	2	1	3	2	9	Medium
Handling of sensitive information	3	3	3	3	2	14	High
Phishing	3	3	3	3	3	15	High
Authentication and authorization	2	2	2	3	1	10	Medium
Cloud flooding	1	1	1	3	2	8	Medium
Cloud malware injection	1	1	1	3	2	8	Medium
Signature wrap	1	1	1	1	1	5	Low
SQL injection	3	2	1	2	3	11	Medium
Denial of service	3	3	2	3	3	14	High
Sybil attack	2	2	2	2	1	9	Medium
Sinkhole attack	1	1	1	2	1	6	Low
Sniffing attack	2	3	2	2	2	11	Medium
Traffic analysis attack	2	3	2	2	2	11	Medium
Replay attack	1	1	1	2	1	6	Low
Man in the middle attack	1	1	1	2	2	7	Low
Sleep deprivation attack	1	3	2	2	2	10	Medium

Finally, the Kappa (K) coefficient was calculated resulting in 0.5, meaning that the strength of agreement of the two evaluators is moderate, reinforcing the results obtained through the surveys conducted. However, it should be noted that they may vary to some extent according to the criteria and experience of each evaluator.

$$\kappa = \frac{\Pr(\alpha) - \Pr(\mathfrak{e})}{1 - \Pr(\mathfrak{e})}$$

$$\kappa = \frac{0.67 - 0.33}{1 - 0.33} = 0.5$$

3.2 STRIDE Model Based Analysis

The table shown below is based on the STRIDE model, this will allow to classify attacks according to its objective and purpose, which will help to recommend some security

Table 5. DREAD model based analysis Evaluator 2

Attack	D	R	E	A	D	Total	Risk
Bluetooth pairing attacks	3	3	2	2	2	12	High
Code injection	2	3	3	3	2	13	High
Buffer overflow	1	1	2	2	3	9	Medium
Handling of sensitive information	3	3	2	3	3	14	High
Phishing	3	3	3	3	3	15	High
Authentication and authorization	2	2	2	2	1	9	Medium
Cloud flooding	1	1	1	2	2	7	Low
Cloud malware injection	1	1	1	3	2	8	Medium
Signature wrap	1	1	1	1	1	5	Low
SQL injection	3	2	2	2	3	12	High
Denial of service	3	3	2	3	2	13	High
Sybil attack	1	2	2	1	2	8	Medium
Sinkhole attack	1	1	2	1	1	6	Low
Sniffing attack	2	1	2	3	3	11	Medium
Traffic analysis attack	3	2	2	2	3	12	High
Replay attack	1	1	2	2	1	7	Low
Man in the middle attack	2	2	3	2	1	10	Medium
Sleep deprivation attack	3	2	2	3	2	12	High

Table 6. STRIDE model based analysis

Evaluator 1	Evaluator 2			
	Low	Medium	High	Total
Low	3	1	0	4
Medium	1	5	4	10
High	0	0	4	4
Total	4	6	8	18

policies. This table will present an X in the category or categories to which each one of the 18 analyzed attacks belongs. The threats observed in the columns of the table

Table 7. Agreement probability calculation

Observer agreement Pr(a)	0.67
Expected agreement Pr(e)	0.33

are: Spoofing (S), Tampering (T), Repudiation (R), Information Disclosure (I), Denial of Service (D) and Elevation of Privilege (E).

Table 8 shows how each type of attack analyzed can belong to more than one category of threat to which wearable devices are vulnerable. In this case, it has been observed that the most important threat to which wearable devices are vulnerable is the disclosure of information, this category violates a very important security property that is the confidentiality of information, disclosing it to unauthorized users and allowing them access to unauthorized files or data.

Table 8. STRIDE model based analysis

Attack	S	T	R	I	D	E
Bluetooth pairing attacks	X			X	X	X
Code injection		X	X	X	X	X
Buffer overflow		X	X	X	X	X
Handling of sensitive information		X		X		X
Phishing	X	X		X		
Authentication and authorization	X	X		X		X
Cloud flooding					X	
Cloud malware injection	X	X		X	X	X
Signature wrap	X	X	X	X		X
SQL injection		X		X	X	X
Denial of service				X	X	
Sybil attack	X			X	X	
Sinkhole attack	X	X		X		X
Sniffing attack			X			
Traffic analysis attack			X			
Replay attack	X			X	X	X
Man in the middle attack	X		X	X		
Sleep deprivation attack					X	

3.3 Security Recommendations

Having analyzed each of the threats, vulnerabilities, risks and types of attacks that wearable devices are victims of, we proceed to make security recommendations so that these terminals used by millions of users daily and the networks to which they are connect can be more secure. The above-mentioned recommendations are observed in Table 9.

Table 9. Security recommendations

Threat	Countermeasure
Spoofing	Do not allow sensitive information to be transferred in plain text -Prevent credentials from being stored in plain text
Tampering	Use more restrictive authentication Use of digital signatures
Repudiation	Make hash code comparisons Use systems that record logs Use digital signatures
Information disclosure	Use secure encryption algorithms Do not allow sensitive information to be transmitted in plain text
Denial of service	Use bandwidth control or load balancing to disperse the flow of information
Elevation of privilege	Follow the principle of least privilege

Additionally, it is important to mention that (PII) personally identifiable information can be assigned to a specific factor in the human system and in various jurisdictions it has legal protection [17]. Transmission of a fixed MAC address tied to an individual's identity would fail the PII test when creating a unique user signature. For example, the Directive on the Protection of Personal Data of the European Parliament promulgated on May 5, 2016 and introduced by its member states in their national laws until May 6, 2018, extends the definition of personal data to include those that can be identified, directly or indirectly [18]. A fixed MAC address runs the risk of invisible surveillance, violating this requirement.

4 Conclusions

The emergence of wearable devices has resulted in a large amount of sensitive information traveling through the digital world. Thus, all this data is exposed to the risk of unjustified privacy violations, threatening the lives of users and their security. Consequently, protecting people's privacy has become a major challenge in the development of wearable technology. The implementation and use of wearable technology using sensor devices have become an integral part of daily life and its different activities. The growing interest in these devices has raised concerns about their security and privacy. This document has been reviewed and analyzed using the DREAD model, the different attacks that wearable devices can suffer and the risk that this entails. Thus, it has been possible to confirm the high importance of information security when working with these devices. Wearable technology despite its rapid growth is in an early stage of its generation and promises a great future in terms of development. Likewise, their vulnerabilities, threats, and risks to which these devices are exposed, are in an initial and research stage that will continue to evolve at the same or higher pace than technology.

Safety hazards in wearable devices could cause a lot of damage. This is because these devices are part of people's daily lives, they are friendly and have several sensors that carry sensitive information. Two examples of attacks that have a high risk on wearable devices are: Bluetooth pairing attacks and code injection, both are easy to implement and can cause a lot of damage with the information acquired. Although the risk is low in cloud flooding attacks, signature wrap sinkhole attack, replay attack, attention should be paid to them. If these attacks materialize on the devices, they can cause the same damage as a high risk attack.

References

1. Ching, K., Singh, M.: Wearable technology devices security and privacy vulnerability analysis. Int. J. Netw. Secur. Its Appl. (IJNSA) **8**(3), 19–30 (2016)
2. HP. http://www8.hp.com/us/en/hp-news/pressrelease.html?id=2037386#.Vi18G7crLIU. Accessed 03 July (2020)
3. Tseng, T., Wu, C., Lai, F.: Threat analysis for wearable health devices and environment monitoring internet of things integration system. IEEE Access **7**, 144983–144994 (2019)
4. Andalusian Society of Intensive Medicine and Coronary Units. http://www.samiuc.es/estadisticas-variables-binarias/medidas-de-concordancia/kappa-de-cohen/#:~:text=Kappa%20de%20Cohen%20(%CE%BA),que%20podr%C3%ADaocurrir%20por%20mero%20azar.&text=Si%20K%20es%20cero%2C%20ello,que%20ocurrir%C3%ADa%20por%20puro%20azar
5. Lotfy, K., Hale, M.L.: Assessing pairing and data exchange mechanism security in the wearable internet of things. In: 2016 IEEE International Conference on Mobile Services (2016)
6. Farooq, M., Waseem, M., Khairi, A., Mazhar, S.: A critical analysis on the security concerns of Internet of Things (IoT). Int. J. Comput. Appl. **111**, 1–6 (2016). https://doi.org/10.5120/19547-1280
7. Zhu, B., Joseph, A., Sastry, S.: A taxonomy of cyber attacks on SCADA systems. In: International Conference on Internet of Things and 4th International Conference on Cyber, Physical and Social Computing, vol. 2011, pp. 380–388 (2011)
8. Nan, Y., Yang, M., Yang, Z., Zhou, S., Gu, G., Wang, X.: UIPicker: user-input privacy identification in mobile applications. In: USENIX Security Symposium, pp. 993–1008 (2015)
9. Simmons, C., Ellis, C., Shiva, S., Dasgupta, D., Wu, Q.: AVOIDIT: a cyber-attack taxonomy (2009)
10. Gruschka, N., Jensen, M.: Attack surfaces: a taxonomy for attacks on cloud services. In: IEEE 3rd International Conference on Cloud Computing (CLOUD), pp. 276–279 (2010). https://doi.org/10.1109/CLOUD.2010.23
11. Padhy, R., Patra, M., Satapathy, S.: Cloud computing: security issues and research challenges. Int. J. Comput. Sci. Inform. Technol. Secur. (IJCSITS) **1**(2), 136–146 (2011)
12. Jensen, M., Schwenk, J., Gruschka, N., Iacono, L.L.: On technical security issues in cloud computing. In: IEEE International Conference on Cloud Computing, pp. 109–116 (2009)

13. Zhang, Q., Wang, X.: SQL injections through back-end of RFID system. In: International Symposium on Computer Network and Multimedia Technology, pp. 1–4 (2009)
14. Dorai, R., Kannan, V.: SQL injection—database attack revolution and prevention. J. Int.'l Com. L. Tech. **6**(4), 224 (2011)
15. Hassan, S.S., Bibon, S.D., Hossain, M.S., Atiquzzaman, M.: Security threats in Bluetooth technology. Comput. Secur. **74**, 308–322 (2017)
16. Network. International Journal of Computer Applications, vol. 40, no. 15, pp. 19–25 (2012). https://doi.org/10.5120/5056-7374
17. Boyle, R.J., Panko, R.R.: Corporate Computer Security. Prentice Hall Press, Essex (2014)
18. European Parliament. https://eur-lex.europa.eu/legal-content/EN/TXT/?uri=uriserv%3AOJ.L_.2016.119.01.0089.01.ENG&toc=OJ%3AL%3A2016%3A119%3ATOC. Accessed 03 July (2020)

A VR-System to Assess Stereopsis with Visual Stimulation: A Pilot Study of System Configuration

Sonia Cárdenas-Delgado(✉), Mauricio Loachamín-Valencia, Pablo Guanoluisa-Atiaga, and Xavier Monar-Mejía

Departamento de Ciencias de la Computación, Universidad de las Fuerzas Armadas ESPE, Av. General Rumiñahui s/n, 171-5-231B Sangolquí, Ecuador
{secardenas,mrloachamin,psguanoluisa,jxmonar}@espe.edu.ec

Abstract. Stereopsis is the binocular perception of depth. The lack of accurate depth perception can have a high impact on life, from early childhood learning to the jobs and daily tasks we are able to do. One of the human factors that can influence in the aviation safety is the lack of stereoscopic vision. Our main aim was to develop a virtual system that includes interaction and immersion using a VR device of visualization. It included a virtual environment, several 3D objects and a virtual task to assess stereoscopic vision in adults. Two groups of military pilots participated in the experimental study (N = 20). Group 1 were participants without experience (N = 10) and Group 2 were participants with experience (N = 10). The components of our VR-system SAS have allowed to assess the stereopsis through visualization, recognition and number of 3D objects seen. The HMD used, allowed the participants of both groups to feel a more immersive experience, have better depth perception and see 3D objects inside the virtual environment. Regarding to the 3D sensations and satisfaction, the participants found it easier to use the VR-system SAS, and they would also participate again and use this system. Furthermore, with both, the LANG Stereo-test and the VR-system, the participants were able to see the 3D objects shown. The results suggested that our system is valid to assess stereopsis. Moreover, our VR-system could be a very useful technological tool in the field of ophthalmological evaluation and pilot training.

Keywords: Stereopsis · Virtual reality · Human factor · depth perception · Ophthalmology

1 Introduction

The technologies and methodologies implemented in the aviation industry have made it possible to guarantee aviation safety to the world [23]. However, Human Factors and their interaction have become major challenges for air operations. These factors include the aspects like their sensory perception, human capabilities and limitations [2,11]. Some errors are produced by physiological and

M. Botto-Tobar et al. (Eds.): CIT 2020, AISC 1326, pp. 328–342, 2021.
https://doi.org/10.1007/978-3-030-68080-0_25

psychological limitations such as the perception, that is received through the senses, cognitive overload, fatigue, and workload [1,2]. Therefore, to fly an aircraft a lot of dexterity, practice, series of skills, psychological, neurological and visual capabilities such as stereopsis are required [21].

Stereopsis is the binocular perception of depth, which is the sensitivity obtained by visual disparity of images formed in the retinas of both eyes. Each eye captures 2D images of the objects around us and brain uses these images to rebuild the visual world with depth dimension or 3D environment. In other words, if one of the two eyes is not functioning, the depth perception lacks of accuracy. The lack of depth perception can affect a child's ability to learn, it can impede the normal development of occupational activities of adults, limit an athlete's potential or cause problems when driving, piloting and navigating [13,30].

The general population can suffer different alterations in binocular vision but they are unaware of it and can go unnoticed, yet this visual capacity confers functional benefits in the development of daily tasks that are related to eye-hand coordination, for example, pour water into a glass, catch a ball, park a car or motorcycle (calculate distances), pilot an aircraft, recognize how far or how close an object is (calculate depth), see space in three dimensions. Above all, it allows the brain to compare images of a scene and estimate the relative depth with high accuracy [10].

Due to the importance of stereopsis in the performance of human activities, which involve the safety of other people and themselves, such as pilots, surgeons or drivers. The need arises to develop a technological tool which allows to assess stereoscopic vision.

In previous works related to ophthalmology and optometry, they have used different traditional methods and tests to assess stereoscopic vision [6,14]. However, to our knowledge, virtual reality devices and convergent technologies have not been used to assess visual impairments such as stereoscopic vision.

Our hypothesis consists of a system using virtual reality with visual stimulation which could be a useful and complementary tool in the stereoscopic vision assessment process in adults. Our contribution is to perform early detection of visual impairments related to the binocular perception of depth to mitigate incidents during air operations performed by pilots in their missions in each Armed Forces of Ecuador.

In this study, we aim to assess stereopsis through a virtual and functional task under an experimental condition which consisted in the visual stimulation while the participants go through a maze. Also, they must recognize the objects in the environment. For the visualization, we used a HMD that allows to feel a more immersive experience and tracking the head movements. For the interaction and navigation into the virtual environment, we used a device with motion control technology. Before each session, the participants were checked for stereopsis using the LANG Stereo-test cards I and II. After using the VR-system, the participants evaluated 3D sensations and satisfaction.

The paper is structured as follows. Section 2 a review of related studies with different visualization systems, VR head-mounted displays, binocular depth perception, and standard tests widely used for this type of studies. Section 3 describes methods and materials such as development, methodology, hardware and software. Section 4 includes participants, measurements and a brief description of the procedure carried out. Section 5 describes the results analysis. Section 6 draws the conclusions and discussions related to this study. Finally, Sect. 7 provides a general idea future works.

2 Background

Stereoscopic visual impairment is related to the difficulty of seeing objects in three dimensions and the lack of depth perception of objects in the real world, it is not commonly assessed [26]. Primary works related with assess stereo vision have been studied with psychophysical techniques, and a few years ago physiological mechanisms have been applied to these techniques [10]. Moreover, other studies suggest that the problem of stereopsis is that it affects the quality of life, in some aspects such as information processing, motor skills and reading competence in children.

Regarding human-computer interaction, virtual environments (VE) are an active field of research. VE offer immersion, are intuitive and due to them, the applications development with virtual reality have increased. These applications are developed mainly for the entertainment industry, however, in other disciplines virtual reality has become a useful tool, for example, Psychology, Medicine, Ophthalmology, Education and Security, among others [6,27].

In the last years, VR technology has received special attention due to the use of a variety of technological tools and virtual reality devices to navigate or interact with virtual worlds, especially HMDs for visual input. Head-mounted displays (HMD) like the Oculus Rift, HTC Vive and similar technologies [17,19, 24] have extended the accessibility of Virtual Reality (VR) to a wide segment of the population and the research community [3,5,6,8,16].

HMDs provide an interactive and immersive virtual reality environment by generating high-resolution 3D graphics and objects, stereoscopic images, and the ability to look around in the virtual environment. Also, HMDs generate a computerized representation focused on first person in the real world, where an individual is immersed, can see as a first person from different angles, navigate, interact and manipulate it, and generate a sense of presence [7]. Furthermore, the manufacture of a large number of device models (virtual reality headsets) and low costs have made them increasingly available devices to experience immersion and presence within different fields.

Regarding the works carried out in the field of ophthalmic diagnosis of various vision functions such as screening, stereopsis, astigmatism and others, significant advances have been made in the creation of computer equipment and systems [6,14]. However, in most cases these systems generally use video display screens to present visual tests, tablets or screens that provide stereoscopic images based on spatial resolution and display geometry.

Concerning applications in the field of stereopsis, studies have been found in which devices showed magnified images to create a binocular field and allow the viewer to look around the virtual environment [20]. Other study carried out in people with lost field centers used both eyes to perform a grasping task. This study demonstrated the correlation with stereopsis, contrast sensitivity, and binocular visual field [29]. Also, there are several works based on the information from the National Eye Institute and other which simulated the effects of the visual diseases implementing different post-processing effects. However, to our knowledge, virtual reality devices and convergent technologies have not been used to assess visual impairments such as stereoscopic vision [25,28].

Stereoscopic vision and depth perception tests help identify conditions such as: amblyopia, strabismus, suppression, and stereopsis. Stereovision is very important in human beings and it is the way each eye can see an object from different angles, combining these two angles we can see a 3D image. There are several depth perception tests on the market; for example some ask the patient to identify the letter, shape or animal to measure depth perception. This allows to identify if the patient's eyes are working together and can identify an image [30].

Usually, stereopsis has been assessed using several traditional tests, the LANG Stereo-test, Titmus test, Ramdom Dot E, among others. The tests measure stereoacuteness (measured in arc seconds) through cards, images and polarized glasses. In addition, there are studies that suggest the effectiveness of the use of the LANG Stereo-test to assess stereopsis [6,22], micro strabismus, depth perception, the thick stereopsis, sensitivity and visual binocular in children and adults. LANG Stereo-test is based on pantographic presentation of random dot pattern. This test does not require glasses to recognize the stereoscopic images embedded in random dots on the test cards [4,18].

Various tests and computer applications have been developed to allow detecting deficiencies or disorders for many visual tasks that humans perform, however, to check stereoscopic vision, they have been surprisingly arduous and limited. For this reason, in the last years, the study of stereopsis disorder has received more attention.

In this work, we developed a virtual system that includes an interactive and immersive environment using devices of visualization to assess stereoscopic vision in adults.

3 Methods and Materials

3.1 Development

The Virtual Reality System to Assess Stereopsis (SAS) includes hardware and software. The components of the hardware are the VR engine, the input and output devices. The software is the application and database. The application involves tools and software for modeling, designing, developing. The database is a repository where the information of the task was stored. Furthermore, it includes a virtual environment, 3D objects and a virtual task for the evaluation of the elements related to stereoscopic vision. Figure 1 shows the architecture of the *VR-System SAS*.

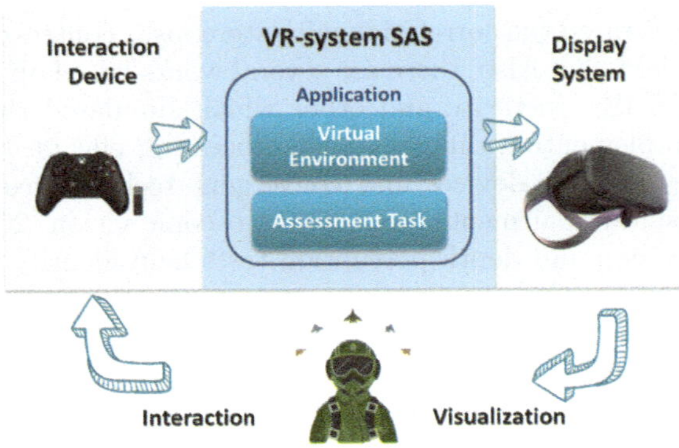

Fig. 1. Architecture of the VR-System SAS.

3.2 Methodology

The design and development of the virtual environment was based on the methodology proposed by Fencott [9]. The phases are (Fig. 2):

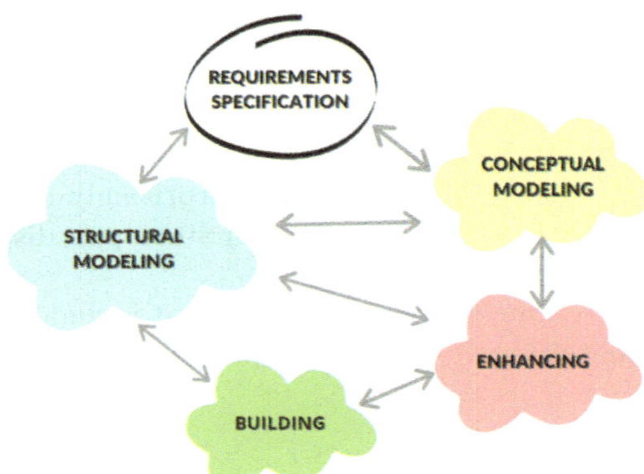

Fig. 2. Phases of the methodology for VE design

- **Requirements specification.** The requirements of the virtual environment were established.
- **Conceptual modelling.** All the necessary objects from the real model that will integrate the virtual environment have been defined and gathered.

- **Structural modelling.** The graphical model was structured including the 3D objects and the environment with their perceptual characteristics and their defined interrelations.
- **Building.** In this phase we created the virtual environment, 3D objects and positioned them in the VE.
- **Enhancing.** We enhanced our VE with texture, lighting, sound and interaction, and we integrated devices of visualization and navigation.

3.3 Virtual Environment

The research team of the "VisioStereo-Diagnostic" project was in charge of the design and development of the virtual environment. The VE consists of a virtual maze with twelve intersections. The maze wall is two meters high and made of exposed brick. Its pathways from brown stone are two meters wide (see Fig. 3). Several virtual objects were created for the virtual environment: a bomb, a flashlight, munitions, a military helmet, a bat, light blue and yellow arrows (see Fig. 4). The design of the virtual environment included five 3D objects. They were placed following the sequence on the right - left side of the path. A first person avatar was designed, this represents the participant in the virtual environment. Through the avatar's eyes, the participant can see the environment. The movements are also personified by the avatar.

(a) External view (b) Internal view

Fig. 3. View of the virtual environment.

The stages of this task were two: habituation and testing.

- ***The habituation stage*** this is a training stage to adapt the participant to the virtual environment. The habituation stage consists of an environment where the participant follows a guided route with light blue arrows (see Fig. 5a).
- ***The test stage*** in this stage the elements related with the human stereopsis: visualization, recognition and number of 3D objects seen; depth perception and immersion were used to assess. The participants must go through the guided route for the yellow arrows and look at 3D objects within the virtual environment (see Fig. 5b).

Fig. 4. Location of 3D objects in the maze

(a) The habituation stage.

(b) The test stage.

Fig. 5. Virtual maze.

3.4 Hardware and Software

The *VR-system SAS* is being developed on a computer with an Intel Core i7 processor, 32 GB of RAM and with a Windows 10 operating system, the equipment includes a 4 GB NVIDIA GeForce GTX-1080 video card. The visualisation device used was an HMD Oculus Rift. This VR device includes motion sensors, surrounding sound and 360° vision, these elements allow to feel immersion within the virtual environment. The interaction with the virtual environment was using a wireless control or gamepad compatible with Windows 10. The virtual scenes were developed using Unity Personal and programs with C# and Javascript. To model the 3D objects we used Blender.

4 Description of the Study

The study consisted in two experiments. Each was carried out in different days. Both had the same virtual maze and way of navigating in the environment. Participants performed two stages in each experimental task. In the experimental study participated two groups of military pilots belonging to the Ecuadorian Air Force and Terrestrial Force.

4.1 Participants

The necessary administrative and logistical procedures were made for two groups of military pilots to participate in the experimental study. The total of participants for the experimental study were 20. Group 1 were participants without experience (N = 10), (Group 1: mean age 22.4 ± 1.43) and Group 2 were participants with experience (N = 10), (Group 2: mean age 42.7 ± 2.75). In the Fig. 6 can be see to participants performing the virtual test.

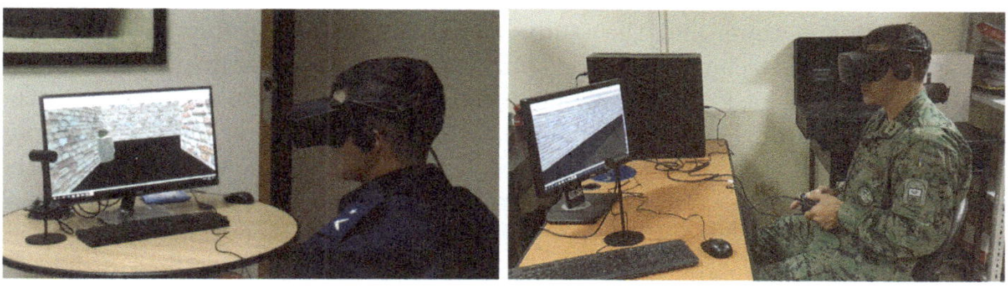

(a) Group 1 (b) Group 2

Fig. 6. Participants using the VR-System.

4.2 Measurements

The assess was carried out with the LANG Stereo-test. This test has been used due to the ease to analyse stereopsis in children and adults, as suggested by [4,6]. The first card measures thick stereopsis and the second card measures fine stereopsis. Each card shows three stereoscopic objects, therefore, 6 objects were evaluated. Maximum score of the traditional test is 6/6.

Only four questions from The Presence Questionnaire [32] were selected, these were related to the depth of objects, sense of moving around inside the virtual environment, and perception of the participants about the immersive experience. Two questions about satisfaction were employed. The questions applied can be seen in Table 1.

In addition, the data that allows assessing the performance of the virtual task was recorded. This consisted of counting the number of objects seen within the virtual environment. Maximum score of the virtual test is 5/5. These data was compared with objects seen with the LANG Stereo-test.

Table 1. Questions about 3D sensations and satisfaction.

Q	3D sensations
Q1	At times it seems to me that objects have depth?
Q2	Have you felt within the 3D environment?
Q3	Has the virtual 3D environment been realistic?
Q4	How involved were you in the 3D virtual environment experience?
Q	Satisfaction
Q5	Would you participate again?
Q6	Did I have a good time?

4.3 Procedure

The scheme of the experimental study can be seen in Fig. 7. Prior to each session, all participants were informed in writing about the aims and procedures of the study, they also signed an informed consent form. They were fully free to leave the study at any time, and the study was conducted according to the principles stated in the Declaration of Helsinki. Participants performed two stages in each experimental task. Each participant was instructed about how to use the devices, such as the HMD and gamepad. The participant was also urged to pay attention at each stage of the task, as well as, to remember the objects of the environment. The first questionnaire administered was the Personal Data Questionnaire (PDQ) and was filled out per each participant previous to the task. Afterwards, they were checked for stereopsis using the LANG Stereo-test card I and II. Then, the participants performed the experimental task. Furthermore, the assess variables were recorded. Finally, when they had finished the task, they completed the questionnaire to know their perceptions about 3D sensations and satisfaction.

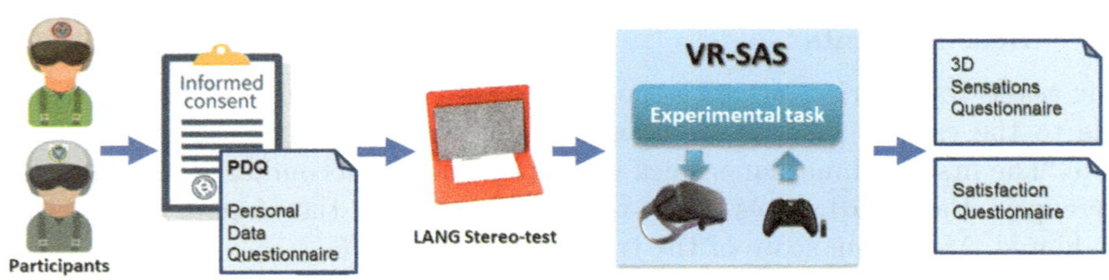

Fig. 7. Scheme of the experimental study.

5 Results

This section presents the analysis of the data collected from this study. Data normality was checked and the pertinent statistical tests were carried out based on those results. The Likert scale was applied in the questionnaires. The group was divided into two groups of pilots, those participants who had years of experience and those participants who did not have experience. There were two groups: Group 1 and Group 2. They used the HMD and gamepad. The data from the study was analysed using the statistical open source toolkit R1 with the R-Studio. The results of the questionnaire were grouped by 3D Sensations, and Satisfaction.

5.1 Outcomes Comparison Between LANG Stereo-Test vs. VR-System SAS-Test

In the traditional test (LANG Stereo-test), the results showed that 2 participants from Group 1 and 7 from Group 2 observed 5 to 6 objects (maximum score of the traditional test 6/6). This means that inexperienced participants saw fewer objects even in the LANG Stereo-test. Whereas in the virtual test (VR-system), 7 participants from Group 1 and 9 from Group 2 observed 4 to 5 objects (maximum score of the virtual test 5/5). This indicates that experienced participants can see more objects. Moreover, in the virtual environment the number of objects in both groups increases. This means that the virtual test and the traditional test correlate their results and demonstrate the validity of the VR-system SAS. Statistically significant differences were not found between both groups when the performance between the LANG Stereo-test and VR-system SAS was compared (see Fig. 8).

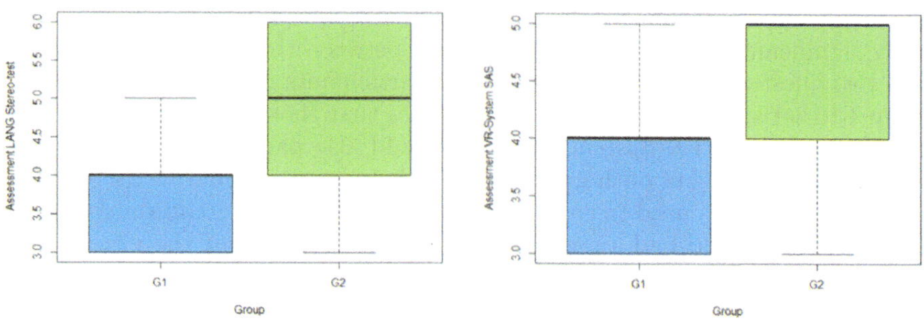

(a) LANG Stereo-test Outcomes. (b) VR-system SAS Outcomes.

Fig. 8. Boxplot with results obtained in the traditional test and the virtual test.

5.2 3D Sensations Outcomes

To determine the outcomes for 3D sensations, the participants answered questions Q1-Q4 after their exposure to the virtual environment using visual stimulation with the HMD. Medians and standard deviation of questions from Group 1 were Q1 (3.9 ± 0.74), Q2 (3.7 ± 0.68), Q3 (3.6 ± 0.52), Q4 (3.8 ± 0.63). Medians and standard deviation of questions from Group 2 were Q1 (4.2 ± 0.42),

Q2 (4.7 ± 0.68), Q3 (4.6 ± 0.52), Q4 (4.6 ± 0.52). Also, statistically significant differences were not found between the two groups in all four questions about 3D sensations. Overall, the HMD Oculus Rift allowed the participants both groups to feel a more immersive experience, have better depth perception and view 3D objects inside the virtual environment (See Fig. 9).

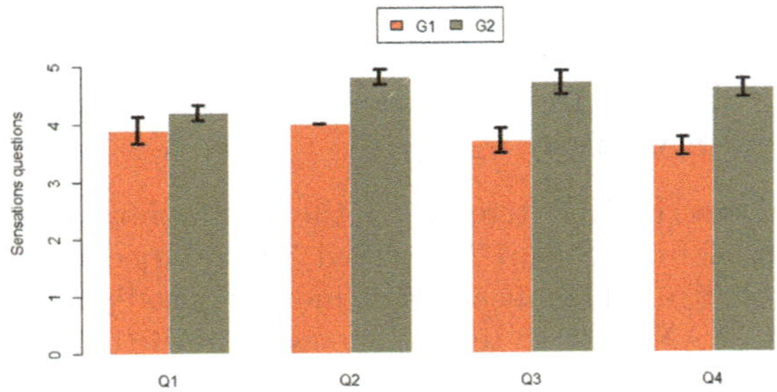

Fig. 9. 3D sensations outcomes

5.3 Satisfaction Outcomes

To determine the outcomes for satisfaction, the participants answered questions Q5-Q6 after their exposure to the virtual environment using visual stimulation with the HMD Oculus Rift. Medians and standard deviation of questions from Group 1 were Q5 (3.9 ± 0.32), Q6 (3.9 ± 0.32). Medians and standard deviation of questions from Group 2 were Q5 (4.7 ± 0.48), Q6 (4.9 ± 0.32). Also, statistically significant differences were not found between the Group 1 and Group 2 in all two questions (See Fig. 10). Group 2 participants gave a higher score than Group 1 to satisfaction questions. This implies that those with experience found it easier to use the *VR-System SAS* and would also participate again. In addition, the participants of both groups stated that this tool could be useful for the training which pilots need to carry out permanently. 19 out 20 affirmed they will participate again and had a good time when they carried out the virtual test.

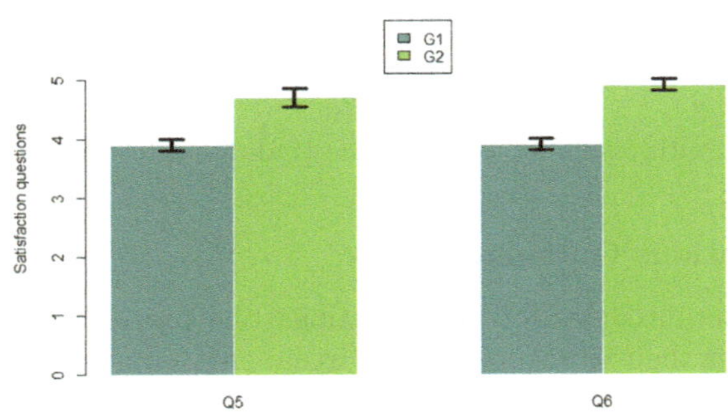

Fig. 10. Satisfaction outcomes

6 Conclusions and Discussions

A VR-system combining immersion and interaction was designed and implemented. It was designed to assess depth perception. Several virtual 3D objects were created. A virtual task was defined, which included two stages (habituation and testing). The features of the HMD (Oculus Rift) used are very significant because it allows to feel immersion within the virtual environment. The immersion was possible due to the properties that it has, these are head tracking, built-in headphones, a wide view, low weight and more ergonomic. The components which have allowed us to assess stereopsis using VR-system were the depth of objects, sense of moving around inside the virtual environment, perception of the participants about the immersive experience and the number of 3D objects seen within the virtual environment.

Regarding the results obtained in the performance comparison between the traditional test and virtual test; the experienced participants can see more objects with the virtual test than the traditional tests. In both tests, the participants were able to see the 3D objects shown. This means that the results of the VR-system SAS and the LANG Stereo-test correlate and suggest that our system is valid to assess stereopsis. The results demonstrate the validity and usefulness of our VR-system SAS to assess deficiencies in stereoscopic vision. These results corroborate our hypothesis.

In relation to the results of the 3D sensations and satisfaction, statistically significant differences not found in all questions were in favor of the VR-system in both groups. The participants were satisfied with our system. More participants in the two groups scored full marks in the virtual task than with the traditional test. Since the participants were satisfied with the virtual task, this satisfaction could positively influence the virtual task outcome. This proves what other authors have suggested the positive relationship between satisfaction, usability and virtual task outcomes [6,15].

Those participants with experience found more easy to use the VR-system SAS than whose without experience. All of them affirmed they will participate again because they had a good time. Besides, they had better depth perception and viewed 3D objects inside the virtual environment. This study has shown that our system with visual stimulation could be a useful and complementary tool in the process of assessing stereoscopic vision in adults because the participants in both groups favorably rated the immersive experience.

Moreover, the use of virtual reality in this work is a technological safe alternative because it allows to assess the stereopsis in a simulated environment like if it was real, without compromising the users' safety. The lack of accurate depth perception can have a high impact on life, from early childhood learning to the jobs and daily tasks we are able to do [31]. The highest form of depth perception is an essential visual ability that pilots must have to avoid any risky situation that threatens the safety of passengers, the aircraft and their own one [12]. In this context, our VR-system can be a very useful technological tool in the field of ophthalmological assessment and pilot training. Even this could contribute to perform early detection of visual impairments related to the binocular perception of depth and mitigate incidents during air operations performed by pilots in their missions.

According to the obtained results, this work also suggests that people who have carried out activities related to eye-hand coordination and visual exercises, like catch a ball, park a car or a motorcycle, pilot an aircraft or to drive a vehicle for a longer time have developed a greater capacity in their stereoscopic vision.

7 Future Works

The main work for the future is to continue with the development of the VR-system and carry out other studies that include the comparison between HMDs of other models and brands; study stereopsis in other professions such as surgeons and drivers; and adapt different types of locomotion and interaction.

Acknowledgment. This work was mainly funded by the project researchers. Moreover, we thank to the Universidad de las Fuerzas Armadas ESPE for providing the facilities to use the CICTE Simulation Laboratory and the of the Integrated Health System. Also, we are grateful with the pilots of the Ecuador Armed Forces for their collaboration as participants.

References

1. Amalberti, R.: La conduite de systèmes à risques: le travail à l'hôpital. Presses universitaires de France (1996)
2. Billings, C.E.: Aviation Automation: The Search for a Human-Centered Approach (2018)
3. Biocca, F.A., Rolland, J.: Virtual eyes can rearrange your body: adaptation to visual displacement in see-through, head-mounted displays. Presence Teleoperators Virtual Environ. **7**(3), 262–277 (1998)
4. Brown, S., Weih, L., Mukesh, N., McCarty, C., Taylor, H.: Assessment of adult stereopsis using the Lang 1 Stereotest: a pilot study. Binocular Vis. Strabismus Q. **16**(2), 91–98 (2001)
5. Bruder, G., Steinicke, F., Rothaus, K., Hinrichs, K.: Enhancing presence in head-mounted display environments by visual body feedback using head-mounted cameras. In: International Conference on CyberWorlds, pp. 43–50. IEEE (2009)
6. Cárdenas-Delgado, S., Juan, M.C., Méndez-López, M., Pérez-Hernández, E.: Could people with stereo-deficiencies have a rich 3d experience using HMDS? In: Lecture Notes in Computer Science, vol. 10513, pp. 97–116 (2017)
7. Cárdenas-Delgado, S., Loachamín-Valencia, M., Calderón, M.P.: VR-system ViPoS: VR system with visual and postural stimulation using hmds for assessment cybersickness in military personnel. In: Smart Innovation, Systems and Technologies, pp. 71–84. Springer (2018)
8. Donalek, C., Djorgovski, S.G., Cioc, A., Wang, A., Zhang, J., Lawler, E., Yeh, S., Mahabal, A., Graham, M., Drake, A., Davidoff, S., Norris, J.S., Longo, G.: Immersive and collaborative data visualization using virtual reality platforms. In: Proceedings International Conference on Big Data, pp. 609–614. IEEE (2014)

9. Fencott, C.: Towards a design methodology for virtual environments. In: Proceedings of the International Workshop on User Friendly Design of Virtual Environments (1999)
10. Hadjidimitrakis, K.: Coupling of head and hand movements during eye-head-hand coordination: there is more to reaching than meets eye. J. Neurophysiol. **123**(5), 1579–1582 (2020)
11. Helmreich, R.L.: On error management: lessons from aviation. BMJ **320**(7237), 781–785 (2000)
12. Helmreich, R.L., Merritt, A.C.: Culture at Work in Aviation and Medicine: National, Organizational and Professional Influences. Routledge (2019)
13. Howard, I.P., Rogers, B.J.: Binocular Vision and Stereopsis. Oxford University Press (2008)
14. Jeon, H.S., Choi, H.Y.: Binocular function test. In: Primary Eye Examination, pp. 71–82. Springer, Singapore (2019)
15. Koenig, S.T., Crucian, G.P., Dünser, A., Bartneck, C., Dalrymple-Alford, J.C.: Validity evaluation of a spatial memory task in virtual environments. Int. J. Des. Innov. Res. **6**(1), 1–13 (2011)
16. Korsgaard, D., Nilsson, N.C., Bjørner, T.: Immersive eating: evaluating the use of head-mounted displays for mixed reality meal sessions. In: Workshop on Everyday Virtual Reality, pp. 1–4. IEEE (2017)
17. Kushner, D.: Virtual reality's moment. IEEE Spectr. **51**(1), 34–37 (2014)
18. Lang, J.: Lang-Stereotest (2019). https://www.lang-stereotest.com/pages/video
19. McGill, M., Boland, D., Murray-Smith, R., Brewster, S.: A dose of reality. In: Conference on Human Factors in Computing Systems, pp. 2143–2152. ACM Press (2015)
20. Mon-Williams, M., Warm, J.P., Rushton, S.: Binocular vision in a virtual world: visual deficits following the wearing of a head-mounted display. Ophthalmic Physiol. Opt. **13**(4), 387–391 (1993)
21. Muñoz-Marrón, D.: Factores Humanos en Aviación: CRM (Crew Resource Management). Psychologist Papers **39**(3) (2018)
22. Nüssgens, Z., Czerwonka, B., Roggenkämper, P.: Examinations on the new lang test. Strabismus **1**(2), 69–73 (1993)
23. Raju, D., Eleswarapu, L., Saiv, R., Nath, V.: Study and design of smart embedded system for aviation system: a review. In: Nanoelectronics, Circuits and Communication Systems, pp. 573–590. Springer, Singapore (2019)
24. Rodríguez-Andrés, D., Cárdenas-Delgado, S., del Carmen Juan Lizandra, M., Pérez-Hernández, E., Méndez-López, M., Lluch, J.: Stereoscopic visualization systems: comparison between a large passive display and a head mounted display. In: Proceedings of International Joint Conference on Computer Vision, Imaging and Computer Graphics Theory and Applications. SciTePress (2015)
25. Saggio, G., Ferrari, M.: New trends in virtual reality visualization of 3D scenarios. In: Virtual Reality - Human Computer Interaction. InTech (2012)
26. Stidwill, D., Fletcher, R.: Normal Binocular Vision: Theory. Investigation and Practical Aspects. Wiley (2011)
27. Syrimi, M., Ali, N.: The role of stereopsis (three-dimensional vision) in dentistry: review of the current literature. Br. Dent. J. **218**(10), 597–598 (2015)
28. Tang, Y., Li, L., Wang, C., Chen, M., Feng, W., Zou, X., Huang, K.: Real-time detection of surface deformation and strain in recycled aggregate concrete-filled steel tubular columns via four-ocular vision. Robot. Comput. Integr. Manuf. **59**, 36–46 (2019)

29. Verghese, P., Tyson, T.L., Ghahghaei, S., Fletcher, D.C.: Depth perception and grasp in central field loss. Invest. Opthalmology Vis. Sci. **57**(3), 1476–1487 (2016)
30. Vishwanath, D.: Advancing a new theory of stereopsis: Reply to Rogers. Psychol. Rev**127**(1), 146–152 (2020)
31. Webber, A.L.: The functional impact of amblyopia. Clin. Exp. Optom. **101**(4), 443–450 (2018)
32. Witmer, B.G., Singer, M.J.: Measuring presence in virtual environments: a presence questionnaire. Presence **7**(3), 225–240 (1998)

Introducing Gamification in Professional Training

Henry Alvear[1], Hugo Arias-Flores[2], Carlos Ramos-Galarza[2,3], and Janio Jadán-Guerrero[2(✉)]

[1] Maestría en Educación Mención Innovación y Liderazgo Educativo - MEILE, Universidad Tecnológica Indoamérica, Av. Machala y Sabanilla, Quito, Ecuador
hfernando1088@hotmail.com

[2] Centro de Investigación en Mecatrónica y Sistemas Interactivos - MIST, Universidad Tecnológica Indoamérica, Av. Machala y Sabanilla, Quito, Ecuador
{hugoarias,janiojadan}@uti.edu.ec

[3] Facultad de Psicología, Pontificia Universidad Católica del Ecuador, Av. 12 de Octubre y Roca, Quito, Ecuador
caramos@puce.edu.ec

Abstract. The incorporation of game elements in non-game contexts is a subject of interest in various sectors, such as business, marketing, engineering, medicine, and military. Gamification and serious games have been involved in the rapid adaptation of professional training. The main goal of game thinking is to motivate and engage learners by using game-like techniques. This paper introduces gamification as an innovative approach for police officers to develop the required competencies for upholding the ethics, principles, and values that the Ecuadorian institution and society demand of them. The aim of gamification is to improve the quality and efficiency of the transfer from theory to real-life practice in police training. The paper focuses on the use of technology in a gamified virtual learning environment that simulates scenarios based upon common incidents reported to Ecuadorian police. The evaluation was conducted with 50 police officers. Our initial evaluation demonstrates that gamification has the potential to be scaled to force-wide proportions.

Keywords: Gamification · Serious games · Ludic strategies · Professional training · Police officers

1 Introduction

Technology plays an important role in different day-to-day activities, which is why, today, we speak of intelligent learning environments. These environments must adopt pedagogical approaches and intelligent technologies [1] that motivate users and improve their participation in activities, adding recreational elements according to the interactions of each user [2]. This leads to the following research question: How can we promote gamification in the training of professionals to enhance their commitment?

M. Botto-Tobar et al. (Eds.): CIT 2020, AISC 1326, pp. 343–353, 2021.
https://doi.org/10.1007/978-3-030-68080-0_26

This interaction (human–computer) modifies professional training to make the learning process more interactive as more concepts of intelligent education are introduced through gamification as an effective and attractive learning method [3]. This intelligent learning environment must promote the commitment of users in different fields of knowledge to foster meaningful learning [4, 5].

In these intelligent environments, technology is only one part in addition to the didactic strategy, pedagogy, processes, human talent, knowledge, attitudes, and behaviors [6]. In this sense, a design with a philosophical approach helps to improve motivation and enhances collaborative learning in a gamified environment, since participants present a better perception, unlike environments that maintain a unique approach [7]. Establishing the role of the player and elements of the game according to an individual motivational strategy and preferences of the player in the gamification design can enable the sensitization and internalization of competencies that are established against different doctrinal cultures to promote commitment. This commitment depends on the participant's motivation, which relies on an innovative and creative environment, since professionals learn in a dynamic way. Needs, interests, experiences, and preferences are aspects to consider for group cohesion and performance in the training process.

The needs of both users and organizations can be implemented strategically in the dynamics of digital environments, thus providing training opportunities to improve work commitment and transform the business with game solutions in work situations. Sophisticated applications of gamification have been used in areas such as financial services on business websites, employee training, business intelligence, market research, electronic commerce, customer relations, among others [8].

A changing world demands that organizations stay competitive and provide rewarding experiences to their customers. Companies in almost all industries must continually foster commitment among their employees and maintain digital products and services as well as processes that support those products and services. Traditional organizations must embrace this modernity to design and manage a business architecture that enables continuous change and a modular design of business capabilities as well as the technologies behind them [9].

The rest of this document is organized as follows: Sect. 2 discusses related work; Sect. 3 outlines the research method; Sect. 4 presents the design of the platform; Sect. 5 provides the results and discussion; and Sect. 6 offers the conclusions and suggestions for future work.

2 Related Work

Gamification has two objectives: (1) To make users play the game to derive extrinsic value (e.g., simulation of scenarios with some activities or missions) and (2) to increase intrinsic value by achieving the desired result (e.g., increasing the user's skills or knowledge). Understanding these methods and how they are related is the basis for creating effective training tools that motivate learning and user participation [1]. The commitment seen from the cognitive and motivational perspectives through learning and gamification seeks to create interactive experiences in the participants through the game.

In the context of health education, two games were developed to guide the training of professionals and the empowerment of patients. The first game simulated clinical cases and was aimed at training professionals in treatment protocols and diagnostic procedures. The game featured a problem-solving approach for saving the patient through diagnosis. The second game, called Edu game, was aimed at caregivers and patients with a mission to control the disease and promote the adoption of a healthy lifestyle and avoid hospitalization. The game involved role play using an avatar. The results show that patients and caregivers increased their motivation to acquire knowledge (mean value of 0.55), and for professionals, the value was 0.3. Regarding the knowledge acquired in the two groups, the effectiveness value was minimal—$K = 0.25$ for patients and caregivers and $K = 0.15$ for professional workers. Thus, the researchers concluded that the results are promising and advocated experiments with a larger sample [10].

In reference to this topic, an investigation carried out with nurses from the intensive care unit of a hospital sought to raise awareness about compliance with hand hygiene, individual behavior changes, and performance optimization through a gamification solution that collected data and provided feedback in real time in a fun and engaging way. A design science research methodology was used following a development cycle made up of six activities. The results are promising for improving the awareness of nurses. The participating nurses enjoyed the concept and felt that it provided a unique opportunity for feedback on their performance [11].

An investigation carried out in the public sector, in which a game called RE-PROVO was developed to evaluate the discussion of requirements along the themes of risk aversion and innovation, sought to encourage participation and creativity in the analysis of business requirements during the replacement projects of legacy systems. The results show that the way in which the game was introduced in the two participating entities affected active participation. In terms of creativity and engagement, the results highlight the paramount importance of anonymity and competence. Creativity and commitment increase when organizational hierarchies do not impede discussion; ideas are presented in a competition setting and there is immediate feedback on the actions of participants; and feedback is given through rewards or badges or by informing players of the reactions of others to their actions. These serious gaming techniques show the greatest promise in reinforcing the commitment and creativity required to deal with an inherited problem [12].

In the field of human resources and hiring, a concept test was carried out, in which managerial skills were analyzed using the Civilization strategy game, to explore whether strategy video games are indicative of managerial skills. Forty business students participated. Students who scored high in the game had better problem-solving and organization and planning skills than students who scored low. Results of the preliminary analysis of the data suggest that strategy games, such as Civilization, can be used to assess problem-solving skills and organizational and planning skills—skills that are highly relevant to managerial professions [13].

The integration of gamification and learning was presented in applied research in the training of engineers through the development of a learning platform for a biometric technologies course. This platform was based on facial recognition and followed the standard process of facial recognition systems. It combined elements of standard gamification with role-playing games, allowing students to find themselves in the roles of both biometric engineers and attackers of the system. The results show an improvement in student motivation and learning outcomes. In addition, teamwork and competition were important factors in the investigation. The results underline the significant impact that the gamification approach has on improving student motivation in the learning process [14].

In the design of a prototype, the training process for launching a lifeboat was imitated. The game used 3D techniques with the support of virtual reality equipment involving both tractor maneuvers and boat operations. The main objective of the game was to understand the sequence of operations and manage the potential risks that can occur during the launch process. Two case studies were conducted for the two purposes of the game: One for training volunteers without launching experience and the other for sensitizing the public to potential dangers with young children. The results show that people learned better when they performed the activities in the game, unlike those who only read the launch instruction manual. The game holds promise for professional training and serves the educational purpose of raising awareness of safety issues for the general public [15].

Recognizing the contribution of new technological trends to the training of professionals through a gamified environment can be a good motivator for acquiring new skills. The entertainment dimension highlights the emotional experience of the user, increasing the attractiveness of learning regardless of the field in which it takes place. For these reasons, the present study aimed to implement a game-based platform as a tool for training police officers.

3 Methodology

The study focused on an environment based on gamification and role play for the training of police officers through the implementation of a virtual platform for the Police Doctrine course. It is important to note that this tool allows for delivering more effective training in a significantly reduced amount of time to more staff and without the need for long, extensive sessions in the classroom. Improving a set of knowledge, principles, and values in an orderly and systematic way will guarantee discipline, respect, and work, which will serve as a well constructed didactic and instructional order oriented to the ideology of being police and supporting the daily activity of a police officer.

3.1 Participants

The population targeted to evaluate the platform comprised 50 individuals in the police department, including managers and operational technicians. Participants ranged in age from 21 to 50 years, with a mean age of 32 years. They belonged to different police units involving the same conceptualization of what would regulate their police activity. Two groups of 25 participants were randomly generated—one participating in traditional training and the other participating in the traditional training and the gamified environment.

3.2 Materials and Instruments

To develop the technological resource, and in compliance with this article, the document titled "Institutional Doctrine Module" (IDM) [16] was used. The technology used to design a gamified virtual learning environment was the Moodle platform integrated with external web 2.0 resources, such as Genially.

3.3 Procedure

The study involved testing an interactive educational resource called "Internalization of My Institutional Doctrine," a digital platform designed to develop meaningful learning that police officers internalize so that it becomes part of their being. The platform is based on a gamified virtual learning environment comprising different learning modules with interactive resources, activities, and evaluation tools. To evaluate this platform, a group of 25 police officers agreed to participate in the training using this online methodology. The evaluation considered various cognitive, procedural, and attitudinal aspects that are necessary in the daily life of a police officer as well as the acquisition of new proposals combining classroom teaching resources with technology. The first phase consisted of the design of the learning strategy—that is, the creation of activities for participants to perform during the training. The second phase consisted of the development of the educational strategy in the gamified virtual learning environment. The rules, mechanics, and objectives of the scenario were revealed during the training. Finally, in the evaluation phase, a survey based on a 5-point Likert scale was applied to ask each police officer about their experience and knowledge of IDM.

4 Design of the Gamification Platform

The platform is aimed at executives and operational technicians in the police department. The learning objective is to understand how staff can be motivated to respect the basic rules of the institution, coexistence in work environments, and how to minimize critical situations. Different scenarios are proposed for each context: police identity, legal basis, principles, and values. The platform was designed as a single player role-playing game and developed in Moodle on the basis of the problem-solving methodology. The user's mission is to give the correct answers to solve the problem posed.

The game begins with an explanation of the rules and how the interactions develop, as shown in Fig. 1.

The avatar gives an introduction to the objectives and scenarios involved in the application of the game. The game presents a menu screen displaying the activity options. These options are configured in graphic elements of the scenario, as will be seen later in the Scenarios section.

4.1 Objectives and Rules

The most complex step in gamification is transforming the learning objective into gamified activities. The main learning objective was to identify the most relevant aspects of IDM. In this sense, scenarios were designed with situations in which the police officers would need to make decisions. Gamified activities would be evaluated, and three objectives were defined to evaluate the gamification components:

Fig. 1. Avatar giving instructions

- Objective 1: Evaluate the motivation of users to acquire knowledge.
- Objective 2: Evaluate the knowledge acquired by users after using the game.
- Objective 3: Evaluate user satisfaction with the use of the platform.

4.2 Scenarios

The designed scenarios represent different situations that can arise in general in the work environment to cover most of the common problems that the police face. For each game interaction, the avatar describes a question or problem to the player. For example, in the main scene of Fig. 2, the player can download the IDM document. In addition, the avatar shows a video with the main synthesized aspects of the doctrine.

Fig. 2. Main scenario of the platform

Each player can visualize their responses and evaluate which rules they broke, according to IDM, using synthesized feedback. The three categories (actions, analysis, and recommendations) of possible responses are shown on the same screen. Next, the player must select a category of answers. For each category, the police officer must choose the correct answer. The available answers are defined according to the competencies and abilities that the player must acquire. In the following example, in the Actions category, a scenario of a city is presented (see Fig. 3).

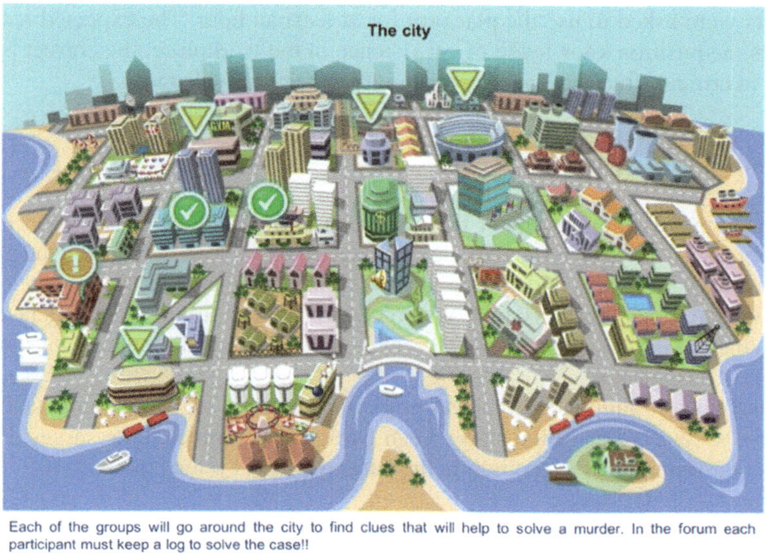

Fig. 3. Scenario of the city

Participants had to find clues to solve the problem posed. The scenario was designed in Genially, in which there are some hidden objects in different places in the city.

By correctly answering questions, the participant earns points and receives clues. This record is used to generate indirect leaderboards and monitor the players' progress. In addition, each player has a representation of the number of lives with the concept of gamification. When the player provides an incorrect answer, they lose a life. If they lose all three of their lives, the player can no longer play unless their coach authorizes it. The strategy is intended to monitor the player and measure their progress, so the coach can assess whether the strategy is correct or seek differentiated initiatives to support the player. When the game is over, a leaderboard is presented showing the final score. A user can share scores with other users and compete with each other (Competition). A range of the best scores is also presented to generate challenges (Social dimension).

4.3 Pilot Study

The application was evaluated through a user test aimed at measuring user satisfaction, motivation, and learning gain within a real context of use. Due to the limitations generated by the sensitive context, only a user test was performed on a very small sample.

To measure motivation (Objective 1) and knowledge acquired (Objective 2), a pre-test and post-test were conducted, respectively. To assess user satisfaction with the solutions proposed in Objective 3, a 10-item questionnaire was used. Using a Likert scale, the items measured system interaction, learning content, terminology, and appearance.

Users were asked to use the platform for at least an hour. The expected result was the users' acquisition knowledge of coexistence in the workplace, the correct behavior to react in critical situations, and motivation to acquire knowledge.

5 Results and Impact

To evaluate the learning results of the module, a knowledge test was carried out. The mean score was 81.2 with a standard deviation of 10.042 for participants in the gamified environment. Participants undergoing the traditional training had a mean score of 73.65 with a standard deviation of 10.674. We proceeded with a t test of independent samples and obtained a significance below 0.05, evincing a statistically significant difference between the two groups.

To evaluate the gamified virtual learning environment, a survey was conducted at the end of the second phase of the study. Each participant completed the survey. Figure 4 shows a question related to promoting the participation of police officers.

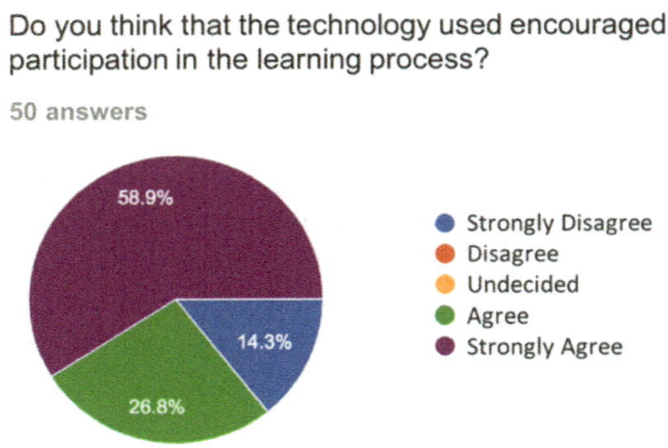

Fig. 4. Question related to the police officers' encouragement to participate

The results show that it was possible to generate interaction and motivation in the participants, since the mean value of the motivational questionnaire for the gamified group was 5.5917. The mean value for the traditional group was 4.918. In general, the participants accepted the gamified environment, thus identifying that gamification is one of the best techniques applied in the design of online training courses with a gamified and motivating methodological base.

To determine if the use of metaphor was effective, participants were asked whether the narration of the metaphor was helpful and provided a rewarding experience. The results in Fig. 5 are encouraging, since 70.9% strongly agreed that the metaphor was useful and generated interest and curiosity. Furthermore, similar to the rest of the participants (29.1%), they also answered that they agree with the use of metaphor. These statistics show that participants identified with the simulated setting and the narrative that motivated the exploration.

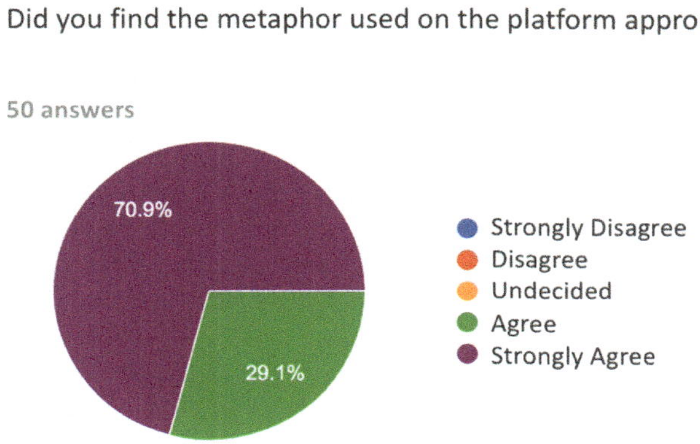

Fig. 5. Question related to metaphor and digital storytelling

Collaborative learning is another benefit of the gamification technique that involves a group of participants working together to impact learning in a positive way. In this sense, the questionnaire included whether the activities carried out collaboratively helped the participants' development. The results show that 58.9% of participants strongly agreed that gamification encourages collaborative work. Moreover, 41.1% considered it suitable for a collaborative environment (see Fig. 6).

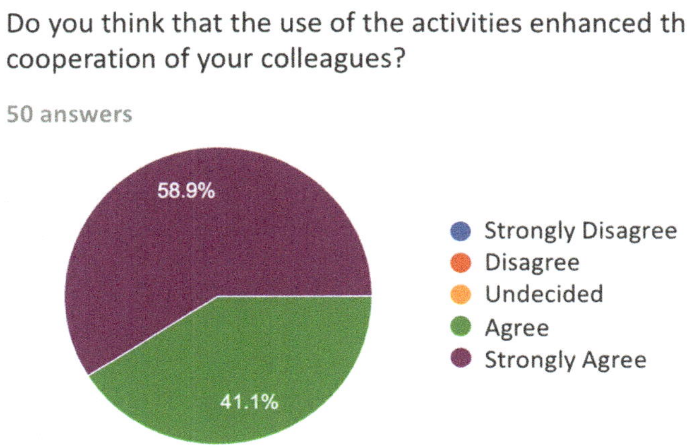

Fig. 6. Question related to collaborative learning

The results indicate that the police officers appropriated the knowledge of IDM based on the different scenarios experienced on the platform and the post-test evaluation. Consequently, there is potential to escalate the application of gamification techniques in the training process at all job levels. Gamification allows the acquisition of complex or complicated concepts through practice and repetition as well as opportunities to make and learn from mistakes. The possibility of going back and repeating establishes the user's own rhythm, which leads to the acquisition of knowledge.

6 Conclusions

The use of technology in the teaching/learning process through gamification is highly feasible, since meaningful learning in apprentices emphasizes active participation through pedagogies that include play as the primary basis of the acquisition of new knowledge, generating skills through the relationship with technology employed as a training tool. Interactive learning environments present an opportunity to evolve the teaching process through the incorporation of game elements that have been shown to attract users' attention, motivate users to achieve objectives, and promote competition, teamwork, and communication.

The analysis of the results verified and validated that the use of technologies resources and gamification through the teaching and learning of police officers will allow the officers to internalize and improve the acquisition and development of competencies and skills, thereby increasing the motivation of those trained in different police educational settings. The implementation of this research will allow for advancing innovative proposals that will generate the construction of knowledge, provide police officers with meaningful, cooperative, and dynamic situations, which make possible the investigation, transmission, and elaboration of information, and learning by discovery, thus improving educational quality.

In this sense, we believe that these tools should be elements that allow and promote continuous interaction between permanent training within the police institution. With all this, the main challenge that we face is tackling a methodological change in the educational field that departs from traditional methodologies based on the memorization and forgetting of contents.

Regarding study limitations, it should be emphasized that the sample was very small, which prevents the generalization of our results. Future research should involve carrying out such training with a larger number of participants.

Expressions of Gratitude. The authors thank Universidad Tecnológica Indoamérica for financing this study through the project "Impacto en la Cognición y Comportamiento Humano dentro de la Interacción con la Tecnología - ICCHIT" and the staff of the Ecuadorian Police.

References

1. Redhead, A., Saunders, J.: Gamification and simulation. In: Akhgar, B. (ed.) Serious Games for Enhancing Law Enforcement Agencies. Security Informatics and Law Enforcement. Springer, Cham (2019). https://doi.org/10.1007/978-3-030-29926-2_5H1.

2. Monterrat, B., Lavoué, É., George, S.: Toward an adaptive gamification system for learning environments. In: Zvacek, S., Restivo, M., Uhomoibhi, J., Helfert, M. (eds.) Computer Supported Education. CSEDU 2014. Communications in Computer and Information Science, vol. 510. Springer, Cham (2015). https://doi.org/10.1007/978-3-319-25768-6_8
3. Rutkauskiene D., Gudoniene D., Maskeliunas R., Blazauskas T.: The Gamification Model for E-Learning Participants Engagement. In: Uskov V., Howlett R., Jain L. (eds.) Smart Education and e-Learning 2016. Smart Innovation, Systems and Technologies, vol. 59. Springer, Cham (2016). https://doi.org/https://doi.org/10.1007/978-3-319-39690-3_26
4. Zyngier, D.: (Re) conceptualising student engagement: doing education not doing time. Teach. Teacher Educ. **24**(7), 1765–1776 (2008)
5. Espín-Álvarez, E., Freire-Muñoz, I.: Relación entre el uso de internet para el entretenimiento y el aprendizaje escolar en estudiantes adolescentes del Ecuador. CienciAmérica **8**(1), 138–157 (2019). https://doi.org/10.33210/ca.v8i1.209.
6. Arias-Flores, H., Jadán-Guerrero, j., Gómez-Luna, L.: Innovación Educativa en el Aula mediante Design Thinking y Game Thinking. Hamut´ay, **6**(1), 82–95 (2019)
7. Challco, G.C., Bittencourt, I.I., Isotani, S.: Can ontologies support the gamification of scripted collaborative learning sessions? In: Bittencourt, I., Cukurova, M., Muldner, K., Luckin, R., Millán, E. (eds.) Artificial Intelligence in Education. AIED 2020. Lecture Notes in Computer Science, vol. 12163. Springer, Cham (2020). https://doi.org/10.1007/978-3-030-52237-7_7
8. Chen, E.T.: The gamification as a resourceful tool to improve work performance. In: Reiners T., Wood, L. (eds.) Gamification in Education and Business. Springer, Cham (2015). https://doi.org/10.1007/978-3-319-10208-5_24
9. Bossert, O., Feldmann, S.: Perpetual evolution—rethinking the way digital transformations are managed. In: Zimmermann, A., Schmidt, R., Jain, L. (eds.) Architecting the Digital Transformation. Intelligent Systems Reference Library, vol. 188. Springer, Cham (2021). https://doi.org/10.1007/978-3-030-49640-1_3
10. Pesare, E., Roselli, T., Corriero, N., et al.: Game-based learning and Gamification to promote engagement and motivation in medical learning contexts. Smart Learn. Environ. **3**, 5 (2016). https://doi.org/10.1186/s40561-016-0028-0
11. Marques, R., Gregório, J., Pinheiro, F., et al.: How can information systems provide support to nurses' hand hygiene performance? Using gamification and indoor location to improve hand hygiene awareness and reduce hospital infections. BMC Med. Inform. Decis. Mak. **17**, 15 (2017). https://doi.org/10.1186/s12911-017-0410-z
12. Alexandrova, A., Rapanotti, L.: Requirements analysis gamification in legacy system replacement projects. Requirements Eng. **25**, 131–151 (2020). https://doi.org/10.1007/s00766-019-00311-2
13. Simons, A., Wohlgenannt, I., Weinmann, M., et al.: Good gamers, good managers? A proof-of-concept study with Sid Meier's civilization. Rev. Manage. Sci. (2020). https://doi.org/10.1007/s11846-020-00378-0
14. Milenković, I., Šošević, U., Simić, D., et al.: Improving student engagement in a biometric classroom: the contribution of gamification. Univ. Access Inf. Soc. **18**, 523–532 (2019). https://doi.org/10.1007/s10209-019-00676-9
15. Jiang, M., Lan, W., Chang, J., et al.: A game prototype for understanding the safety issues of a lifeboat launch. Virtual Reality **22**, 137–148 (2018). https://doi.org/10.1007/s10055-018-0334-7
16. Policía nacional del Ecuador: Módulo de doctrina institucional. Dirección nacional de educación, Departamento de doctrina (2018)

Identifying Technological Alternatives Focused on Early Alert or Detection of Forest Fires: Results Derived from an Empirical Study

Efraín R. Fonseca C.(✉), Diego Marcillo, Santiago P. Jácome-Guerrero, Tatiana Gualotuña, and Henry Cruz

Universidad de las Fuerzas Armadas ESPE, Ambato s/n y Av. General Rumiñahui, Sangolqui, Ecuador
{erfonseca,dmmarcillo,psjacome,tmgualotunia,hocruz}@espe.edu.ec
https://www.espe.edu.ec

Abstract. *Context:* Forests are home to 80% of terrestrial life; however, they have been seriously threatened as millions of its hectares are lost each year due to forest fires. *Aim:* To elicit the requirements, methods, techniques, technological elements, situations, and specifications to structure a current, compact, generic, and low-cost technological model that allows early alert or detection of forest fires and provide timely support in decision-making. *Method:* We carried out a Preliminary Literature Review (PLR) on Scopus digital database strictly following the guidelines for systematic literature reviews proposed by B. Kitchenham with some improvements. *Results:* The PLR process obtained 15 primary studies as of July 2020, which highlight emerging technologies to address the forest fires' issues. *Conclusions:* Forest fires represent an unprecedented danger to life, so there is a broad range of possibilities for their early detection since time ago. Emerging technologies give the science new alternatives to develop compact, economic, accessible, and functional proposals for early warning about this phenomenon.

Keywords: Forest fires · Emerging technologies · Early alert · Preliminary literature review

1 Introduction

Forests are home to 80% of terrestrial life, distributed in an abundant variety of creatures that play a crucial role in sustaining life, allowing trees and other essential vegetal species to grow. Sometimes they are even inadvertently the heroes of the life circle who keep the planet healthy. However, this biodiversity is seriously threatened, as millions of hectares of forests are lost each year due to forest fires (FF) [8].

M. Botto-Tobar et al. (Eds.): CIT 2020, AISC 1326, pp. 354–368, 2021.
https://doi.org/10.1007/978-3-030-68080-0_27

According to the World Health Organization (WHO)[1], the FF represent a recurrent phenomenon caused by humans and nature. Although it is not possible to ascertain FF's origin, human negligence is the most frequent cause. The effects of the FF phenomenon are generally varied and devastating since people and animals die, and resources and the environment are destroyed. Likewise, there are collateral effects due to the FF, among which are traffic, air, and maritime accidents; damage to humans, animals, and vegetation due to smoke, in places near and even far from the disaster area; and high concentrations of other gases and particles in the air for prolonged periods, which pollute the environment.

Nevertheless, the WHO also indicates that FFs are predictable and avoidable in many cases by applying appropriate policies that address different factors that influence their origin, such as:

- Dryness of vegetation.
- Climatic variations that influence the spread and intensity of the FF.
- Inappropriate use of fuels.
- Prolonged droughts, and
- Lack of early warning systems.

We will focus on the lack of early warning systems for FF in this study, for which we carried out a Preliminary Literature Review (PLR). The results obtained from the research made it possible to structure a hybrid technological model that prevents potential FFs, and alerts early on their start, for timely decision-making. More specifically, the proposed model is characterized by the following elements:

- Wireless sensors network to monitor environmental parameters and the forest as such.
- Centralized public repository to store data from the wireless sensor network.
- Alternative long-distance communication network to transmit the information from the wireless sensor network.
- Unmanned aerial vehicles outfitted with sensors to corroborate the data obtained from the wireless sensor network.

The remainder of the article is organized as follows. In Sect. 2, related works to early warning of forest fires are analyzed. Section 3 details the preliminary literature review process performed. Next, Sect. 4 presents the results obtained after the research, which provides answers to the research questions. Section 5 details the model obtained after the synthesis of results. Finally, Sect. 6 reports the conclusions and future work.

2 Background and Related Works

Forests involve a life context for many animal and plant species; likewise, they provide oxygen, purify water, enrich the soil, and regulate climate [28]. However, several factors can cause the destruction of forests, such as the agricultural frontier's expansion, deforestation policies, and forest fires (FF) [23].

[1] Source: https://www.who.int/hac/techguidance/ems/vegetation_fires/es/.

FF implies a potential threat to living beings, the environment, and resources. FFs are forecast to destroy half of the world's forests by 2030 [27].

This problem has aroused great interest in the scientific community, which is visible in several proposals focused on addressing FF's problem. For example, there are works focused on early alerts of FFs based on Satellite Remote Sensing (SRS), Wireless Sensor Network (WSN), Internet of Things (IoT), Artificial Intelligence (AI), Unmanned Aerial Vehicle (UAV), Big Data Analytics (BDA); likewise, there are proposals based on different network architectures, routing algorithms and protocols.

Below we show an analysis of the technological trend of solutions focused on FF, which motivates our research giving us different alternatives.

Khan et al. [15] studied emerging technologies applied to natural disasters. The authors determine the need to early detect an FF disaster based on monitoring systems using different technologies such as SRS, WSN, IoT, Mobile AdHoc Network (MANET), AI, UAV, and BDA. The solution proposed includes architectures for detection and monitoring scheme based on WSN and UAV.

Yuan et al. [30] conducted a review of a conceptual system's development and architecture of unmanned aerial vehicles for monitoring, detection, and fighting against forest fires. The authors determined the essential elements and functions of a general UAV-based forest fire surveillance and fighting system: (1) monitoring (identifying a potential fire), (2) detection (activating an alarm to inform firefighting operators o initialize additional diagnostics and forecasts), (3) diagnostics (determine the location and extent of the fire and track its evolution) and (4) forecast (predict future fire evolution based on fire extinguishing conditions and wind in real-time). A single UAV performs those functions with different sensors and a central ground station for monitoring, which could mean excessive battery charge.

Sendra et al. [24] propose a system to assess the risk level of a forest fire and the presence of one in rural areas, based on a low-cost network architecture with Long Range technology (LoRa). This system includes several nodes with sensors to acquire data and measure temperature, relative humidity, wind speed, and CO2 in the environment. The data is stored and processed on a The Things Network (TTN) server and presented through graphic web visualization.

Jin et al. [12] propose a method of forest fire detection combining image processing with machine learning on video sequences. The method consists of three functionalities: (1) Detection of moving objects to separate the foreground from the background of the video images and thus significantly reduce the operation time of forest fire detection and improve detection accuracy; (2) extraction of characteristics in two steps: in the first instance, the color is processed and segmented to filter irrelevant background information, and then the characteristics of color, texture, and shape of the image are extracted to perform changes of images from high to low dimension; and (3) classification of the extracted characteristics to place them in high or low dimension (through these classifiers, the image can be judged as a flame, and consequently, the algorithm will give a fire alarm).

Fan et al. [7] propose a system of smoke detection and monitoring, based on a rotating camera that takes panoramic backgrounds. This system consists of four modules: (1) Video frame location module through an algorithm that accurately and quickly locates video frames on a panoramic background, (2) foreground extraction module through algorithm optimization of real-time extract, (3) smoke detection module by optimizing real-time algorithms merged with the previous stage to improve efficiency and stability, and (4) background update.

Serna et al. [25] propose a system of disaster management focused on forest fire extinguishing operations. The authors based their proposal on the Equipment Destined for Orientation and Safety (EIDOS) system based on a large and dense WSN composed of random nodes represented by UAVs, which fly over the surroundings of the area affected by a forest fire. The system provides critical information to improve the safety of the people who will act on the disaster. The network nodes process the data collected in a distributed and collaborative way to track the fire's position and current limits, which is notified to the population.

After analyzing the related literature, we found that most of the proposals focus on early detection of forest fires and base their solutions on emerging technologies in various ways, emphasizing WSN, IoT, and UAV, which motivated us to establish the following research objective.

Elicit the requirements, method, techniques, technological elements, situations, and specifications to structure a current, compact, generic, and low-cost technological model that allows early warning of forest fires and provide timely support in decision-making.

We posed four research questions (RQs) to accurately define the scope of this study:

RQ1: What technologies are used to predict and give early alerts of forest fires?

RQ2: What systems are used to predict and give early alerts of forest fires?

RQ3: Under what criteria can a technological platform be designed to predict and give early alerts of forest fires?

RQ4: What is the frequency with which proposals of early alerts and prediction of forest fires have been submitted?

3 Research Method

We carried out a rigorous Preliminary Literature Review (PLR) in order to achieve the stated research objective, based on the protocol for Systematic Literature Reviews (SLR) proposed by Kitchenham et al. [16], but with certain modifications aimed at improvement. Our research protocol includes (1) Approach to the review, (2) planning the review, (3) Studies selection, (4) Quality evaluation, (5) information extraction, and (6) synthesis of results. We present phases 5 and 6 in different sections, given their magnitude and significance.

3.1 Approach to the Review

The review protocol was structured based on the objective and the research questions (RQs) detailed in Sects. 2. SLRs propose RQs generally in PICOC format [21], which could be a complication, depending on the approach of the RQs. In this case, expressing RQs in PICOC format is not possible. PICOC is defined in the SLRs' framework to compare two treatments, such as applying penicillin against a placebo on treating an infection. In this study, we look for specific information in particular articles; therefore, propose PICOC questions does not make sense.

3.2 Planning the Review

We planned the review through a search strategy that includes: (1) Establishment of inclusion and exclusion criteria, (2) Shaping a control group, and (3) Composing the search string.

Inclusion and Exclusion Criteria. We consider the establishment of inclusion and exclusion criteria essential since it is necessary to explicitly make the characteristics of the studies that we wish to identify and those of studies that we do not require. We also used such criteria to structure the control group. It is necessary to clarify that we settled down a *group of 3 researchers* (three full-time Ph.D. professors) to avoid bias when making decisions. For example, to establish the inclusion and exclusion criteria and the rest of the review process.

Inclusion Criteria

- Current studies focused on improving the forests and their surroundings' security, based on the acquisition or data processing for the forest fires' control.

- Current studies that present low-cost proposals of models or technological frameworks that are easy to implement; mainly focused on the forest fires' control.
- Current studies focused on the early prediction of potential forest fires, based on the information and communications technologies.
- Current studies focused on the early discovery of forest fires, based on the information and communications technologies.

Exclusion Criteria

- Studies that refer only to merely technological aspects oriented to implementation, but not to improve forest fires control.
- Studies whose purpose is to improve the forest ecosystem, but not the control of forest fires.
- Studies that present proposals for collateral damage from forest fires, but not alerts about them.
- Articles published before 2015 or published in unrepresentative conferences or magazines, or written in languages distinct from English.

Control Group. The control group (CG) is a set of studies characterized by being intimately related to the review's aim and strictly responding to the inclusion criteria. We used the CG as the source of terms of which emerged the search string.

The control group's shaping starts with a literature review performed by the researchers. Each researcher proposes papers that strictly comply with inclusion criteria and that cam from reliable sources. Afterward, we made a focus group discussion to validate the researchers' proposals and forming the CG. More specifically, we shape a CG of *5 studies*.

Search String. The search string was structured based on a detailed review of the CG studies carried out by each researcher. Through this activity, we identify and extract the most relevant terms related to the review's objective. Afterward, we organize a discussion group to categorize the terms by their relevance and frequency in each study and between studies.

The categorization of terms outlined *3 categories*: Problem, purpose, and technology. We selected the search string showed below after an iterative trial-and-error piloting on the SCOPUS digital database[2] with the most relevant terms in each category:

(("forest fires" OR "wildfires" OR "flames") AND ("early detection" OR "early alert") AND ("emerging technologies" OR "technological breakthroughs"))

3.3 Selection of Studies

The selection process includes searching for: (1) Candidate studies, (2) Relevant studies, and (3) Primary studies.

[2] Source: https://www.scopus.com/home.uri.

Candidate Studies. We got the candidate studies during the search string definition process. The selected search string obtained *79 candidate studies.*

Relevant Studies. The process of identifying relevant studies involves applying inclusion and exclusion criteria to the candidate studies at the title, abstract, keywords, and introduction level, by each researcher.

Afterward, we established a discussion group to settle the discrepancies between the researchers who made the selection. After the discussion, a consensus was reached, and *18 relevant studies* were selected.

Primary Studies. The selection process of primary studies starts downloading the full text of all *relevant studies.* Next, each researcher reviews and applies inclusion and exclusion criteria in each relevant study's full text. Finally, the researchers in consensus selected *15 primary studies.*

We have omitted process graphs and summary tables related to the findings due to a lack of space. You can visit the following *link* if more detail is required.

3.4 Quality Evaluation

The studies' selection process includes activities that could be vulnerable to potential threats to validity; however, we added contingency plans in the most vulnerable activities to mitigate the impact on our research's validity. The main threats to validity recognized in this type of study are (1) literature coverage, (2) bibliographic sources coverage, and (3) reliability on the studies' selection. The threats and plans in place are detailed below.

Literature Coverage. The literature coverage includes determining the inclusion and exclusion criteria, shaping the control group, and structuring the search string. We added the following quality elements to guarantee the validity of such activities: Guide from an expert on the research topic and in conducting literature reviews; reliable sources of information, that is, articles from high-impact journals or conferences; the consensus in decision making; constant approach towards the aim of the investigation. We found that the control group's composition supports the origin of the terms used in the search string and, therefore, the literature's coverage, based on the activities carried out.

Bibliographic Sources Coverage. We consider articles from journals and conferences, books, and book chapters to obtain the highest coverage of BDD sources. We also considered a digital database where works related to our research topic are usually indexed or published. In this case, we used SCOPUS for our research.

Studies Selection Quality. The selection process has a potential threat to validity in terms of bias that researchers could represent. In this research, there were three stages of selection: (1) Selection of candidate studies, (2) Selection of relevant studies, and (3) Selection of primary studies.

We minimized the threat of bias in selecting candidate studies starting from a structured search string with relevant terms obtained exclusively from the CG studies; that is, our search string does not contain improvised terms suggested by some researchers. Therefore, the candidate studies do not respond to any bias and are those obtained from the search string used in the SCOPUS digital database.

Regarding the threat of bias in selecting relevant studies, it was minimized by performing a systematic analysis and candidate studies filtering. The cross-validation of the candidate studies' reviews guided by an expert in conducting PLR reached 83% agreement between the researchers.

To complement the cross-validation of relevant studies, we performed another quality analysis using the statistical method Cohen's Kappa coefficient (k) [19], with which we measure the degree of agreement between two criteria. According to this method, k values higher than 0.75 show excellent concordance; while, values less than 0.40 indicate little agreement. We obtained $k = 0.57$, which indicates an acceptable agreement between the two researchers.

We based the quality of the primary studies selection process on the guidelines proposed by Kitchenham et al. [16], which consider nine quality criteria to evaluate each study. Two researchers individually reviewed each primary study with such criteria and the guidance of an expert performing PLRs. Finally, the researchers reached a consensus in the selected primary studies.

4 Results

The process of extracting information from the primary studies, based on their exhaustive reading, allowed us to answer the research questions. We describe in detail the answer given to each RQ below.

RQ1: What technologies are used to predict and give early alerts of forest fires?

We identified different categories of technological elements used in forest fires detection. Some proposals recommend its integration. Each category identified is described below.

Satellite Remote Sensing: Satellite remote sensing is a valuable tool when natural disasters occur since it allows measuring the impact of the disaster over a large geographic area. High-resolution spatial images provide critical information for proper disaster management [5,6,22].

Unmanned Aerial Vehicles (UAV): Such devices can capture the entire disaster environment from a privileged perspective to route the actions into being carried out, such as the collection of information, the location of the victims, and the optimization of the rescue [4,9,20].

Internet of Things (IoT): It is a technological paradigm used worldwide to interconnect a varied type of device that provides a particular service. Among other things, this paradigm can offer solutions for real-time monitoring of possible disaster scenarios [3,14,32]. Despite the devices' limitations, such as severe operating conditions, or the cost of maintenance and implementation, IoT is highly adaptable and integrable [11]; for example, with web/mobile technologies, Artificial Intelligence (AI) techniques, machine learning, and Big Data analysis.

Big Data: It is a field that allows real-time analysis of large volumes of data obtained from different forest fire detection and monitoring devices. The analysis results can facilitate decision-making in order to manage the disaster. Some researchers use AI algorithms to perform data analysis, which generates predictive and prescriptive models to extract patterns from possible forest fire scenarios. Kansal et al. [14] point out some machine learning techniques used in the prediction of forest fires: regression techniques, neural networks, SVM, decision trees, among others.

Alternative Communication Networks: In general, when disaster beats, traditional communication networks are disrupted, as these too can be affected. Therefore, it is essential to use alternative networks that allow maintaining the operation and connectivity between the different devices in these scenarios. Among the most used communication technologies, we can point out: WSN, LoRa, Ad hoc, ZigBee, SigFox, NB-IoT, LTE, WiFi, Infrared, WiMax, GSM, Satellite [15,24,31].

Wireless Sensor Network (WSN): WSN is a widely used network for forest fire monitoring and detection through short-range, low-power, and low-cost nodes. In general, the nodes are deployed at random and collaborate to carry out a particular action. Each node can have calculation, detection, and communication capacity through sensors [1,18]. Forest fires monitoring and detection are carried out by detecting parameters such as humidity, temperature, rainfall, smoke, wind speed, solar radiation, CO2, and other gases. The main challenges surrounding the use of WSNs in forest fires detection are determining the proper location of nodes, power consumption, synchronization, and setting the temperature threshold for activating the sensors (a situation usually solved using fuzzy logic).

Mobile Ad-Hoc Networks (MANET): A decentralized network that allows the use of active, simple, and economical techniques for monitoring and detecting forest fires. MANET is a wireless network structured through multiple interconnected mobile nodes, which can move independently [10].

As can be seen, the technological elements used in the proposals focused on the detection and monitoring of forest fires are varied and characterized by their permanent evolution. However, we noted that the most significant efforts in recent years aim to develop technologies related to IoT, Artificial Intelligence, Big Data, and lower-cost emerging communications networks.

RQ2: What systems are used to predict and give early alerts of forest fires?

Several authors proposed forest fires detection systems around the technological elements identified to give a response to RQ1, some more complex than others. Below, we describe some of these systems.

Kang et al. [13] propose a model based on edge computing for intelligent data acquisition generated by sensors in an IoT environment, through edge nodes in forest fire scenarios, where rapid response and analysis is required. The authors indicate that big data and smart devices' analysis increases the processing load on the edge nodes in traditional scenarios, which is related to the latency and power consumption of the edge processing.

Vikram et al. [29] introduce a low-cost proposal that uses a Wireless Sensor Network (WSN) through a sensor location technique applying a Support Vector Machine (SVM) as an alternative to traditional sensor systems with GPS. The proposal is based on a semi-supervised classification model, with which it divides the forest area into High Active, Medium Active, and Low Active zones to predict the state of the fire zone with high precision.

Aksamovic et al. [2] present a system based on the modeling of a WSN through a network design simulator for a specific application domain. The simulator makes it easier to understand the benefits and limitations of implementing a WSN in the early detection of forest fires and their monitoring. The simulator addresses essential design issues, such as the coverage of the area under surveillance and the number of sensors required for specific deployments.

[6] proposes a system as a model based on a forest fires detection algorithm that processes satellite images according to their intensity level to determine the region affected by the fire. The model allows detecting the fire and its behavior from the forest fire's images and the wind direction.

According to Son et al. [26], in South Korea, a project was proposed using WSN hybrid camera surveillance for forest fire detection based on cluster topology network. Each primary node of the cluster performs calculations such as the level of fire risk by measuring humidity, temperature, rainfall, smoke, wind speed, and solar radiation.

Lee et al. [17] present a UAV-based forest fire detection system, using Deep Convolution Neural Networks, for a wide range of aerial photography. Five deep convolution neural networks were analyzed to evaluate the system: AlexNet, GoogleNet, modified GoogleNet, modified VGG13, and VGG13.

The primary studies mostly report proposals of systems for predicting and detecting forest fires, characterized by being varied, efficient, sophisticated, and complete from the perspective of the identified need. Many of these systems integrate several of the technological elements referred to above. We believe that many of these systems could be complemented or combined to structure new ones more efficiently and effectively in predicting and detecting forest fires.

RQ3: Under what criteria can a technological platform be designed to predict and give early alerts of forest fires?

Based on the identified systems to predict and alert about forest fires, the technological elements used for them, and the context in which they would be or were used, it was possible to determine the criteria to design solutions to predict and alert about forest fires, which are divided into three categories: (1) Hardware devices, (2) Communication devices, and (3) Information processing.

Hardware Devices: Concerning hardware devices, those that have stood out the most are the sensors since it is necessary to measure different parameters that vary around the FF. The most common are sensors for humidity, temperature, rainfall, smoke, wind speed, solar radiation, and dangerous gases. Among the dangerous gases are carbon monoxide, butane, propane, methane, and carbon dioxide sensors.

We cannot but consider TTN Servers, image/video cameras, UAV, Satellite, GPS, GSM (Global System for Mobile Communication), among others, within the hardware devices.

Communications Devices: Communication has proven to be essential in the systems to predict and alert about forest fires, primarily to transmit the data obtained through the sensors. Depending on the circumstances that characterize the forest context under study, different network technology could be used, such as WSN, LoRa, Ad hoc, ZigBee, SigFox, NB-IoT, LTE, WiFi, Infrared, WiMax, GSM, Satellite, Mobile Ad hoc Network (MANET).

Information Processing: The third consideration, no less important than the previous ones that must be taken to design systems to predict and alert about forest fires, is the type of information processing to be carried out and, therefore, the selection of the most appropriate technique, among which are IoT, Big Data, Fuzzy Logic, AI, Deep Learning, Neural Networks, Web/mobile applications, among others.

Once we had clear the criteria for designing technological platforms to predict and give early alerts of forest fires, we also believe that it is appropriate to highlight the considerations for this type of system:

- Recognition of events related to a fire.
- The information processing of these events, and
- Suggest actions that could be taken during decision making.

RQ4: What is the frequency with which proposals of early alerts and prediction of forest fires have been submitted?

After carrying out an analysis of the primary studies' general characteristics, such as year of publication and types of proposals reported in them, we found significant findings, as shown in Fig. 1.

The results indicate that 2015 was a year in which the FF topic attracted much attention, and the most significant number of proposals were received. However, the proposals decreased from that year until 2017, even where they stabilized. Finally, an exponential increase was noted in 2019.

Regarding the type of technology, data management, sensor-based monitoring, and image processing stand out without neglecting communication, which is fundamental in this proposal type.

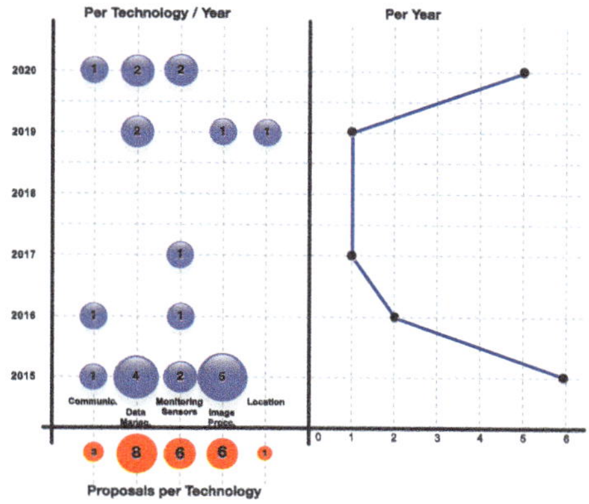

Fig. 1. Timeline of primary studies

We summarize each primary study's contribution to the RQs through a table (accessible through this *link*) if a more graphic appreciation is required.

5 Synthesis of Results

This study is part of a research project aiming to create wireless sensor networks with nodes focus on the early alert or detection of forest fires complemented by a UAVs network equipped with the same functionalities, which constituted the selection criteria for each element that conforms a compact, economical, and functional model derived from a synthesis of results, as seen in Fig. 2.

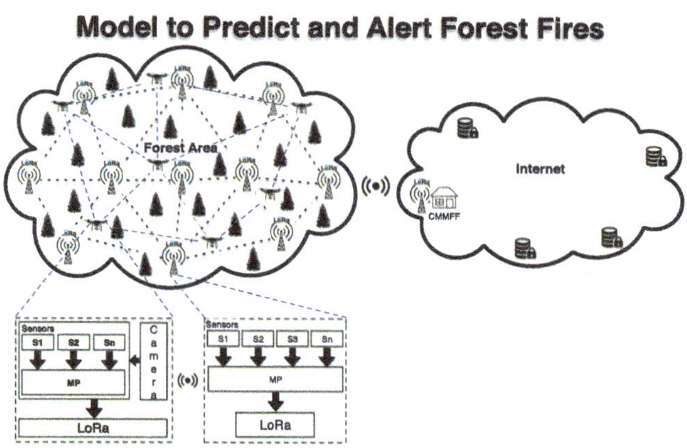

Fig. 2. Model to predict and alert forest fires

The model blends three modules: (1) WSN connected to a microprocessor (MP) and a LoRA device; (2) UAV network to validate FF events through a set of sensors, a microprocessor, and a camera; and (3) Forest fire control and monitoring module (FFCMM).

The WSN comprises nodes distributed throughout the monitored forest area. Each node includes sensors to acquire data of humidity, CO2, temperature, and wind speed. Additionally, each node has a microprocessor that manages the sensors' data, and a LoRa communication device of low-power and wide-area coverage. Once the network is working, the nodes exchange information on events that occur in the monitored area. Specific nodes extend the network coverage and connect the WSN to the internet to alert about a possible event detected until reaching the control center, where is the FFCMM to processes the information.

A temporary network of Unmanned Aerial Vehicles (UAVs) is extended in the monitored area to confirm the event alerted by the WSN. Each UAV constitutes a node of such network and includes a set of sensors, a microprocessor that validates the WSN alert, and a high definition camera that captures and processes frames of different objects in the monitored area corroborating the possible event or event in progress. Likewise, this information is sent to the control center through the WSN to be managed by the MCMIF. All the time, the information from the WSN and the UAVs network is processed and stored. The result of the information processing is the activation or not of FF alerts.

6 Conclusions and Future Works

FF has always meant an unprecedented danger to life, so there have been regular attempts for their early detection. This study state that emerging technologies have given new alternatives to science to develop improved, economical, and efficient proposals for early warning about FF more intensive in 2019. The results stand out the technologies of data management, sensor-based monitoring, and communication.

Several technologies are used to propose early warning systems on FF, indicating no defined standards for their use. We found that each forest area's characteristics and needs to determine the use of specific technologies to develop this kind of system. Therefore, the systems built or proposed are also varied.

We believe that it is possible to build early warning systems for FF that include other solutions to cover possible gaps not considered. Therefore, we estimate that our model is feasible, and in our future work, we will implement a prototype.

Acknowledgments. The authors would like to thank to the Universidad de las Fuerzas Armadas ESPE of Ecuador for the financial support given to the PIS-04 research project titled "Detección temprana de incendios forestales basado en aprendizaje automático y en algoritmos de clasificación aplicados al espectro visible", and additionally to Corporación Ecuatoriana para el Desarrollo de la Investigación y Academia - CEDIA for the financial support given to the present research, development, and innovation work through its Grupos de Trabajo, especially for the "Grupo de Trabajo en Internet de las Cosas y Ciudades Inteligentes" fund.

References

1. Afzaal, H., Zafar, N.: Robot-based forest fire detection and extinguishing model, pp. 112–117 (2016)
2. Aksamovic, A., Hebibovic, M., Boskovic, D.: Forest fire early detection system design utilising the WSN simulator, vol. 2017, pp. 1–5 (2017)
3. Akter, S., Wamba, S.: Big data and disaster management: a systematic review and agenda for future research. Ann. Oper. Res. **283**(1–2), 939–959 (2019)
4. Asatryan, D., Hovsepyan, S.: Method for fire and smoke detection in monitored forest areas, pp. 77–81 (2015)
5. Baruah, S., Boruah, G., Sharma, S., Hoque, W., Chetia, T., Dey, C., Gogoi, D., Das, P., Baruah, S., Basumatari, D., Pathak, J., Barua, A., Choudhury, S.: Seismic vulnerability assessment of earthquake-prone mega-city Shillong, India using geophysical mapping and remote sensing. Georisk **14**(2), 112–127 (2020)
6. Divya, T., Vijayalakshmi, M.: Analysis of wild fire behaviour in wild conservation area using image data mining (2015)
7. Fan, Y., Ma, H.: Video based forest-fire smoke detection. Qinghua Daxue Xuebao/J. Tsinghua Univ. **55**, 243–250, 256 (2015)
8. Food and Agriculture Organization of the United Nations: International day of forests 2020: Forests and biodiversity
9. Gallego, A.J., Pertusa, A., Gil, P., Fisher, R.: Detection of bodies in maritime rescue operations using unmanned aerial vehicles with multispectral cameras. J. Field Rob. **36**(4), 782–796 (2019)
10. Gupta, N., Gupta, R.: Route-discovery optimization in LAR: a review. In: Advances in Intelligent and Soft Computing, AINSC, vol. 131, pp. 877–884 (2012)
11. Herutomo, A., Abdurohman, M., Suwastika, N., Prabowo, S., Wijiutomo, C.: Forest fire detection system reliability test using wireless sensor network and OpenMTC communication platform, pp. 87–91 (2015)
12. Jin, S., Lu, X.: Vision-based forest fire detection using machine learning. In: Proceedings of the 3rd International Conference on Computer Science and Application Engineering, pp. 1–6 (2019)
13. Kang, J., Kim, S., Kim, J., Sung, N., Yoon, Y.: Dynamic offloading model for distributed collaboration in edge computing: a use case on forest fires management. Appl. Sci. (Switzerland) **10**(7), 2334 (2020)
14. Kansal, A., Singh, Y., Kumar, N., Mohindru, V.: Detection of forest fires using machine learning technique: a perspective, pp. 241–245 (2016)
15. Khan, A., Gupta, S., Gupta, S.K.: Multi-hazard disaster studies: monitoring, detection, recovery, and management, based on emerging technologies and optimal techniques. Int. J. Disaster Risk Reduct. **47**, 101642 (2020)
16. Kitchenham, B.A., Budgen, D., Brereton, P.: Evidence-Based Software Engineering and Systematic Reviews. Chapman & Hall/CRC (2015)
17. Lee, W., Kim, S., Lee, Y.T., Lee, H.W., Choi, M.: Deep neural networks for wild fire detection with unmanned aerial vehicle, pp. 252–253 (2017)
18. Lin, R., Wang, Z., Sun, Y.: Energy efficient medium access control protocols for wireless sensor networks and its state-of-art, vol. 1, pp. 669–674 (2004)
19. McHugh, M.L.: Interrater reliability: the kappa statistic. Biochem. Med. **22**(3), 276–282 (2012). PMID: 23092060
20. McRae, J., Gay, C., Nielsen, B., Hunt, A.: Using an unmanned aircraft system (drone) to conduct a complex high altitude search and rescue operation: a case study. Wilderness Environ. Med. **30**(3), 287–290 (2019)

21. Petticrew, M., Roberts, H.: Systematic Reviews in the Social Sciences: A Practical Guide. Wiley-Blackwell (2005)
22. Qu, T., Xu, Q., Liu, C., Li, Z., Chen, B., Dai, K.: Radar remote sensing applications in landslide monitoring with multi-platform InSAR observations: a case study from china. Int. Arch. Photogramm. Remote Sens. Spatial Inf. Sci. **42**, 1939–1943 (2019)
23. Rowell, A., Moore, P.F.: Global review of forest fires. Citeseer (2000)
24. Sendra, S., García, L., Lloret, J., Bosch, I., Vega-Rodríguez, R.: Lorawan network for fire monitoring in rural environments. Electronics **9**(3), 531 (2020)
25. Serna, M.Á., Casado, R., Bermúdez, A., Pereira, N., Tennina, S.: Distributed forest fire monitoring using wireless sensor networks. Int. J. Distrib. Sens. Netw. **11**(10), 964564 (2015)
26. Son, B., Her, Y.s., Kim, J.G.: A design and implementation of forest-fires surveillance system based on wireless sensor networks for South Korea mountains. Int. J. Comput. Sci. Netw. Secur. (IJCSNS) **6**(9), 124–130 (2006)
27. Stipaničev, D., Vuko, T., Krstinić, D., Štula, M., Bodrožić, L.: Forest fire protection by advanced video detection system-croatian experiences. In: 3rd TIEMS Workshop-Improvement of Disaster Managment Systems-Local and Global Trends (2006)
28. Sukhodolov, A.P., Sorokina, P.G., Fedotov, A.P.: Numerical analysis of ecology-economic model for forest fire fighting in Baikal region. Discrete Contin. Models Appl. Comput. Sci. **27**(2), 154–164 (2019)
29. Vikram, R., Sinha, D., De, D., Das, A.: EEFFL: energy efficient data forwarding for forest fire detection using localization technique in wireless sensor network. Wireless Netw. **26** (2020)
30. Yuan, C., Zhang, Y., Liu, Z.: A survey on technologies for automatic forest fire monitoring, detection, and fighting using unmanned aerial vehicles and remote sensing techniques. Can. J. For. Res. **45**(7), 783–792 (2015)
31. Yusoff, A., Shafiril, S., Wills, G., Gilbert, L., Crowder, R.: Utilizing the application of sensors to develop low-cost remote sensing system for monitoring forest environmental activity, vol. 2017, pp. 1–6 (2017)
32. Zhang, Q.X., Lin, G.H., Zhang, Y.M., Xu, G., Wang, J.J.: Wildland forest fire smoke detection based on faster R-CNN using synthetic smoke images. Procedia Eng. **211**, 441–446 (2018)

Computational Modeling and Characterization of Composites

Chitosan-Based Hydrogels as Ibuprofen Controlled Delivery System

Juliana María Patiño Medina[1], Angélica María Candela Soto[1](✉), Olga Lucía Bayona Ayala[1], Lizeth J. Alvarado Rueda[1], and Hernando Alberto Camargo García[2]

[1] Universidad Santo Tomás, Bucaramanga, Colombia
angelica.candela01@ustabuca.edu.co
[2] Servicio Geológico Colombiano, Bogotá, Colombia

Abstract. The goal of this research was to synthesize a compound from a hybrid method of in situ synthesis of magnetic nanoparticles aggregation to the chitosan hydrogel (MCH), as well as to synthesize a chitosan-glutaraldehyde (HCG) hydrogel by shaking and gelling, these for controlled release of ibuprofen. The hydrogels were chemically crosslinked, their swelling capacity allows the exchange of fluids. Therefore were studied as possible alternative materials in the controlled release of ibu-profen. The ibuprofen was extracted from the conventional tablet by solid-liquid extraction. The controlled release was carried out at pH 7.4, simulating the pH of the intestine of the human body, and was monitored by UV-Vis spectroscopy. The MCH is the zero-order with a correlation coefficient of 0.92, and HCG is Korsmeyer-Peppas model with a correlation coefficient of 0.99 and a diffusional coefficient of 0.49. In this way, the HCG showed better performance. In the characterization of the materials, the analytical techniques were: FTIR, SEM, TGA, and UV-Vis.

Keywords: Chitosan · Hydrogel · Ibuprofen · Controlled release

1 Introduction

Chitosan is a natural origin polymer compound by glucosamine and n-acetylglucosamine and derived from chitin. Chitin is the second most abundant polymer on earth and is present in fungus and exoskeleton of insects and crustaceans. Chitosan has interesting characteristics such as its non-toxic character, biocompatibility, biodegradability. It has different physicochemical properties such as molecular weight, degree of deacetylation, degree of crystallinity, and degree of ionization [1]. This polymer has a high number of hydrophilic groups capable of crosslinking for the formation of networks, these networks are called hydrogels. Hydrogels are flexible chains that have elasticity and high affinity for water, but the chemical or physical bonds formed between the chains of the polymer prevent their dissolution [2]. Through the networks that form, they enter water or biological fluids allowing the swelling of the material, which occurs by the osmotic forces between the incoming water and the polymer chains which are opposed to the

M. Botto-Tobar et al. (Eds.): CIT 2020, AISC 1326, pp. 371–385, 2021.
https://doi.org/10.1007/978-3-030-68080-0_28

expansion. One of the most important characteristics in hydrogels is the low interfacial tension between the surface of the hydrogel and the body fluid since this condition allows us to minimize protein adsorption and cell adhesion, therefore, the possibility of a negative immune reaction is reduced. Magnetic hydrogels or ferrogel are named in this way for the presence of magnetic nanoparticles in them. They are materials sensitive to the presence of a magnetic field to which they react with a quick response, remote action, and no contact. They have been used in biomedical applications, some advantages are that they can be controlled in such a way that they are non-invasive and in turn cannot alter the biological environment [3, 4]. Thus, the hydrogels are a good alternative for the controlled release of drugs, which consists of administration systems used to transport therapeutic drugs in the body, safely, at an adequate speed, over time, and with the desired therapeutic effect. The design of these systems is focused on improving the aqueous solubility and chemical stability of the active agents, increasing pharmacological activity, and reducing side effects. In this way, the desired therapeutic concentration of the drug is provided and maintained at the target biological site [5]. The release of a drug in the system can take place by diffusion and relaxation or by swelling of the polymer chains. The diffusion process occurs when the drug passes through the polymer that forms the release system, influenced by the concentration gradient between the system and the surrounding medium. The swelling of polymeric chains is related to the diffusion of the drug from and through the polymer, under countercurrent diffusion with the medium fluid towards the interior of the polymer [6]. The release kinetic data must be processed by different mathematical models to establish which of them the material follows. The models studied are the zero-order model, single order model, Higuchi model, and Korsmeyer-Peppas model [7].

The preliminary study is prepared taking into account that ibuprofen ((RS) -2- (4-isobutylphenyl) propanoic acid) is a high-consumption drug used for the symptomatic relief of occasional mild or moderate pain, such as headaches, dental, menstrual, mus cular, lumbar pain or febrile states, but the interest is that it can cause kidney problems due to high concentrations and even be toxic [8]. From a chemical point of view, it is a weak acid with high permeability, it is a relatively simple and small molecule. It has low solubility in water and in acidic pH < 4 but it is soluble in alkaline pH. It has R and S enantiomers, but the (S) enantiomer is the most biologically active. Furthermore, the presence of the carboxyl group in its structure limits its solubility in the stomach [9–11]. Belonging to the group of non-steroidal anti-inflammatory analgesics (NSAIDs), these drugs have the mechanism of action to block the synthesis of prostaglandins by inhibiting cyclooxygenase (COX) isoforms [12]. The objective of this research was the synthesis of two alternative and novel hydrogels for the controlled release of ibuprofen at a pH similar to that of the human intestine.

2 Materials and Methods

2.1 Materials

Chemical Reagents: Sodium hydroxide $\geq$97% (Merck), 37% hydrochloric acid (Carlo Erba), acetic acid 99.5–100.5% (PanReac Applichem), iron (III) anhydrous chloride

≥98% (Merck), iron (II) tetrahydrate chloride ≥98 % (Merck), 98% dipotassium phosphate (Carlo Erba), sodium chloride (Honeywell), potassium chloride ≥99.5%, potassium dihydrogen phosphate ≥99.5% (Merck), potassium bipthalate 99.99% (JT Baker), glutaraldehyde 50% (Suquin), hexane ≥60% (Merck), chitosan extracted according to Escobar Sierra [13].

Minor Equipment: HQ40E multi-parameter (Hach), Fusion meter (Stuart SMP 20), balance (OHAUS Pioneer), Carousel Tech stirring heating plate (RADLEYS), drying and heating oven (BINDER).

3 Methods

3.1 Infrared Spectroscopy

Infrared spectra were obtained on the Shimadzu IRAffinity-1 spectrometer. The spectra were made using the ATR MIRacle support; and they were processed in ADC Spec-Manager. The most characteristic absorption bands, expressed in cm-1, are reported in Table 1.

3.2 Ultraviolet-Visible Spectroscopy

UV-Vis spectra were performed on Agilent Technologies Cary 60 G6860A equipment. A 1 cm optical path quartz cell was used, and the spectra were processed in Cary WinUV v5.0.0.999.

3.3 Scanning Electron Microscopy

The analyzed sample was placed on a metallic stub with carbon adhesive tape, and then it was coated with gold in a Quorum 150ES coating equipment. Imaging was performed on a QUANTA FEG 650 Field Emission Gun (FEG) Scanning Electron Microscope under the following characteristics: high vacuum and 25kV acceleration voltage. Detectors used are secondary electrons (SE): Everhart Thornley detector ETD and backscattered electron (BSE): Backscattered electron detector (BSED) type SSD (to observe compositional variations). To carry out the chemical analysis, an acceleration voltage of 25kV was used with an EDAX APOLO X detector with a resolution of 126.1 eV (in. Mn Kα) to carry out EDS (Energy-Dispersive Spectroscopy) analysis. The EDX Genesis Software was used for the semi-quantitative information of the chemical elements.

3.4 Thermogravimetric Analysis

The analysis by thermogravimetry was carried out on a TGA Discovery 5500 brand TA Instruments as follows for each of the samples: the MCH sample was analyzed under a flow of 20 mL/min of N2, with a heating ramp of 10 °C/min up to 800 °C and with a mass of 28,078 mg. The data from the HCG sample were collected at a flow of 20 mL/min of N2, with a heating ramp from 10 °C/min to 800 °C and with a mass of 39,235 mg.

3.5 Synthesis of the Magnetic Chitosan Hydrogel

In the synthesis of the magnetic chitosan hydrogel, a solution of 2% w/v chitosan in 2% acetic acid was prepared. Subsequently, a solution with 0.97 g of anhydrous iron (III) chloride and 0.42 g of iron (II) chloride tetrahydrate in 25 mL of deionized water was added. The mixture of chitosan and iron salts was stirred for two hours, after this time 22 mL of 1.25M sodium hydroxide were added. This process was maintained under an inert atmosphere. Finally, it was filtered under vacuum, washed until pH = 7 and the solid was dried at 70 °C [4].

Infrared spectra were obtained on the Shimadzu IRAffinity-1 spectrometer. The spectra were made using the ATR MIRacle support; and they were processed in ADC SpecManager. The most characteristic absorption bands, expressed in cm^{-1}, are reported in the Table 1.

3.6 Synthesis of the Chitosan-Glutaraldehyde Hydrogel

The chitosan hydrogel was chemically crosslinked with glutaraldehyde, a 2% chitosan solution was prepared which was stirred for 24 h, then 4 mL of 25% glutaraldehyde was added and homogenized for 30 min. Subsequently, it gelled for 30 min and was taken to the oven to evaporate the solvent for 3 h, finally, it was taken to the desiccator [14, 15].

3.7 Liquid-Liquid Separation and Purification of Ibuprofen from a Commercial Tablet

Two tablets of the medicine were used, weighed, and macerated until a fine powder was obtained. Subsequently, the tablets were dissolved in 30 mL of hexane and 50 mL of distilled water in a settling funnel. The phases were separated, and 30 mL of hexane were added again, the phases were separated, and the organic phase was removed. Then, the solvent was slowly evaporated at 60 °C. The IBU was subsequently purified by dissolving the solid in 15 mL of hexane and evaporating the solvent until crystals were obtained [16].

3.8 Buffer Solutions pH = 7.4 and pH = 2

The phosphate buffer solution pH = 7.4 was made by adding 8.03 mmol of K2HPO4.3H2O, 2.00 mmol of KH2PO4, 13.03 mmol of NaCl, and 2.79 mmol of KCl in approximately 900 mL of deionized water, subsequently its pH was adjusted to 7.40 with the addition of 1M HCl and finally, it was added to 1000 mL with deionized water [16].

3.9 Ibuprofen Fit Curve

An IBU fit curve was drawn up where a 500 μg/L stock solution was made with a buffer solution pH = 7.4. Subsequently, the respective solutions were made to prepare the concentrations of 400, 250, 125, 100, 50, 40, 30, 20, 10, and 5 μg/L. Subsequently, its absorbance was recorded in the UV-Vis spectrophotometer at a wavelength of 265 nm and the fit curve was elaborated with the collected data [18].

3.10 Drug Loading to Hydrogels

A 500 μg/L ibuprofen solution was prepared, 30 mL were taken from it and added to an erlenmeyer where the hydrogel was introduced, this process was performed for both MCH and HCG. The drug was loaded for 72 h [4].

3.11 Ibuprofen Release Kinetics

The same process described below was performed for MCH and HCG. 100 mL of the buffer solution was taken, and the temperature was increased to 37 ± 1 °C. The pH of the solution was verified, and the loaded material was added to the solution. Subsequently, 3 ml of the solution were taken at the minute of contact, and every 5 min until completing 30 min and every 10 min until completing one hour, then aliquots were taken every hour for 8 h and every 24 h for 4 days. At each extraction, the same amount of buffer solution was added and finally, the samples were read by the UV-Vis spectrophotometer at 220 nm [19].

4 Results and Discussion

4.1 Characterization Magnetic Chitosan Hydrogel

Figure 1 shows the micrograph of the chitosan magnetic hydrogel analyzed by the SEM enlarged to 200 μm and 500 nm. In micrograph a) It is possible to see in the upper right part the porosity of the material, and its morphology. Hydrogels are porous materials, the pores of which are formed by the crosslinking of polymer chains or by magnetic particles, observed in Fig. 1 b). The qualitative elemental analysis evidenced the presence of carbon, oxygen, sodium, silicon, gold, and iron.

In Fig. 2 the thermogram of the magnetic chitosan hydrogel is presented, where the temperature vs. Percentage of mass and its respective derivative is displayed where important changes are observed.

A change of around 100 °C equivalent to the loss of humidity is perceived, then after 250 °C, a dominant and a progressive change occurs that would be associated with the loss of crosslinking of the polymer with the magnetic particles and between the chains thereof since in this temperature range the depolymerization of chitosan occurs. The pyrolysis of chitosan occurs from 600 °C so that chitosan decomposes and the mass that is constant from 700 °C corresponds to the magnetic particles. The MCH thermogram is similar to that of chitosan, given the presence of the polymer in it, however, it presents changes corresponding to the fact that the structure of the chitosan was modified. The thermogram is similar to that published by Tandekar et al. 2020, Hernández et al. 2009 and Reddy and Lee, 2013 [19–21].

4.2 Characterization Chitosan-Glutaraldehyde Hydrogel

The thermogram of the HCG is presented in the Fig. 3, where the % of mass is related to its respective derivative as the temperature increases.

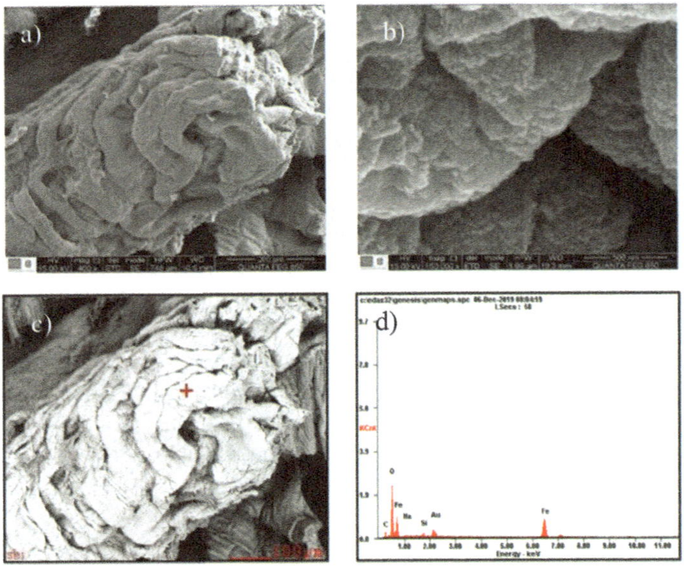

Fig. 1. SEM micrographics of MCH with amplification at a) 200 μm. b) 500 nm c) reference point of EDS data collection and d) elemental analysis graph of chitosan.

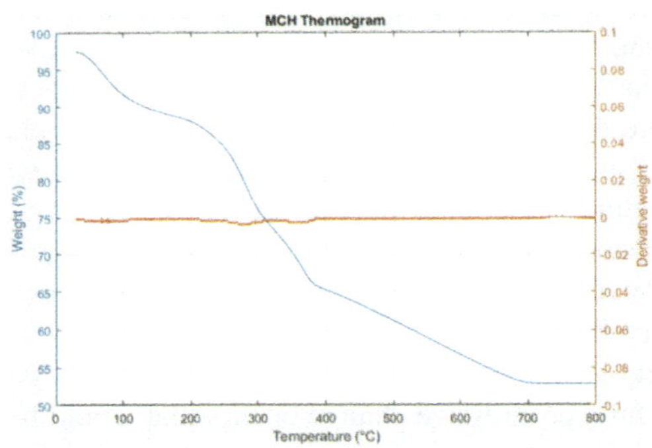

Fig. 2. Thermogram of the magnetic chitosan hydrogel

In this case, a significant change between 50 °C and 150 °C is observed, which agrees with the loss of humidity, acetic acid and glutaraldehyde, whose boiling points correspond to 100 °C, 118 °C and 187 °C respectively, as several compounds, the decrease is high. A small change is observed around 250 °C which can be influenced is the depolymerization of chitosan and its subsequent decrease in mass until pyrolysis. The thermogram is analogous to that reported by Submitted et al. 2014 and Ray et al. 2010 [22, 23].

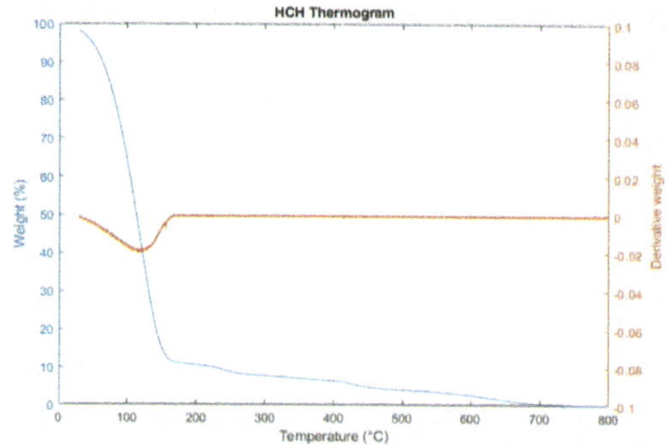

Fig. 3. Thermogram of the Chitosan-glutaraldehyde hydrogel

4.3 Ibuprofen

Through liquid-liquid extraction, an extraction yield of ibuprofen of 72.83 ± 3.76E-3% was obtained. Subsequently, the melting point was determined where a range of 77–78 °C was obtained, this is similar to that reported in the literature, which corresponds to 75–77 °C [9]. Similarly, a visible infrared spectrum was obtained in Fig. 4.

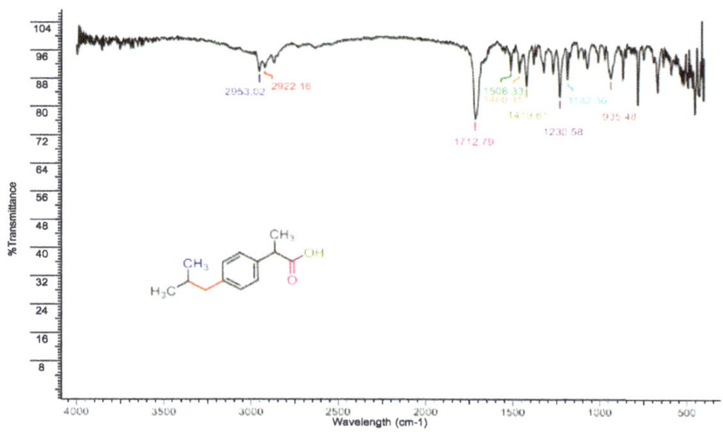

Fig. 4. Infrared spectrum of the Ibuprofen

The main peaks in the infrared spectrum were identified considering the functional groups present in the molecule (Table 1)

Then in the Fig. 5 shows the UV-Vis spectrum performed on ibuprofen. Ibuprofen has conjugated double bonds in the benzene ring and a carboxyl which are chromophoric groups, these are responsible for emitting a signal in the ultraviolet-visible region [26].

The literature reported that ibuprofen at a pH close to neutral, evaluated in a buffer solution pH = 7.4, has a wavelength between 260 and 270 nm [27] and as observed in this case, ibuprofen absorbed at a wavelength of 265 nm.

Table 1. The main peaks in the infrared of the Ibuprofen

Región	Vibración	Nota
2953 cm^{-1}	C-H	υas
2922 cm^{-1}	-CH2-	υas
1712 cm^{-1}	C = O	Carbonyl stretch
1230 cm^{-1}	C-O-OH	Carboxyl twist/stretch
1508 y 1460 cm^{-1}	Benceno	Condensed benene core system
1419 cm^{-1}	C-O-H	δ del carboxyl
935 cm^{-1}	Carboxyl dimer	

Note: *υas asymmetric stress, δ reflection in the plane. Vibrations of the bonds present in the extracted ibuprofen according to the analysis of the infrared spectrum. The assignment of each type of vibration was granted according to Conley, Avendaño and Calderón, 1979 [24]. In addition, it agrees with that reported in the SDBS database [25]

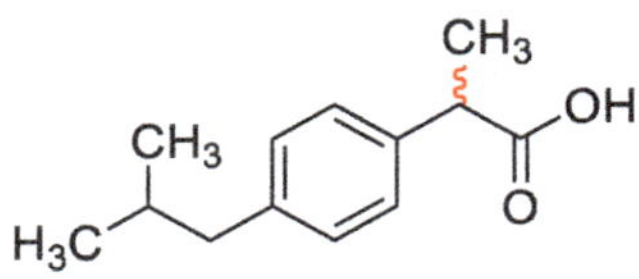

Fig. 5. Ibuprofen structure

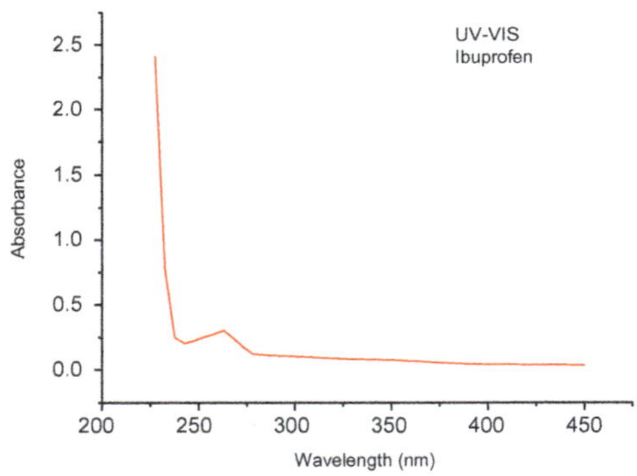

Fig. 6. UV-Vis spectrum of Ibuprofen

4.4 Ibuprofen Fit Curve

Subsequently, the Fig. 6 shows the ibuprofen fit curve, which was performed at a wavelength of 265 nm.

The adjustment curve for ibuprofen presents a strong positive correlation with a 99% confidence percentage, so that the absorbance and concentration variables are linearly

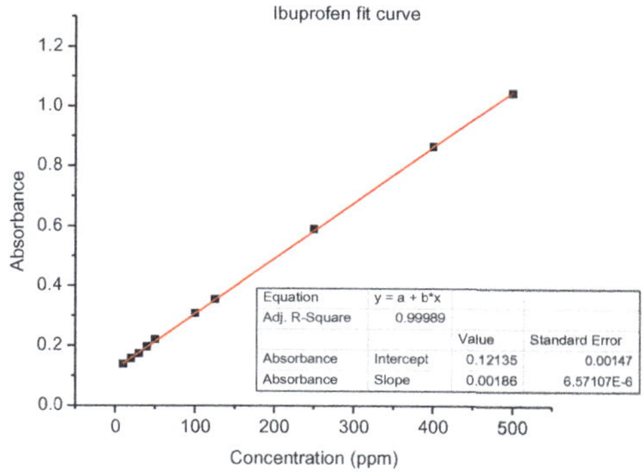

Fig. 7. Ibuprofen fit curve

related, likewise the deviation it presents is very low and allows us to determine data with certainty. From it, in this case the concentration of ibuprofen released by each hydrogel analyzed.

4.5 Preliminary Study of the Kinetics of Ibuprofen Release

Hydrogels are capable of responding to stimuli from the environment where they are found, such as pH, temperature, magnetic field, among others [28], therefore one of these factors was chosen to carry out the preliminary study of the release kinetics of ibuprofen this was the pH, therefore a phosphate buffer solution pH 7.4 was prepared to simulate the pH in the intestine [29]. The analyzes were carried out at a temperature of 37 °C since this is the normal human body temperature [30].

There are three different ways to load a hydrogel, these are by permeation, entrapment, and covalent bonding. In the case of the two hydrogels evaluated, neither could perform covalent bonds to ibuprofen, since their hydrophilic groups are crosslinked, therefore, permeation and entrapment are possible. The permeation is influenced because when the hydrogel comes into contact with the simulation of the intestinal or gastric fluid, the level of ibuprofen released is high. Entrapment occurs due to the swelling of the hydrogel, to the fact that the ibuprofen molecule is small, and to how the medicine is released over time [2].

The hydrogels evaluated were chemically crosslinked, the chitosan magnetic hydrogel, which is crosslinked by means of chelation, which is carried out by the magnetite with the hydrophilic groups of the chitosan. In the chitosan-glutaraldehyde hydrogel, crosslinking is effected by glutaraldehyde by means of covalent bonds as seen in Fig. 7.

Next, the analysis of the release of ibuprofen was carried out using the two hydrogels evaluated: MCH and HCG. Both were studied, according to the models of zero-order, order one, Higuchi and Korsmeyer-Peppas. The correlation coefficient is the one that allows us to establish to which model the material fits, therefore only the models to which each material is fitted are presented.

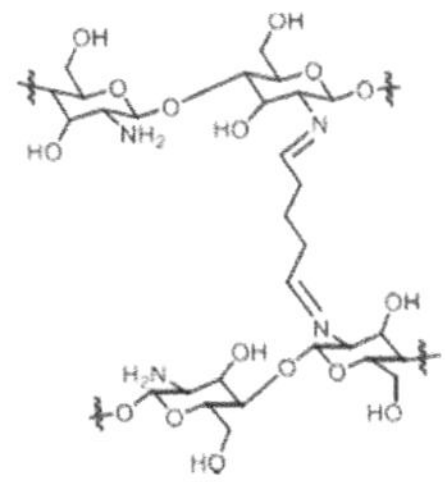

Fig. 8. Representation of the crosslinking between chitosan and glutaraldehyde for the formation of HCG. Taken and adapted from [14].

4.6 Zero Order Model

To understand this model, the following equations must be considered:

$$C_0 - C_t = K_0 t \quad (1)$$

$$C_t = C_0 + Kt \quad (2)$$

Where, C_t is the amount of drug released at time t and C_0 is the initial concentration of the drug at time t = 0 and K_0 is the zero order rate constant. Thus, it is observed that according to zero-order kinetics, the drug administration process is constant. Through the data analysis using the straight line, the zero order speed constant is provided and according to its correlation coefficient whether the release of the drug used follows the model or not [7].

4.7 First Order Model

First order kinetics is described by:

$$\frac{DC}{dt} = -K_1 C \quad (3)$$

Where K1 represents the first-order rate constant where the first-order model is one whose speed is directly proportional to the variation of the concentration of the drug over time. This relationship implies that the higher concentration, the faster the reaction, and the linear kinetics would follow. By integrating and reorganizing the equation obtained was [7]:

$$\log\ C = \log C_0 - \frac{K_1 t}{2.303} \quad (4)$$

In Eq. (3), K1 is the first-order rate equation expressed in time per hour, C0 is the initial drug concentration, C is the percentage of drug remaining in time t. Graphically, log C vs time is performed, where the slope of the graph provides the constant first-order rate, in the same way the correlation coefficient indicates whether or not the drug release follows the first-order model [7].

4.8 Higuchi Model

Higuchi's kinetic model is described by dissolution and diffusion. Today this model is widely recognized in studies of drug dissolution. The classical Higuchi equation is [7].

$$Q = A\sqrt{D(2C_0 - C_s)C_s t} \tag{5}$$

Where Q is the cumulative amount of drug released over time t per unit area, C0 is the initial concentration of the drug, Cs is the solubility of the drug in the matrix, and D is the diffusion coefficient of the drug molecule in the matrix. However, this formula can be simplified as follows [7]:

$$Q = K_H * t^{\frac{1}{2}} \tag{6}$$

In it, KH is Higuchi's dissolution constant. To identify if the release follows the Higuchi model, the cumulative percentage of drug release is plotted vs. The square root of time, the line graph generates the equation of the line and the correlation coefficient. In the equation of the line, the slope corresponds to KH and the correlation coefficient allows us to interpret whether the release kinetics adapt to this model or not, in the same way if the release mechanism is controlled by diffusion [7].

4.9 Korsmeyer-Peppas Model

The Korsmeyer-Peppas model is governed by the following equation:

$$\frac{\mathrm{M_t}}{M_\sigma} = K_{kp}\mathrm{t}^{\mathrm{n}} \tag{7}$$

Where the relation $\frac{\mathrm{M_t}}{M\infty}$ is the fraction of the drug released at time t. Equation 7 can be linearized by applying logarithm to it, as follows [7]:

$$\log(\frac{\mathrm{M_t}}{M_\infty}) = \log(K_{kp}) + n\log(t) \tag{8}$$

Mt is the amount of drug released at time t, M∞ is the amount of drug released after time ∞, n is the diffusion exponent or exponent of drug release, Kkp is the Korsmeyer release rate constant. So, the percentage of drug release is $\log(\frac{\mathrm{M_t}}{M\infty})$ vs. The logarithm of time (log t) [7].

The diffusional exponent n is of utmost importance since it allows identifying the material's release mechanism [7]. If the value of n is 0.5, the release of the drug follows a Fickian-type diffusion or case I transport. A value of n greater than 0.5 but less than 1 causes an abnormal or non-Fickian diffusion. If the value of n is greater than 1, the kinetics of the release system are zero-order and the process is controlled by the relaxation of the polymeric chains [6].

Given the presence of porous materials, the diffusion exponent can take values less than 0.5, which indicates the combination of the mechanisms of partial diffusion through the swollen matrix and through the pores filled with water or the solution of the medium in which is found [6].

4.10 Magnetic Chitosan Hydrogel Kinetics

Figure 9 shows the study of the kinetics of IBU release with MCH at said pH, with the zero-order model which consists of the linear relationship between the accumulated concentration of ibuprofen released with respect to time in the which was released.

The release kinetics presents a strong positive correlation with a 92% reliability percentage, so that the variables accumulated % of ibuprofen and time are linearly related, however, the deviation presented by the values of the equation of the line is high. Equivalent to 1.62 for the intercept and is low for the slope whose value is 9.34E−4.

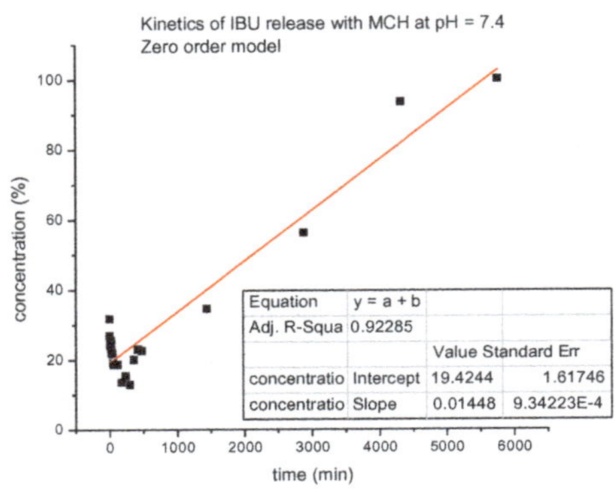

Fig. 9. Kinetics of IBU release with MCH at pH = 7.4 under the zero order model

Zero-order helps us describe the concentration of the drug being administered and what the level of ibuprofen would be in the blood, as it is released where it is expected to be constant [7]. According to Eq. (3) described for the zero order model, when linearized, it is obtained that the zero order cup constant is the slope of the line, so K = 1.44E−2 ppm * min^{-1} corresponding to speed.

The correlation coefficient indicates whether or not the release process follows the zero-order kinetic model, based on this, the R^2 obtained was 0.92, for its value closest to 1 with respect to the other models evaluated, the release is slightly close to this model.

Ibuprofen release does not occur in the same way when chitosan magnetic hydrogel comes into contact with the buffer solution compared to when it is in contact for a long time. During the first minutes, the release occurs quickly and disproportionately, while over time it is proportional.

4.11 Chitosan-Glutaraldehyde Hydrogel Kinetics

The study of IBU release with HCG at pH = 7.4 is observed, with the Korsmeyer- Peppas model which consists of the linear relationship between the log of the concentration of ibuprofen released with respect to the log of the time in which it was released.

The release of ibuprofen using the chitosan-glutaraldehyde hydrogel presents a strong positive correlation with a reliability percentage of 99%, thus the logarithm variables of the ibuprofen concentration and the logarithm of time are strongly linearly related.

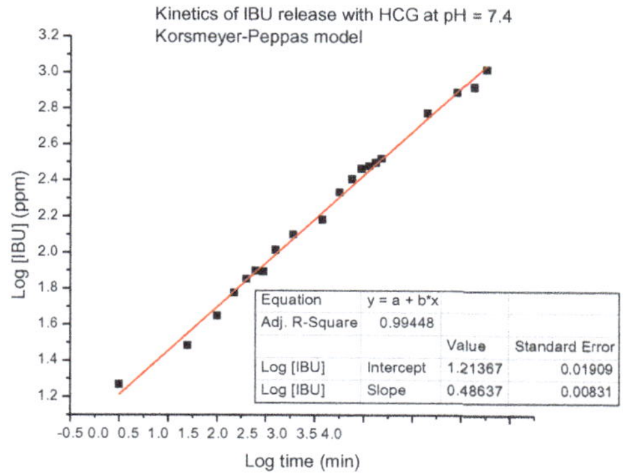

Fig. 10. IBU release kinetics with HCG at pH = 7.4 under the Korsmeyer-Peppas model

Similarly, R^2 being close to 1 allows us to determine that the kinetics of HCG follow the K-P model. The equation of the line presented a low error, this is important since it is used to determine the Korsmeyer release rate constant and the diffusional exponent.

The rate (KKP) at which the drug was released was determined to be 16.35 ppm * min^{-1}. In the case of n, the value is 0.49, being a number very close to 0.5, we infer that the chitosan-glutaraldehyde hydrogel follows a Fickian-type diffusion or case I transport.

5 Conclusions

The chitosan magnetic hydrogel was synthesized in situ by means of the hybrid method of aggregation of magnetite particles to the chitosan hydrogel, where the morphology, porosity, and existence of particles in the chitosan magnetic hydrogel were visi- ble according to the SEM, as well as the analysis elemental micro of this. Ibuprofen was also identified thanks to the FTIR analytical technique, where the peaks corresponded to the functional groups of the molecule.

UV-Vis spectroscopy is a suitable technique for the quantification of ibuprofen in the preliminary study of the controlled release of this drug since the molecule has visible chromophoric groups at a maximum absorbance of 265 nm.

The preliminary controlled release study of ibuprofen for MCH showed that pH 7.4 follows the zero-order model, where release occurs by relaxation of the polymer chains. In the same way, it describes how the level of ibuprofen in the blood would be as it is released, in addition, a constant increase in the concentration of medication over time was visualized.

The HCG in the preliminary ibuprofen controlled-release study responded positively to the Korsmeyer-Peppas model for analysis at pH 7.4, where it presented a correlation coefficient of 0.99. According to the diffusion coefficient, it is indicated that the liberation is given by diffusion of the Fickian type or type I case.

The hydrogel that responds best as an access route of the active ingredient to the body is the chitosan-glutaraldehyde hydrogel, mainly for the release of ibuprofen.

References

1. Bakshi, P.S., Selvakumar, D., Kadirvelu, K., Kumar, N.S.: Chitosan as an environment friendly biomaterial – a review on recent modifications and applications. International Journal of Biological Macromolecules (2019)
2. Bhattarai, N., Gunn, J., Zhang, M.: Chitosan-based hydrogels for controlled, localized drug delivery. Adv. Drug Deliv. Rev. **62**(1), 83–99 (2010)
3. Liu, Q., Liu, M., Li, H., Lam, K.Y.: Multiphysics modeling of responsive deformation of dual magnetic-pH-sensitive hydrogel. Int. J. Solids Struct. **190**, 76–92 (2019)
4. Wang, Y., Li, B., Xu, F., Han, Z., Wei, D., Jia, D., Zhou, Y.: Tough Magnetic Chitosan Hydrogel Nanocomposites for Remotely Stimulated Drug Release [Research-article]. Biomacromolecules **19**(8), 3351–3360 (2018).
5. Li, C., Wang, J., Wang, Y., Gao, H., Wei, G., Huang, Y., Yu, H., Gan, Y., Wang, Y., Mei, L., Chen, H., Hu, H., Zhang, Z., Jin, Y.: Recent progress in drug delivery. Acta Pharma- ceutica Sinica B **9**(6), 1145–1162 (2019)
6. Aragón Fernández, J., González Santos, R., Fuentes Estévez, G.: Estudio in vitro de liberación de fármacos desde un biomaterial compuesto. Revista CENIC. Ciencias Químicas **41**, 1–8 (2010)
7. Gouda, R., Baishya, H., Qing, S.: Application of mathematical models in drug release kinetics of carbidopa and levodopa ER tablets. J. Developing Drugs **06**(02), 1–8 (2017)
8. Álamo, C.: Guía Farmacológica. Arán (2005)
9. American Chemical Society. https://www.acs.org/content/acs/en/molecule-of-the-week/archive/i/ibuprofen.html. Accessed 23 Apr 2020
10. Matiz-Melo, G.E., Rodríguez-Cavallo, E., Osorio, del Rosario Osorio, M.: Estudio comparativo de la calidad biofarmacéutica de marcas comerciales y multifuente de tabletas de ibuprofeno en el mercado colombiano. Revista Colombiana de Ciencias Químico-Farmacéuticas **46**(1), 61–83 (2017).
11. Yang, Z., Hu, Y., Tang, G., Dong, M., Liu, Q., Lin, X.: Development of ibuprofen dry suspensions by hot melt extrusion: characterization, physical stability and pharmacokinetic studies. J. Drug Delivery Sci. Technol. **54**(June), 101313 (2019)
12. Lorenzo, P., Moreno, A., Lizasoain, I., Leza, J.C., Moro, M.A., Portolés, A.V.: Farmacología Básica y Clínica (18th ed.). Médica Panamericana. (2015)
13. Escobar Sierra, D., Ossa Orozco, C., Alexander Ospina, W.: Optimización de un proto- colo de extracción de quitina y quitosano desde caparazones de crustáceos. Scientia et Technica **18**(1), 260–266 (2013)
14. Akakuru, O.U., Isiuku, B.O.: Chitosan Hydrogels and their Glutaraldehyde-Crosslinked Counterparts as Potential Drug Release and Tissue Engineering Systems - Synthesis, Characterization, Swelling Kinetics and Mechanism. Journal of Physical Chemistry & Bi- ophysics, 07(03) (2017)
15. Padilla Quintero, B.C., Cardona Trujillo, V., Muñoz Ruiz, G.A., Lizcano Valbuena, W.H., Zuluaga Corrales, H.F., Giraldo Parra, W.L., Rodríguez Marmolejo, A.F., Padilla Quintero, B.C., Cardona Trujillo, V., Muñoz Ruiz, G.A., Lizcano Valbuena, W.H., Zuluaga Corrales, H.F., Giraldo Parra, W.L., Rodríguez Marmolejo, A.F.: Preparation and Characterization of Chitosan Based Gels for Possible Biomedical Applications. Revista de Ciencias **21**(1), 91–99 (2017)
16. Cuesta Hoyos, S.A.: Estudio químico, computacional y farmacológico de Ibuprofeno. Pontificia Universidad Católica del Ecuador (2014)
17. Aragón Fernández, J., González Santos, R., Fuentes Estévez, G.: Cinética de liberación de cefalexiana desde un biomaterial compuesto por HAP-200/POVIAC/CaCO3. Anales de La Real Academia Nacional de Farmacia **75**(3), 345–363 (2009)

18. Al-Gharabli, S., Al-Omari, B., Kujawski, W., Kujawa, J.: Biomimetic hybrid membranes with covalently anchored chitosan – material design, transport and separation. Desalination **491**(May), 114550 (2020)
19. Tandekar, S., Saravanan, D., Korde, S., Jugade, R.: Gamma degraded chitosan-Fe(III) beads for defluoridation of water. Materials Today: Proceedings (2020)
20. Hernández, R., Zamora-Mora, V., Sibaja-Ballestero, M., Vega-Baudrit, J., López, D., Mijangos, C.: Influence of iron oxide nanoparticles on the rheological properties of hybrid chitosan ferrogels. J. Colloid Interface Sci. **339**(1), 53–59 (2009)
21. Reddy, D.H.K., Lee, S.M.: Application of magnetic chitosan composites for the re- moval of toxic metal and dyes from aqueous solutions. Adv. Coll. Interface. Sci. **201–202**, 68–93 (2013)
22. Submitted, T., Fulfilment, R., For, R.: Synthesis and Characterization of Chitosan-Glutaraldehyde Sorbent Materials For the Removal of Arsenate Shaguftah Younus Permission to Use. August (2014)
23. Ray, M., Pal, K., Anis, A., Banthia, A.K.: Development and characterization of chi- tosan-based polymeric hydrogel membranes. Des. Monomers Polym. **13**(3), 193–206 (2010)
24. Conley, R., Avendaño Ruiz, R., Calderón Martínez, J.: Espectroscopía Infrarroja (1st ed.). Alhambra (1979)
25. SDBS Information. Ibuprofen (2-(p-isobutylphenyl)propionic acid). https://sdbs.db.aist.go.jp/sdbs/cgi-bin/landingpage?sdbsno=10570, Accessed 08 June 2020
26. del Carmen Silva García, M., García Bermejo, M.J:. Técnico Especialista en Laboratorio de Atención Primaria (Primera). MAD (2006)
27. Campos Hernández, T.: Caracterización química del ibuprofeno en medio acuoso [Universidad Autónoma Metropolitana Unidad Iztapalapa] (2010). https://www.cdc.gov/nccdphp/
28. Arredondo Peñaranda, A., Londoño López, M.: Hidrogeles: potenciales biomateriales para la liberación controlada de medicamentos. Revista Ingeniería Biomédica **3**(5), 83–94 (2009)
29. Kumar Thakur, V., Kumar Thakur, M.: Funtional Biopolymers. Springer (2018)
30. Williams, M.: Para la salud, la condición física y el deporte. Paidotribo (2002)

Modern Vehicle Technology

Influence of Failures not Detected by the On-Board Diagnostic System on the Performance and Pollutant Emissions of a Spark Ignition Engine

Jairo Castillo-Calderón[1(✉)], Diego Díaz Sinche[1], Rubén Carrión Jaura[1], and Diego Rivera Campoverde[2]

[1] Universidad Nacional de Loja, Facultad de la Energía, las Industrias y los Recursos Naturales no Renovables, Ciudad Universitaria Guillermo Falconí, Loja, Ecuador
jdcastilloc@unl.edu.ec

[2] Universidad Politécnica Salesiana, Calle Vieja 1230 y Elia Liut, Cuenca, Ecuador

Abstract. This research is aimed at determining the influence on the performance and concentration of exhaust gas emissions of a spark ignition (SI) engine by inducing controlled failures not detected by the on-board diagnostic system (OBD). Through the Analytic Hierarchy Process (AHP), two defects are selected to cause, obtaining the air filter in poor condition and excessive spark gap. The experimental unit is a Hyundai Accent vehicle that is subjected to 16 treatments on the dynamometric bench with the engine at full load and constant speed, both in normal and defective operating conditions; two commercial fuels from the region are used with a RON of 87 and 92, respectively, whose preference of the Ecuadorian population is conditioned by the worrying increase in fuel prices. The results obtained indicate that the highest fuel consumption is 11.75 L/100 km caused by the clogged air filter and lower octane naphtha. Power and torque behave similarly; their lowest values are 14.2 kW and 67.8 Nm, respectively, due to the clogged air filter combined with the faulty spark plug and 87 octane gasoline. The highest concentration of CO is 5.21%, due to the defective spark plug together with the clogged air filter, and the HC emissions amount to 179 ppm with the clogged air filter; both polluting gases using lower quality naphtha.

Keywords: Analytic Hierarchy Process · Engine performance · Induced failures · On-board diagnosis · Pollutant emissions

1 Introduction

In the last two decades, global climate policy is the main topic on the international agenda, even more so knowing that the damages of climate change are latent, since, in many countries, as in emerging ones, they have severe consequences on health and security of power supply, among others. Thus, part of a rapid mitigation of this reality arises within the United Nations Framework Convention on Climate Change (UNFCCC),

M. Botto-Tobar et al. (Eds.): CIT 2020, AISC 1326, pp. 389–403, 2021.
https://doi.org/10.1007/978-3-030-68080-0_29

where substantial financial support is provided to developing countries [1]. One of the main factors, and of growing concern, that contribute to air pollution and climate change is traffic [2, 3], for around 95% of transport energy comes from liquid petroleum fuels and around 60% of all oil produced is destined for transport fuels [4–8]. It is estimated that 1.3 billion trucks, buses, and automobiles circulate worldwide [9–12] and the harmful gases emitted by these motor vehicles cause around 9 million deaths a year [13]; For light vehicles, the main pollutants emitted are carbon monoxide CO and hydrocarbons (HC) [14, 15]. Particularly in developing countries urban growth is, among other concerns, associated with increases in air pollution levels [16], Thus, for example, a metric of the number of hours per year that the average traveler spends on congestion shows that Ecuador has a significant position since Quito occupies 15th place and Guayaquil 22nd among more than 975 cities in 43 countries studied [17], which implies that urban mobility and its effect by vehicular pollution is alarming.

Although it is undeniable that to reduce polluting emissions and increase the sustainability of energy production, and the efficient use of it, it is a priority to procure renewable sources [18], there is still a long way to go until sustainable energy systems meet most of humanity's energy needs, including transport, so research can necessarily be directed towards the use of vehicles with internal combustion engines (MCI) and their emissions. Even with the strict regulations on pollutants and especially on CO2 carbon dioxide emissions that could restrict the use of MCI in automotive applications in the near future, it is clear that this will continue to be the dominant technology driving transportation for decades to come [19]. According to [20], research at MCI will focus on managing subsystem requirements, operating limits, and their interaction, providing a systematic calibration procedure, and adapting the control system to variations in the environment and to engine aging. On the other hand, [19] concludes that to significantly mitigate greenhouse gases (GHG) and other transport impacts, combustion engines assisted by partial electrification must be improved and control and post-treatment systems must be optimized.

In this context, numerous technological advances have been developed in the vehicle to address the various aspects of pollution control and mitigation [21]. The electronic engine control system that manages injection, ignition, and gas treatment has evolved over time, improving vehicle performance and reducing emissions [22]. However, as mileage increases, maintenance of the different subsystems is imperative to avoid failures, preserve vehicle performance, and comply with emissions regulations [23].

There is some research on simulation of failures in specific vehicle components that have gas measurement as a common factor. The impact on driving capacity, emissions and fuel consumption is evaluated by [24], by simulating deterioration or failure of the air intake, fuel supply, ignition, and exhaust of a vehicle that uses liquefied petroleum gas (LPG). In [25] the damage that a three-way catalyst (TWC) can suffer due to misfires and concentrations of rich and poor mixtures is studied. In [26] the ignition timing is varied to assess the impact on exhaust emissions and the performance of a gasoline engine. Similarly, [27] studies the effect of the variation of the ignition and injection time on the performance and emissions of a spark ignition engine using compressed natural gas (CNG) and a mixture of hydrogen with CNG as fuels, where experiments were carried out at constant 3000 rpm. In [28] the MAP sensor, the oxygen sensor, and a fuel injector are disconnected, to simulate faults that are recognized by the OBD system. Tests carried out on a dynamometric bench at various operating regimes on four vehicles equipped with the same engine show that, if an injector does not work, the emission level increases significantly more than the other failures; this study does not evaluate engine performance. In [29, 30] the failure of the exhaust gas recirculation valve (EGR) is related to the concentration of HC, CO, and nitrogen oxides (NOx). In [31] the HC, CO, and NOx are measured in three taxis with the TWC deactivated and compared with those that have it active, demonstrating that there are significantly higher air pollutant emissions when the TWC is not working. The study presented in [32] consists of the generation of failures in a 16-ton diesel truck that is subjected to various transitory tests on a chassis dynamometer. Simulated failures included those that would commonly occur in intake, fuel injection, exhaust after-treatment, and other systems; do not include misfire. The results showed that all failures increased fuel consumption, except the EGR malfunction when fully closed; HC and CO emissions were negligible, while nitrogen monoxide (NO) increased markedly. Most defects or faults that affect the level of polluting emissions in the vehicle are reported to the driver by the OBD system through the Malfunction Indicator Lamp (MIL) or Check Engine on the instrument panel. However, some faults are not recognized by the OBD system and the MIL lighting is not carried out, consequently, the driver continues to operate the car without noticing an eventual reduction in engine performance and increased pollutant emissions [28].

Precisely, the objective of this research lies in simulating engine failures not detected by the OBD system that would commonly occur by not performing adequate preventive maintenance, opportunely, on the air induction system and ignition system, and establishing their influences in polluting emissions, engine torque, power, and specific fuel consumption. In addition, the experiment considers the use of two commercial fuels from the region with a RON of 87 and 92, respectively. It is presumed that the vehicle will show an increase in emissions and fuel consumption, and a reduction in power and torque, being exposed to different failures, and will be more noticeable with lower octane fuel.

2 Materials and Methods

2.1 Test Rig and Procedures

In 2018, the number of motorized vehicles registered in Ecuador was 2.4 million, with an increase of 7.4% compared to the previous year; of these, 69,424 corresponded to the province of Loja [33]. The test vehicle selected is the Hyundai Accent, for having the highest frequency of registration records, including all categories and subcategories of motor vehicles, in the city of Loja, where the study was conducted. Its predominance in the local automotive fleet would indicate a significant incidence in polluting emissions. According to the Ecuadorian Institute for Standardization (INEN), the selected vehicle corresponds to subcategory M1: four-wheel motor vehicles designed and built for the transportation of passengers, with a capacity of no more than eight seats, excluding the driver's seat [34]. Considering the current technological validity, the Accent model is used, version 1.4 L of gasoline, year of manufacture 2019. The power and maximum nominal torque are 99 HP @ 6000 rpm and 133.4 Nm @ 4000 rpm respectively.

A total of 16 treatments were carried out, implicitly one repetition for each engine configuration, that is, under normal operating and faulty conditions to estimate the measurement uncertainties. A MAHA LPS 3000 chassis dynamometer is used with a single set of rollers with a diameter of 318 mm, where the vehicle is anchored (see Fig. 1). For each experimental condition, where engine performance and polluting emissions are evaluated, the vehicle is subjected to a protocol covered by the NTE INEN 960, SAE J1349, ISO 1585: 1992 and IS 14599: 1999 standards, which establish the method of general test for the performance evaluation of the motorized vehicle to determine the values of torque, power and specific fuel consumption at full load as a function of the speed of the machine.

For the study, the parameters involve a total load obtained of 85.50%, engine running speed of 2000 min^{-1} with a tolerance of ± 10 min^{-1}, stabilized speed of 60 km/h in fourth gear, engine coolant temperature of 90 °C $\pm$ 5 °C, intake air temperature of 23 °C $\pm$ 1 °C. As the standards mention, no data is taken unless the speed has been held constant for at least one minute with the accelerator pedal fully depressed.

Beyond the failures caused, the vehicle was previously reviewed and prepared to guarantee ideal operating conditions, that is, without presenting mechanical, electrical, or electronic problems that could affect the measurements. To measure the gaseous concentrations of CO and HC, a BRAIN BEE AGS-688 gas analyzer was used, which has INEN certification to be used as a measuring instrument for the technical review of vehicles in Ecuador. In all treatments, the sampling probe was inserted 400 mm into the car's exhaust pipe; measurements were recorded after stabilization of the engine running speed in the established interval. The data obtained are compared with each other and with the emission limits allowed for mobile sources with gasoline engines (CO $<$ 1% by volume; HC $<$ 200 ppm), defined in the NTE INEN 2204 [35].

Fig. 1. Vehicle placed on the chassis dynamometer.

2.2 Test Fuels

In Ecuador, the largest shipments of oil derivatives have been made to the automotive sector with 55.85%, that is, 15,300,982 barrels, between the months of January to May 2020 [36], where there are three types of gasoline: Super, ECOPAÍS, and Extra, however, the latter is not sold in some areas of the country due to its replacement by ECOPAÍS gasoline [37]. In this context, the fuels used for the research are Super and ECOPAÍS with a RON of 92 and 87 respectively, established by the NTE INEN 935 [38]. ECOPAÍS gasoline is a biofuel made up of 5% bioethanol and 95% base gasoline, furthermore, it has the same octane rating and price as Extra gasoline [39]; it has been marketed since 2010 in order to reduce the importation and subsidy of high-octane naphtha that is used for the preparation or refining of regular gasoline, and to contribute to the reduction of polluting gas emissions through the use of nationally produced ethanol. The Hydrocarbon Regulation and Control Agency (ARCH) is in charge of controlling the quantity and quality of the fuel that is sold in the distribution centers complies with current legal regulations to guarantee the right of consumers to receive optimal fuel conditions for use in your vehicles. The considerable rise in the price of Super gasoline through Executive Decree 490 of August 27, 2018, caused a reduction in its demand of 39.2% when comparing the period from January to July 2019 with respect to 2018 [40, 41]; The fact that users prefer the use of the other two cheaper and lower octane gasoline, raises concerns about the possible repercussions on engine performance and emissions of polluting gases.

2.3 Selection of Faults not Detected by the OBD

Initially, through an analysis of experts, the eight most typical faults in specific elements of the vehicle have been identified, where five are associated with the fuel control system and three with the ignition control system: pressure regulator, injector, fuel pump, fuel filter, and air filter, ignition coil, electrodes and spark plug wire. From these eight, two failures are selected to be caused in the experiment through the help of Expert Choice, multi-criteria decision-making software. Expert Choice implements the Analytical Hierarchy Process (AHP) which is a structured technique for organizing and analyzing complex decisions. Instead of prescribing a correct decision, AHP helps decision-makers find the one that best suits their goal and their understanding of the problem. As explained [42], the problem can be subdivided into sub-problems that start from the bottom of the hierarchy in order to identify: the alternatives; the criteria; sub-criteria and the objective to be achieved. The objective is the selection of failures, there are eight alternatives and four criteria to consider: item most likely to fail, increased severity of failure (sub-criteria: higher cost and greater impact), lower cost of fault generation, and shorter generation time of the fault for the experiment. The AHP offers a structured way to include and make the qualitative elements comparable [43], immediately the experts systematically evaluate the various elements of the structure comparing them with each other two by two, with respect to their impact on an item that is above them in the hierarchy. That is, the criteria are compared in pairs with the importance objective and the alternatives are compared in pairs with each of the preference criteria. To make the comparison, the "Saaty semantic scale" is used, which allows transforming qualitative judgments into quantitative and objective numerical judgments. This scale comprises the first nine numbers (1 = equally important and 9 = much more important) with which each of the judgments is weighted [44]. In making the comparisons, experts can use hard data about the items, but they generally use their judgments about the relative meaning and importance of the items. The essence of AHP is that human judgments, and not just the underlying information, can be used to make the assessments [45].

For the current problem, each of the eight alternatives has a default priority of 0.125, however, as the AHP progresses, the priorities change from their default values as experts enter information on the importance of the various nodes of the hierarchical structure, through the series of pairwise comparisons. For the specific case, with respect to the objective "Selection of failures", it was obtained that the air filter and the spark plug electrode, with priorities of 0.179 and 0.152 respectively, were the failures selected for the experiment. The level of inconsistency is located at 0.09, below the 0.1 allowed, giving the results an acceptable status, since this shows that there are no subjectivities derived from comparisons between pairs through the judgments issued. The Fig. 2 is showing the sensitivity graph, which allows to have a better understanding of the attributes presented by the failures evaluated, based on the selection criteria.

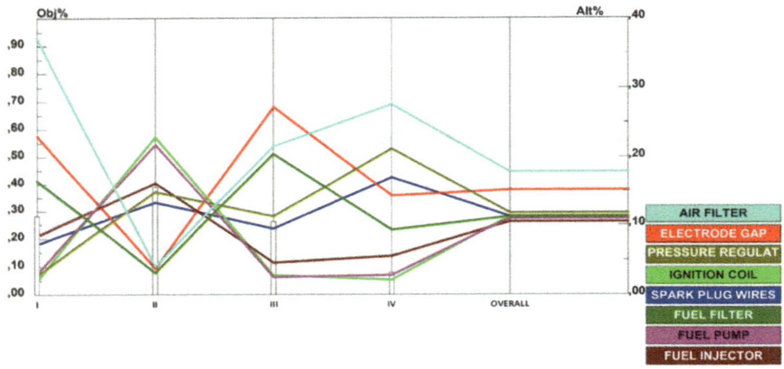

Fig. 2. Performance sensitivity for target: fault selection. Criteria: I. Item most likely to fail; II. Increased severity of failure; III. Lower cost of fault generation; IV. Shorter generation time of the fault.

2.4 Simulation of Malfunction

The vehicle in normal operating conditions is the starting reference for comparison with simulated engine failures, which are not detected by the OBD system. Defects that would commonly occur by failing to perform proper preventive maintenance on the air induction and ignition system are simulated. The simulated malfunctions consider failures individually and in combination, where a repetition is performed for each treatment. The defects caused are described below.

Clogged Air Filter. The air filter prevents dust penetration into the combustion chamber and thus protects the kinematic coupling against abrasive wear [46]. By dispensing with the air filter, engine wear would be accelerated and combustion efficiency would be reduced [32]. To simulate the failure, a used and blocked air filter is installed that has been in use for 18,000 km, where its mass is 6 g greater than a new one. A clogged air filter decreases the volumetric performance of the engine and increases pumping losses, leading to higher fuel consumption [32].

Faulty Spark Plugs. The spark plug ignites the working fluid that is under pressure in the combustion chamber by means of an electric spark that jumps between its electrodes. Over time the electrodes wear out causing an increase in the disruptive distance, this means that the amount of energy required to form the spark is also greater, which can affect the duration of the spark, which is insufficient, the engine performance, fuel consumption, and pollutant emissions [24, 47]. To simulate the failure worn spark plugs are used, the electrodes of which have an inadequate gap of 1.3 mm compared to 1 mm for new spark plugs.

3 Results and Discussion

The experimental results and discussion are presented in two sections. The first covers the effect of the engine malfunction on its performance, and the second covers the effect of the malfunction on polluting gas emissions, including the comparison with the current Ecuadorian regulation. In both cases, the use of the country's two commercial gasoline is evaluated, as well as the failures generated, and their combinations, with respect to the engine operating under normal conditions. In addition, the behavior of the engine performance is evaluated in the first section when carrying out the test of the net power measurement.

3.1 Engine Performance

Figure 3 shows the effect of engine malfunction, due to induced faults, on the fuel consumption rate (L/100 km). The same trend exists for both types of gasoline, both under engine malfunction conditions and under normal reference conditions. It is noteworthy that all failures increase the rate of fuel consumption; it should be noted that, when using higher-octane gasoline, consumption is reduced [48]. Thus, the lowest consumption obtained is 10.90 L/100 km, corresponding to the faultless engine that uses 92 RON gasoline. When the engine is subjected to high loads, as in all the tests in this study, the amount of fuel injected increases [49, 50], this suggests the engine runs on rich mixtures [28]. This enrichment ($\lambda < 1$) is even more significant with a clogged air filter since the volumetric efficiency and the intake pressure are reduced, causing a greater pumping loss that affects the air intake, justifying the consumption increases [51]. For the specific case, the highest consumption is 11.75 L/100 km, caused by the clogged air filter and the use of 87 RON gasoline. On the other hand, Fig. 4 shows the box plot where the behavior of the variable specific fuel consumption of the four groups corresponding to the operating condition of the engine is evaluated, obtained through the measurement of the net power, with 92 octane gasoline. The average specific consumptions, as well as the variability, of the groups are similar. It is denoted that the median specific consumption is slightly the lowest for the non-fault engine with 183.19 g/kWh, and the highest is for the malfunction condition with clogged air filter, this is 185.50 g/kWh. These data suggest having a similar trend to those obtained in Fig. 3.

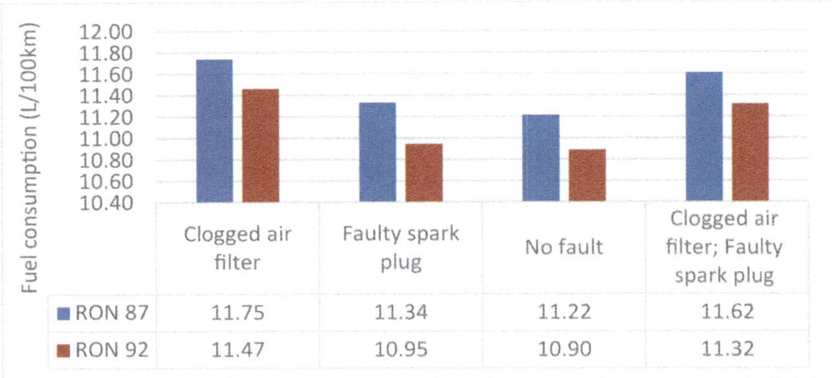

Fig. 3. Influence of engine failures on fuel consumption (L/100km).

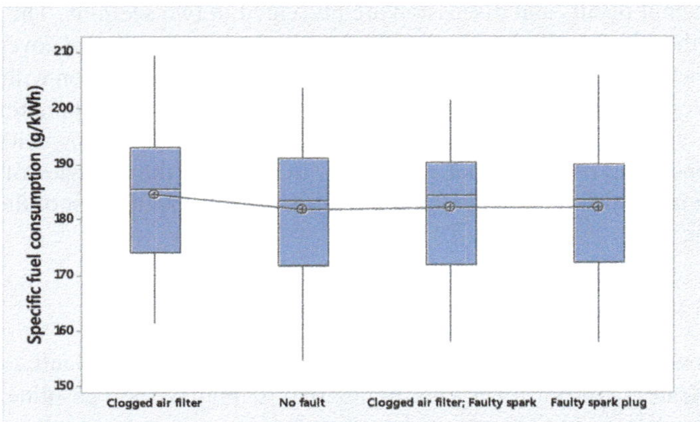

Fig. 4. Box plot of specific fuel consumption at different engine conditions.

As regards the motor power (kW), in Fig. 5 its variation is observed concerning the failures caused. For all operating conditions there is an increase in power due to the use of higher-octane gasoline [48]. The lowest power obtained is 14.2 kW, concerning the engine that uses 87 octane gasoline in malfunctioning conditions, both with the clogged air filter, which produces a loss of load, and the excessive disruptive distance of the electrodes spark plugs, which do not offer an optimal ignition that ensures complete combustion of the working fluid, conditioning the engine performance [52]. In the other three configurations, there is considerable similarity in the maximum power values achieved with the highest octane fuel. On the other hand, Fig. 6 shows the box plot evaluating the behavior of the power variable for each engine operating condition, obtained through the measurement of net power, with 92 octane gasoline. A similar behavior is observed with respect to the data in Fig. 5, since the mean powers, as well as the variability, of the groups are analogous, where it could be mentioned that the lowest median is 38.28 kW for the fault condition with clogged air filter and faulty spark plug.

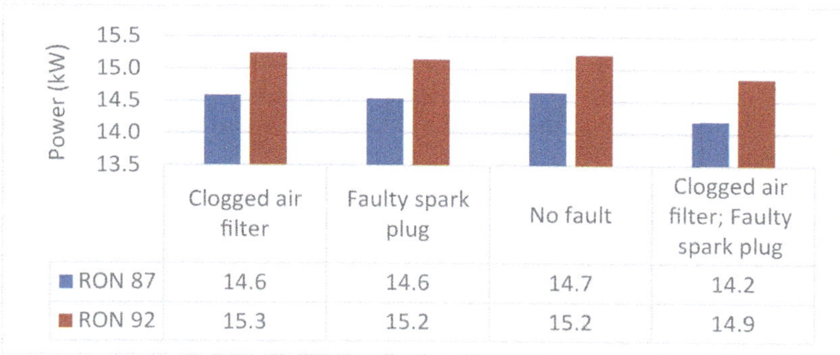

Fig. 5. Influence of engine failures on power (kW).

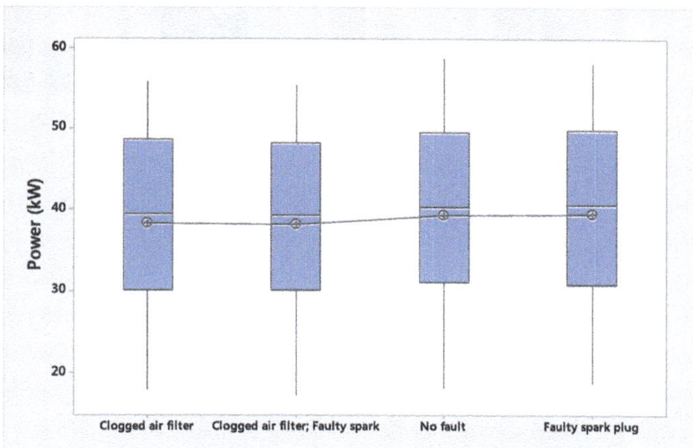

Fig. 6. Box plot of power at different engine conditions.

Engine torque (Nm) shows a behavior similar to power; Fig. 7 shows its variation concerning the failures caused. With 92 octane gasoline the torque is higher in all engine operating configurations compared to lower octane gasoline, in contrast to what was demonstrated in [53]. The higher engine performance observed with 92 RON gasoline is attributed to its optimized manufacturing to work with a high octane, however, if the fuel used has an octane higher than that stipulated by the manufacturer, the performance would not improve, according to the results of [54], because some engine modification should be accompanied for this. The lowest torque obtained is 67.8 Nm, with the engine using 87 octane gasoline in malfunctioning conditions, that means, with clogged air filter and faulty spark plug. In the remaining conditions there is similarity in the maximum torque values, where the engine without failures stands out slightly with the 92 octane fuel. Figure 8 shows the box plot evaluating the behavior of the torque variable of the four groups corresponding to the operating condition of the engine, obtained through the measurement of net power, with 92 octane gasoline. The average torques of the groups are similar, but the torques of some of the groups are more variable than others. The median torque is highest for the faultless motor (132.3 Nm), however, this operating condition demonstrates the greatest variability, with an interquartile range of 6.47 Nm; furthermore, the distribution has a negative skewness, with a minimum torque of approximately 117 Nm. The faulty spark plug malfunction condition shows the least variability, with an interquartile range of 3.36 Nm and has a median torque similar to the malfunction condition with clogged air filter and faulty spark plug, it means, 127.9 and 128.2 respectively. Distribution of engine operating condition data with clogged air filter and faulty spark plug shows the lowest torque values. The data made with this test maintain a trend or pattern similar to that obtained in Fig. 7.

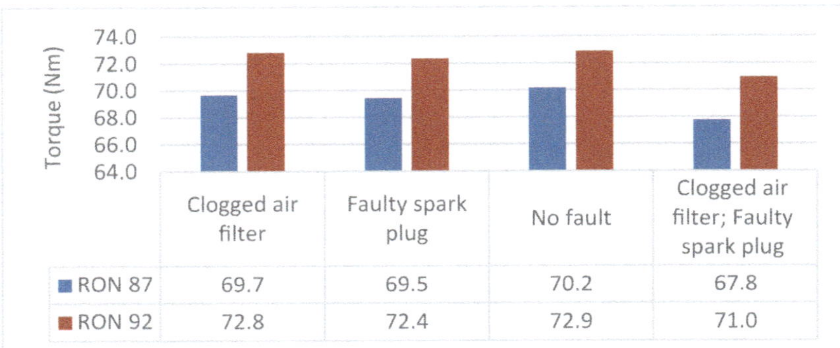

Fig. 7. Influence of engine failures on torque (Nm).

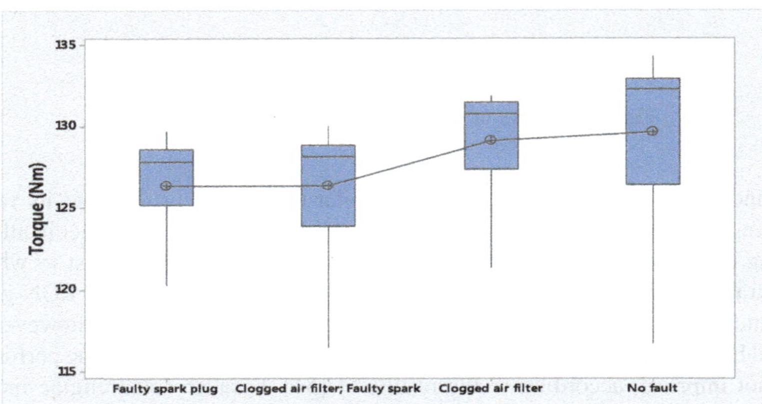

Fig. 8. Box plot of torque at different engine conditions.

3.2 Pollutant Emissions

Based on Fig. 9, it can be seen that the HC emissions increase with a lower RON, this corresponds with the findings of [48]; except the engine in fault conditions with clogged air filter and defective spark plug. Likewise, according to Fig. 10, CO emissions are slightly lower with a RON of 92 compared to RON of 87, in agreement with the results of [54]. The impact of the simulated dysfunctions does not show major differences in HC and CO emissions with respect to the standard condition of the engine, in fact, in some cases, they are lower, corresponding to the findings of [24]. It is highlighted that the lowest HC emissions are obtained with a defective spark plug, following the results of [47], where it is justified that the greater separation of electrodes involves a greater volume of the mixture exposed to the spark that benefits its ignition. If the electrode gap is small, the core of the flame may go out due to the possible transfer of heat from the flame to the electrode; besides, the closer proximity of the electrodes would result in a loss of the turbulence benefits of the mixture between them, causing a less efficient combustion process and more HC emissions. On the other hand, it is verified that, having the clogged air filter, the engine will run with a rich mixture, which leads to the maximum values of HC, specifically 179 ppm with a RON of 87.

Regarding CO concentrations, the lowest are 3.59% during the operation of the engine without failures and 87 octane gasoline; a close behavior is observed with a RON of 92. The highest CO emissions, 5.21%, are achieved under the conditions of a clogged air filter and a defective spark plug, using lower octane naphtha. This is due to incomplete combustion as the proportion of oxygen to the engine is lower due to the restriction caused, that is, having a rich mixture, added to it the effect of excessive separation of the electrodes from the spark plug, where the similar findings of [55], explain a detriment in the quality of combustion of the working fluid, revealing a greater increase in CO emissions. In general, HC emission concentrations are relatively low for all engine operating conditions, remaining below the maximum limit allowed in current national regulation, that is, HC < 200 ppm. The opposite occurs with CO emission concentrations, where in all engine operating conditions the values are higher than those stipulated in the standard, CO <1% by volume. However, the maximum limits set in the standard are due to a static test and the tests carried out in this investigation evaluate the pollutant emissions at full load and constant engine operating speed.

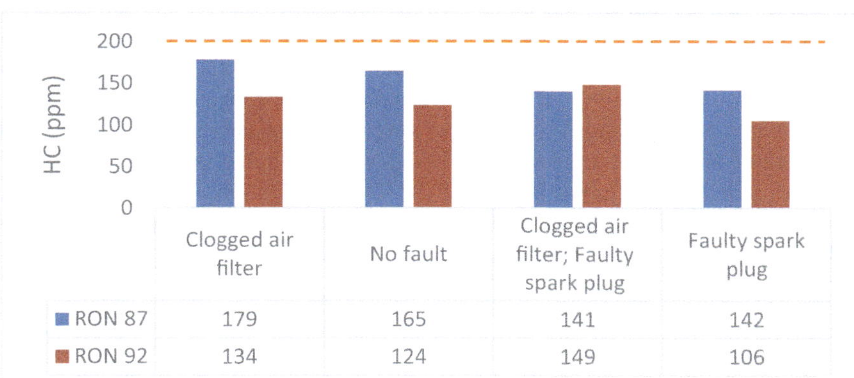

Fig. 9. Influence of failures on HC emissions. The dashed line indicates the limit established by NTE INEN 2204.

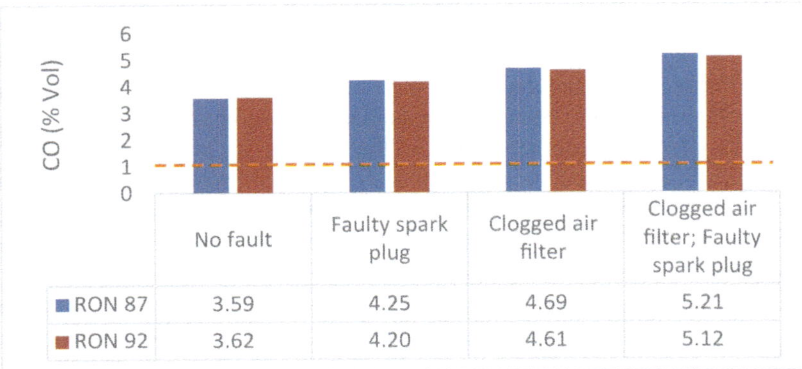

Fig. 10. Influence of failures on CO emissions. The dashed line indicates the limit established by NTE INEN 2204.

4 Conclusions

It is concluded that the lowest rate of fuel consumption is 10.90 L/100 km, corresponding to the faultless engine that uses gasoline with a RON of 92, and the highest consumption is 11.75 L/100 km caused by the clogged air filter and use of gasoline with a RON of 87. On the other hand, the lowest power obtained is 14.2 kW, relative to the engine in malfunctioning conditions, that is, with the clogged air filter, excessive spark gap, and gasoline of 87 octanes. In the other three engine configurations, there is considerable agreement in the maximum power values reached with the highest octane fuel. Engine torque has a behavior similar to power, where the lowest is 67.8 Nm, with the use of 87 octane gasoline, a clogged air filter, and a defective spark plug. In the remaining conditions, there is a similarity in the maximum torque values, where the faultless engine with the 92 octane fuel slightly stands out, which gave 72.9 Nm. The lowest concentration of HC emissions is obtained with a faulty spark plug, specifically 106 ppm with a RON of 92. Meanwhile, the clogged air filter leads to the maximum HC values, which is 179 ppm with a RON of 87; however, it remains below the maximum limit allowed in current national regulation. All engine operating conditions give CO values above those stipulated in the standard. Thus, the lowest CO emissions are 3.59% with the faultless engine and 87 octane gasoline; although with the higher-octane naphtha, 3.62% is obtained, indicating a similar behavior. The highest CO emissions are 5.21% under the conditions of a clogged air filter and faulty spark plug, using lower octane gasoline. Clearly, the best engine performance and emissions are produced with the use of better quality gasoline.

References

1. Michaelowa, K., Betzold, C., Weiler, F.: Development Aid and Adaptation to Climate Change in Developing Countries (Cham, CH: Palgrave MacMillan). The Review of International Organizations, vol. 13(4), pp. 621–625. Springer (2018)

2. Laña, I., Del Ser, J., Padró, A., Vélez, M., Casanova-Mateo, C.: The role of local urban traffic and meteorological conditions in air pollution: a data-based case study in Madrid. Spain. Atmos. Environ. **145**, 424–438 (2016)
3. Weber, C., Sundvor, I., Figenbaum, E.: Comparison of regulated emission factors of Euro 6 LDV in Nordic temperatures and cold start conditions: diesel- and gasoline direct-injection. Atmos. Environ. **206**, 208–217 (2019)
4. BP Energy outlook (2017). https://www.bp.com/content/dam/bp/business-sites/en/global/corporate/pdfs/energy-economics/energy-outlook/bp-energy-outlook-2017.pdf. Accessed 03 June 2020
5. Exxonmobil, 2017 Outlook for energy: a view to 2040. https://safety4sea.com/wp-content/uploads/2017/10/ExxonMobil-2017-Outlook-for-Energy-A-View-to-2040-2017_10.pdf. Accessed 4 June 2020
6. World Energy Council, Global Transport Scenarios 2050. https://www.worldenergy.org/assets/downloads/wec_transport_scenarios_2050.pdf. Accessed 14 June 2020
7. OPEC. World oil outlook (2013). https://www.opec.org/opec_web/static_files_project/media/downloads/publications/WOO_2013.pdf. Accessed 14 June 2020
8. U.S. Energy Information Administration (EIA). International energy outlook (2016). https://www.eia.gov/outlooks/ieo/pdf/0484(2016).pdf. Accessed 17 June 2020
9. Nakashima, Y., Sadanaga, Y., Saito, S., Hoshi, J., Ueno, H.: Contributions of vehicular emissions and secondary formation to nitrous acid concentrations in ambient urban air in Tokyo in the winter. Sci. Total Environ. **592**, 178–186 (2017)
10. Qin, Y.M., Tan, H.B., Li, Y.J., Schurman, M.I., Li, F., Canonaco, F., Prévôt, A.S.H., Chan, C.K.: Impacts of traffic emissions on atmospheric particulate nitrate and organics at a downwind site on the periphery of Guangzhou. China. Atmos. Chem. Phys. **17**, 10245–10258 (2017)
11. Singh, V., Sahu, S.K., Kesarkar, A.P., Biswal, A.: Estimation of high-resolution emissions from road transport sector in a megacity Delhi. Urban Climate **26**, 109–120 (2018)
12. Ma, F., Liang, Y., Yuen, K.F., Sun, Q., Zhu, Y., Wang, Y., Shi, W.: Assessing the vulnerability of urban rail transit network under heavy air pollution: a dynamic vehicle restriction perspective. Sustain. Cities Soc. **52**, 101851 (2020)
13. Lelieveld, J., Klingmüller, K., Pozzer, A., Pöschl, U., Fnais, M., Daiber, A., Münzel, T.: Cardiovascular disease burden from ambient air pollution in Europe reassessed using novel hazard ratio functions. Eur. Heart J. **40**, 1590–1596 (2019)
14. Lopes, T.F.A., Policarpo, N.A., Vasconcelos, V.M.R., Oliveira, M.L.M.: Vehicular emissions estimate in the Fortaleza, Ceará, Brazil, metropolitan region in 2010. Engenharia Sanit. Ambiental **23**(5), 1013–1025 (2018)
15. Teixeira, E.C., Feltes, S., Santana, E.R.R.: Study of the emissions from moving sources in the metropolitan area of Porto Alegre, RS. Brazil. Química. Nova **31**(2), 244–248 (2008)
16. Shen, L., Shuai, C., Jiao, L., Tan, Y., Song, X.: A global perspective on the sustainable performance of urbanization. Sustain. **8**, 1–16 (2016)
17. INRIX Global Traffic Scorecard (2019). https://inrix.com/scorecard/
18. Caetano, N.S., Mata, T.M., Martins, A.A., Felgueiras, M.C.: New trends in energy production and utilization. Energy Procedia **107**, 7–14 (2017)
19. Kalghatgi, G.: Is it really the end of internal combustion engines and petroleum in transport? Appl. Energy **225**, 965–974 (2018)
20. Payri, F., Luján, J.M., Guardiola, C., Pla, B.: A challenging future for the IC engine: new technologies and the control role. Oil & Gas Science and Technology–Rev. IFP Energies nouvelles **70**(1), 15–30 (2015)
21. Chandra, S., Camal, F.: A simulation-based evaluation of connected vehicle technology for emissions and fuel consumption. Procedia Eng. **145**, 296–303 (2016)

22. Zhang, Q., Fan, J., Yang, W., Ying, F., Bao, Z., Sheng, Y., et al.: Influences of accumulated mileage and technological changes on emissions of regulated pollutants from gasoline passenger vehicles. J. Environ. Sci. **71**, 197–206 (2018)
23. He, L., Hu, J., Yang, L., Li, Z., Zheng, X., Xie, S., et al.: Real-world gaseous emissions of high-mileage taxi fleets in China. Sci. Total Environ. **659**, 267–274 (2019)
24. Organ, B., Huang, Y., Zhou, J., Yam, Y., Mok, W., Chan, E.: Simulation of engine faults and their impact on emissions and vehicle performance for a liquefied petroleum gas taxi. Sci. Total Environ. **716**, 137066 (2020)
25. Lee, S., Bae, C., Lee, Y., Han, T.: Effects of engine operating conditions on catalytic converter temperature in an SI engine. SAE International (2002)
26. Zareei, J., Kakaee, A.: Study and the effects of ignition timing on gasoline engine performance and emissions. Eur. Transp. Res. Rev. **5**(2), 109–116 (2013)
27. Khan, M., Qidwai, M., Chausalkar, A.: Effect of Variable Ignition and Injection Timing on Emission Characteristics of SI Engine Using CNG and HCNG as Fuel. Int. J. Adv. Mech. Eng. **4**(1), 83–94 (2014)
28. Toma, M., Micu, D., Andreescu, C.: Influences of engine faults on pollutant emission. Procedia Manuf. **32**, 529–536 (2019)
29. Nevius, T., Rooney, R., Zummer, R.: Speciation of nitrogen oxides in a light duty diesel engine during an EGR system failure. SAE International (2012)
30. Rajesh, B., Saravanan, S.: Effect of exhaust gas recirculation (EGR) on performance and emissions of a constant speed DI diesel engine fueled with pentanol/diesel blends. Fuel **160**, 217–226 (2015)
31. Zheng, X., Wu, Y., Zhang, S., He, L., Hao, J.: Evaluating real-world emissions of lightduty gasoline vehicles with deactivated three-way catalyst converters. Atmospheric Pollut. Res. **9**, 126–132 (2018)
32. Huang, Y., Ng, E.C.Y., Yam, Y., Lee, C.K.C., Surawski, N.C., Mok, W., et al.: Impact of potential engine malfunctions on fuel consumption and gaseous emissions of a euro VI diesel truck. Energy Convers. Manag. **184**, 521–529 (2019)
33. INEC, Anuario de Estadísticas de Transporte 2018. https://www.ecuadorencifras.gob.ec/documentos/web-inec/Estadisticas_Economicas/Estadistica%20de%20Transporte/2018/2018_ANET_BOLETIN.pdf. Accessed 22 June 2020
34. NTE INEN-ISO 3833:2008: Vehículos Automotores. Tipos. Términos y Definiciones, Quito-Ecuador (2008)
35. NTE INEN 2204:2016: Gestión Ambiental. Aire. Vehículos Automotores. Límites Permitidos de Emisiones Producidos por Fuentes Móviles Terrestres que Utilizan Gasolina, Quito-Ecuador (2016)
36. EP PETROECUDOR. Informe Estadístico Enero–Mayo 2020. https://www.eppetroecuador.ec/wp-content/uploads/downloads/2020/06/INFORME-ESTADISTICO-MAYO-2020.pdf, Accessed 02 July 2020
37. EP PETROECUDOR. Informe Estadístico Enero 2018. https://www.eppetroecuador.ec/wp-content/uploads/downloads/2018/02/INFORME-ESTADISTICO-ENERO-2018-V1-23-02-2018.pdf. Accessed 03 July 2020.
38. NTE INEN 935: Productos derivados de petróleo. Gasolina. Requisitos, Quito-Ecuador (2016)
39. Agencia de Regulación y Control Hidrocarburífero, ARCH: Recurso, La Revista Hidrocarburífera del Ecuador (2017). https://www.controlhidrocarburos.gob.ec/wp-content/uploads/2017/11/Revista-ARCH-2017.pdf. Accessed 07 July 2020
40. EP PETROECUDOR. Informe Estadístico Enero–Diciembre 2019. https://www.eppetroecuador.ec/wp-content/uploads/downloads/2020/01/INFORME-ESTADISTICO-DICIEMBRE-2019.pdf. Accessed 04 June 2020

41. EP PETROECUDOR. Informe Estadístico Enero–Diciembre 2018. https://www.eppetroecuador.ec/wp-content/uploads/downloads/2019/01/INFORME-ESTADISTICO-DICIEMBRE-2018-16-01-2019.pdf. Accessed 04 June 2020
42. De Luca, S., Di Pace, R., Bruno, F.: Accounting for attitudes and perceptions influencing users' willingness to purchase Electric Vehicles through a Hybrid Choice Modeling approach based on Analytic Hierarchy Process. In: AIIT 2nd International Congress on Transport Infrastructure and Systems in a changing world, Rome-Italy (2019)
43. Heindl, A., Liefner, I.: The Analytic Hierarchy Process as a methodological contribution to improve regional innovation system research: explored through comparative research in China. Technol. Soc. **59**, 101197 (2019)
44. Ji, P., Jiang, R.: Scale transitivity in the AHP. J. Oper. Res. Soc. **54**(8), 896–905 (2003)
45. Saaty, T.: Relative Measurement and its Generalization in Decision Making: Why Pairwise Comparisons are Central in Mathematics for the Measurement of Intangible Factors. The Analytic Hierarchy/Network Process. Rev. R. Acad. Cien. Serie A. Mat. **102**(2), 251–318 (2008)
46. Toma, M., Bobâlcă, C.: Research on drivers' perception on the maintenance of air filters for internal combustion engines. Procedia Technol. **22**, 961–968 (2016)
47. Badawy, T., Bao, X., Xu, H.: Impact of spark plug gap on flame kernel propagation and engine performance. Appl. Energy **191**, 311–327 (2017)
48. Stradling, R., Williams, J., Hamje, H., Rickeard, D.: Effect of octane on performance, energy consumption and emissions of Two Euro 4 passenger cars. Transp. Res. Procedia **14**, 3159–3168 (2016)
49. Kumar, D., Kumar, A., Datta, A., Kumar, R.: Advances in Internal Combustion Engine Research: Energy, Environment, and Sustainability. 1st edn. Springer, Cham (2017)
50. Sharma, N., Kumar, A.: Effect of fuel injection pressure and engine speed on performance, emissions, combustion, and particulate investigations of gasohols fuelled gasoline direct injection engine. Energy Resour. Technol. **142**(4), 042201 (2020)
51. Basshuysen, Rv., Schäfer, F.: Internal combustion engine handbook – basics, components, system, and perspectives. 2nd ed. SAE International (2016).
52. Bhaskar, H.: Effect of spark plug gap on cycle-by-cycle fluctuations in four stroke spark ignition engine. Int. J. Innovative Res. Dev. **5**(11), 85–90 (2016)
53. Rashid, A., Mansor, M., Ghopa, W., Harun, Z., Wan, W.: An experimental study of the performance and emissions of spark ignition gasoline engine. Int. J. Automot. Mech. Eng. **13**(3), 3540–3554 (2016)
54. Rashid, A.K., Abu Mansor, M.R., Racovitza, A., Chiriac, R.: Combustion characteristics of various octane rating fuels for automotive thermal engines efficiency requirements. Energy Procedia **157**, 763–772 (2019)
55. Daut, A., Bahrom, M., Abdul, R., Nadirah, J.: Effect of side gapping spark plug on engine performance and emission. Int. J. Recent Technol. Eng. **8**(4), 6145–6148 (2019)

Characterization of Mechanical Properties of Materials Composed of Epoxy Resin Reinforced with Carbon Fiber and Polyurethane Foam, Epoxy Resin Reinforced with Carbon Fiber and Aluminum Honey-Comb and Polyester Resin Reinforced with Glass Fiber with Polyurethane Foam for an Electric Sports Vehicle for Academic Purposes

D. C. Dulcey-Diaz[1](✉), O. Lengerke-Pérez[1], A. D. Rincón-Quintero[1,2], C. L. Sandoval-Rodriguez[1,2], C. G. Cardenas-Arias[1,2], and C. A. Castañeda-Flórez[1]

[1] Unidades Tecnológicas de Santander UTS, Bucaramanga, Santander 680005, Colombia
{ddulcey,olengerke,arincon,csandoval,ccardenas}@correo.uts.edu.co, cjosecastaneda@uts.edu.co
[2] University of the Basque Country UPV/EHU, 48013 Leioa, Bizkaia, Spain

Abstract. In the present work, the main objective is to characterize the mechanical properties of composites materials such as epoxy resin reinforced with carbon fiber and polyurethane foam (ER-CP), epoxy resin reinforced with carbon fiber and aluminum honeycomb (ER-CA), and polyester resin reinforced with glass fiber and polyurethane foam (PR-GP), that will allow selecting the most accurate material for the future construction of parts for a prototype electric vehicle for academic competitions, taking into account the impact resistance, deformation for flexure charges and the cost. To obtain the variables, identification is required by the performance of mechanic tests like flexion and impact. The investigation begins with the manufacture of all test tubes for mechanical flexural and impact tests according to ASTM C393 and ASTM D6110 respectively. Finally, the analysis of each test is carried out according to the criteria of resistance, weight, and cost to select the most precise material. The experimental evaluation found that the composite material with the best performance as a proposal for use in applications of an electric sports vehicle in academic competitions is PR-GP.

Keywords: Bodywork · Composite · Electric vehicle · Flexure · Impact

M. Botto-Tobar et al. (Eds.): CIT 2020, AISC 1326, pp. 404–417, 2021.
https://doi.org/10.1007/978-3-030-68080-0_30

1 Introduction

Composite Materials for Automotive Applications Overview

In the present, the composite materials are born by the need to bond two or more materials to obtain a combination of properties not found individually in the materials. In the fabrication of bodyworks is common to find metallic materials like steel and aluminum alloys. In the manufacture of chassis, aluminum is the second most used material, due to its advantages of low weight and good resistance to fluctuating loads, torsion, and impact loads, and it is not corrosive in atmospheric condition [1, 2]. These materials make the vehicles heavier and consume more fuel because of that; in the last decades, the automotive industry has been set in composite materials since they can preserve mechanical properties and reduce the weight of the vehicle, saving fuel consumption. Greenhouse gases continue to increase, to reverse it is necessary to replace high-strength steel, aluminum, and CFRP with new materials of lower density that allow to improve fuel efficiency and increase the life-long drive distance [3]. Composite materials reinforced with fiberglass and carbon fiber have an excellent performance to impact loads; due to their high levels of specific energy absorption, especially those whose thermoplastic matrix provides benefits in terms of toughness [4, 5]. Wang et al. [6] use composite materials of Carbon Fiber-Reinforced Composite Sandwich Structures with Aluminum Honeycomb Cores through the effects of core thickness and density on the laminate material properties studying the three-point bending and panel peeling tests.

The challenges to design a composite material are often part of the process. One of the main characteristics of a composite material is its ability to be lighten without losing good mechanical properties but also the performance and productions costs are important too, the methodology that couples weight-optimization and technical cost modeling through an application-bound cost. Lightweight designs and implementation of electric motors are strategies of the automotive industry to avoid high demand for fuel and reduce emissions, respectively, both strategies could be combined to reduce environmental impacts even more [7, 8]. The material systems method and application-bound comparative value of weight have led to a solid discussion for future design recommendations; the sandwich structures are weight and cost efficient, the recycled carbon fiber prepeg provide up to a 50% on cost reductions [9], but this structures are susceptible to failures like de-bonding of skin and core, core shear failure, buckling [10, 11].

The carbon fiber reinforced polymer composites are used in many industries such as the vehicle industry because can reduce the weight of structures and increase the strength, durability, and stiffness due to its good mechanical properties. It is well known that the aerospace industry has used composite materials for many decades and in consequence, its prices have risen, but now thanks to the new materials the inclusion on the market can be easier [12].

In these changing times, the need to improve and be more efficiently has led to the automobile transportation and industry to grow fast in the past few decades especially, in the urban area, considering composites materials to be the new alternative due to its good performance in such good mechanical properties like fatigue resistance and impact. Customer demands and global warning for carbon saving and green mobility have been the principal motive in this investigation and how the new composite materials will provide a very important approaching a new era for the automotive industry [13, 14].

Before using a composite material for a racing propose or an efficiently competition it is important to know how will the vehicle react on a collision situation, the energy absorption is an important test to implement when designing a prototype. Based on the energy absorption characteristics of composites, the competitions rules and the overall vehicle arrangement, the most accurate material for the components have to be selected, in order to protect the driver. The fiber reinforced epoxy resin composite is used for its good mechanical properties [15].

Carlstedt, D., & Asp, L. E. [16] presents an approach in the use of composite materials to analyze the performance for a framework on a structural battery on an electric vehicle (EV). The mechanical and electrical properties are known from the material. The framework is to evaluate the estimation of the electric vehicle drive range for existing batteries in EV. The study shows that weight saving potential doesn't affect the drive range performance and also increases by up to 70%. The application of CFRP unidirectional plies on batteries for EV presented good results, due to the high specific mechanical and electrical performance; the mass reduce especially in this area helps to increase the vehicle drive range and provides the battery for a better life span. This investigation provides a better understanding of the use of CFRP on battery systems and the environmental impact.

There are many international academic competitions focused on the development of prototype vehicles such as the Shell Eco-Marathon and The Solar Challenge that seek energy efficiency through the use of composite materials but most of their resistance providing a high increase compared to traditional materials in their mechanical properties. The University of Lisbon Formula Student Team developed a crash absorber for their vehicle prototype to replace the aluminum one using composite materials [17, 18].The World Solar Challenge is a competition of solar vehicles; the University of Bologna designed a strategy based on changing the prototype vehicle form metal to a fiber-reinforced composite, the new prototype with CFRP lightening the roof as a whole. This integration between structural and non-structural parts in the composite roof prototype also allowed other elements to be lightening, in particular, those locally supporting the solar cells, reducing the overall weight by roughly 15 kg [19]. The Shell Eco-Marathon competition is about energetic efficiency more than a velocity competition, for that, the materials used in the tests named above are composites materials ER-CP, ER-CA, and PR-GP, resulting in the selection of the most suitable material for the future construction of the bodywork for the prototype vehicle.

Among the activities to justify the use of composite materials in the automotive industry is to achieve an analysis of the fuel consumption induced by the weight of the automobile to determine the fuel reduction coefficient, seeking the lowest process benefit amount of fuel and reduce the cost of its operation [7]. For this reason, this investigation is focused on the characterization of composites sandwich materials by performing mechanical tests of flexion and impact for the future use of bodywork for a vehicle in an academic energy efficiency competition.

2 Methods and Materials Selection

In this research, analyzed three materials were worked, ER-CP, ER-CA, and PR-GP, the volumetric fractions for the work are: 80% matrix and 20% reinforcement, and the density of the composite material was determined by the mix rule for reinforced fiber materials [19].

The sandwich-structured composite is chosen on the bases of material properties. The reinforcement and core materials acquired for the construction of the specimens are the carbon fiber Fig. 1(a) hat is made up of fine carbon composite filaments, the glass fiber Fig. 1(b) consists of numerous polymeric filaments based on silicon dioxide are used as reinforcement and the aluminum honeycomb Fig. 1(c) is constructed by honeycomb aluminum panels used as the core. Other materials such as isocyanate and polyol are chemical compounds used to make polyurethane foam, the 856 polyester resin is the main compound of the polyester matrix, the Mek (methyl ethyl ketone) is the catalyst for polyester resin and the styrene is used that serves to dissolve the polyester resin and acts in the catalyzing time, the rigid epoxy resin (A) is the main compound of the matrix of the ER-CP and ER-CA materials, together with the epoxy resin is used as the catalyst (B).

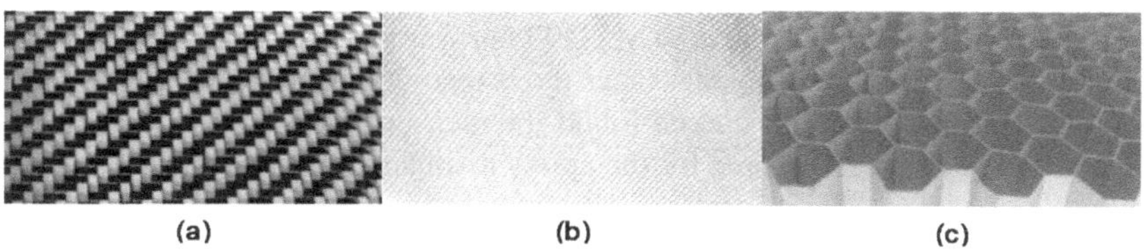

Fig. 1. Reinforcement materials

2.1 Test Tubes for Flexure Testing

The manual stratification method was used to elaborate the test tubes for the flexure testing, due to its simple application. But before starting the PU core for the material has to be made.

a) *Elaborations of the molds for PU foam for the flexure test tubes.*
   Two wooden-molds are considered for the elaboration for the PU foam core used on test tubes for the flexure testing, each mold is up to 4 test tubes, due to the ASTM C393 norm. The molds have a tolerance in its dimensions.
b) *Elaboration of the PU foam core for the flexure testing.*
   Wooden was the chosen material due to the low cost and easy way to handle. The specifications of the mold are listed on the Table 1.
c) *PU foam elaboration.*
   The PU foam has two components, an A part called Isocyanate and a B part called polyol. It is necessary to know the mold volume, because the PU foam grows up to 40 times its original volume, on the Table 2 there are the quantities for the A and B parts.

Table 1. Flex test mold specifications.

Specification	Dimensions (cm)
Long	41
Width	16
Depth	2.1
Volume	1337.6 cm^3

Table 2. Proportions for polyurethane foam for bending specimens

Impact Test	Proportion (cm^3)
Isocyanate	34,4
Polyol	34,4

d) *Measurements and work volume for the flexure test tubes.*
For the flexure test tubes, the ASTM C393 norm was used, which says the specific measurements for the testing, also says the is recommended to test 5 specimens for the material. In Table 3, it can be found the measurements for the specimens and detailed the new work volume for the epoxy resin and polyester resin.

Table 3. Measurements plane and work volume for the flexion test.

Specification dimensions	Measurements and plane	Work volume epoxy and polyester resin
Large (cm)	20	41
Width (cm)	7	16
Thickness (cm)	1.5	0.2
Volume (cm^3)	210	131.2

When the volumetric fractions are established and the work volume, on the Table 4 the volume and weight for the matrix needed.

2.2 Test Tubes for Impact Testing

The manual stratification method was used to elaborate the test tubes for the flexure testing, due to its simple application. But before starting the PU core for the material has to be made.

a) *Elaborations of the molds for PU foam for the impact test tubes.*

Table 4. Necessary amounts of matrix and reinforcement for polyurethane foam for bending specimens.

Matrix		Reinforcement	
Volume (cm^3)	Weight (g)	Volume (cm^3)	Weight (g)
104,96	114,4	26,24	49,85

A wooden-mold is considered for the elaboration for the PU foam core for test tubes for the impact testing, the mold is up to 7 test tubes, due to the ASTM D6110 norm. The molds have a tolerance in its dimensions.

b) *Elaboration of the PU foam core for the impact testing.*
Wooden was the chosen material due to the low cost and easy way to handle. The specifications of the mold are listed on the Table 5.

Table 5. Impact test mold specifications.

Specification	Dimensions (cm)
Long	18
Width	13
Depth	2.1
Volume	491.4 cm^3

3 Experimental Set Up

3.1 Three-Point Flexure Testing Developing

For the three-point flexure testing, the Universal MTS Bionix 515 machine is used, located on the mechanical engineering building at Universidad Industrial de Santander in Bucaramanga, Santander. In conformity whit, ASTM C393 specification, Fig. 2, Fig. 3, and Fig. 4 shows before test samples.

Fig. 2. Test sample ER-CP for flexure testing

Fig. 3. Test sample ER-CA for flexure testing

Fig. 4. Test sample PR-GP for flexure testing

3.2 Charpy Impact Testing Developing

For the Charpy Impact Testing, the Treble Charpy Impact Machine is used, located on the materials characterization building of the civil engineering faculty at Universidad Industrial de Santander in Bucaramanga. In conformity with ASTM D6110 specification, Fig. 5, Fig. 6 and Fig. 7 show before the samples.

Fig. 5. Test sample ER-CP for impact testing

Fig. 6. Test sample ER-CA for impact testing

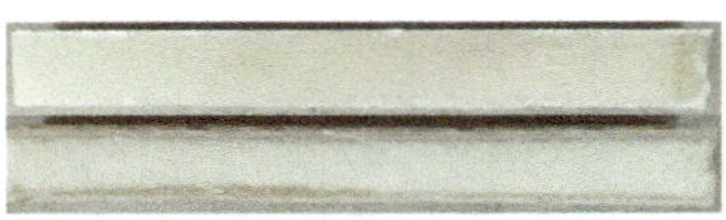

Fig. 7. Test sample PR-GP for impact testing

Before starting the impact testing, a rub test should be performed in which the pendulum free falls without placing any specimen on the support beam; the energy, angle, and stress for rubbing readings are taken.

4 Results and Discussions

4.1 Flexure Testing Results

The core shear ultimate stress (Fs), was calculated from Eq. (1), with respect to the facing stress is needed to know the core shear ultimate stress(σr), and Eq. (3) use in the calculation of ultimate flexion (σf), the flexion module (E) takin into account the maximum deflection (δ) and the spacing between supports calculated from the thickness of the specimen, all of the above items using the values of the maximum load (Pm), the sandwich thickness (d), the core thickness (tn),the width of the specimen(b), the length of support span (s), and the reinforce thickness (t).

$$F_s = \frac{P_m}{(d + t_n)b} \tag{1}$$

$$\sigma_r = \frac{Fs}{2t} \tag{2}$$

$$\sigma_f = \frac{3P_m s}{2bd^2} \tag{3}$$

$$E = \frac{s^3 P_m}{4bd^3 \delta} \tag{4}$$

On Fig. 8(a) it is observed that the core shear ultimate stress average by each group mate-rial, the ER-CA core has the higher core shear ultimate stress resistance of all materials, similarly on Fig. 8(b) is indicating the facing stress (average), the ER-CA has the high up facing stress resistance of all materials. The contrast in the ultimate flexion stress average shown in Fig. 8(c) the ER-CA core has the higher ultimate flexion stress resistance of all materials. And Fig. 8(d) provides adequate information for the study flexion module average by each group material, the ER-CA core has the higher flex-ion module resistance of all materials with 492 MPa.

The unitary deformation (ε) is needed to know δ were used to calculate the relation shown in the Eq. (5), the ER-CA has ε maximum of 9%, that is superior to ER-CP with 8.2% and PR-GP with 4.1% of unitary deformation.

$$\varepsilon = \frac{6\delta d}{s^2} \tag{5}$$

The failure mode of 5 specimens under flexural test of the composite sandwich for the ER-CA, ER-CP and PR-GP shown in the Fig. 9. The results of the experiments shown the relation of the flexion stress vs unitary deformation is observed in the Fig. 10, Fig. 11 and Fig. 12. The ER-CA has more stress than deformation on the elastic zone, the PR-GP has the intermediate value on the deformation – stress. Lastly, the material with the lowest results respecting stress and deformation is the ER-CP.

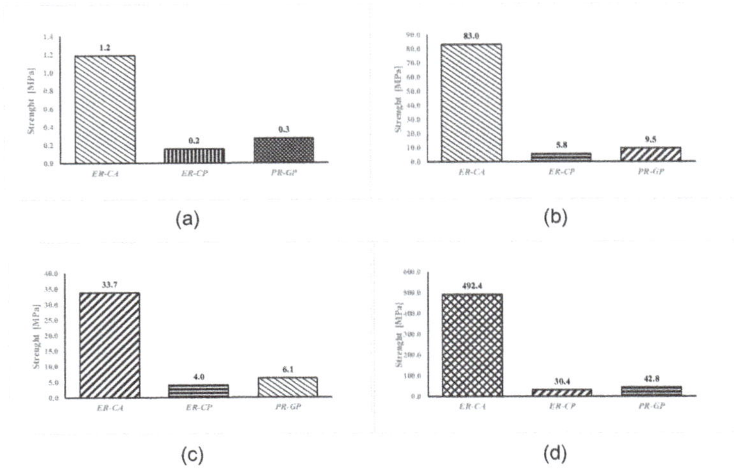

Fig. 8. Comparison maximum core shear strength (a), maximum facing stress average (b), ultimate flexion stress average (c) and the flexion module (d).

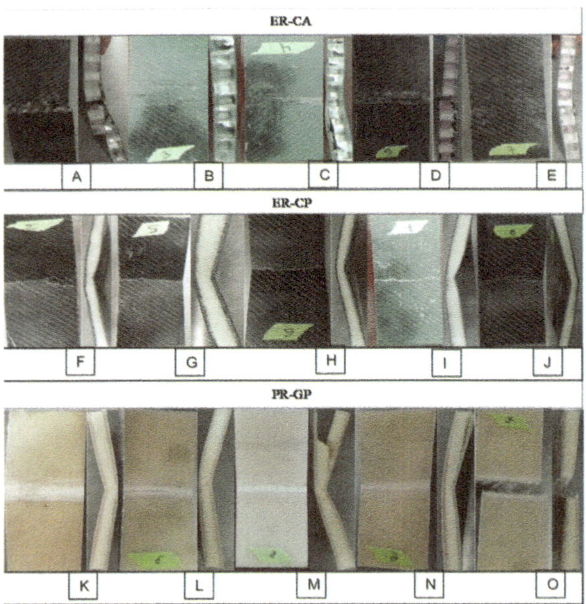

Fig. 9. Failure mode of specimen the flexure test; A, B, C, D, E specimen test ER-CA; F, G, H, I, J specimen test ER-CP, K, L, M, N, O specimen test PR-GP.

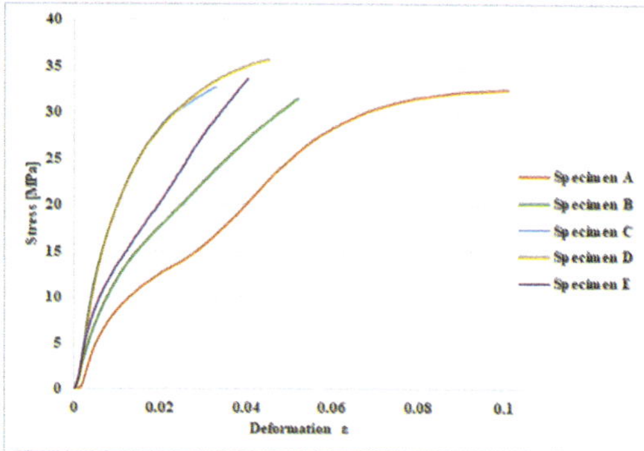

Fig. 10. The stress versus strain curve A, B, C, D, E specimen test ER-CA.

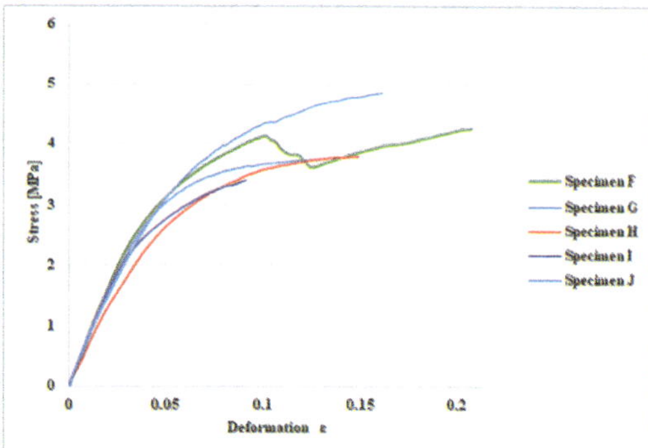

Fig. 11. The stress versus strain curve F, G, H, I, J specimen test ER-CP.

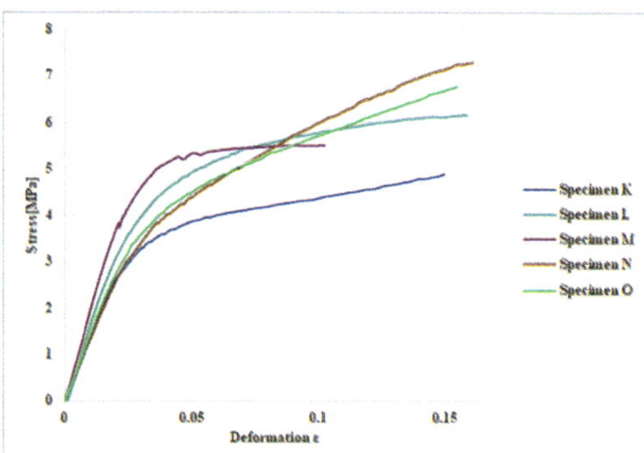

Fig. 12. The stress versus strain curve K, L, M, N, O specimen test PR-GP.

4.2 Impact Testing Results

The results obtained from the Charpy impact test are used to determine the energy absorbed by the specimens when it is impacted, must know the energy impact and friction energy. To calculate it, Eq. (6) was used where impact energy absorption is E_i, test energy is E_e, and friction energy is E_f. For determination of the impact resistance (I_s) the E_i and the specimen width (t) has to be known were use it Eq. (7).

$$E_i = E_e - E_f \tag{6}$$

$$I_s = \frac{E_i}{t} \tag{7}$$

Figure 13(a) shows the comparison of typical average absorbed energy by group of materials, the PR-GP is the material that most absorbed energy of all. From Fig. 13(b) explains the comparison of the average impact resistance by group of materials is observed, PR-GP is the material with the highest resistance to impact among the other materials. There is a perceptual increasing in both properties of 28.6% regarding the ER-CA and a perceptual increasing of 41.5% regarding the ER-CP.

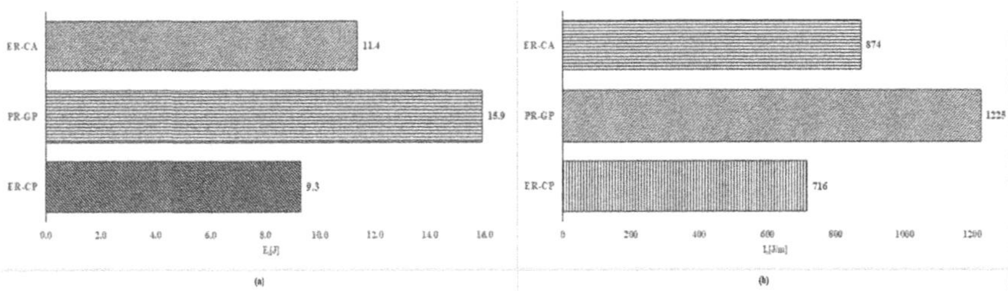

Fig. 13. Comparison of average absorbed energy(a); comparison of the average impact resistance (b).

4.3 Results of the Application of Selection Criteria According to Performance

Production costs and the properties of the materials according to the behavior of the three-point flexure test and the Charpy impact test are used to apply the selection methodology for materials and processes by Ph.D. Michael Ashby [21].

In this methodology the selection of materials and the elaboration processes work together to bring an organized scheme in order to identify all the variables that interfere with a selection process when a material is a design, using a taxonomy to classify the different materials and processes. To support this method, materials properties charts are used to graphically select the accurate materials for the product according to its properties, another important aspect to take into account is the elaboration costs.

On the Fig. 14(a) it is observable that ER-CA is the material with the best properties according to the three-point flexure test and the PR-GP presents the best relation between impact energy absorbed, the impact resistance, and elaboration besides, and the Fig. 14(b) the costs for minimum quantities of raw material that are available in the market are U$D 21.17, making this material the cheapest material to elaborate, thus the most suitable composite material for use in an electric sports vehicle in academic competitions is PR-GP.

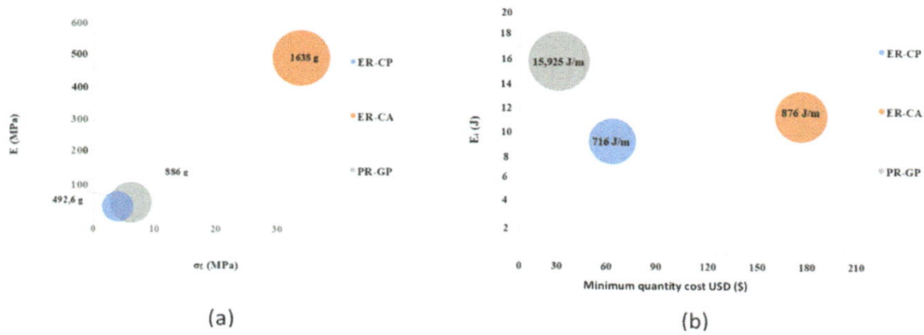

Fig. 14. Material map with properties

The selection methodology for materials elaboration and processes by Ph.D. Michael Ashby related many mechanical properties such: the flexion module, the ultimate flexion stress, the weight, the impact energy absorbed, the impact resistance, and the elaboration cost; these properties allowed to construct graphics that show the relations between them and finally select the best material for use in an electric sports vehicle in academic competition.

5 Conclusion

The inclusion of lightweight materials in the manufacturing vehicle process can present good results in terms of high-efficiency fuel consumption, some parts like bumpers, seats, dashboards, car roofs, seatbelts, and some electronic components such as sensors. Other applications can be on the brake system and suspension system, however more investigation and tests have to be developed to validate the mechanical properties and compare the results to traditional materials, properties such as thermoelectrically behavior, magnetism behavior, and fatigue behavior have to be tested as well. One important aspect that has to be considered on future investigations is the energy efficiency and how the lightweight materials not also increase the autonomy and life span of the battery but also the reduction for environmental impact due to low greenhouse gases emission, composite materials has the advantage to reduce considerately the weight-maintaining the mechanical properties.

References

1. Kiran, L., Kakkeri, S., Deshpande, S.: Proposal of hybrid composite material for light commercial vehicle chassis. Mater. Today Proc. **5**(11), 24258–24267 (2018)

2. Ascanio-Villabona, J.G., Rincón-Quintero, A.D., Cardenas-Arias, C.G., Sandoval-Rodriguez, C.L.: Incidence of corrosion in low voltage electrical conductor. IOP Conf. Ser. Mater. Sci. Eng. **844**(1), 012055 (2020)
3. Kawajiri, K., Kobayashi, M., Sakamoto, K.: Lightweight materials equal lightweight greenhouse gas emissions?: a historical analysis of greenhouse gases of vehicle material substitution. J. Clean. Prod. **253**, 119805 (2020)
4. Brooks, R., Shanmuga Ramanan, S.M., Arun, S.: Composites in Automotive Applications: Design. Elsevier Ltd. (2017)
5. Rincón-Quintero, A.D., et al.: Manufacture of hybrid pieces using recycled R-PET, polypropylene PP and cocoa pod husks ash CPHA, by pneumatic injection controlled with LabVIEW Software and Arduino Hardware. IOP Conf. Ser. Mater. Sci. Eng. **844**(1), 012054 (2020)
6. Wang, J., Shi, C., Yang, N., Sun, H., Liu, Y., Song, B.: Strength, stiffness, and panel peeling strength of carbon fiber-reinforced composite sandwich structures with aluminum honeycomb cores for vehicle body. Compos. Struct. **184**, 1189–1196 (2018)
7. Delogu, M., Zanchi, L., Dattilo, C.A., Pierini, M.: Innovative composites and hybrid materials for electric vehicles lightweight design in a sustainability perspective. Mater. Today Commun. **13**, 192–209 (2017)
8. Ascanio-Villabona, J.G., Sandoval-Rodriguez, C.L., Rincón-Quintero, A.D., Tarazona-Romero, B.E., Paez-Castro, R.E.: Building a prototype for functional analysis of the energy potential of the water flow in pipe 1/2 using microturbines applied to Unidades Tecnológicas de Santander. IOP Conf. Ser. Mater. Sci. Eng. **844**(1), 0–26 (2020)
9. Hagnell, M.K., Kumaraswamy, S., Nyman, T., Åkermo, M.: From aviation to automotive - a study on material selection and its implication on cost and weight efficient structural composite and sandwich designs. Heliyon **6**(3), e03716 (2020)
10. Selwyn Jebadurai, D., Razal Rose, A., Aatthisugan, I.: A novel approach to enhance mechanical characteristics of sandwich composites. Mater. Today Proc. **24**, 618–627 (2018)
11. Romero-Tarazona, B.E., Rodriguez-Sandoval, C.L., Villabonai-Ascanio, J.G., Rincón-Quintero, A.D.: Development of an artificial vision system that allows non-destructive testing on flat concrete slabs for surface crack detection by processing of digital images in MATLAB. IOP Conf. Ser. Mater. Sci. Eng. **844**(1), 012058 (2020)
12. Hiremath, N., Young, S., Ghossein, H., Penumadu, D., Vaidya, U., Theodore, M.: Low cost textile-grade carbon-fiber epoxy composites for automotive and wind energy applications. Compos. Part B Eng. **198**, 108156 (2020)
13. Kumar, S., Bharj, R.S.: Emerging composite material use in current electric vehicle: a review. Mater. Today Proc. **5**(14), 27946–27954 (2018)
14. Cárdenas-Arias, C.G., Rincón-Quintero, A.D., Santos-Jaimes, A., Sandoval-Rodriguez, C.L., Rojas-Gomez, D.F., Ardila-Galvis, S.J.: Elasticity modulus variation of the AISI SAE 1045 steel subjected to corrosion process by chloride using tension test destructive. IOP Conf. Ser. Mater. Sci. Eng. **844**(1), 012059 (2020)
15. Pavlović, A., Sintoni, D., Minak, G., Fragassa, C.: On the modal behaviour of ultralight composite sandwich automotive panels. Compos. Struct. **248**, 1–4 (2020)
16. Carlstedt, D., Asp, L.E.: Performance analysis framework for structural battery composites in electric vehicles. Compos. Part B Eng. **186**, 107822 (2020)
17. Fontana, M., Araujo, A.L., Madeira, J.F.A.: Optimization of a thin-walled composite crash absorber. Thin-Walled Struct. **155**, 106826 (2020)
18. Sandoval-Rodriguez, C.L., Ascanio Villabona, J.G., Cárdenas-Arias, C.G., Rincon-Quintero, A.D., Tarazona-Romero, B.E.: Characterization of the mechanical vibration signals associated with unbalance and misalignment in rotating machines, using the cepstrum transformation and the principal component analysis. IOP Conf. Ser. Mater. Sci. Eng. **844**(1), 012057 (2020)

19. Fragassa, C., Pavlovic, A., Minak, G.: On the structural behaviour of a CFRP safety cage in a solar powered electric vehicle. Compos. Struct. **252**, 112698 (2020)
20. Askeland, D.R., Sánchez García, G.: Ciencia e ingeniería de los materiales. International Thomson Editores (1998)
21. Ashby, M.F., Jones, D.R.H.: Engineering Materials 2: An Introduction to Microstructures, Processing and Design: Second Edition, 3rd edn. Oxford (2013)

Defense Engineering

Airworthiness and Safety in Air Operations in Ecuadorian Public Institutions

Brian Flores[1(✉)] and Rosalba Rodríguez[2]

[1] Fuerza Aérea Ecuatoriana, Guayaquil, Ecuador
floresbriand@gmail.com
[2] Universidad de las Fuerzas Armadas - ESPE, Sangolquí, Ecuador
rmrodriguez5@espe.edu.ec

Abstract. With the expansion of civil and public aviation, all aircraft that share Ecuadorian airspace must be properly regulated. The objective is to minimize air accidents through a single standard to raise safety levels in air operations through an airworthiness regulation that extends specific rules for this important sector and led by the Ecuadorian Air Force, to issue a corresponding aeronautical certification.

This research is mixed in nature, applying the qualitative approach with expert consultation through the Delphi method and quantitative approach using bibliographic research techniques. It is analyzed the applicability of the international standards: ISO IEC 17021–1;2015 and ISO IEC 17021–3:2017 for accreditation of the Airworthiness Directorate of the Ecuadorian Air Force (FAE) with the National Accreditation Service (SAE), which is the highest certifying body at Ecuador level; by identifying the requirements that the Ecuadorian Air Force meets and what should be worked on for its improvement, to propose changes to the Organization's Handbook based on the ISO standard.

The result of the investigation following the analysis of the documentation, procedures and specific training of the personnel working in the Airworthiness Directorate shows that the FAE meets 39% of the requirements, it is still up to 55% of them to work and that 6% are not applicable to our reality.

Keywords: Safety · Air operations · Airworthiness · Public aviation · Aeronautical certifications

1 Introduction

Airworthiness in both the Ecuadorian military and other state-owned aviation has been diminished for many years due to the lack of state-level laws and a framework in the Joint Staff (CCFFAA) to implement aeronautical certification activities in both flight operations and airworthiness. Also, Ecuador does not have a quality management system that ensures long-term sustainability of its regulations and standards for the certification of its aeronautical programs.

In Ecuador, the legal accreditation body for certification under international standards is the Ecuadorian Accreditation Service (SAE). The SAE is the signatory to international

M. Botto-Tobar et al. (Eds.): CIT 2020, AISC 1326, pp. 421–432, 2021.
https://doi.org/10.1007/978-3-030-68080-0_31

recognition agreements such as the Inter-American Accreditation Cooperation (IAAC) since March 2018 [1] and the International Accreditation Forum (IAF) since 2011.

Both public and private bodies implementing a certification process are regulated by the SAE. The SAE has the expertise to validate the application of the international standard ISO IEC 17021. As a certifying body, the SAE ensures, through specific regulations and competencies, the standards for certification. This research focuses on the applicability analysis of the international standards ISO IEC 17021–1;2015 and ISO IEC 17021–3:2017 for the accreditation of the Airworthiness Directorate (DIRAER) of the Ecuadorian Air Force (FAE) with the SAE. The DIRAER then could serve as the SAE-accredited certifying body to standardize air operations safety. The following hypothesis is presented: The absence of an international standard applied to DIRAER processes negatively impacts the quality of airworthiness inspections performed in maintenance organizations.

It is important to note that the Chicago Convention on Civil Aviation signed on August 20, 1954 [2], established mandatory standards for international civil aviation flight and maintenance operations. With the advances in aeronautic technology, these standards are continuously updated. At the regional level, regulations are applied according to the Regional Cooperation System for Operational Security Surveillance (SRVSOP) and through the Latin American Aeronautical Regulations [3], and at the national level, through the Directorate-General for Civil Aviation (DGAC), which is responsible for the civil aeronautical sector standards through the Technical Regulations (RDAC).

In contrast to civil aviation, there are no international agreements regulating or standardizing the airworthiness and air operations of public and military aircraft. In Ecuador, each institution maintains, or not, a quality control system to ensure compliance with air operations and maintenance tasks for flight safety.

1.1 Public Aviation in Ecuador and Its Regulations

In Ecuador, public aviation not regulated by the DGAC such as the Armed Forces and Police aircraft organizations, have individually established systems to ensure airworthiness without state-level standards. Throughout the modern world, countries such as the United States, United Kingdom, Australia, and the European Union maintain a compilation of specific regulations for public aviation. Within the region, Argentina, Brazil, Chile, and Colombia have already implemented airworthiness regulations and standards for public aviation safety and operations.

As an example, the Argentine military's airworthiness had to be standardized in record time after several accident investigations determined a lack of control in the airworthiness processes. This was necessary to avoid being controlled by civil aviation rules and regulations [4]. Also, in the European Union, after a series of accidents with military aircraft, they established the Military Airworthiness Authorities (MAWA) to regulate military aviation maintenance and flight safety operations [5].

A statement from Filippo De Florio, a leading airworthiness expert, the airworthiness processes must integrate three main elements: safe conditions, consider requirements according to the aircraft, and considering tolerance limits for its operation [6]. However, military airworthiness has unique components that differentiate it from civilian, these being: the high-risk operational environment, the use of weapons, and the tolerances for its operation, as shown in the following (Fig. 1):

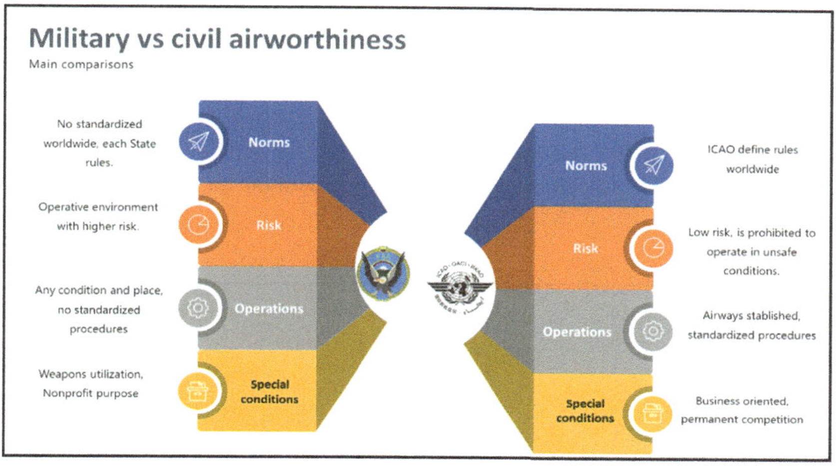

Fig. 1. Military airworthiness and civil airworthiness

Civil aviation authorities in all states are regulated through inspections of the International Civil Aviation Organization (ICAO), for the quality standards of maintenance and operations of their airports and services.

The generic quality standards used globally are recognized under International Organization for Standardization (ISO). These standards are established by committees comprised of different states establishing specific regulations for both certification of organizations and accrediting entities at the country level.

The Ecuadorian Air Force, faced with an urgent need to standardize airworthiness activities, took the first steps in 2009 by forming the FAE Airworthiness Directorate, which was officially established on April 30, 2018; its mission: "Verify and certify technical and legal capability of aircraft through airworthiness certifications, specialized engineering studies, airworthiness assurance, and continuous surveillance…" [7]. This Directorate is the highest-level organization established in public aviation, certifying more than 70 aircraft and 500 maintenance technicians. Further, their goal was to certify all Maintenance Organizations (OMAs), through their regulations and standards that parallel to RDAC 145, LAR 145 and FAR 145 but, specific to military aviation.

To optimize DIRAER as a certification body in the aeronautical field, internal processes should be closely related to a quality program. One certification body, SAE, would ensure continuous improvement, internal control of its processes, and validated not only at the national but international level with safety in flight operations. Validating flight safety processes through an external accredited organization gives DIRAER a long-term sustainable and credible program.

The specific standard for achieving accreditation from a quality certifying body corresponds to ISO 17021 whose latest version 2015 details:

"…the standard ISO/IEC 17021 specifies requirements for bodies that perform audit and certify management systems. It presents generic requirements for such bodies to carry out audits and certifications in the field of quality management, environmental, and other types of management systems. These bodies are known as certification bodies" [8].

The ISO standard 1702 corresponds to the international standard used for the accreditation of certifying bodies. This is precisely the role DIRAER fulfills by requiring it to be validated by the national certification authorities of the country. This standard has been chosen due to its applicability and compatibility with the current management system. Also, DIRAER processes are in accordance with the audit standard ISO 19011 (Fig. 2).

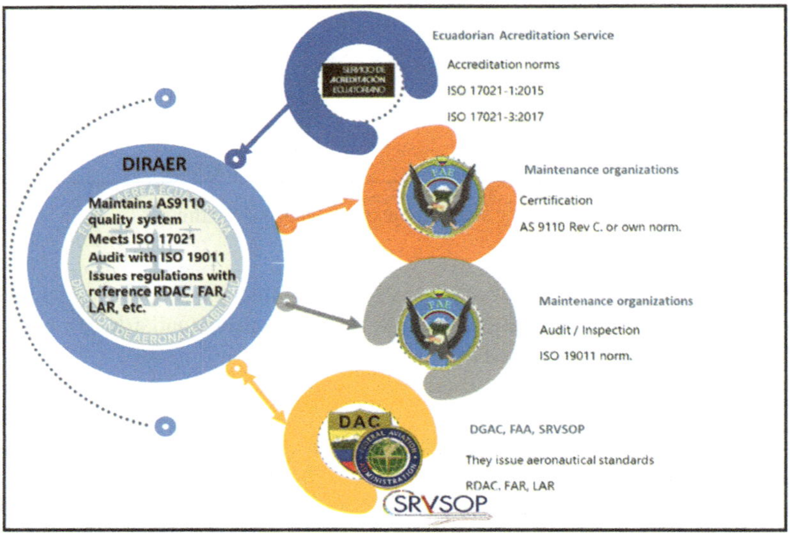

Fig. 2. DIRAER's relationship with regulations

As an accreditation requirement, the certification body must maintain a quality system the same as ISO 9001 or a specific standard as AS 9110 Rev. C, corresponding to quality systems for aeronautical maintenance organizations.

The process began by researching all best practices and available literature on the topic. Items reviewed were the structures and operation of military airworthiness systems of the countries within the region, locating multiple regulations and using standards such as ISO 9001 for management systems; accreditation was not a consideration in this review. Noted in the review of scientific documents on the specific topic of military airworthiness were scarce and more focused on peculiarities and differences with civil airworthiness, but not to a standard that allows systems to be accredited. The indexed articles of the sixth International Scientific Conference on Defense Technology, as well as publications in Science Direct and the Federal Aviation Administration (FAA), provided significant information. The studies revealed the experiences in the implementation of AS 9110 and its future inclusion as an ISO standard.

2 Methodology

This research was diverse, employing a qualitative approach applied with expert consultation using the Delphi method. The most experienced military personnel in airworthiness and aeronautical maintenance experts were consulted. In the first quarter of 2020 the research program used three rounds of queries.

The quantitative approach was also used when conducting a comparative analysis to verify the applicability of the international standards ISO IEC 17021–1:2015 and ISO IEC 17021–3:2017 for accreditation of DIRAER with the SAE (The highest accreditation body level in Ecuador). Experts identified the ISO requirements met by the Ecuadorian Air Force; they proposed improvements to those ISO requirements needing work in the organization's manual based on the ISO standard.

To visualize the results of the study, primary information was quantified, focusing on the use of objective indicators [9]. This allowed the interpretation and collection of information through numerical data representing the performance variables.

In addition, to document research, standards in ISO 17021–1:2015 and ISO 17021–3:2017 were fully analyzed, allowing for comparison with what is working within DIRAER, generating a compliance gap analysis.

Standards in ISO 19011 and AS-9110 Rev C were considered as an additional reference concerning audit processes and quality management systems for aeronautical maintenance organizations, respectively (Fig. 3).

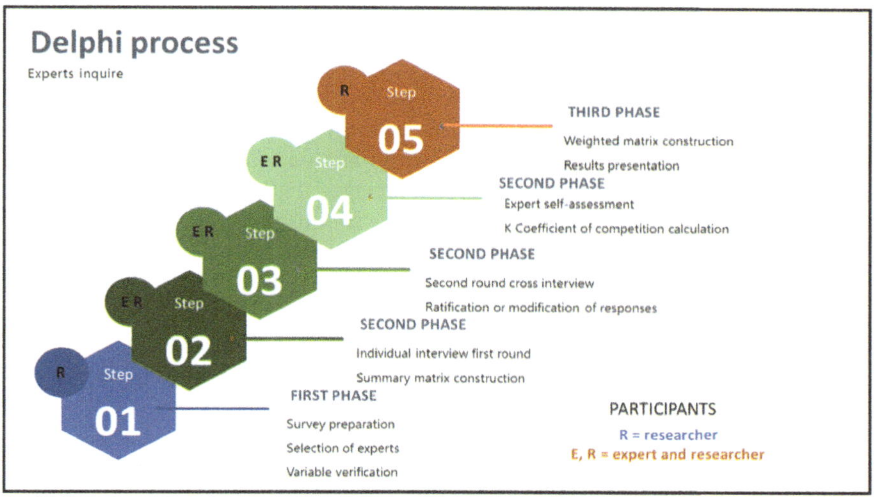

Fig. 3. Delphi process Fuente: Reguant, M., & Torrado, M. 2016.

Data collection tools used a matrix for each expert, allowing information for comparison to the study hypothesis, both for the independent and dependent variables (Fig. 4).

The variables operationalization is detailed in Tables 1 and 2.

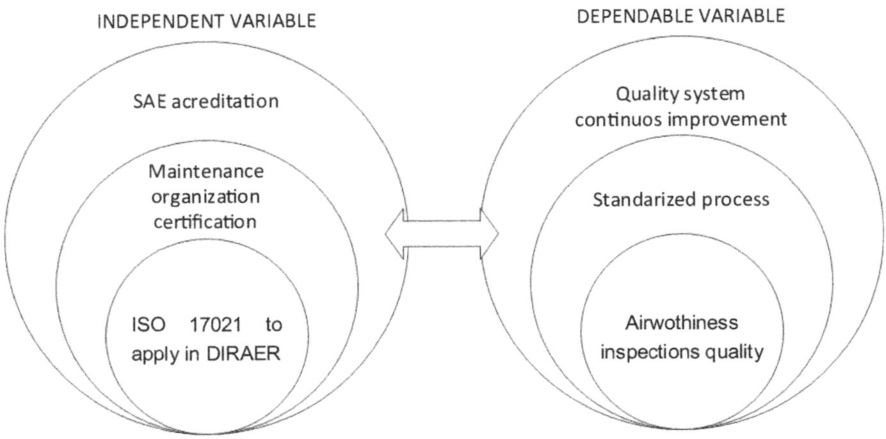

Fig. 4. Research variables

The research used one of the variants of the Delphi method, originally created by the U.S. Air Force for the development of their nuclear program (declassified in the 1960s [10]) as a systematic method for expert consultation. The method is used in multiple research areas, especially under the following considerations: a) When information is insufficient and a non-specialized, collective criterion may deflect the investigation, b) When the problem can benefit from the judgements of the experts and the anonymity of each is required. Based on these premises and considering that the research subject requires specialized knowledge for valid data collection, results involve the application of a regulation, using experts to collect specific information and those with organizational decision-making authority to apply suggested changes. Further, it was determined that the use of a generalized survey would be effective for this project.

For Delphi method application and based on the methodology presented by Torrado and Reguant, the following phases are considered in this study, which have been adapted according to the type of research and organization [11]:

FIRST PHASE: Interview question preparation and expert selection. Development of the questions related to investigation variables were defined, as well as the identification of the expert group selection.

SECOND PHASE: First round, application of individual interviews and the construction of a summary matrix. Second round, analysis by each expert to ratify or modify their own responses and identify those considered in alignment. Third round, with the self-assessment of each expert based on organizational knowledge, field experience, and answer argumentation.

THIRD PHASE: Construction of the weighted matrix with the consensual responses, considering the self-assessment for the final valuation and the use of the competency ratio K, which allows weighing the answers according to expert-specific knowledge, using the following formula:

K: competitive ratio:

$$K = \frac{kc + ka}{2} \quad [12]$$

Table 1. Evaluation criteria for knowledge ratio kc

Feature	Expert	High	Middle	Low
Organization knowledge	0.5	0.4	0.3	0.2
Field experience consulted at tactical level	0.25	0.2	0.1	0.05
Field experience consulted al strategic level	0.25	0.2	0.1	0.05

Table 2. Evaluative criteria for argumentation coefficient (ka)

Feature	Expert	High	Middle	Low
Answer argumentation	0.8	0.7	0.55	0.4
Research evaluation	0.2	0.1	0.05	

3 Results

With the end of the first round, the criteria expressed was analyzed. As noted in Table 3, questions B, D, and E resulted in a close alignment of thoughts from the experts, a new round was not necessary for those questions.

To generate the weighted matrix of the responses considering the coefficient of competence K, the coefficients corresponding to each expert were calculated based on self-assessments. The illustration in Table 5 shows the procedure carried out with Expert No. 1.

In addition, Table 6 details the results of the calculation of the K coefficient for all the experts consulted (Table 4).

Weighting table No. 5 with No. 6 by using the competency coefficient, the answers to questions A, C, F, and G in both decimal and percentage form are:

The interpretation of the most important findings before each question is as follows:

Regarding question A, 82% of the experts recognize the necessity to adopt a quality standard that accredits DIRAER as a certifying body, obliging itself to standardize its procedures for performing inspections, reducing inspector discretion. It recognizes that there is no authority for establishing procedures as a certifying body. Noting that it

Table 3. First-round match matrix

Question	Yes	Partial	No	Observation
B. Do you think it is important to implement an accreditation standard in DIRAER?	10			Perfect match It is recommended to standardize the maintenance organizations it is necessary that DIRAER rule
D. Do you consider that DIRAER's airworthiness inspections are of quality?	6	4		Majority match It is recommended to standardize inspector
E. Do you think that implementing an international quality standard would improve the quality of inspections?	9		1	Majority match Those who think differently do not consider a specific standard but to adapt and create their own standard

Table 4. Calculation of the proficiency coefficient K for Expert 1

	E1	Expert	High	Middle	Low
kc	Organization knowledge		0,4		
	Field experience consulted at tactical level			0,1	
	Field experience consulted al strategic level	0,25			
ka	Organization knowledge	0,8			
	Field experience consulted at tactical level		0,1		
K	Competition coefficient	0,825			

Table 5. Expert competency coefficient

Expert	1	2	3	4	5	6	7	8	9	10
K	0.825	0.825	0.625	0.750	0.800	0.725	0.875	0.625	0.750	0.750

Table 6. Weighting responses

Question	Yes	Partial	No	Total
A	0,75	0,63	6,23	7,6
C	5,28	0,83	1,5	7,6
F	0,83	6,03	0,75	7,6
G	1,58	6,03	0	7,6

Table 7. Percentage weighting of responses

Question	Yes	Partial	No	Total
A	10%	8%	82%	100%
C	69%	11%	20%	100%
F	11%	79%	10%	100%
G	21%	79%	0%	100%

would be necessary to establish this regulation at the level of the Joint Command or even a verification process by collaboration with other aeronautical authorities.

Question B was in full agreement with all experts. It highlights the need to implement a quality standard to accredit DIRAER as a body that executes certifications under international standards with a quality process in place for continuous improvement.

The results for question C are more diverse from the experts. Most of the experts, 69%, believe accreditation to a higher body is feasible; however, 20% believe peer accreditation with other military aeronautical authorities would be sufficient or create a state-level law.

In question D the results does not differ further from the criteria because 60% considered inspections carried out by DIRAER are a quality process. However, 40% mentioned, although adequately working, there would be a significant benefit to future improvements from standardized procedures in the execution of inspections by reducing dependency of specific training and experience of each inspector. In this regard, it recommends that DIRAER increase its measurement process of inspections to evaluate the effectiveness and identify opportunities for improvement.

The results in question E show a high confidence correlation with 90% in agreement, with the only one exception. One expert believes the most useful option would be to create and adopt existing regulations without considering a higher standard certification. Also noted, if a quality standard is implemented, it must be focused on aeronautical maintenance processes, as well as the AS 9110 Rev. C.

Question F shows 79% of respondents believe there are training deficiencies with DIRAER inspectors. In addition, 10% believe relevant training is not available and should be a priority. It recommends implementing a training and ongoing education plan for airworthiness inspectors, as well as training in specialized areas.

The answer in question G shows 79% of experts believe checklists will likely improve as maintenance organizations apply them during aircraft inspections. However, they currently do not cover all the aspects required when compared to the checklists of the Civil Aviation Airworthiness Inspector Manual.

Based on the results, the question that supports the verification of the hypothesis for the dependent variable is B, "Do you consider that it is important to implement an accreditation standard in DIRAER?". For the independent variable question E, "Do you consider that implementing an international quality standard would improve the quality of inspections?".

Regarding question B, there is a perfect correlation in the expert's criteria, while for question E, you have an almost perfect correlation with a single decent. However, in the second round, the same expert expressed "there is also a lack of regulation for DIRAER certification", a fact that strongly supports the acceptance of question E.

Figure 5 shows the result of the questions that needed a second round by the experts to obtain a standardized criterion.

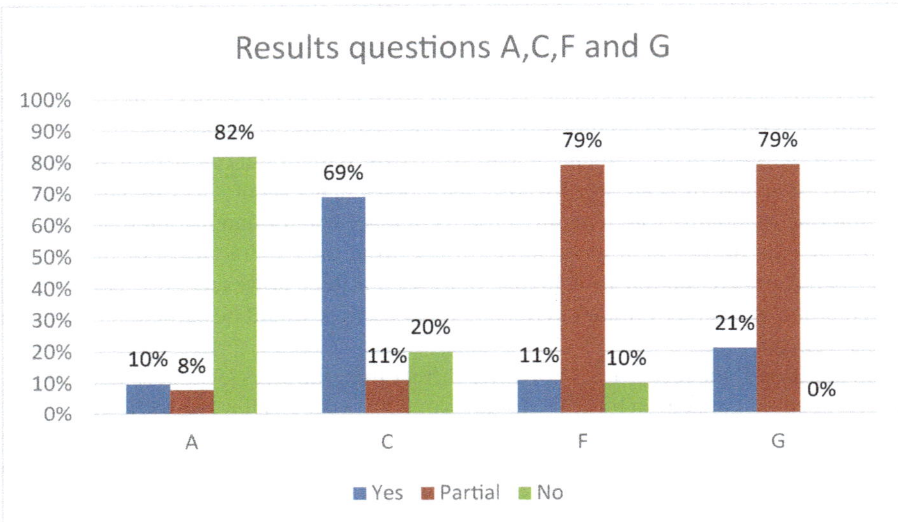

Fig. 5. Quality regulations implemented

At this point, the first stage of the investigation was completed and complemented by the second phase, corresponding to a point-by-point gap analysis of ISO IEC 17021–1:2015 and ISO IEC 17021–3:2017 concerning the documentation, procedures, and specific training of the staff working at DIRAER. The results are as follows (Table 8):

Table 8. Compliance with requirements according to ISO 17021–1:2015 and ISO 17021–3:2017.

Requirements	Number	Percentage
Fulfilled	82	39%
No fulfilled	105	55%
No applicable	13	6%

With these results, the hypothesis is verified, as the percentage of non-compliance with quality standards at the international level exceeds 50%.

Among the identified requirements for compliance, the most prominent are:

- implementation of procedures for complaints and appeals
- declaration and implementation of the impartiality policy
- limitation of DIRAER in certification-only functions and not advisory
- organizational independence for decision-making
- training of DIRAER staff as internal auditors, leading auditors, as well as evaluation and monitoring according to their competencies
- establishment of the auditors' development program (DIRAER airworthiness inspectors) with defined profiles and competencies
- execution of internal audit processes
- alignment of audit procedures with the provisions of the standard

4 Discussion and Conclusions

As a result of the study, accrediting DIRAER with international standards represents an advantage by placing the Air Force in a position to lead an initiative to standardize public aviation safety in Ecuador. The Air Force has the technical and legal support to exercise its authority and is viable and sustainable.

The DIRAER's ISO standards were tested and are appropriate for a certification body, although its implementation will require effort, the objective is to minimize risk in flight operations and provide for continuous improvement of the organization. This action can serve as an example for all public sector oragizations lacking standardization and an environment of continous improvement.

Considering the expansive growth of both civil and public aviation, all aircraft that share common airspace must be adequately regulated to minimize accidents and improve flight safety. This condition is the basis for designating the Air Force as the institution to exercise air space control, taking the lead in front of other public institutions that have aviation and are not regulated by the DGAC due to the characteristics of their operation.

The Air Force's experience gained over the last ten years in matters of airworthiness, quality inspections, training of maintenance technicians, and specialized engineering in the aeronautical field, should be considered by political authorities when deciding the lead of an integrated airworthiness regulation.

It is necessary to consider the journey other countries have taken to achieve public aeronautical regulations and standards. As with Australian, Argentine, or European cases, initial obstacles concluded a need for specific standards for public aviation. However, despite a strong regulatory system, its effectiveness depends on how it is managed and how each service implements regulatory requirements [13].

At the global and regional levels, each country creates its aeronautical authority for public aviation in order to maintain the safety of air operations. Equipped with an international accreditation such as ISO 17021, under the AS9100 standard, with which the Air Force has experience, gives it a definite advantage in leading the aeronautical authority, initially for the Armed Forces, and then for all public aviation in Ecuador.

The implementation of a standard for certifications will facilitate the permanent improvement of quality in logistics and airworthiness systems, directly improving maintenance and safe operation of Air Force aircraft. Using the Air Force as a model under international standards can be imulated by all Armed Forces and state aviation organizations, delivering meaningful contributions to service Ecuadorian society.

For the development of common standards, under an accredited certifying body, it is necessary to determine the specific characteristics of the operation and maintenance of each State institution with public aviation.

References

1. SAE, 2019. [En línea]. https://www.acreditacion.gob.ec/historia-2018/
2. ICAO, 2014. [En línea]. https://www.icao.int/secretariat/legal/List%20of%20Parties/Chicago_ES.pdf
3. SRVSOP, 2019. [En línea]. https://www.srvsop.aero/biblioteca/reglamentos/lar/
4. Flores, B.: La normativa internacional de calidad AS 9110 en los escuadrones de mantenimiento aeronáutico. Quito (2014)
5. Jankovic, M., Ilic, Z., Dronjak y, M., Parezanovic, V.: Military airworthiness. In: International Scientific Conference of Defensive Technologies, 10 Octubre (2014)
6. De Florio, F.: Airworthiness, An Introduction to Aircraft Certification. Burlington. Elsevier, MA (2006)
7. FAE: Manual de la Organización DIRAER. Quito (2018)
8. ISO, International Organization for Standardization 17021–1. 2015. [En línea]. https://www.iso.org/obp/ui/#iso:std:iso-iec:17021:-1:ed-1:v1:es
9. Cáceres, P.: Análisis cualitativo de contenido: un alternativa metodológica alcanzable, pp. 53–81 (2003)
10. Cabrero y, J., Infante, A.: Empleo del método Delphi y su empleo en la investigción en comunicación y educación. EDUTEC, vol. 48, p. a272 (2014)
11. Reguant y, M., Torrado, M.: El método Delphi. REIRE (2016)
12. Zharta, J., Montes, J., Toro y, I., Villada, S.: Método Delphi - Propuesta para el cálculo del número de expertos en un estudio Delphi sobre empaques biodegradables al 2032. REVISTA ESPACIOS, **35**(13), p. 10 (2014)
13. Bushell, T.: The Never Ending Story of Airworthiness versus Murphys Law (2016)

Fires in World Heritage Buildings

Diego Venegas[1](✉), Oswaldo Erazo[1], Óscar Farías[2], César Ayabaca[3], and Ana Medina[4]

[1] Bio-Bio University, 4051381 Concepción, Chile
diego.venegas1801@alumnos.ubiobio.cl
[2] Concepcion University, 160-C Concepción, Chile
[3] National Polytechnic School, Quito E11-253, Ecuador
[4] Consultant in Occupational Health and Safety Management Systems, Quito, Ecuador

Abstract. Heritage, cultural and historical buildings have been built all over the world and they represent part of the history and wealth of the people. Most of these buildings are hundreds of years old. Minimum criteria of prevention and protection against fires were not taken into account during their construction. This makes them vulnerable to catastrophes that cause incalculable losses. This paper investigates fires in heritage buildings around the world during the period 1990–2019. Results showed that the five main causes for these events to occur were: vandalism 26%, during restoration: 21%, electrical problems and short circuits: 17%, accidents: 16% and undetermined causes: 20%. Based on these results, safety recommendations are provided to prevent these events and to minimize both human and material damages if a fire were to occur.

Keywords: Heritage buildings · Fire · Preservation · Maintenance

1 Introduction

"World heritage sites are those that, due to various reasons, are considered that without them, the places where they are located, would no longer be what they represent to all Humanity" [1], UNESCO defines cultural heritage as the "heritage of physical artifacts and intangible attributes including tangible culture, intangible culture and natural heritage" [2]. A heritage building gives us a sense of wonder. It also makes us want to know more about the culture and people who have created it. It has architectural, aesthetic, historical, documentary, archaeological, economic, social and even political and spiritual or symbolic values. However, the main impact is always emotional, since it is a symbol of our cultural identity and continuity, a part of our heritage [3]. Here is the importance of the buildings that represent part of the people's culture, which also have unique pieces of art that identify their social environment. If a heritage structure with more than 100 years of use has survived destruction, undoubtedly, it has the right to be called a historic or heritage building [4]. History has recorded the destruction by fire of world heritage buildings and cities such as the Library of Alexandria (48 BC), Rome (64 AD), St. Paul's Cathedral in London (1666), the Cathedral of León (1966), El Chiado de Lisboa (1988) [5] or the Iglesia Compañía de Jesús Santiago de Chile in 1863

M. Botto-Tobar et al. (Eds.): CIT 2020, AISC 1326, pp. 433–447, 2021.
https://doi.org/10.1007/978-3-030-68080-0_32

[6]. Despite the current technology in the science of preventing and extinguishing fires, we have seen desperately how some of the heritage buildings have been burnt down, from 1975 to 2005, the amount of disasters has increased by about 400% and the annual number of affected people has nearly quadrupled, also depending on the rise in the population and urbanization [7], and the most recent events such as the Al-Aqsa Mosque in Jerusalem [8], the Notre Dame Cathedral in Paris [9], the National Museum in Rio de Janeiro, Brazil [10], from the start of 2010 to the end of 2013, 52 fires endangering historic buildings have been reported by the Chinese administration [11], among the most recent ones.

Regarding what happened in above, several questions arise. Prior the construction of these buildings: were elements and devices for fire prevention and mitigation planned? In places of tourist attractions that gather large numbers of people worldwide: are there any contingency and escape plans in case of accidents? People who visit these buildings: do they have the basic knowledge to act if an emergency such as a fire were to occur? Are there any necessary measures for the preservation of these buildings and are they fulfilled when maintenance works are carried out?

Upon carrying out the corresponding analysis for the causes of the fires, it is presumed that most of these could have been avoided if meticulous plans and measures for prevention and mitigation of fires had been made available [12]. Some factors that contributed to fires in heritage buildings throughout the world have been identified, therefore in this paper a study of accidents that occurred since 1990 was made. Their causes are investigated, and some minimum safety considerations are issued, which can be implemented immediately in these types of buildings in order to prevent, mitigate and, in the worst-case scenario, minimize the fire disaster in heritage buildings.

2 Methodology

In this work, a qualitative methodology was used to determine the causes of fires in heritage buildings, based on scientific reports and press articles. The causes of the fires were determined according to an occurrence pattern This comparative analysis found the main problems in the security management of fires in heritage buildings. Following a quantitative methodology, the origin and number of fires were recorded. Finally, a descriptive qualitative methodology to suggest the best-recommended practices given by the prevention and mitigation of fire experts was used. Online database such as IEEE, Emerald, and Scopus was used. The selection of keywords was down to fires in heritage buildings, security management of fires, security systems of fires, and security fires in buildings.

3 Development

A selection of topics was narrowed down to keywords such as: fire in heritage building; fire safety management and system; and fire safety in buildings. Table 1 shows a series of fires worldwide that occurred in heritage buildings from 1990 to 2019, and Fig. 1 shows quantitative results of the most probable causes of fires.

Table 1. Fires in heritage buildings and their causes.

Incident date	World heritage	City/Country	Cause
April 2019	St. Anthony's Chuch	Colombo, Sri Lanka	Vandalism [13]
April 2019	Notre Dame Cathedral	Paris, France	Restoration [14]
April 2019	Al-Aqsa Mosque	Jerusalen, Israel	Short circuit [8]
September 2018	National Museum	Río de Janeiro, Brasil	Short circuit [15]
August 2018	Primak Building	Belfast, Irland	Restoration [16]
June 2018	School of Art	Glasgow, Scotland	Restoration [17]
June 2018	Anba Makar	Cairo, Egypt	Short circuit [18]
January 2017	San Nicolás de los Arroyos Cathedral	San Nicolás, Argentine	Short circuit [19]
November 2017	Sagrado Corazón de Jesús Chapel	Buenos Aires, Argentine	Vandalism [20]
May 2016	Serbian Orthodox Cathedral	New York, USA	Vandalism [21]
October 2014	Sosnowiec Cathedral	Sosnowiec, Poland	Unknown [22]
November 2013	Municipal theatre	Santiago de Chile, Chile	Short circuit [23]
April 2010	Evangelical church	Bitrista, Romania	Unknown [24]
May 2010	Santa Maria dei Derelitti Church	Venice, Italy	Accident [24]
September 2010	Phillips Collection	Washington, USA	Restoration [25]
February 2009	Holy Name Cathedral	Chicago, USA	Restoration [24]
February 2008	Namdaemun Gate	Seoul, South Korea	Vandalism [26]
April 2008	Castello di Moncalieri	Turin, Italy	Restoration [27]
May 2008	Alma College	Ontario, Canada	Vandalism [28]
January 2007	Taltal church	Taltal, Chile	Short circuit [29]
January 2006	Casa San Cristobal de la Laguna	Tenerife, Spain	Short circuit [25]
March 2006	Beaulieu Tide Mill	New Forrest, England	Vandalism [25]
May 2006	Porvoo Cathedral	Porvoo, Finland	Vandalism [24]
June 2006	Old Town of Films	Graubunden, Swiss	Unknown [25]
August 2006	Trinity Cathedral	Saint Petersburg, Russia	Restoration [30]
September 2006	Eastern Ortodox Church	Kamienica, Poland	Unknown [24]
January 2005	S. Johannis Church	Gotteingen, Germany	Vandalism [25]
April 2005	Naksansa Temple	Yangyang, South Korea	Accident [25]

(continued)

Table 1. *(continued)*

Incident date	World heritage	City/Country	Cause
June 2005	Rand Club	Johannesburg, South Africa	Short circuit [25]
June 2005	Northam Library	Devon, England	Accident [24]
June 2005	Museum of Biblical Art	Dallas, USA	Unknown [25]
August 2005	Schloss Elmau	Kruen, Germany	Short circuit [25]
August 2005	St Mary´s Lodge	London, England	Unknown [25]
November 2005	Gasthofs Lowen	Oberrohrdorf, Swiss	Vandalism [25]
December 2005	The Redwood Library	Dedham, USA	Accident [25]
December 2005	Peterhof Palace	St. Petersburgo, Russia	Unknown [25]
March 2004	16 Churches	Kosovo	Vandalism [24]
September 2004	Duchess Anna Amalia Library	Weimar, Germany	Short circuit [31]
October 2004	Hafodunos Hall	Llangernyw, North Wales	Vandalism [25]
November 2004	Sinsheim Mezquite	Heidelberg, Germany	Vandalism [24]
November 2004	Harbin Synagogue	Harbin, China	Restoration [25]
December 2004	Turin Cathedral	Turin, Italy	Short circuit [24]
January 2003	The Chariot of Glory	St. Petersburgo, Russia	Accident [24]
January 2003	Mount Stromlo Observatory	Camberra, Australia	Accident [25]
January 2003	Luneville Chateau	Luneville, France	Short circuit [24]
April 2003	Glienicke Jagschloss	Berlin, Germany	Restoration [24]
May 2003	Brighton West Pier	England	Vandalism [24]
June 2003	Nacional Library	Bagdad, Irak	Vandalism [32]
July 2003	North Carolina State Capitol	Charlotte, USA	Restoration [24]
October 2003	National Motorcycle Museum	Solihull, England	Accident [24]
October 2003	Babintong House Spa	Somerset, England	Short circuit [24]
November 2003	Presbyterian Church	Exeter, USA	Accident [24]
January 2002	Cinematheque	Paris, France	Unknown [24]
September 2002	Burakuden Hall	Muko, Japan	Unknown [24]
1999–2002	60 Museums	USA	Short circuit [25]
July 2001	Mosque-Cathedral of Córdoba	Córdova, Spain	Accident [33]

(continued)

Table 1. (*continued*)

Incident date	World heritage	City/Country	Cause
August 2001	Salem United Methodist Church	Allentow, USA	Restoration [24]
September 2001	Twin Towers, Pentagon	New York, Washington, USA	Vandalism [34]
October 2001	Sodra Rada Church	Värmland, Sweden	Accident [24]
November 2001	Peterborough Cathedral	Peterborough, England	Vandalism [24]
December 2001	St. John the Divine Cathedral	New York, USA	Unknown [35]
May 2000	St. Paul Church	Deptford, England	Restoration [24]
July 2000	Presbyterian Church	Lexington, USA	Restoration [24]
August 2000	Sofiensaal	Vienna, Austria	Restoration [24]
1999–2000	50 Churches	Poland	Vandalism [24]
September 1998	St Brandon Church	Durham, England	Unknown [24]
January 1997	Old City, Junkerngasse	Berna, Swiss	Unknown [24]
April 1997	Jean Baptiste Cathedral, Royal Palace	Turin, Italy	Short circuit [36]
January 1996	La Fenice Opera	Venice, Italy	Vandalism [37]
February 1996	The Company of Jesus Church	Quito, Ecuador	Restoration [38]
June 1996	The Treasury Building	Washington, USA	Restoration [39]
December 1996	Madonna della Grazie Church	Millan, Italy	Unknown [40]
July 1994	Halifax Church	Halifax, Canada	Vandalism [25]
October 1994	Teruel Cathedral	Teruel, Spain	Accident [41]
October 1994	Berlin Cathedral	Berlin, Germany	Restoration [42]
January 1994	Gran Teatre del Liceu	Barcelona, Spain	Restoration [43]
February 1993	Lundby Church	Goteborg, Sweden	Vandalism [24]
Summer 1993	Parish Church of St Michael	Newquay, Cornwall. England	Accident [44]
August 1993	Pont de la Chapelle	Lucerna, Swiss	Unknown [45]
September 1993	Esztergom Cathedral	Budapest, Hungary	Unknown [42]
February 1992	Proveantgarden	Copenhague, Denmark	Accident [24]
April 1992	Odd Fellow Palace	Copenhague, Denmark	Accident [24]

(*continued*)

Table 1. (*continued*)

Incident date	World heritage	City/Country	Cause
June 1992	Christianborg Palace Chapel	Copenhague, Denmark	Accident [24]
November 1992	Windsor Castle	Windsor, England	Restoration [46]
November 1992	Redoutensal Hofburg Palace	Viena, Austria	Unknown [24]
August 1992	Sarajevo Library	Sarajevo, Bosnia	Vandalism [47]
1992–1994	40 Churches	Noruega	Vandalism [24]
September 1991	San Francisco Javier Cathedral	Yamaguchi, Japan	Unknown [48]
October 1991	Petruzzelli Theatre	Bari, Italy	Vandalism [49]

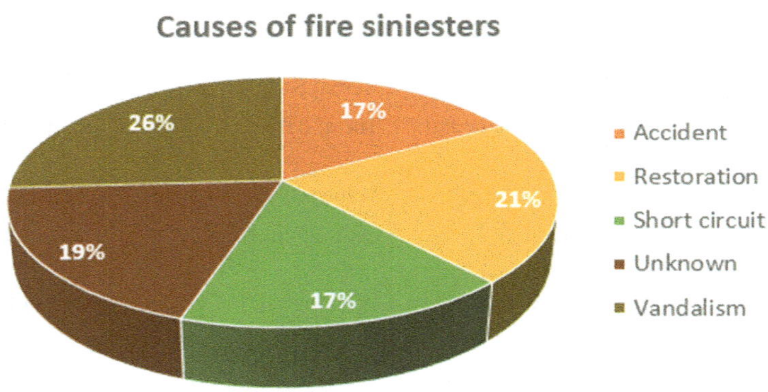

Fig. 1. Events in heritage buildings and their most likely causes.

4 Results and Discussion

Heritage buildings around the world were built several decades or centuries ago. Their designs and construction did not consider principles of fire protection and prevention [50]. In some cases, electrical lighting systems were not foreseen. Over time, heritage buildings have been remodeled, expanded, even converted into museums, since they are considered a tourist attraction. As a result, new electrical equipment such as lighting fixtures, amplification, and air conditioning, among others have been installed. On the other hand, the vast majority of religious and civil patrimonial structures are made out of wood due to its strength, aesthetics properties, and availability [51, 52]. Furthermore, curtains and canvases are used to decorate their structures [53], thereby they become more vulnerable to the presence of fire [54]. The main groups causing fires in heritage buildings are shown below.

4.1 Vandalism/Terrorism

Terrorism and religious intolerance have caused much destruction around the world. Heritage buildings such as churches are especially vulnerable to fire, as their structures are made up of wood, fabric, candles, and empty structural elements, all of these components spread the fire. Criminal activities such as smuggling or theft can lead to arson to cover up major crimes [24].

4.2 Restoration

New needs regarding energy efficiency standards, reduction of carbon footprint [55], improvement of interior comfort, integration of renewable energies in remodeling and retrofitting of historic buildings should be considered [56]. Heritage buildings have been subjected to anthropic action since their construction, causing serious damage and even total destruction. Being old buildings, they are subjected to restoration processes by experts in this field, but with little to no knowledge on preventing and extinguishing fires. Additionally, they do not have the necessary equipment in the event of a fire and generally perform inadequate interventions in fire outbreaks causing serious incidents.

4.3 Electricity/Short Circuits

Overloading of cables and circuits can cause a fire in heritage buildings. Especially the installations of equipment which were not included in the initial calculations and were incorporated into the buildings to meet their new needs (lighting, air conditioning, amplification, among others). Inadequate maintenance of electrical connections has contributed to fires incidents.

4.4 Accidents

The influence of sunlight on easily flammable elements such as curtains or candles near a windows have caused fires with dire consequences. On the other hand, the use of fireworks during festivities in the surroundings of the heritage buildings cause serious damage to their structures.

4.5 Common Causes

In addition to the causes above, there are common causes that have been identified in many reports of fires in patrimonial sites, which maximize the damage caused to these types of buildings, including:

- The heritage buildings are located in areas with high vehicular traffic, which makes it difficult to access the Fire Department and firefighting authorities.
- There are few special requirements established in the law of heritage buildings, especially concerning fire protection. Financial and economic priorities sometimes limit investments in mitigation of an active or passive fire defense systems [24].

- Conservation studies in heritage buildings are usually of an architectural nature but do not consider legal and engineering regulations for fire prevention [57].
- Several of the reports and subsequent investigations have determined that extinguishing agents used during the mitigation efforts have caused greater damage to the patrimonial sites [25].
- Those fires where people have been inside the building, have resulted in an innumerable loss of life because people have a lack of knowledge in evacuation protocols. Moreover, most buildings do not have emergency plans involving visitors' control. Which confirms the fact that there is no civic culture in terms of civil protection, especially regarding the prevention and response to fires [58], even worse, administrators in heritage buildings do not have basic instructions to act in the event of a fire [23].
- The adverse climate where heritage buildings are located affects their design and structure for indoor comfort. This increases the risk of electrical overloads [25].

5 Safety Actions to Take

5.1 General Actions

Fire safety management is defined as the application of policies, standards, tools, information and practices to analyze, evaluate and control fire safety [59], becoming a safety strategy that requires management policies and procedures. A safety strategy must be continuous with constant reviews to guarantee its effective application [60, 61].

Four factors acting simultaneously trigger a fire: a source of spark or heat (activation energy), combustible material, a sufficient concentration of oxygen (oxidizer), and a chain reaction (propagation). If one of these factors is controlled, there is no fire [31]. An effective fire prevention policy must put forward objectives, as shown in Fig. 2.

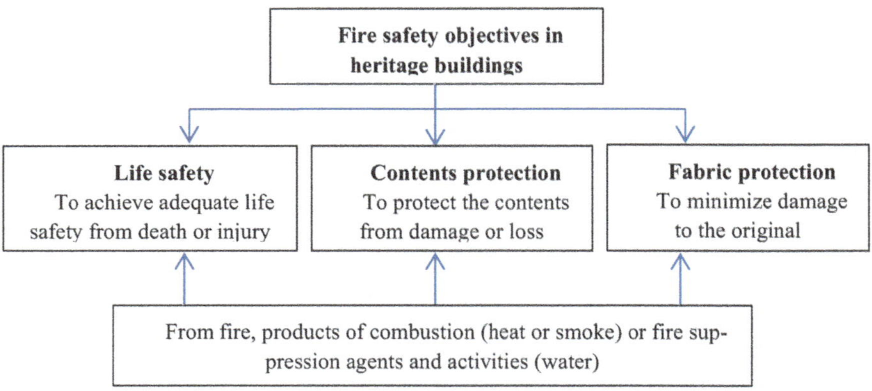

Fig. 2. Fire safety objectives in heritage buildings. Adapted from [62].

The safety and fire protection objectives must be coordinated and agreed upon in conjunction with current authorities, which based on the local legislation, must require the minimum protection measures. Furthermore, authorities must check that the legal requirements for building consent, building permits, building regulations, fire protection standards and certification procedures are fully fulfilled [63].

If a country or region does not have legislation for this type of buildings, one alternative is to adapt or adopt the guidelines of organizations with a lot of experience such as the National Fire Protection Association (NFPA), which among others has: "NFPA 914: Code for Fire Protection of Historic Structures" [64] or "NFPA 909: Code for the Protection of Cultural Resource Properties - Museums, Libraries, and Places of Worship" [65]. On the other hand, the world organizations for the defense of heritage are responsible for developing sustainable strategies and urge governments and international organizations to help preserve the national heritage of each country [24]. Likewise, the authorities, organizations, and administrators of heritage buildings are responsible for hiring qualified personnel, including architects and engineers for preserving heritage, security and fire prevention respectively. These professionals must also coordinate aesthetic actions with protection and mitigation strategies [66]. Passive and active systems, building characteristics must be considered in protection action plans to ensure heritage buildings' integrity. Actions for the protection of heritage buildings and their assets must consider passive and active systems [67]. Protection systems are summarized in Fig. 3.

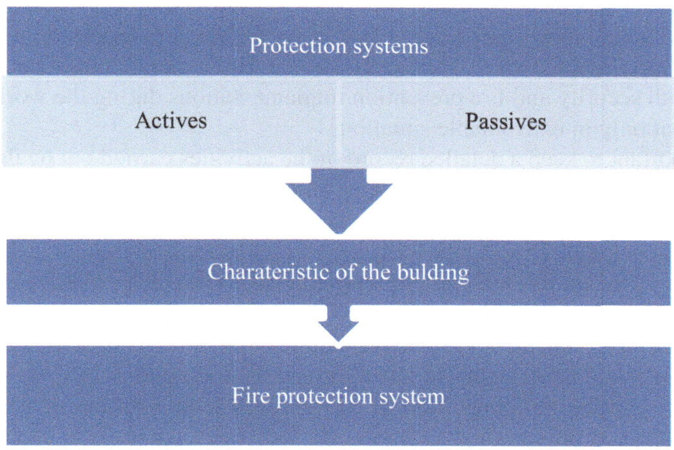

Fig. 3. Action to be taken for preventing fires in heritage buildings. Adapted from [68].

5.2 Restoration/Maintenance

Maintenance is the most important preservation action. It must be carried out continuously to protect and maintain a building or property to a certain level of standard [69]. Maintenance should include all of the technical and practical elements that ensure heritage buildings are properly maintained. This way, heritage buildings will not loss their value and significance [70]. Restoration works in heritage buildings must be carried out with a minimal impact to preserve its historical character [71], and should follow the steps suggested in Fig. 4.

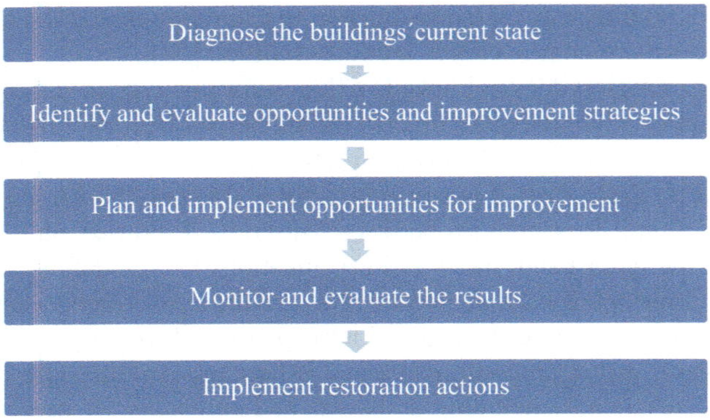

Fig. 4. Steps to follow in a remodel of heritage buildings. Adapted from [72].

Preservation and protection plans should include a detailed study of the materials used in heritage buildings (wood, fabric, metal, among others), in order to determine the appropriate fire extinguishing agents according to their type [73].

Special emphasis should be placed on the surveillance and supervision of restoration works including welding equipment or heat sources that emit sparks or generate overloads in the electrical network. This supervision must be carried out by an expert team equipped with security and fire prevention implementations during the works and must be carried out only in unavoidable situations.

It is important to keep a detailed record of the activities carried out by the personnel in charge of the heritage buildings (training, emergency rescue services, restoration and conservation) and a document must be prepared with the assignments and responsibilities during emergencies [25].

A modern concept for building conservation is sustainability. In this sense, restorations and maintenance must be made more flexible and adapted with new techniques and ingenious ideas to achieve sustainable maintenance in heritage buildings [74], especially those where it is necessary to install fire protection systems [75]. In addition, the conditions of existing buildings need to be improved to adapt them to new functions, new protection and preservation technologies, and strict supervision and surveillance policies [76].

5.3 Electricity

Short circuits normally occur due to excess loads, when consumption is greater than the capacity of the electrical installations. Electrical load studies should be carried out by qualified electricians. When necessary, actions such as increasing the capacity of transformers and generators, changing the wiring, among others in heritage buildings should be taken [80]. Churches, mosques, or buildings with courtyards should have lightning protection systems, especially if they are built in areas with a high probability of electrical storms.

5.4 Terrorism/Vandalism

Unfortunately, there is no much to do when it comes to terrorism. Terrorist attacks are normally unexpected, and the consequences are usually devastating. However, authorities, in those areas with the presence of extreme religious and political groups, may consider prevention and mitigation plans to defend their heritage buildings.

5.5 Additional Actions

Here are some additional basic suggestions to protect heritage buildings against fires:

- Actions taken by governments must become a security strategy that requires management policies and procedures of the highest level [77].
- For an effective protection of buildings, it is necessary to know in detail about the type of materials used in buildings. This allows getting appropriate extinguishing systems and agents [78].
- Many of the heritage buildings have atriums, which are decorative structures that requires special attention with regards to fire prevention systems. Their shapes can generate additional problems in the case of a fire [79, 80].
- Prevention and safety analyzes should include studies of climate change as a catalyst for fires [81], and special care should be considered in heritage buildings located in areas prone to forest fires [82].
- Finally, if a fire has already occurred in a heritage building, the causes must be investigated thoroughly, in order to avoid future damages by fire. This should be done with expert advice, and based on the guidelines, for example NFPA in the "NFPA 921: Guide for fire and explosion investigations" [83] or "NFPA 1033: Standard for Professional Qualifications for Fire Investigators" [84].

6 Conclusions

Many of the incidents with the presence of fire reported in this paper could have been avoided if the storage of flammable materials had been managed technically and responsibly. Also if fire safety management and fire emergency plans had been available. This should leave a lesson for the future so that governments can plan investments for prevention and mitigation programs.

The way of protecting heritage buildings around the world is striking. There is a clear lack of concrete measures for emergency situations.

There must be a global agreement between governments and international organizations for the protection of heritage buildings that limits events where fire is allowed.

Many of the reported incidents could have been avoided if basic security measures had been taken. An appeal is made to those responsible for the care of these buildings to preserve and maintain them.

Engineers and architects have a joint task to protect heritage buildings. On the one hand, adequate plans must be maintained or elaborated in order to keep unaltered the historical and artistic character of the treasures in the buildings. On the other hand, equipment that guarantees a reasonable level of security for its occupants, visitors, and materials of great historical value must be provided.

Finally, it should be noted the importance of safeguarding the historical heritage of humanity because they constitute an asset of incalculable value for society. They cannot be restored if they are completely affected by the fire.

References

1. Plataforma Urbana. Incendios: la constante amenaza a nuestro patrimonio arquitectónico (2014)
2. UNESCO. Illicit Trafficking of Cultural Property (2019)
3. Okba, E, Embaby, M.: Sustainability and heritage buildings. Int. J. Eng. Res. Technol. **2**, 1682–1690 (2013)
4. Bernard, M.: Conservation of historic buildings. Archit Press, 1–12 (2003)
5. Viajar El periódico. Notre Dame y otros importantes patrimonios culturales que han sido devorados por las llamas (2019)
6. Archivo Nacional de Chile. 8 de Diciembre de 1863: La Tragedia del Incendio en la Iglesia de la Compañía de Jesús (2019)
7. Parisi, F., Augenti, N.: Earthquake damages to cultural heritage constructions and simplified assessment of artworks. Eng. Fail. Anal. **34**, 735–760 (2013). https://doi.org/10.1016/j.engfailanal.2013.01.005
8. La Vanguardia. La mezquita de Al-Aqsa se ha incendiado a la vez que Notre Dame (2019). https://www.lavanguardia.com/internacional/20190416/461699084409/incendio-mezquita-al-aqsa-jerusalen-a-la-vez-notre-dame-paris.html
9. BiobioChile. Incendio en Notre Dame: cómo el fuego atacó uno de los monumentos del mundo (2019). https://www.biobiochile.cl/noticias/internacional/europa/2019/04/16/incendio-en-notre-dame-como-el-fuego-ataco-uno-de-los-monumentos-del-mundo.shtml
10. El País. Un incendio consume el Museo Nacional de Río, uno de los más antiguos de Brasil (2018). https://elpais.com/internacional/2018/09/03/america/1535935296_562657.html
11. Yuan, C., He, Y., Feng, Y., Wang, P.: Fire hazards in heritage villages: a case study on Dangjia village in China. Int. J. Disaster Risk Reduct. **28**, 748–757 (2018). https://doi.org/10.1016/j.ijdrr.2018.02.002
12. Verzoni, A.: Salvaguardando la Historia. NFPA J Latinoam, 4 (2018)
13. El Mundo. Sri Lanka sufre una devastadora cadena de atentados contra iglesias y hoteles de lujo (2019)
14. La Tercera. Las secuelas del incendio en Notre Dame miradas desde un drone (2019). https://www.latercera.com/la-tercera-tv/noticia/las-secuelas-del-incendio-notre-dame-miradas-desde-drone/618563/
15. El País. Catástrofe cultural El incendio del Museo Nacional revela que Brasil renunció a velar por el saber (2018). https://elpais.com/elpais/2018/09/04/opinion/1536086612_654125.html
16. El Economista. El edificio histórico que alberga Primark en Belfast (Irlanda) en riesgo de derrumbe por un incendio (2018)
17. Fire risk heritage. Second Fire almost Destroys the Glasgow School of Art (2018)
18. Aciprensa. Egipto: Obispo agradece a musulmán por pedir ayuda para apagar incendio en iglesia (2018)
19. Infobae. Se incendió la catedral de San Nicolás (2017)
20. Aciprensa. Incendian capilla dedicada al Sagrado Corazón de Jesús en Argentina (2017)
21. EFE. Nueva York investiga las causas de un gran incendio en una catedral ortodoxa (2016)

22. La Vanguardia. Un incendio daña gravemente una catedral en la Silesia polaca (2014)
23. Puro Chile. Infierno patrimonial: la incultura chilena en prevención de incendios (2013)
24. Fire risk heritage. Church fires (2019)
25. Fire risk heritage. Museum fires (2018)
26. Reuters. National monument burns down (2008)
27. Cronache. Incendio nel Castello di Moncalieri (2008)
28. Fire Heritage. Fire Safety in Heritage Buildings (2019)
29. Mercurio Antofagasta. Incendio arrasó con histórica iglesia de Taltal (2007)
30. El Mundo. Un incendio causa enormes daños en la catedral de la Santa Trinidad de San Petersburgo (2006)
31. Tacón, J.: Los desastres en archivos y bibliotecas: Causas y efectos, protección y recuperación. Bibl Histórica Marqués Vald, 1–22 (2010)
32. El País. Más de un millón de libros, destruidos en Irak (2003)
33. El País. Un incendio en la Mezquita de Córdoba destruye una parte de su archivo histórico (2001)
34. Simensen, C.: A theory for the collapse of the world trade center. Eng. Fail. Anal. **41**, 39–47 (2014)
35. West Side Rag. Fire During Palm Sunday Service Damages Items at Cathedral of St. John the Divine (2019)
36. El País. La Sábana Santa se salva del incendio de la catedral de Turín (1997)
37. El País. Un pavoroso incendio reduce a cenizas el teatro de ópera La Fenice de Venecia (1996)
38. Explored. Incendio Consume Un Altar De La Iglesia De La Compañía (1996)
39. Fire risk heritage. 1996 Treasury Building Fire – Washington DC (USA) (2019)
40. Ticino. Chiesa di Santa Maria delle Grazie (2019)
41. Cadena SER Radio Zaragoza. Los dos incendios que sufrió la catedral de Teruel en el siglo XX (2019)
42. La Razón. Otras catedrales en el mundo que ardieron en llamas (2019)
43. La Vanguardia. Lo que se esconde detrás del incendio del Liceu (2016)
44. Chakrabarti, B., Yates, T., Lewry, A.: Effect of fire damage on natural stonework in buildings. Constr. Build. Mater. **10,** 539–544 (1996). https://doi.org/10.1016/0950-0618(95)00076-3
45. Arcinfo. Incendie du pont de la Chapelle à Lucerne: vingt ans après, un point de discorde subsiste (2013)
46. El País. Un incendio devasta el castillo real de Windsor y amenaza el tesoro artístico de Isabel II (1992)
47. Radio Cooperativa. No sólo es Notre Dame: Los patrimonios que sufren por incendios, guerras y terremotos (2019)
48. El Confidencial. El incendio de Notre-Dame, un desastre que se repite en la historia del mundo (2019)
49. La Gazzetta del Mezzogiorno. Petruzzelli, 20 anni fa l'incendio (2011)
50. Uzun, Z., Köse, C., Nesibe, C.K.: multidisciplinary study to reveal the historical value of wooden structures and to develop a conservation approach: Dere and Karlı Mosques in Samsun, Turkey. J. Cult. Herit. **32**, 60–72 (2018)
51. Arun, G.: Evolution of timber construction in Turkey, archi-cultural translations through the silk road. In: Proceedings of 2nd International Conference, pp. 225–30 (2012)
52. Irbe, I., Karadelev, M., Andersone, I., Andersons, B.: Biodeterioration of external wooden structures of the Latvian cultural heritage. Cult. Herit. **13**, 79–84 (2012)
53. D'Orazio, M., Bernardini, G., Tacconi, S., Arteconi, V., Quagliarini, E.: Fire safety in Italian-style historical theatres: How photoluminescent wayfinding can improve occupants' evacuation with no architecture modifications. J. Cult. Herit. **19**, 492–501 (2016)
54. Othuman, M., Sani, N., Abas, N., Khaw, Y.: Evaluation of fire hazard and safety management of heritage buildings in Georgetown, Penang. In: MATEC Web Conference, vol. 3 (2014)

55. Yung, E., Chan, E.: Implementation challenges to the adaptive reuse of heritage buildings: towards the goals of sustainable, low carbon cities. Habit. Int. **36**, 352–361 (2012). https://doi.org/10.1016/j.habitatint.2011.11.001
56. Cabeza, L.F., de Gracia, A., Pisello, A.L.: Integration of renewable technologies in historical and heritage buildings: a review. Energy Build. **177**, 96–111 (2018). https://doi.org/10.1016/j.enbuild.2018.07.058
57. Fazli, M.F., Gupta, S.K., Mehta, S.B.: Statistical analysis of variables in heritage buildings. Int. J. Archit. Infrastruct. Plan. **1**(2), 10–7 (2017)
58. Loria, J., Cruz, F., Bautista, A:. Características de incendios-conato de incendio en unidades hospitalarias del Instituto Mexicano del Seguro Social, 2012–2015 2016:2012–5
59. Roslan, R., Said, S.Y.: Fire safety management system for heritage buildings in Malaysia. In: 3rd ABRA Int. Conf. Qual. Life "Quality Life 3," 221–226 (2017)
60. Salleh, N., Ahmad, A.: Fire management in heritage buildings: the current scenario in Malaysia. In: 22nd CIPA Symp, p. 6 (2009)
61. Jácome, N.: Historic Building Conservation (2015)
62. Salleh, N.H.: Fire Safety in Heritage Buildings: Life vs Property Safety. Herit. study Muslim World (2017). https://doi.org/10.13140/2.1.3955.5848
63. Grupo Español de Ciudades Patrimonio de la Humanidad. Manual Básico de Seguridad y Protección contra Incendios en Ciudades Patrimonio de la Humanidad (2006)
64. NFPA. NFPA 914: Code for Fire Protection of Historic Structures (2007)
65. NFPA. NFPA 909: Code for the Protection of Cultural Resource Properties - Museums, Libraries, and Places of Worship (2017)
66. Watts, J., Solomon, R.: Fire safety code for historic structures. Fire Technol. **38**, 301–310 (2002)
67. Kasza, Z., Kovacs, K.: Risk analysis about lightning protection for buildings focusing on risk of loss of human life. Procedia Manuf. **32**, 458–465 (2019). https://doi.org/10.1016/j.promfg.2019.02.240
68. Ibrahim, M., Abdul-Hamid, K., Ibrahim, M., Mohd-din, A., Yunus, R., Yahya, M.: The development of fire risk assessment method for heritage building. Procedia Eng. **20**, 317–324 (2011). https://doi.org/10.1016/j.proeng.2011.11.172
69. Kalsum, U., Zakaria, N., Mohammad, A., Shah, A.: Maintenance work for heritage buildings in Malaysia: owners' perspectives. Int. J. Build. Pathol. Adapt. (2018). https://doi.org/10.1108/IJBPA-07-2018-0062
70. Musso, S., De Marco, L.: Teaching Conservation/Restoration of the Architectural Heritage Goals, Contents and Methods. (2008)
71. Kidd, S.: Fire Safety Management in Heritage Buildings. Historic S, Edinburgh (2005)
72. Ezzat, A., Elsaay, H.: Adaptive reuse: an innovative approach for generating sustainable values for historic buildings in developing countries. Organ Technol Manag Constr. 10, 1704–1718. https://doi.org/10.2478/otmcj-2018-0002
73. COWI. Manual Fire Extinguishing Equipment for Protection of Heritage (2006)
74. Hasnain, H., Mohseni, F.: Creative ideation and adaptive reuse: a solution to sustainable urban heritage conservation. Earth Environ. Sci. 8 (2018). https://doi.org/10.1088/1755-1315/
75. Central Public Works Department. Handbook of Conservation of Heritage Buildings (2013)
76. Hapizah, S., Suhaiza, N., Ahmad, Y., Azidah, N., Mohd, F.: Conceptual design of GIS database for heritage building in melaka. J. Eng. Technol. **3**, 114–128 (2018)
77. He, Y., Park, L.A.F.: A statistical analysis of occurrence and association between structural fire hazards in heritage housing. Fire Saf. J. **90**, 169–180 (2017). https://doi.org/10.1016/j.firesaf.2017.04.021
78. Macari, R.: Criterios que determinan los requerimientos de resistencia al fuego de elementos estructurales (2015)

79. Faller, G.: Seguridad contra incendios en edificios con atrio. Inf La Construcción n.d, 55, 27–32
80. Riggio, M., D'Ayala, D., Parisi, M.A., Tardini, C.: Assessment of heritage timber structures: Review of standards, guidelines and procedures. J. Cult. Herit. **31**, 220–235 (2018). https://doi.org/10.1016/j.culher.2017.11.007
81. Ankersmit, B., Stappers, M.: Managing Indoor Climate Risks in Museums. n.d
82. Fire risk heritage. Vegetation Fire and Cultural Heritage buildings: the Paul Getty Museum case study (2017)
83. NFPA. NFPA 921: Gude for fire and explosion investigations (2017)
84. NFPA. NFPA 921: Norma para Calificaciones Profesionales para Investigadores de Incendio (2014)

Mobile Robot Teleoperated Through Wireless Communication and Video Acquisition in Real Time for Hazard Applications

Mélany J. Yarad Jácome[1(✉)] and José B. Carguachi Caizatoa[2]

[1] Universidad de Las Fuerzas Armadas ESPE, Av. El Progreso S/N, Sangolquí, Ecuador
mela6327@yahoo.com

[2] Instituto Superior Tecnológico Sucre, Teodoro Gómez de la Torre S14-72 y Joaquín Gutiérrez, Quito, Ecuador
joselocc69@gmail.com

Abstract. This research is focused on providing security by the acquisition of images in real time to be implemented on hazardous situations, through a teleoperated robot. Robot can be implemented in activities of any kinds; such as military or police missions, expeditions to places with radiation, rescues. To begin, the robot is programmed to be able to move and to send images which are processed in streaming, keeping the safety on the communication network. After a statistical analysis, the preferences of usage of these tools in function of the elements which compose the robot were determined. Since the targets of the robot are explorations, it's must be autonomous, for this, it's possesses batteries that provide energy to the Electronic and Mechanical parts. The gathered results depicts the importance operators assign to determined applications, making expedition enterprises to remote places even easier. To implement this project, free software was utilized to the programming. The robot counts with xBee shields which provokes communication with the robot and the computer. Additionally, the video acquisition application is included. This application is composed by a LAN network; that links an IP wireless Wi- Fi camera, generating data communication in streaming mode with the computer.

Keywords: Teleoperated robot · Streaming · xBee shields · LAN · Wi- Fi · IP camera

1 Introduction

Within human development, countless activities have been presented that day by day have put at risk the personnel who execute them, from explorations in inhospitable places, to the deactivation of detonators, among many others that with the advancement of technology, have also become more harmful and dangerous, so it is necessary that in this research process regarding the importance of developing aid systems for operations in high places risk, the background and needs to be able to generate the respective objectives that must be met for an appropriate design of a Teleoperated Robot. In addition

M. Botto-Tobar et al. (Eds.): CIT 2020, AISC 1326, pp. 448–461, 2021.
https://doi.org/10.1007/978-3-030-68080-0_33

to establishing the guidelines by which research should be investigated, it is essential to have the respective theoretical support and the necessary foundations. In addition to establishing the guidelines by which research should be investigated, it is essential to have the respective theoretical support and the necessary foundations. Knowledge of the types of networks, the programming languages used for the respective control, mechanical and electronic equipment and elements as well as the design and selection of parts, pieces, elements and spare parts are concepts necessary to develop the prototype according to the needs of avoiding the unnecessary loss of human life in the various activities in hazard environments or in missions in which they may be exposed to any kind of risk that can be can threaten their integrity.

2 Background

The missions carried out by police personnel, the army or scientific explorations, are often high risk, where it is not convenient to use the human hand in some work, due to the high risk that these can be exposed, it is for this reason that different tools are developed that can meet the objectives set, allowing them to be replaced by a teleoperated mobile equipment that is called a teleoperated robot.

Efforts have been made to safeguard the lives of people in charge of high-risk operations, but often failed processes have been resulted, it is for this reason that a streaming manipulation robot is created, thanks to the technology that currently exists, can achieve research aspirations, involving electronics, communications, control, computer vision and network security that is incorporated into the acquisition of the video.

Presented with the problem that arises in high-risk scans, this research focuses on the effective application of wireless communication and real-time video acquisition, taking into account the different types of programming that exist and network protections to avoid infiltrations in signal acquisition. To do this, an analysis of free software was performed (raspberry linux, Arduino), then proceeded to design and build, on the one hand, an instrument that allows to manipulate elements (GRIPPER) and on the other the application of the streaming viewer.

After data stratification, the preferences of the use of this robot were determined according to each application.

3 Statement of the Problem

The research work was carried out after detecting a number of inconveniences that the army and the national police have in carrying out missions that made access difficult, under specialist specifications, a teleoperated robot was developed, capable of solving problems in high-risk operations. Generating the next problem.

Is it possible to develop a mobile teleoperated robotic arm with protected wireless communication and video streaming acquisition, using mechatronic elements, in order to safeguard the lives of people in charge of high-risk operations?

4 Limitations of Study

The use of artificial intelligence in a robot, is verified in the decisions that it makes, by means of a program established by the user, which depends on the application that will develop, within this is also considered computer vision, which is most used in teleoperated robots, that is when the tasks of the robot involve the processing of images coming from video cameras that can be located in it.

The transcendental part of this work is to identify the means of teleoperation and protections to wireless networks that are used for the mobilization and acquisition of streaming images, and which considerations should be taken into account at the time of use.

5 Objectives

5.1 Main

The objective of this research is to develop a mobile teleoperated robot with protected wireless communication and video streaming acquisition, using mechatronic elements, in order to safeguard the lives of people in charge of high-risk operations.

5.2 Specifics

- Design and apply the teleoperated robot in high-risk operations for in charge.
- Set the mechatronic elements to be used in a teleoperated robot.
- Identify the operations the robot may be exposed to.
- valuate the results.

6 Hypothesis

Specialized people use the teleoperated robot in such a way that they do not influence high-risk research.

7 Arguments

This part presents the technical-scientific compendium of the research carried out, as well as the design, application and synergy of the elements that were used. In relation to synergy, Fig. 1, indicates the reciprocity of each of the studies used.

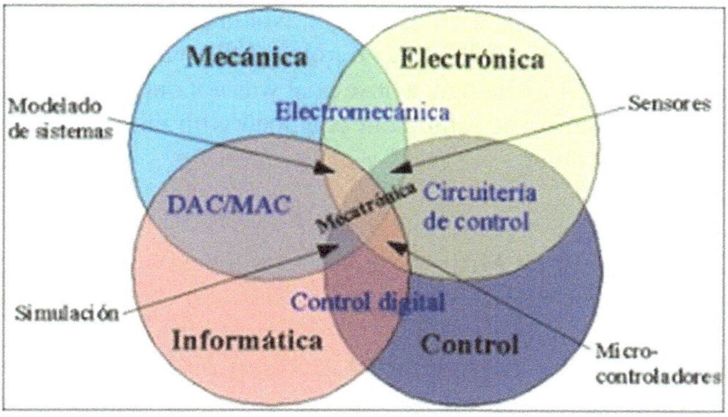

Fig. 1. Synergy of compendium that make up MECATRÓNICA [1]

7.1 Teleoperated Robot

It is understood that teleoperation is a set of technologies that achieve long-distance operation of a device, which is manipulated by a human; therefore, teleoperating is the action that a human performs to operate or remotely govern a device [2].

By definitions of NASA as Gonzales mentions, teleoperated robots are defined as robotic devices with manipulative arms and sensors with a certain degree of mobility, remotely controlled by a human operator directly, or by a computer [3].

The military area provides many possibilities for teleoperated systems, most of the mobile teleoperation technologies were developed for applications in space, underwater, medical, in the nuclear industry, military, the technologies used here range from remote monitoring systems, to the use of UAV[1]. An example of this type of vehicle is the US Air Force Predator [4]. Teleoperation has proven to be a very useful tool to allow humans to perform certain tasks that otherwise would not have been possible [5].

7.2 Wireless Networks

a) Bluetooth

Bluetooth technology appeared more than a decade ago. After its emergence on the market, Bluetooth technology has revolutionized the way electronic devices are connected to each other. The technology behind Bluetooth continues to advance and can be found on a wide range of electronic devices, including cell phones, game controllers, audio and video equipment.

b) WIFI

Wireless technology has been living with us for many years, no less than since the early 1990s, albeit in a disorderly manner, because each manufacturer developed its own models, thus generating difficulties for others.

[1] Unmanned Air Vehicles.

In the late 1990s, companies such as Lucent, Nokia and Symbol Technologies met to create a partnership known as WECA (Wireless Ethernet Compatibility), which in 2003 was renamed the Wi-Fi Alliance, whose goal was not only the promotion of Wi-Fi technology, but set standards for equipment equipped with this wireless technology to be compatible with each other.

c) WiMAX

Worldwide Interoperability for Microwave Access is a data transmission standard that uses radio waves at frequencies from 2.3 to 3.5 GHz and can have coverage of up to 50 km.

It is a technology within the so-referred to as last mile technologies, also known as local looping that allows microwave data reception and radio wave relaying. The standard that defines this technology is IEEE 802.16. One of its advantages is to provide broadband services in areas where the deployment of cable or fiber by low population density has very high costs per user (rural areas).

You can define a wireless network as the set of computers or computer devices that communicate with each other by elements that do not require connection cables for their operation, these networks are classified into several categories according to the geographical area in which the user is connected to the network, such as LAN[2], MAN[3] and WLAN[4]; called coverage areas (Fig. 2).

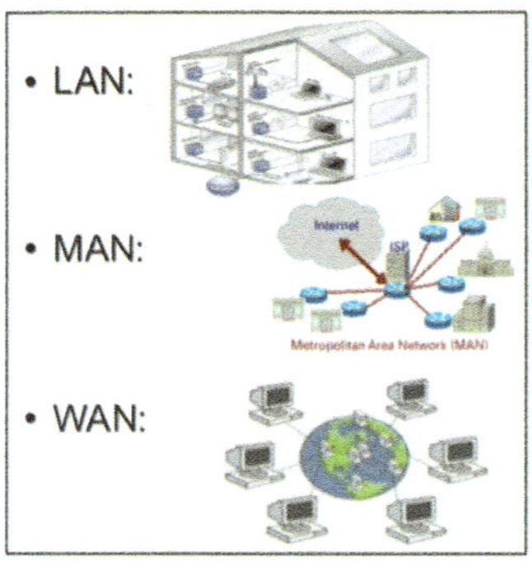

Fig. 2. Wireless networks types [6]

The advantage in wireless networks is the compatibility between the different equipment that connect to each other, the security of a network is very important, since they have encrypted packets, which must be centralized, depending on the protocol to be used.

Among the disadvantages that exist within a wireless network is that it has more chance of infiltration to wired networks, and if they are not protected in some way they can suffer computer attacks since they are visible to any compatible device.

[2] Local Area Network.

[3] Metropolitan Area Network.

[4] World Area Network.

7.3 Processing

It is a flexible programming language, and with free access for software development in the creation of virtual environments, the environment is based on JAVA, but developed by IDE (Integrated Development Environment), which is easy to use and compatible with several programs, since it has a graphic environment of pixels that have an axis of coordinates that help in the displacement in rectangular coordinates (right, left, (x); up, down, (y) (Fig. 3).

Fig. 3. Processing program presentation [7]

The creation of designs in processing requires several commands which execute an order when they are programmed, that is to say, the program is based on a computer language so that the processing understands the type of design that is intended to be made.

7.4 Arduino

"It is a free hardware board that incorporates a microcontroller and a series of female pins (which are internally attached to the pins of the microcontroller that allow to connect there in a very simple and comfortable way different sensors and actuators)" [8].

Arduino was created as a centralized open-coding digital and electronic platform basically in software and hardware, designed for easy learning of both programming and electronic circuits.

It is compatible with a wide variety of sensors thanks to the inputs with which it was created just as it has the feature of manipulating different elements such as motors, LED lights, Arduino programming is based on C++ language and the environment is based entirely on processing with some variations.

The current market has an extended range of Arduino for various applications, as the technology advances in great strides and applications become more demanding, that is why manufacturers must develop better technologies that are easy to use and economically accessible to all users.

7.5 Streaming (Acquisition if Image)

"Security management should include both personal security and heritage security; in addition, a security system must take different functions that ensure the three basic security tasks: prevention (before the attack occurs to avoid it), the alarm (at the time of the attack, warning) and the reaction (once the attack has occurred to alleviate its effects)" [9].

The fundamental thing about these cameras is that they always need to be connected to an internet network and have the characteristic of being remotely controlled by the user.

In some cases these cameras can turn on when they detect motion and send messages to a user with the sequences or the entire video to report why the camera has been activated.

According to José Huidrobo the operation of the cameras basically depends on an address since it needs static routing on the A P[5] for video and audio transmission, once this is done the image reaches a receiver over a WAN, it does not need cable as there are cameras that work with WI-FI with the 802.11 standard, the user can access at any time the acquisition of real-time video called Streaming by accessing the I[6] address of the installed computer.

7.6 Mechatronic System

7.6.1 Wireless Module x-bee

"ZigBee is a wireless communications protocol based on the 802.15.4 standard, it is designed for low-speed communications between two or more devices, networks can be formed with thousands of devices communicating with each other, it is perfect for many applications" [10].

For the Computer, Robotics and Electronics Club, X-BEE wireless modules are devices consisting of a transmitter, ZigBee receiver and processor, so it allows users to perform simpler tasks with little data transmission and is part of a PAN[7] network the standard used is 802.15.4. They are classified as a PAN network as they work with radio frequency and do not need to connect to an internet network.

7.6.2 Servo Motors

"Servos are a special type of DC motor. Which are characterized by their ability to immediately position themselves in any position within their operating range. For this, the servomotor waits for a train of pulses that corresponds to the movement to be carried out. They are generally made up of an amplifier, a motor, a reduction system made up of toothed wheels and a feedback circuit, all in the same small box" [11].

For Iván T. and Omar O., a servo motor is a type of Vcc motor, which is defined by its ability to move at different angles in a range from 0° to 180° immediately, to perform such action the servo motor receives a train of PWM pulses that indicates the position or movement to be carried out, servo motors are generally made up of amplifiers, a speed reduction system made up of toothed wheels (gears) and a re-feeder circuit.

Like all electronic elements, there are also types of servo motors, among the most representative are the Servo motors of direct current, alternating current, Brushless permanent magnet, step by step.

[5] Access Point.

[6] Internet Protocol.

[7] Personal Area Network.

7.6.3 Bateries

"An electric cell or electric battery is the industrialized and commercial format of the galvanic or voltaic cell. It is a device that converts chemical energy into electrical energy by a transitory chemical process, after which its activity ceases and its constituent elements have to be renewed, since its characteristics are altered during it" [12].

According to IDEA, a battery is a chemical vessel that transmits electrons, that is to say, an energy supply based on chemical reactions. The batteries consist of two poles, positive (+) and one negative (−).

Electrons move from the negative pole to the positive pole, completing the chemical reaction, that is to say, electricity is produced by connecting a charge to the battery. Batteries consist of electrodes, electrolytes, separators and medium. Giving rise to the types of batteries that are Ni-Cd[8], Ni-MH[9] and Lithium batteries.

7.6.4 Mechanic System

"The mechanical system is composed of various joints. It is usually distinguished between the arm and the terminal organ or final effector that can be interchangeable, using tweezers or specific devices for different tasks" [13].

According to Yhoana Yirley the manipulating robots are those that are composed of several joints which can be interchangeable for the application to which it was intended.

Manipulating robots having joints based on the morphology of a human arm have several degrees of freedom that are listed according to the number of positions that can take the same either a fixed robot arm or one that moves through a terrain.

7.6.5 Drive Design

The design of the drive system refers to the mobility of the robot as such, that is to say, which will be the most suitable system for movement and displacement on the ground, there are 3 types of motor designs today are the track tractor, arthropods and mechanically steering rims.

8 Experiment Configuration

A teleoperated robot was made, which has supervised control, being the robotic arm the slave, is the element that can be part of the risk tasks, while the operator manipulates by means of high-level commands through the software for the manipulator to execute them, these applications are currently the most used in the military field to execute high-risk operations to execute high-risk operations, as they allow the use of virtual models, consider parameters of danger and other things. A clear example of this type of control can be seen in [14].

For the movement of the robotic arm, direct current servo motors were used, which allow the mobility of the mechanical parts assembled according to the generalized application for which the device was created, in the same way the mobility of the robot is done by means of tracks, as presented in the Fig. 4.

[8] Nickel Cadmium.

[9] Nickel Mercury.

Fig. 4. Robot with gripper for servo motors and mobility by means of tracks [15]

In Fig. 4 you can see the teleoperated robot, which counts 4 tracks for mobilization, a robotic arm with terminal function in gripper, with a structure of 6 degrees of freedom, and an IP camera that emits a signal in streaming, all these functions are controlled by the application software, which consists of a distribution of sliders for the movement of each of the parts of the robotic arm, which are shown in Fig. 5 and Fig. 6, Fig. 7 shows the control of the overall displacement of the robot, having the following configuration:

Figure 5 shows the position control of the robotic arm in streaming, these sliders allow to locate the arm, the movements that develop are as follows:

- Clamp: opens and closes for clamping elements, can be replaced as needed.
- Wrist: allows the clockwise and counterclockwise rotation of the clamp.

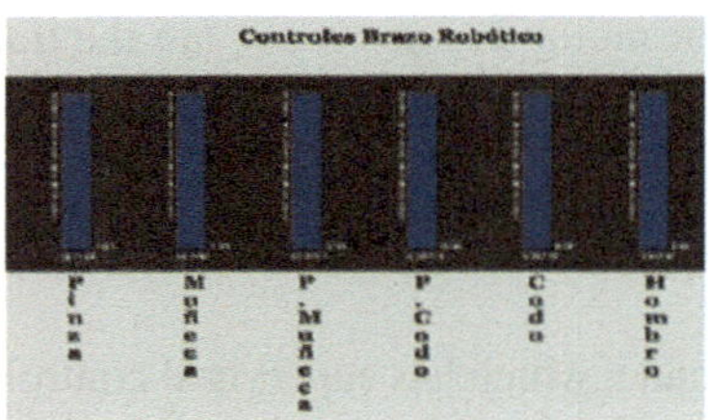

Fig. 5. Mobility controls of robotic arm [15]

- P. wrist: upper and lower position of the wrist of the robotic arm.
- Elbow: upper and lower position of the robotic arm.
- P. elbow: upper and lower position of the forearm of the robotic arm.
- Shoulder: clockwise and counterclockwise of the robotic arm.

Figure 6 shows the position of the DC motors, which allow the movements mentioned in the previous section.

For the control was use dan Arduino Mega card, which allows teleoperated mobilization and also thanks to the Processing you could have the acquisition of image in

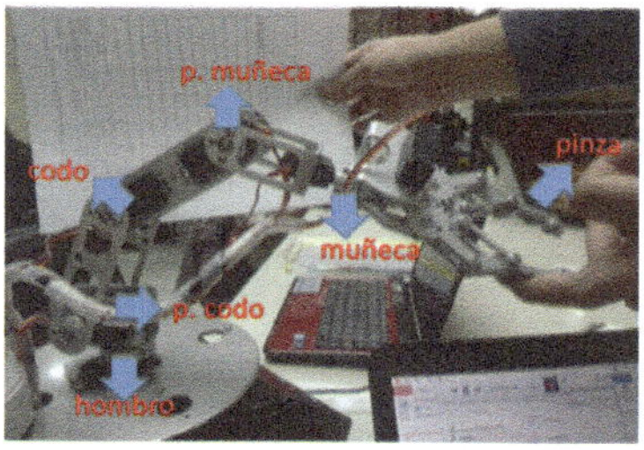

Fig. 6. Controlled parts of robotic arm [15]

real time, for this was necessary an IP camera, which is configured by a wireless router so that it can send the captured information. And the communication to the robot was obtained through the Xbee modules, which allow to direct the robot and give it feed rate, the general interface used can be seen in Fig. 7.

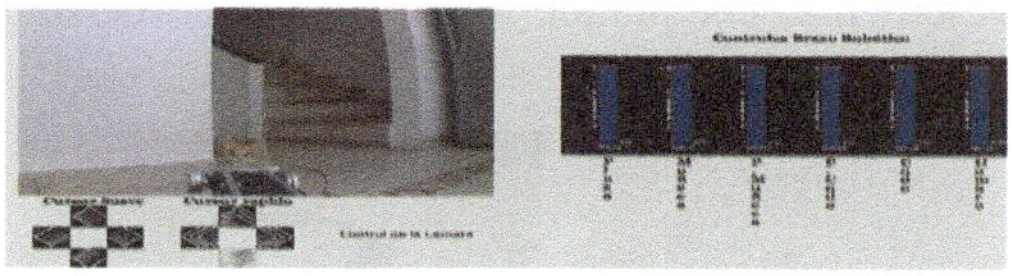

Fig. 7. Control of teleoperated robot [15].

In Fig. 7 you can see the controls of the robotic arm are distributed by sliders or control bars are arranged as follows: clamp, wrist, first wrist, first elbow, elbow and shoulder, while the camera buttons are arranged horizontally for the left and right movement of the camera. The Soft Cursor is to make the camera rotate 5 degrees which makes it more accurate when choosing a specific point of view, as well as the Fast Cursor advances 10 degrees which makes it reach a faster point of view.

The movement control of the robot tracks are arranged by the numerical keypad of the computer, since it is less efficient for the user to control the tracks with the computer cursor, because it would become a more complex process and it would be late the response time by the operator until the button position is placed at the cursor of the computer screen, the control is done with the number 8 (to advance), 2 (for back up), 6 (to turn right) and 7 (to turn left).

The power of the robot is autonomous since it has two lithium-ion batteries, which are intended for control, that is to say, all motors and programming card, the other battery is for the power of the IP camera, it is not necessary to remove the batteries to charge them, it can be done by means of the cables that the robot has on the left side.

To implement the robot, certain parts were imported, since in the country do not exist, these are the tracks, batteries and the arm, each of these has its own implementation manual, so each of the mechanical and electronic parts was assembled according to the indications and specifications established.

Figure 9 shows the mechanical part of the robotic arm, in which the motors that help the movement are implemented, depending on the application you want to use, the position of the motors may vary (Fig. 8).

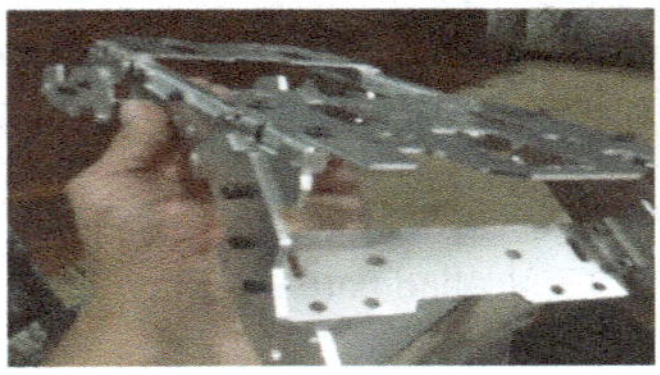

Fig. 8. Mechanical parts of robotic arm and track [15]

Figure 9 shows the electronic board that was developed for the control of the teleoperated robot, in Fig. 9.b you can see the assembly of the Arduino card and the routing of the PCB[10], while in the 9.a indicate the drivers of the cc motors and the assembly of the necessary electronic elements. The plaque was made under the standard IPC[11] 2221.

a b

Fig. 9. Arduino card and motor driver CC [15]

The streaming configuration is done by configuring the processing and router of the robotic arm. (Programming).

8.1 Proposal Pending

As it is a team for the national police or the military, for high-risk processes, it is necessary that existing wireless communication has protections, that is to say, the communication between the slave master (operator – robot) is secure, for this it will be necessary to protect the communication network against attackers, performing a DMZ.[12]

If the robot is used in an espionage process becomes vulnerable, causing threats to the operator, in this way the risk to the activity to be executed occurs.

Given this premise, it is intended that the design of the security policy of the information received by the teleoperated robot, avoid the leakage of such information acquired by the operating entities through social networks, based on the Regulations for the preparation, management, custody, dissemination and security of classified information and controls of ISO/IEC 27002:2013.

[10] Print Circuit Board.

[11] Print Circuit Institute.

[12] Demilitarized zone.

A comparison is made with the work carried out on the subject obtaining the following observations:

The research focuses on the area of development of technologies that improve the interaction of the operation of robotic systems in search and rescue. The guideline that follows this investigation has to do the search and rescue, focused on the service to society in police, military, and other entities involved in public order. Being a particular issue, the requirement of a complex system is less, because the needs are more basic. Therefore, this research generates an important contribution to improve the guidelines of the project and thus complement specific needs [16].

The use of teleoperated robots for infiltration, rescue, exploration in dangerous areas, where it is possible for humans to be in danger, and their physical integrity may be threatened, has allowed the research to generate an important contribution to the improvement of these systems. For this reason, it is important to consider each of the needs when designing a robot, so that it can meet several requirements and meet all possible needs. This research imposes a different field of action in a very dangerous and dangerous environment where this technology can be applied and contribute greatly to the safety, rescue and missions that can save lives or avoid catastrophes [17].

9 Results

9.1 Contributions

In this research it is intended to describe the construction of the robot, through its components which consist of electronic cards intended to send and receive radio signals, a robotic arm consisting of several servo motors to perform the movement, motors located in the tracks of the equipment which will make possible its movement, a camera for real-time video acquisition and its functionality to be used in high-risk work environments, through its assembly of mechanical parts, electronic card programming and the autonomy generated by the power source.

Once programmed and assembled, performance tests were performed as shown in Fig. 10, that is to say, the range of the radio frequency cards and the router's video transmission signal having as 100 m for the XBEE Pro S1 transmit and receive card (approximately) and 30 m for the transmission of video by the router (approximately), these values are adjusted depending on the environment in which the equipment is operating since the values previously expressed can vary by the obstacles that arise before the transmission and reception of the signal such as walls, windows, columns and others.

To get a better idea of what should be developed with a teleoperated robot for risky applications, the interview was used as a method of data collection, in order to analyze and interpret the different opinions expressed by experts on the subject and possible recommendations that may have the same about the implementation of a teleoperated robotic arm with wireless communication and video in real time.

For the conduct of the interview a sample of the people working in the Eplicachima barracks, who are in the area of risk, resulting in a total of 500 people to be interviewed, the instrument consists of 10 questions regarding the subject to clear doubts around the research.

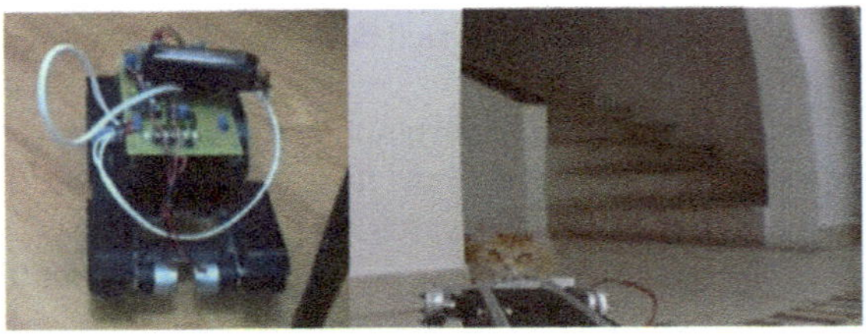

Fig. 10. Teleoperated Robot and imagen acquisition [15]

The Eq. 1 was used to determine the number of people interviewed.

$$n = \frac{k^2 * p * q * N}{\left(e^2 * (N-1)\right) + k^2 * p * q} \tag{1}$$

Where:
N: is the size of the population or universe (total number of potential interviewees).
k: is a constant that depends on the level of trust we assign. The level of trust indicates the probability that the results of our research are true: 95.5% trust is the same as saying that we can be wrong with a probability of 4.5%.
p: is the proportion of individuals who possess the study characteristic in the population. This data is generally unknown and is often assumed to be that p = q = 0.5 is the safest option.
q: is the proportion of individuals who do not possess that characteristic, that is 1-p.
n: is the sample size (number of surveys to take).

Knowing that the results require a 95% trust level, you have to k is 1.96, p and q is 0.5, and the sample error is 5%. N = 500.

Applying the Eq. 1.

$$n = \frac{1.96^2 * 0.5 * 0.5 * 500}{\left(0.05^2 * (500-1)\right) + 1.96^2 * 0.5 * 0.5} \quad n = 217, 54$$

This indicates that the research instrument will apply to 218 people close to the scope of latent risk.

Once these results were obtained, the recommended options for the development of the teleoperated robot were chosen with applications running high-risk tasks to safeguard people's integrity.

It should be emphasized that once the network was created, communication is not unique to a single brand router can be performed with any type of device, and connect to the network provided by the same.

9.2 Conclusions and Recommendations

The usefulness together with the efficiency of a teleoperated system adapted to the needs of carrying out certain types of risk activities, is very high, so it is important to consider each of the needs that these activities demand. It was possible to create a teleoperated robotic arm that works in an automaton manner, working with protected wireless communication allowing to acquire visual information in real time, with the need to harmoniously combine mechanics, electronics, control and programming, whose sole purpose is to allow staff to carry out activities without putting their lives in danger. It is also important to analyze in a very timely manner and define the activities in which the robot will be applied, since according to the risks contemplated by the design, the costs of the design will vary and the more complex or demanding the design, the more technology must entail. The important to consider that to harmonize design processes and generate the appropriate synergy between the mechanical and electronic part, the appropriate knowledge must be available, there are currently mechatronic professionals who have the necessary capacity. In addition, it is necessary to sum the programming processes of the systems. Design with poor knowledge should be considered to impact costs and effective operation. Life precautionary is one of the main approaches of electronic equipment developers, this research, design and construction of the teleoperated robot prototype, maintains the aforementioned approach, guided to be a contribution within the infiltration, rescue, tracking and control activities in places where human life may be at risk.

References

1. Florian, F.: Tecnología del Futuro. [En línea]. https://florian-tecnologiadelfuturo.blogspot.com/. Último acceso 9 Diciembre 2016
2. Escobar y, L.M., Aroca, K.: Robot Manipulador Teleoperado, Sangolquí (2017)
3. González, V.R.: Educa Lab, 03 2004. [En línea]. https://platea.pntic.mec.es/vgonzale/cyr_0204/ctrl_rob/robotica/teleoperado.htm, Último acceso: 6 diciembre 2016
4. Baño, K.: Aplicaciónes en teleobótica (2018)
5. Breñosa, J.M.: Desarrollo de tecnologías de telemanipulación con alto grado de escalado orientadas a la interacción Hombre-Robot (2016)
6. Botoa, M.: SlideShare, 2 Octubre 2014. [En línea]. https://es.slideshare.net/mariobotoa/las-redes-informaticas-39786501, Último acceso: 27 diciembre 2016
7. Fry, B.: Processing, Massachusettss (2011)
8. Vaca, E.: Arduino: Tecnología para todos, Editronix, 17 febrero 2013. [En línea]. https://arduinodhtics.weebly.com/, Último acceso: 28 diciembre 2016
9. Huidrobo, J.M.: Manual de domótica (2010)
10. Club de Informática,R.Y.E.: C.I.r.E, 12 febrero 2012. [En línea]. https://webdelcire.com/wordpress/archives/1714, Último acceso: 12 diciembre 2019
11. Terceros, I., Ormachea, O., Urquidi, Ó., Villazón, A.: Desarrollo de un espectrómetro de bajo costo con una rejilla ajustable y control electrónico inalámbrico. Revista Politécnica 37(1) (2016)
12. IDAE, Circuitos en C.C. Pilas y Acumuladores. 16 Mayo 2013. [En línea]. https://www.rjconsultores.es/pila-electrica/, Último acceso: 28 Septiembre 2020
13. Yirlet, Y.: Sistema Mecánico, 18 Marzo 2013. [En línea]. https://sites.google.com/site/latecnorobotica/sistema-mecanico, Último acceso: 20 Julio 2020
14. Fong, T.: Vehicle teleoperation interface. Auton. Rob. **11**(1), 9–18 (2001)
15. Yarad, M.: Robot teleoperado para riesgos militares, Quito (2019)
16. Baker, M.: Improved interfaces for human-robot interaction in urban search and rescue. Syst. Man Cybern., 2960–2965 (2004)
17. Nagatani, K.: Emergency response to the nuclear accident at the Fukushima Daiichi Nuclear Power Plants using mobile robots. J. Field Rob. **30**, 44–63 (2013)

Low Cost Multitask Mobile Robot for Military Application

Wilbert G. Aguilar(✉), Pablo Albán, Henry Cruz, Alexis Tiuma, and Jean Figueroa

CICTE, DEEL, Universidad de las Fuerzas Armadas ESPE, Sangolquí, Ecuador
wgaguilar@espe.edu.ec

Abstract. In this work, we proposed the design and construction of an unconventional mobile robot. The robot will be able to move in different environments, including stairs, and will have a camera and wireless communication between the robot and the computer with an HMI interface. In the design proposal versatility, robustness and modularity are desired. The designed system works on subsystems ensuring the functionality of the robot modules, and a simple technical support. ROS was selected as the operating system due to its higher performance. Our proposed robot is capable of performing missions in environments similar to those that military personnel have access, but which are high risk. One of the main problems with most unmanned ground vehicles (UGVs) is their inability to climb stairs, which limits access to multiple workspaces. This robot presents an important advance in UGV research with military application, which can be complemented with robotic arms or other tools according to demand.

Keywords: UGV · Mobile Robot · Multitask · Military Application

1 Introduction

In the military industry [1–7], there are several robotic platforms as Stretcher robot, which provides a powerful multi-purpose mobility. Also all of these robots use different mechanisms for mobility, by wheels [7], sliding tracks [9], locomotion limbs [10] and hybridizes configuration [11]. Our proposal offers a UGV robot for military purposes, in environments with high level of risk that could compromise the security and integrity of military personnel. We developed an unmanned ground vehicle, in the CIAM-CICTE Research Center, capable of moving on different surfaces inclinations varied soil rich functionality and proven mastery of the environment, greatly helping in the work field review, demining against possible terrorist attacks.

Recent Ecuadorian technological advances in teleoperated vehicles have been limited, being one of the most notorious work in this area that conducted in 2014 at Universidad de las Fuerzas Armadas ESPE that used the Rocker-Bogie system to traverse different terrains, the main disadvantages of this model the large size and lack of strength in the structure implemented [11] Another important work was done in 2012 by Universidad Politécnica Salesiana, developed as support high-risk operations in the Fire Department, this model used traction by wheels being its main drawback as it is not able

M. Botto-Tobar et al. (Eds.): CIT 2020, AISC 1326, pp. 462–473, 2021.
https://doi.org/10.1007/978-3-030-68080-0_34

to cross obstacles like steps [12]. Notwithstanding, none of these platforms is able to climb stairs, a task that is commonly necessary to access areas of interest.

This paper is organized as: Sect. 2 describes our approach for the design of the multitask robot. Experimentation, results and the respective discussion for different conditions are presented on Sect. 3. Finally, the last two Sections are designated to conclusion, future works and acknowledgement.

2 Our Approach

To support security and defense institutions in Ecuador, it is necessary to have sophisticated technological equipment to perform dangerous tasks that threaten the integrity of the personnel working in these institutions, and the army is no exception. Currently there are few devices that help in the fulfillment of reconnaissance tasks, where access is difficult or risky. Due to this problem, CICTE worked to equip military personnel with a remote control of a road vehicle, which can be mobilized on any ground and under different inclinations or access difficulties. Additionally, the vehicle can be operated by an HMI interface with a computer, providing a rugged and versatile equipment. QFD defines a matrix, which allows prioritizing the needs and technical characteristics of the UGV, setting the guidelines or specifications of the system. The modules of the mobile robot are described on Table 1.

Table 1. Matrix modules for design

N	Module	Subsystem	Function
1	Structure	Extremities	Positioning over obstacles
		Chassis	Integrated vehicle components and provides support and rigidity
2	Locomotion	Locomotion	All-terrain vehicle movement
3	Monitoring	Camera	Video streaming real-time environment
		Interface	Human communication - machine
4	Tele-operation	Communication	Transmission and reception of signals
		I send	Tele-operation of the robot

In the market there are commercial robots similar to those of this project such as the PackBot company iRobot, this robot measures 17.8 cm high, 52.1 cm wide and 88.9cm long with arms extended features, also has a weight of 10.89 kg without including batteries. In this project it was established that the maximum size should be 11 cm high, 55 cm wide and 80 cm long with arms outstretched also set the maximum weight is 15 kg [14].

For material selection or the weight percentage of importance of the parameters is determined. Where 0 indicates that a parameter is more important than another, 2.5

Table 2. Weighting parameter for the selection of the material of the structure

Parameter	Mech. properties	Weldab	Machinab.	cost	Weight	$\sum+1$	%
Mechanical properties	–	2.5	0	5	0	8.5	15.45
Weldability	2.5	–	0	5	2.5	Eleven	Twenty
Machinability	5	5	–	2.5	0	13.5	24.54
Cost	0	0	2.5	–	0	3.5	6,36
Weight	5	2.5	5	5	–	18.5	33.6
					Total	55	100

indicates that both parameters are of equal importance and value 5 it indicates that a parameter is more important than another (Table 2).

Knowing the hierarchical order of the material properties, the performance of the most common materials in the market (aluminum, steel and stainless steel) based on the defendants in the application parameters was evaluated. Aluminum being ideal for project implementation material (Table 3).

Table 3. Results for material selection of the structure

Parameter	% importance	Aluminum		Steel		Stainless steel	
Mechanical properties	15.45	3	0,46	3	0,46	5	0.773
Weldability	Twenty	3	0.6	5	One	5	One
Machinability	24.54	5	1,23	One	0.25	One	0,245
Cost	6,36	5	0.32	3	0,19	One	0,064
Weight	33.6	5	1,68	One	0,34	One	0,336
		Total	4,29	Total	2,24	Total	2,418

The unmanned ground vehicle has two limbs (arms) front to help you position yourself on the obstacle in its path, so that the whole structure adopts the angle needed to avoid it and continue his career, being able to cross different terrain and climb stairs with angles of inclination to 42°, as the inclination usual steps of homes, shops, offices, etc., is between 26° and 42° [15]. To fulfill these functions using three DC motors, one of which raises the robot to position the robot on the stands and the other two engines transmit the motion to the bands that move the robot arms on the surface is made.

The engine raising the robot arms must produce than that generated by the weight of such arms, enabling him to break the inertia and torque to achieve the desired movement (Fig. 1). The estimated weight of the two arms and all its components is 2.6 kg.

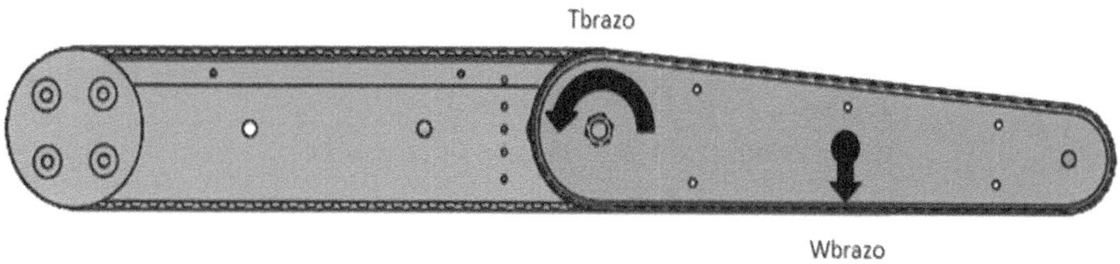

Fig. 1. Side view of the robot with arms outstretched. Tbrazo is the T_{arm} and Wbrazo is the W_{arm}

Torque of the robot arms:

$$T_{arm} = (Number\ of\ arms) * (W_{arm}) * (Distance\ to\ center\ of\ mass)$$

$$T_{arm} = (2) * (1.3\text{kg}) * (17, 5\ \text{cm})$$

$$T_{arm} = 45.5\text{kg cm}$$

Engine torque to the robot arms:

$$T_{arm\,motor} = (T_{arm}) * (1/Motor\ Efficiency)$$

$$T_{arm\,motor} = (45.5\ \text{kg cm}) * (1/75\%)$$

$$T_{arm\,motor} = 60.67\ \text{kg cm}$$

The torque required for the motors is multiplied by a safety factor of 1.6, equivalent to 60% of additional torque to the theoretical torque calculated.

Safety factor motor arm:

$$T_{motor\ de\ brazos} = 60.67\ \text{kg cm} * 1, 6$$

$$T_{motor\ de\ brazos} = 97, 072\ \text{kg cm}$$

For the selection engine use the same procedure carried out to select the material of the structure, considering parameters: price, torque, speed and weight (Table 4).

The design of unmanned ground vehicle has a chassis that allows you to host and set in the proper position for all mechanical and electronic components for operation. Chassis are incorporated DC 2 motors, which require less than the demanded torque arm, which was employed by the same motor.

Batteries for powering the motors and control board (Table 5 and 6):

Fatigue for work demanded by the robot, the mechanical elements to ensure its strength and resistance were designed.

For motor coupling with the axis of the robot arms spur gears they were used, which were designed using the method of the AGMA.

Table 4. Technical characteristics of the engine

Engine	Features	
Tsiny TS-40 GZ868-505	Price	$ 78
	Torque	120 kg.cm
	Weight	1.2 kg
	Supply voltage	12 V
	No load speed	9 RPM
	No-load current	0,46 A
	Nominal speed	RPM 6.8
	Rated current	1.2A
	Peak torque	98 Kg.cm

Table 5. Battery characteristics

Battery TURNIGY XT-90 engines	
Minimum capacity:	5000 mAh
Configuration:	4S1P/14.8 v/4Celdas
Constant discharge:	25C
Maximum discharge (10 s):	35C
Package weight:	552 g
Package size:	147 × 49 × 33 mm
Charging Plug:	JST-XH
Discharge outlet:	XT-90

Bending stress:

$$\sigma_{\mathrm{f}} = \mathrm{W_t\, k_o\, k_v\, k_s} \frac{\mathrm{k_H k_B}}{\mathrm{b\, mt\, Y_J}}$$

$$\sigma_{\mathrm{f}} = 199,76\,\mathrm{MPa}$$

Effort to contact:

$$\sigma_M = Z_e \sqrt{W_t\, k_o\ \ k_v\ \ k_s \frac{k_H\, Z_R}{d_{w1} b\, Z_I}}$$

$$\sigma_M = 266\,\mathrm{MPa}$$

Mechanical strength of Duralon:

$$S_{ut} = 540\,\mathrm{MPa}$$

Table 6. Battery characteristics

SONY Portable Charger CP-V10A/WC control board	
LED	Yes
Connectivity	USB 2.0 output port: 1
Power input	5 V, 1.5 A
Power output	5 V, 1.5 A
Capacity	10000 mAh
Recharge cycles	1000
Type output connector	Micro USB
LED	Yes

$$S_y = 480\,\text{MPa}$$

The shaft design has taken the following characteristics:

- Machining or cold rolling
- Room temperature
- Reliability of 99%
- Management theory Goodman failures.

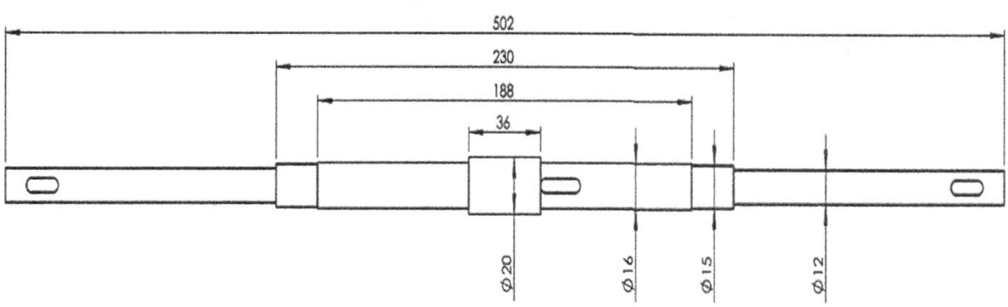

Fig. 2. End shaft design

The shaft (Fig. 2) has 6 bearings, each selected based on the existing load at the point of location. The bearings that were used in the robot are: DIN and DIN 625 6202 625 620 (Fig. 3).

Software using finite element analysis under different meshing convergence of the stresses and strains of the faces of the chassis frame and arms of the robot was analyzed.

Mechanical Analysis chassis for a distributed load of 15 [N] is show on Fig. 4, 5 and 6, and mechanical Analysis of the arms for a point load of 67.17 [N] is show on Fig. 7, 8 and 9.

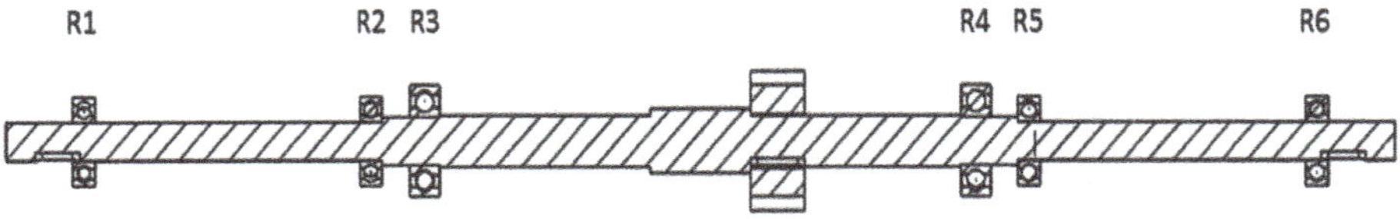

Fig. 3. Distribution of bearings on the shaft

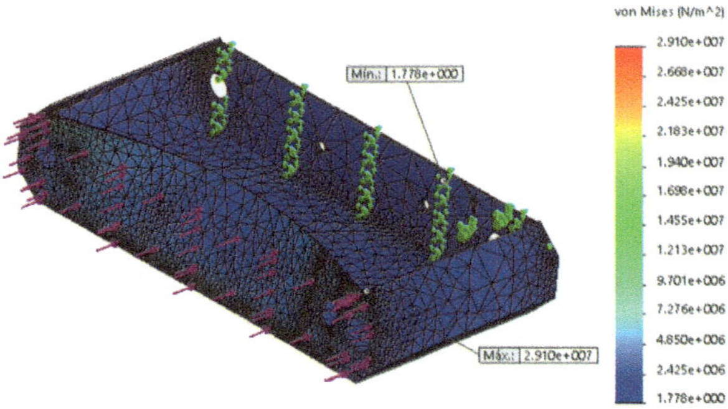

Fig. 4. Von Mises effort 15 [N].

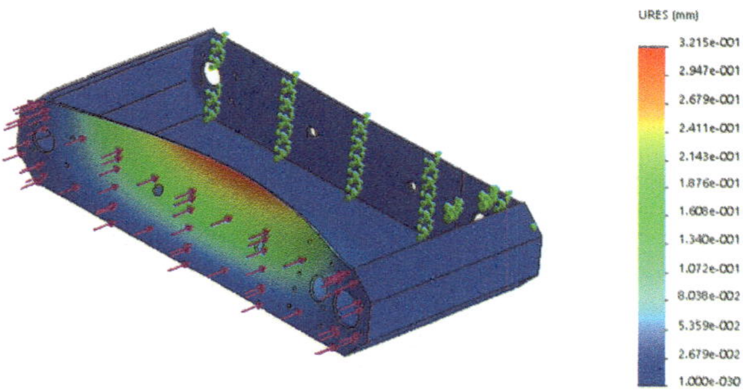

Fig. 5. Deformation 15 [N].

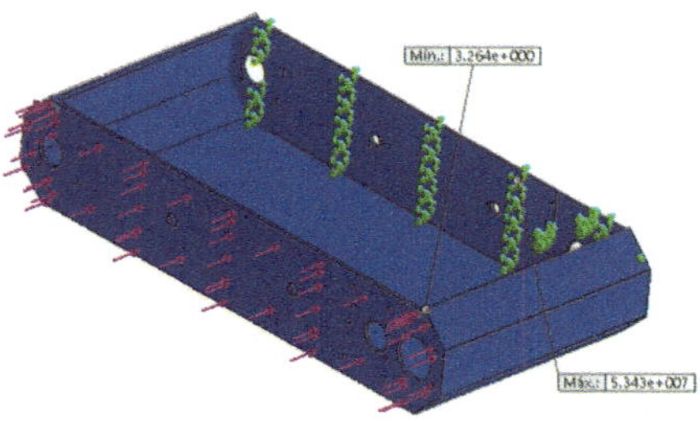

Fig. 6. Security factor 15 [N].

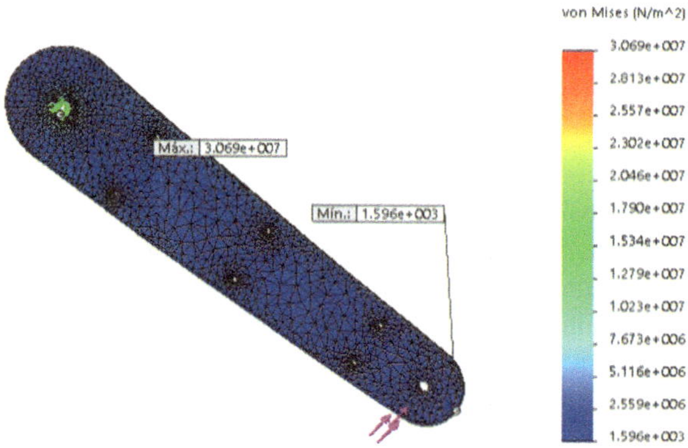

Fig. 7. Effort Von Mises 67.17 [N].

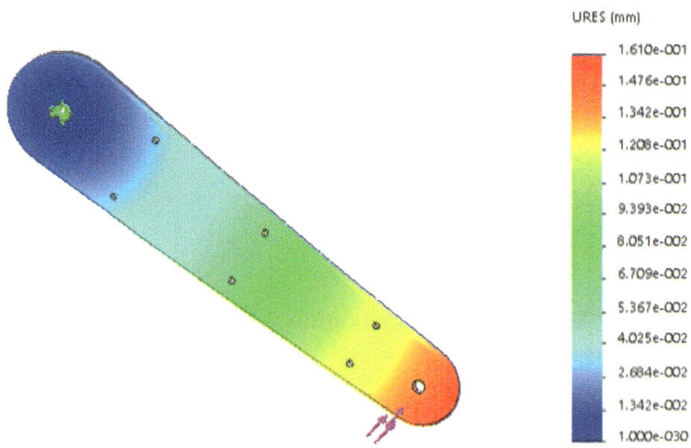

Fig. 8. Deformation 67.17 [N].

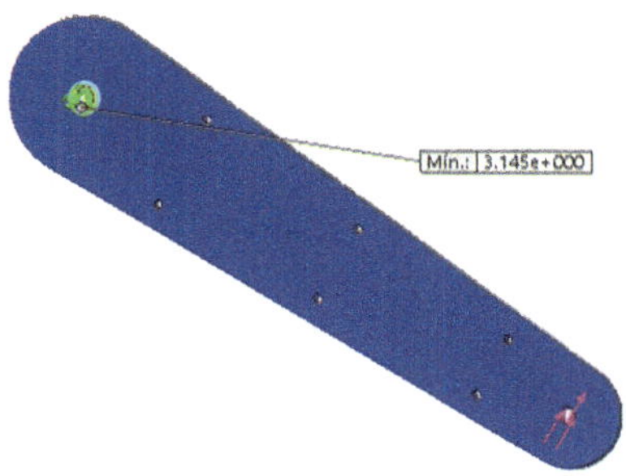

Fig. 9. Safety Factor 67.17 [N].

The program was loaded onto the plate 3 Model B Raspberry Pi under the Lubuntu operating system and using the ROS platform.

Control the robot using the keyboard is done through a node ROS which constantly reads all keys pressed subsequently processed, commands and sends the data to a main node programmatically verifies that access has been made in the HMI, if this is the case sends data via Wi-Fi to the control board in the robot to perform the operations required by the user. Two nodes, one for PC and one for the control board in order to verify all there is a connection between the PC and the robot, if not this case the robot stops until you have connection again when developed with the computer.

Monitoring is performed by a camera allowing visualization of the environment, the data are sent to a computer which via an interface allows the operator to display this environment. This system would be extrapolate to several applications [16–18]. Camera subsystem and HMI interface subsystem are showed on Fig. 10 and 11.

Fig. 10. Camera module V2 Raspberry Pi.

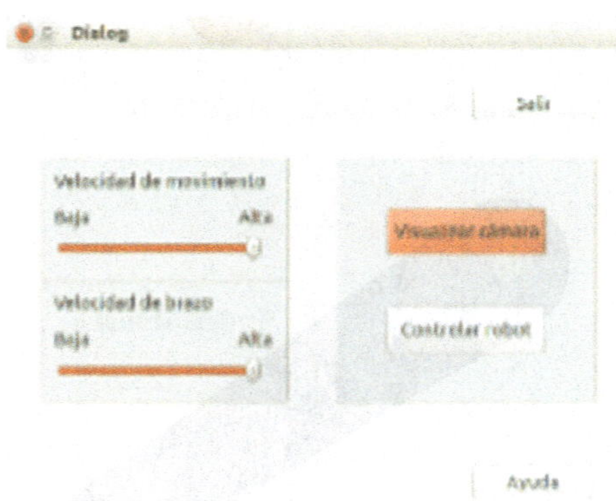

Fig. 11. HMI using QT creator.

3 Results and Discussion

Performance tests were conducted indoors and outdoors in sunny weather with average temperature of 20 °C and average humidity of 62%. The purpose of testing the test operation of the robot on different surfaces, evaluating their performance in speed and travel time for different angles, the speed will be taken to a distance of 200 cm.

Field test on Concrete: Without the presence of water hard and rough that promotes adhesion of the terrain texture robot. In this type of land is not necessary to use the robot arms, only the translational and rotational movement is performed by moving caterpillars (Table 7).

Table 7. Travel time on ground of concrete

Test number	Slope in degrees		
	10	10	30
One	22,5 s	22.61	22.57
Two	22.54	22,55	22.56
3	22.62	22,55	22.62
4	22.52	22.57	22.58
Average	22,55	22.57	22.58

Test on Grounds with Lawn: Wet irregular form with lawn soft and smooth texture that discourages adhesion of the robot to the ground. The limit on this surface tilt is 25°, because for greater inclinations the robot patina and does not advance (Table 8).

Table 8. Travel time on grounds with lawn

Test number	Slope in degrees		
	10	Twenty	25
One	23.55 s	25,21 s	25.68 s
Two	23.45 s	24,98 s	25.87 s
3	23.56 s	24.24 s	26.12 s
4	23.55 s	25.67 s	26.32 s
Average	23.53 s	25.03 s	26,00 s

Test on Mixed Terrain: Formed of sand, grass and small stones you different types of texture, adhesion is not uniform, being good or bad in different sections of the land (Table 9).

Stair Climbing Test: Hard and rough road or tile texture hard, smooth texture, both offer good adhesion between the robot and the ground (Table 10).

Video results are provided on https://www.youtube.com/watch?v=cggtrslkRCM.

Table 9. Travel time on mixed terrain

Test number	Slope in degrees		
	10	Twenty	25
One	22,9 s	23,01 s	23,01 s
Two	22.78 s	23.15 s	23.39 s
3	22.87 s	23.45 s	23.76 s
4	23.09 s	22,99 s	23.47 s
Average	22.91 s	23.15 s	23.41 s

Table 10. Travel time on stairs

Test number	Slope in degrees	
	Fifteen	2.3
One	24.47 s	25.87 s
Two	24.59 s	25.67 s
3	24.01 s	25.31 s
4	25.34 s	26.13 s
Average	24.60 s	25.75 s

4 Conclusions and Future Work

The robot meets its goal of moving through different types of terrain and especially the ascent and descent of stairs as shown on Table 7, 8, 9 and 10. The robust design allows you to scroll through all these areas without suffering structural damage. The design with two robotic arms for positioning on the steps malfunctioned and is highly recommended for this type of robots. These robotic arms offer many advantages when it comes to overcoming obstacles providing versatility in mobility robot.

The robot can operate at its normal position be reversed because the design is the same above and below. Should the ground or some external factor overturn the robot it can continue its role and continue to work normally. The design of the structure prevents most external agents affect the internal components of the robot, however, water and dust can easily enter therefore has no waterproofing is why the robot cannot be used in rain or land as puddles or mud. Although the shape of the structure allows include devices such as a robotic arm it is not recommended as this will adversely affect the strength of the structure and operation of the motors.

Acknowledgement. This work was partially supported by the PIS-04 "Detección temprana de incendios forestales basado en aprendizaje automático y en algoritmos de clasificación aplicados al espectro visible" research project of the Universidad de las Fuerzas Armadas ESPE, Ecuador and partially supported by CIAM-CICTE Research Center.

References

1. Aguilar, W.G., Angulo, C.: Real-time model-based video stabilization for microaerial vehicles. Neural Process. Lett. **43**(2), 459–477 (2016)
2. Aguilar, W.G., Manosalvas, J.F., Guillén, J.A., Collaguazo, B.: Robust motion estimation based on multiple monocular camera for indoor autonomous navigation of micro aerial vehicle. In: International Conference on Augmented Reality, Virtual Reality and Computer Graphics, pp. 547–561 (2018)
3. Orbea, D., Moposita, J., Aguilar, W.G., Paredes, M., León, G., Jara-Olmedo, A.: Math model of UAV multi rotor prototype with fixed wing aerodynamic structure for a flight simulator. In: International Conference on Augmented Reality, Virtual Reality and Computer Graphics, pp. 199–211 (2017)
4. Aguilar, W.G., Casaliglla, V.P., Polit, J.L.: Obstacle avoidance based-visual navigation for micro aerial vehicles. Electronics **6**(1), 10 (2017)
5. Aguilar, W.G., Cobeña, B., Rodriguez, G., Salcedo, V.S., Collaguazo, B.: SVM and RGB-D sensor based gesture recognition for UAV control. In: International Conference on Augmented Reality, Virtual Reality and Computer Graphics, pp. 713–719. (2018).
6. Ceron, A.: . teleoperated robotic systems. monthly Journals Network Latin America and the Caribbean, Spain and Portugal, pp. 62–69. Retrieved September 17, 2018.
7. Direct Industry. (Nd). HiBot. https://www.directindustry.es/prod/hibot/product-177237-1923367.html
8. Arqhys: Arqhys magazine, December 2012https://www.arqhys.com/articulos/robot-hexapodo.html
9. Arias, M., Garcia, R.: virtual prototype of a mobile robot for multi-terrain. Solano, Oaxaca, Mexico, 24 October 2016
10. Quezada, D.: Design and construction of an all-terrain robot using the Rocker-Bogie system and TELEOPERATED inalámbicamente Robotics Laboratory at the University of the Armed Forces - ESPE. Sangloquí, Pichincha, Ecuador, 21 April 2014
11. Argudo, G., Arpi, A.: Design and construction of a mobile robot tele-operated for assistance in operations altoriesgo fire department. Cuenca, Azuay, Ecuador, July 2012
12. Bambino, I.: An introduction to mobile robots. I. In Bambino. Buenos Aires: AADECA (2008)
13. Pot. (Nd). NASA Mars Exploration. https://mars.nasa.gov/programmissions/missions/present/2003/
14. iRobot (2018). PackBot portable multi-mission robot. https://www.endeavorrobotics.com/media/docs/Spanish/Endeavor%20Robotics%20510%20PackBot%20Spec%20Sheet%20SPANISH.pdf
15. Elevate (2011). https://www.eleveescaleras.com.ar/clasificacion-de-'escaleras.htm.
16. Aguilar, W.G., Morales, S.G.: 3D environment mapping using the Kinect V2 and path planning based on RRT algorithms. Electronics **5**(4), 70 (2016)
17. Aguilar, W.G., Salcedo, V.S., Sandoval, D.S., Cobeña, B.: Developing of a video-based model for UAV autonomous navigation. Latin American Workshop on Computational Neuroscience, pp. 94–105 (2017)
18. Aguilar, W.G., Luna, M.A., Ruiz, H., Moya, J.F., Luna, M.P., Abad, V., Parra, H.: Statistical abnormal crowd behavior detection and simulation for real-time applications. In: International Conference on Intelligent Robotics and Applications, pp. 671–682 (2017)

Corruption in the Security Forces. An Isolated Fact or a Generalization? A Brief Study of the Ecuadorian Case

Henry Cruz(✉), Miguel Tapia, and Diego Garcia

Universidad de las Fuerzas Armadas- ESPE, Sangolquí, Ecuador
hocruz@espe.edu.ec

Abstract. With the appearance of new dissident groups of the Revolutionary Armed Forces of Colombia – People's Army (FARC- EP), the perception of insecurity has increased in Ecuador, especially on the northern border. These criminal organizations have tried for a long time to insert themselves into institutions responsible for security to facilitate their criminal activities, mainly associated with transport, collection, custody, and sale of narcotics and illegal substances, kidnapping, illegal mining, among others. This article carries out a generic analysis of the level of corruption and the evil effects generated in the Latin American region and why the irregular armed groups have tried to infiltrate the State security forces and led the Ecuadorian case as a study model. This paper also verifies the situation of participation of the uniformed in acts of common or organized crime, and its analysis on whether cases of non-compliance with the law can be considered as isolated or as the beginning of a process of contamination and low professionalization of these security institutions. The study carried out shows that participated in acts of corruption, or common crime, representing 0.02% of the Ecuadorian Security Institutions' total personnel.

Keywords: National security institutions · Corruption · Drugs trafficking · Irregular Armed Groups (IAG)

1 Introduction

The report prepared by the United Nations Development Program (UNDP) in 2002 indicates that many of the Latin American countries have the highest corruption rates globally, being one of the main obstacles for these countries to achieve economic and social development [1].

We must recall that corruption is not an exclusive problem of developing countries, nor only of the State or public entities, but has also affected countries considered developed and private entities.

As part of each country's public sector, the tentacles of corruption have also affected, including Armed Forces and the Police, control entities, and especially the security forces. In these entities, two types of corruption could be evidenced: An administrative and an operational one. The first could include administrative acts that seek to benefit

M. Botto-Tobar et al. (Eds.): CIT 2020, AISC 1326, pp. 474–485, 2021.
https://doi.org/10.1007/978-3-030-68080-0_35

their own or third parties within internal processes through the convenient management of human and/or economic resources in consideration of the execution activities such as promotions, passes, contracts, purchases, etc. The second type of influence occurs in the operational field, where the security forces carry out their fundamental missions of border control or the commission of crimes, which opens up a series of opportunities for them to carry out acts of corruption [2].

In the Ecuadorian case, the military institution from its origin has enjoyed prestige and trust, due to the approach and inclusion that this institution has shown to Ecuadorian society. This evidence is shown by the variety in recruiting their cadres, supporting state institutions in risk, management missions, and immediate response.

On the other hand, in recent years, the police force has been increasing the level of trust and credibility. This fact is also the product of a communication strategy and of linkage with Ecuadorian society. Undoubtedly, the prestige achieved by the Ecuadorian security forces, which have become public knowledge, could be affected by cases involving elements in criminal activities.

From the signing of the peace agreement between the Colombian State and the Revolutionary Armed Forces of Colombia – People's Army (FARC - EP), new dissident groups of these criminal organizations have emerged. These groups have increased their actions on the northern Ecuadorian border, especially in the province of Esmeraldas, forcing the Ecuadorian State to increase the military and police presence in said region [3].

The FARC - EP has tried for years to infiltrate several Latin American countries' security institutions and bordering the Republic of Colombia to facilitate their criminal activities as part of their operating doctrine. Indeed, part of the strategy of these irregular armed groups is to involve the personnel of the security forces in active service. Besides, affecting the institutional image in the sense of knowing if these institutions are facing a low professionalization effect or are simply isolated events.

The present work tries to establish conclusions to know if corruption has effectively managed to penetrate the Ecuadorian security forces. This study's is a false perception motivated by society's sectors disaffected to institutions that manage agendas under their interests. This work has the following contents:

- Section 2: analyzes the level of relationship of the Latin American security forces with the population.
- Section 3: presents corruption as a national and regional problem.
- Section 4: analyzes FARC-EP dissidents' presence as a cause of corruption in Ecuadorian security forces.
- Section 5: shows the generalization of corruption as a State security problem and finally,
- Section 6: presents the conclusions of this work.

2 Relationship of the Security Forces in Latin America and the Citizens

Hobbes in his work "Leviathan" mentions that the State comes from a pact of submission of men to a monarch, to whom he gives all his individual rights and freedoms to obtain a minimum of security. From this statement, we can deduce that the State, through its coercive power, oversees providing security to the individuals who are part of it [4].

The security that the State provides to its population generates a relationship between them, aiming to guarantee stability and social order, depending on the views; these can be positive or negative. In Latin America, there are various types of relationships between the states' security forces and their respective societies. Some of these are very good since the beginning of their respective republics, and others not very close. Essentially, due to the variety of doctrinal and philosophical schemes that have sought to establish itself in Latin American societies, especially because of the Cold War.

From this Geopolitical-Historical event of humanity, the doctrine of National Security is spread in Latin America through the revolutionary war. This doctrine was supported by various techniques of irregular combat and strategies of collective control of masses and territory. The objective of the State Security Forces is to control the people; in this way, the necessity to fight an "internal enemy" is justified. According to this doctrine, the main objective of these State security institutions is to control the people, as a need to combat an "internal enemy" [5]. In reply, the Latin American States applied the national security doctrine. The confrontation between the state security forces and ideologically opposed sectors of the population was imminent. The results were a negative relationship and rejection from certain sectors of the community population, generating reductions in acceptance level.

On the other hand, the Ecuadorian Armed Forces have written outstanding historical pages regarding safeguarding national sovereignty, such as what happened in the Cenepa War in 1995. After signing peace between the republics of Ecuador and Peru, the Ecuadorian soldier has become a "flexible military" since its role is adapted to the fulfillment of new missions and challenges related to humanitarian support, support for risk management, strengthening civil-military relations, coordination with the National Police for the fight against organized crime, among others. All these activities bring closer and increase the relationship with society; evidence of this is the high acceptance rates of the Ecuadorian armed forces after the Cenepa War.

In April of 2016, acceptance increased the level of approval of the armed forces and the National Police after the earthquake occurred in the provinces of Manabí and Esmeraldas. The deployment capacity and means of the Armed Forces in general and the Land Force were demonstrated through the use of 36 operational and administrative units that carried out a search, rescue, and evacuation operations for victims, logistical support, construction temporary camps, and security [5].

During 2019, large-scale social protests took place in some Latin American countries, specific in Ecuador; there were several acts of violence and destruction. The response of the Ecuadorian Armed Forces to these events was aimed at respect for human rights, control, and progressive use of force; this attitude allowed to avoid an escalation of social shock [6].

This year, the Ecuadorian security forces are playing a fundamental role during the health emergency that occurred due to the global pandemic of COVID-19, executing crisis response operations and allowing compliance with the different measures and provisions adopted by the Government National.

Indeed, it is evident that at present, the security forces and especially the military forces carry out complementary missions that add value to society, and that broadens the conception of its structural doctrinal base where the scenarios are uncertain and the answers must be creative, fast, and attached to the legal order [7].

This new reality of employment of the Ecuadorian security forces has allowed them a close approach with the people, reaching the first places of institutional trust level. The latest survey carried out by CEDATOS in the cities of Quito, Guayaquil, Cuenca, and Manta, closed at the end of 2019, confirms the following: After family with 97.3%, the Armed Forces received the highest response 78.4%, followed by the Ecuadorian Army 76.5%, youth with 75.8%, Ecuadorian Air Force with 73.7%, Navy with 72.04%; Church with 70.02% and the National Police with 53.7% and the private media, 50.1% [8].

3 Corruption: Regional and National Concern

One of the biggest concerns, not only in Ecuador but also throughout the region, is corruption. The Círculo de Estudios Latinoamericanos (CESLA) has updated the corruption report on the region, revealing that 67.65% of the countries that make up Latin America present high levels of corruption, according to the results of the CESLA-2020 indicator. The study shows that Perú and El Salvador place it above the average (69), Bolivia is among the countries with high corruption indicators as it has a scale of 75. Paraguay, the Dominican Republic, Mexico, Guatemala, Honduras, have a high corruption level too. Nicaragua, Haiti, and Venezuela appear on the blacklist. For Ecuador, the synthetic indicator of corruption receives a score of 67/100; in other words, it is a high level of corruption and extreme weakness in anti-corruption policy [9].

As shown in Fig. 1, the corruption index in Ecuador places it near the average for the Latin American region. This figure should be taken as a wake-up call to rethink anti-corruption policies as a society and as a State.

In the Ecuadorian reality, the perception of corruption does not vary much from the regional one. According to CEDATOS, this is one of the social problems that most afflicts Ecuadorian society with 15% occupying the third place among the main issues that worry the population, and added the percentages between corruption, drugs, insecurity, and crime add up to a total of 30.5% [8]. Conceptualizing this problem, corruption is defined as the intentional breach of the principle of impartiality to derive from such behavior a personal or related person's benefit. This definition of corruption can be broken down into three parts:

1. Intentional non-compliance means that someone aware of what they are doing commits the act and knows it is wrong.
2. The principle of impartiality refers to the absence of inclination in favor of or against one or another position.
3. Personal benefit or for related people which is seen as the result of the act committed.

For this reason, corruption is an act committed consciously in favor of some interest to achieve a result or benefit for itself or third parties. Although it is true throughout

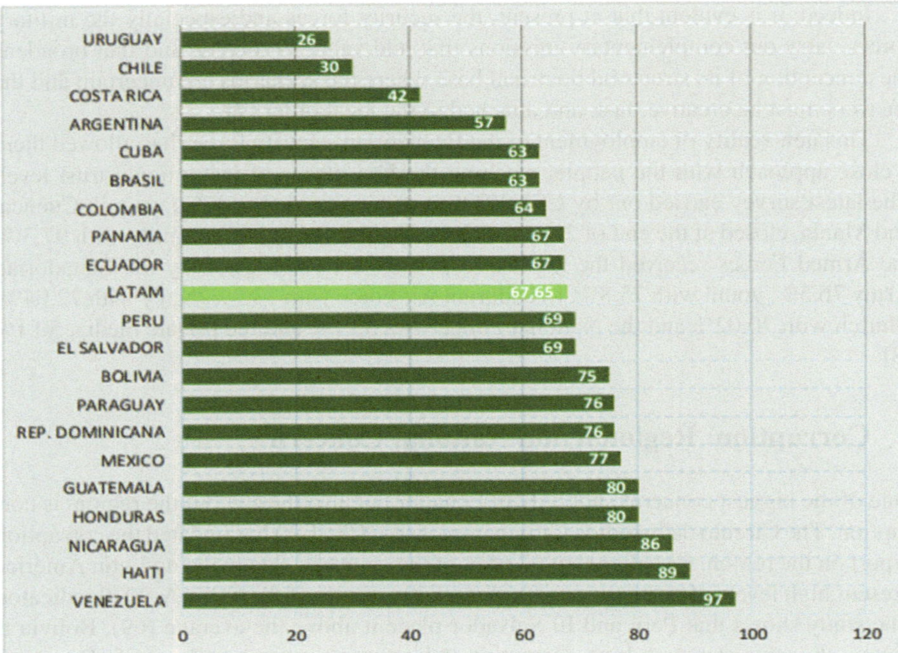

Fig. 1. Corruption in Latin America barometer CESLA 2020 [9].

history, the security forces have been considered and are until today's example of honor, discipline, and loyalty; neither can it be thought that they possess an impenetrable shield to this great social problem of corruption. It is unacceptable for the institutions of order and society in general that active members of the security forces are involved in corruption acts. However, among the causes for the occurrence of these acts, it could be argued that this happens as a logical and, to some extent, expected consequence due to the increase in the number of Armed Irregular Groups (AIG) and their criminal activities within the region.

4 Actions of the FARC - EP Dissidents on the Colombian-Ecuadorian Border, as a Possible Cause of Cases of Corruption in the Ecuadorian Security Forces

During the peace negotiations between Colombia and the FARC - EP, some components did not agree with the terms and agreements, such as the announcement on June 10, 2016, by the Front First Armando Rios FARC. They operated in Vaupés, Guaviare, and Meta's departments, a statement that expressed its decision not to participate in the demobilization process, dereliction of weapons, and continuing hostile activities [10].

These groups were called FARC dissidents, and their actions increased as the negotiations progressed. They reached their peak of criminal activities after the signing of the peace agreement on September 26, 2016. Figure 2 shown the percentage of actions by place of occurrence attributed to dissidents.

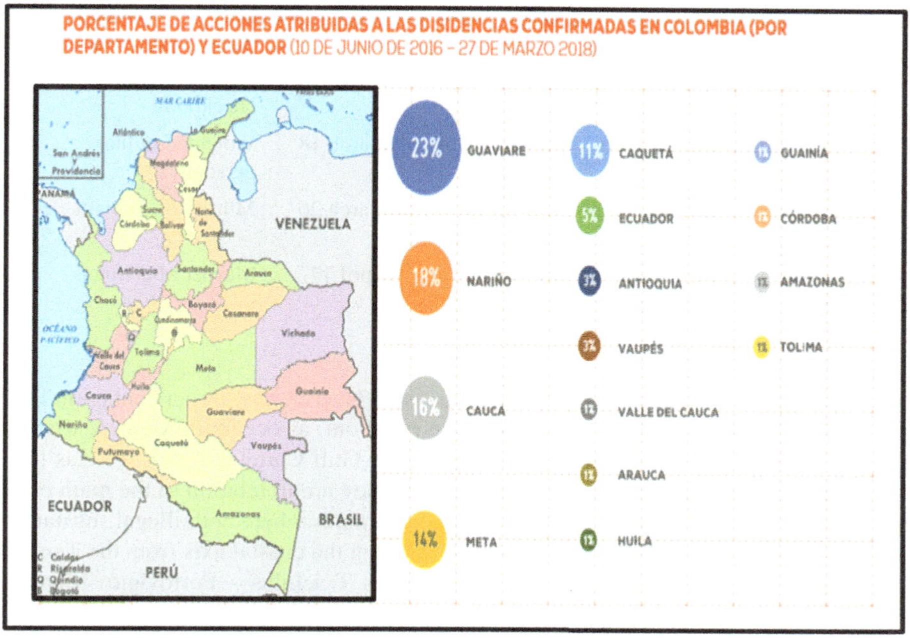

Fig. 2. Percentage of actions by place of occurrence attributed to dissidents [10]

As shown in Fig. 2, 5% of the actions attributed to the confirmed FARC dissidents have been carried out in Ecuadorian territory, which has somehow influenced the decrease in the perception of security in Ecuador, especially on the northern border. This 5% reflects five terrorist attacks that occurred in the first quarter of 2018. Table 1 shown a detail of the attacks that occurred in Ecuador (2018).

For their criminal actions, the AIG must finance themselves with crimes related to drug trafficking. Therefore, they need weapons, which are under the exclusive custody, control, and management of the Armed Forces and the National Police. Besides, they will seek in any case the involvement of its members, infiltrating, penetrating, and corrupting to achieve their objectives.

After the signing of peace with Peru and due to its geographical position, Ecuador became a strategic place for the exchange, transport, and mobility of illegal substances, as well as for arms trafficking. As the UNODC points out, arms trafficking in Ecuador uses land and sea routes and connects a network of national towns that invariably culminates in Colombian territory. In this sense, the State Attorney General's Office has detected light weapons from Chile and Peru, some of them belonging to those countries' military arsenals. The weapons enter through the southern border provinces, which are then dispersed in different Ecuador locations -Quito, Santo Domingo, San Lorenzo, Manta, Guayaquil, Ibarra, Tulcán, Lago Agrio, among others, to reach Colombia as their final

Table 1. Detail of attacks that occurred in Ecuador (2018)

Ord.	Attempt	Location	Date	Effects
1	Car bomb	San Lorenzo	January 27	Destruction of the Police headquarters
2	Explosive charge	Borbón	March 17	Damage to naval checkpoint
3	Attacks on Ecuadorian patrols	Alto Tambo y El Pan	March 18	Wounded military personnel
4	Kidnapping and murder of journalists	No location	March 26	Murder of civilians
5	Kidnapping and murder of civilians	No location	April 17	Murder of civilians

destination. Based on these facts, and as López and Negrete say, direct contagion from institutions of the Ecuadorian State cannot be ruled out, as occurs in Mexico, where a group of ex-military formed the armed wing of the Gulf Cartel called Los Zetas [11]. Exist some routes of arms trafficking in Ecuador, they are in relation to the main cities and regions in this country. Mainly, there are three routes where both illegal substances and materials are trafficked. The first route goes along the coastal axis from the Tumaco (Colombia) – Emeraldas – Santo Domingo de los Tsáchilas – Portoviejo -Manta – Guayaquil – Salinas – Machala - Huaquillas to Zarumilla (Perú). The second route goes through the Inter-Andina region from Machala - Cuenca – Ambato – Quito - Tulcán to Pasto (Colombia). And finally the eastern axis that goes from the cities of Ambato and Quito towards Nueva Loja - Puerto Nuevo - Puerto Maestanza - Puerto Asis (Colombia) [12].

Among the objectives of these AIG is to permeate, contaminate or recruit personnel belonging to the control and State Security Agencies, to facilitate their criminal activities through the infiltration technique, which allows members to of a subversive minority or to people who unconsciously or consciously cooperate, especially to enter public organizations. For the AIGs to obtain the collaboration of the security forces' elements, it will depend on the means of pressure they use and the commitment of the members of these institutions with their ethical and moral responsibility in the fulfillment of their functions.

The Ecuadorian Armed Forces and National Police are diverse institutions made up of all social strata, populations, and ethnic groups that inhabit the country; in other words, the men and women who make it up come from all regions, provinces, cities, towns. As an example of the above, the percentages of admission to the training schools of the Land Force for the admission process of the year 2018 are highlighted, as shown in Table 2.

Indeed, the security forces in Ecuador and specifically the military forces are made up of inhabitants from all its regions; consequently, it is normal for this institution to coexist with the same ties, problems, social and economic needs that afflict its population in general. While the security forces strive to keep their personnel isolated from any form of

Table 2. Employees per Armed Forces School (AFS) in the 2018 admission process [13]

	Races or nationalities					
AFS	Montubio	Half blood	Afro Ecuadorian	White	Other, peoples or nationalities	Total
ESMIL		137		1		138
ESFORSFT	1	596	9	1	41	648
IWIAS		4			35	39
TOTAL	1	737	9	2	76	825
% TOTAL	0,1%	89,3%	1,1%	0,2%	9,2%	100,0%

contamination to their ethics, principles, and institutional values, common or organized crime could use this relationship to promote corruption within security institutions' ranks.

5 The Generalization "An Attack on State Security"

Recent cases related to alleged acts of corruption by personnel are analyzed to determine whether corruption has effectively managed to introduce its tentacles into the Ecuadorian security forces' ranks. These cases are considered by their importance and impact generated in the society.

1. **Crime report in Esmeraldas**. On Monday, January 29, 2018, four soldiers who fled in two vehicles were arrested by members of the National Police, allegedly escaping when they were discovered while trying to attack a house in the Tachina sector, Esmeraldas province [14].
2. **Detained in Arenillas**. On Thursday, March 8, 2018, the National Police dismantled, in the city of Arenillas, El Oro province, an alleged criminal group made up of 10 members, two of whom were active-duty military personnel [15].
3. **Detained in "La Y" Sucumbios.** On Sunday, August 26, 2018, in the sector "The Y" Sucumbios, between Santa Barbara and Carmel, at a military checkpoint a van Nissan brand, which carried white goods appliances stopped, within which was found a substance, presumably a drug, leading to the arrest of two active-duty army members [16].
4. **Cocaine seized at the Manta Air Base.** On September 13, 2018, the National Police seized approximately one ton of cocaine at the Manta Air Base, Manabí province, because of an anti-narcotics operation. Two soldiers of the Ecuadorian Air Force on active duty were arrested and placed under the orders of the competent authorities [17].
5. **Military, former police, and eight civilians arrested for illicit association**. On October 11, 2018, two members of the Ecuadorian Air Force, a former police officer, and eight civilians were arrested for the alleged crime of illicit association while they were transported in four high-end cars to the north of the province of Esmeraldas, to steal money and jewelry [18].

6. **Possible crime of arms and ammunition trafficking.** On October 17, 2018, the Prosecutor's Office, with the National Police's support, carried out a series of proceedings in the country's military units for investigative purposes on the possible crime of trafficking in arms and ammunition. As a result, seven active-duty military personnel were arrested, as well as one public servant [19].

Table 3 shows the data of corruption cases in 2018 in which personnel in active service have been involved. On the other hand, Table 4 has been compared with the total population of Ecuadorian society during the same period of occurrence as follows:

Table 3. Number of military personnel arrested for alleged acts of crime and corruption

Cases raised in 2018	Number of military personnel detained
Military detained in Esmeraldas	4
Military detained in Arenillas	2
Military detained in "La Y" Sucumbios	2
Cocaine seized at the Manta Air Base	2
Possible crime of arms and ammunition trafficking	7
Military, ex-police and 8 civilians arrested for illicit association	2
Total Personnel Detained	19

Table 4. Percentage comparison of detained personnel (year 2018)

Universe	Population	Detained 2018	Percentage (%)
Ecuadorian inhabitants	16´620.000	15.566	0,0934
Ecuadorian security forces personnel	90.000	19	0,0211

As can be seen from Table 3 and Table 4 of the results obtained, the percentage of active duty military and police personnel involved in illicit activities corresponds to less than a quarter of the percentage of citizens detained for the same reasons in the same period. This information supports the claim that the security forces are far below the general population's criminal level.

As analyzed, no one can talk of a generalization of corruption in the Ecuadorian security forces. It is true that in 2018 there were several cases of involvement of active duty personnel in illegal activities; however, it is also true that illegal activities on the part of the AIG that operate on the Ecuadorian borders increased, as one of the main causes of this. In recent years, we must also consider that corruption at all levels has increased in general, to the point of becoming a complex problem on a national and regional scale.

On the other hand, the response of the Ecuadorian security forces authorities at different levels before the public opinion has been forceful in providing all the collaboration so that the alleged crimes are investigated and, if applicable, punished with total impartiality, but at the same time with all the rigor of the law.

It is important to point out the significant effect it produces on the people's perception, the approach with which columnists in all its forms treat these events. It will also depend on their professionalism and ethics to leave in the collective mind a perception of generalized corruption or simply isolated events. Added to this is the negative effect produced by the increasing use of social networks irresponsibly and inappropriately. Thus, in 2018, various statements were made through multiple media that generalized the concept of "corruption in the security forces", negatively affecting the image and prestige of these institutions.

Considering the broad spectrum of participation of the Ecuadorian State's security forces in compliance with their fundamental missions and support for development, to speak of the generalization of corruption in these institutions is an attack on their security.

Cases of participation in criminal actions by military or police personnel do not have a connection between them; in other words, the members who are being investigated for their alleged intervention do not belong to or have formed any permanent, generalized, or articulated criminal organization or group to act with the security forces. These events may have the same causes, but they would not necessarily be related to the action of a few members within the military institution is considered as worrying isolated events.

The facts must be analyzed and investigated in-depth so that the security forces implement the necessary policies that combat and prevent their personnel's involvement in criminal acts related to AIG's activities.

The Ecuadorian Armed Forces, above all, have taken important steps to avoid contamination of their members; proof of this is the new Law on Personnel and Discipline of the Ecuadorian Armed Forces (FF.AA) seeks to increase commitment to society and highlight the ethical and moral values of the members of the armed institution. This is also part of an institutional strategy for society to maintain its trust in military institutions' professionalism, ethics, and integrity.

6 Conclusions

Corruption in Latin America is one of the fundamental factors that impede the development of this region of the world since it leads to the degradation of the social structure and, with it a significant loss of large amounts of public economic and financial resources, which in the end significantly reduces the levels of quality of life and human well-being.

Starting from the signing of the peace agreement between the Colombian State and the Revolutionary Armed Forces of Colombia – People's Army (FARC - EP), new dissident groups of these criminal organizations have emerged. These groups have increased their actions on the northern Ecuadorian border, especially in the province of Esmeraldas, consequently forcing the Ecuadorian State to increase the military and police presence in said region.

Common and organized crime have as a strategy to include recruiting members of the security forces to carry out their illegal actions. The Ecuadorian security forces' members run a high risk of being incited to be part of them because the geostrategic position of the country facilitates the mobility and trafficking of illicit substances and/or weapons.

In the year 2018, the number of members of the security forces who have participated in acts of corruption or common crime represents 0.02% of these institutions' total personnel. The percentage is very low, so it is evident that they are not a generalized act of corruption but rather isolated cases.

The Ecuadorian security forces in general and the Armed Forces specifically enjoy high levels of acceptance and credibility. In the Ecuadorian case, corruption is considered a risk within the defense political agenda, for which strategies have been established and objectives defined to avoid possible contamination and involvement of uniformed personnel in illegal actions.

References

1. El Bautista, O.D.: Problema de la corrupción en América Latina y la incorporación de la ética para su solución. Espacios Públicos **15**(35), 48–62 (2012)
2. García Mejía, M.: Evidencia y marcos conceptuales de la lucha contra la corrupción en la policía de Latinoamérica (Doctoral dissertation, Universidad Complutense de Madrid) (2016)
3. Fierro, V., Villarreal, D.: La influencia del Acuerdo de Paz del Estado con las FARC-EP en las relaciones de intercambio de inteligencia entre Colombia, Ecuador ante potenciales nuevas amenazas a la seguridad 2012–2016 (Bachelor's thesis, QUITO/UIDE/) (2017)
4. Griffiths Spielman, J.: Fuerzas Armadas: ¿preparadas solo para la guerra o efectivo instrumento para, además, asegurar la paz y seguridad estatal? Estud. internacionales (Santiago) **49**(187), 131–161 (2017)
5. Osuna, M.F., Pontoriero, E.: El impacto de la doctrina "de la seguridad nacional" en la argentina durante la guerra fría (1955–1983). Izquierdas **49**, 352–364 (2020)
6. Cruz, H., Reyes, Ch.R.P., Pinillos, M.: Automatic counting of people in crowded scenes, with drones that were applied in internal defense operations on October 20, 2019 in Ecuador. In: Developments and Advances in Defense and Security. MICRADS 2020. Smart Innovation, Systems and Technologies, vol. 181. Springer, Singapore (2020). https://doi.org/10.1007/978-981-15-4875-8_10
7. Cruz, H.: La securitización¿ percepción o realidad tangible? Un análisis multidimensional. Rev. Acad. Guerra Ejército Ecuatoriano **13**(1), 11 (2020)
8. CEDATOS: Investigación cerrada al 5 de mayo de 2018, en Quito, Guayaquil, Cuenca y Manta. https://www.CEDATOS.com.ec/m/detalles. Accessed 21 Nov 2019
9. Círculo de Estudios Latinoamericanos [CESLA]. Indicar de corrupción para América Latina CESLA. https://www.cesla.com/pdfs/Informe-de-corrupcion-en-Latinoamerica.pdf. Accessed 13 June 2020
10. Fundación Ideas para la Paz [FIP] Trayectorias y dinámicas territoriales de las disidencias de las FARC. http://ideaspaz.org, http://ideaspaz.org/media/website/FIP_Disidencias_Final.pdf. Accessed 21 June 2020
11. López, J., y Negrete, D.: Estudio prospectivo al año 2016, sobre el narcotráfico y su incidencia en FF.AA., Master Thesis. https://repositorio.espe.edu.ec/bitstream/21000/7970/1/T-ESPE-047576.pdf. Accessed 02 May 2020
12. Entre Ecuador y Colombia hay 125 pasos fronterizos clandestinos. https://lahora.com.ec/noticia/1101547157/entre-ecuador-y-colombia-hay-125-pasos-fronterizos-clandestinos. Accessed 09 Sep 2018

13. Dirección General de Talento Humano de la Fuerza Terrestre. Informe estadístico de la D.G.T.H. del Ejército (2018)
14. Detienen a militares durante asalto a vivienda. http://www.ecuadorenvivo.com/sociedad/190-sociedad/71584-detienen-a-cuatro-militares-durante-asalto-a-vivienda-en-esmeraldas.html#.X3EV02hKjIU. Accessed 24 Feb 2020
15. Policía desarticuló a banda liderada por dos militares en Arenillas. https://www.elcomercio.com/actualidad/militares-policia-asaltos-autos-ecuador.html. Accessed 07 April 2020
16. Boletín de prensa. Militares detenidos en sector 'La Y' en Sucumbíos. https://www.juiciocrudo.com/articulo/dos-militares-fueron-detenidos-por-presunto-trafico-de-drogas-en-sucumbios/10124. Accessed 01 July 2019
17. Policía Ecuador (13 de septiembre de 2018). Más de una tonelada de cocaína es incautada en la Base Aérea de Manta. http://www.policia.gob.ec/. Accessed 13 Sep 2018
18. Detienen a dos miembros de la FAE por asociación ilícita. https://www.eltelegrafo.com.ec/noticias/sociedad/6/detenidos-miembros-fae-asociacionilicita. Accessed 03 Feb 2020
19. Detienen a militares e indagan sus vínculos con alias Guacho. https://www.eluniverso.com/noticias/2018/10/17/nota/7004009/operativo-camaleon-seis-provincias-investiga-presunta-venta-armas. Accessed 17 Feb 2019

Author Index

M. Botto-Tobar et al. (Eds.): CIT 2020, AISC 1326, pp. 487–489, 2021.
https://doi.org/10.1007/978-3-030-68080-0

The manufacturer's authorised representative in the EU is Springer Nature Customer Service Centre GmbH, Europaplatz 3, 69115 Heidelberg, Germany. If you have any concerns regarding our products, please contact ProductSafety@springernature.com

Printed and bound by CPI Group (UK) Ltd, Croydon, CR0 4YY
28/09/2026
02224057-0006